ISBN 978-0-265-45341-4
PIBN 10630196

THE PHILIPPINE

JOURNAL OF SCIENCE

EDITED BY

PAUL C. FREER, M. D., PH. D.

CO-EDITORS

RICHARD P. STRONG, PH. B., M. D.

E. D. MERRILL, M. S.

PUBLISHED BY

THE BUREAU OF SCIENCE

OF THE

GOVERNMENT OF THE PHILIPPINE ISLANDS

C. BOTANY

VOLUME II

1907

WITH 8 PLATES

MANILA
BUREAU OF PRINTING
1907

No. 1, January, 1907.

No. 2, April, 1907.

No. 3, June 15, 1907.

No. 4, July 15, 1907.

No. 5, October 15, 1907.

No. 6, December, 1907.

ERRATA.

Pages 18, 19, for ***Achrostichum*** read ***Acrostichum.***

Page 119, title ***PTERIDOPHYTA HALCONENSES*** should be ***PTERIDOPHYTA HALCONENSIA.***

Page 120, first paragraph, omit: "and for one period of thirteen days while they were at the higher altitudes they encountered constant heavy rain, day and night, and in this time never saw the sun."

Page 121, line 9, ***Blechnun*** should be ***Blechnum.***

Page 121, lines 20 to 22, omit: "it being a mass granite, marble, white quartz and with some schist-like rock."

Page 122, line 9 should read "related genera such as ***Prosaptia*** and ***Acrosorus,*** and in ***Plagiogyria.***"

Page 125, under **DENNSTAEDTIA,** ***hypolepoid*** should be ***hypolepidioid.***

Page 126, line 10, omit the comma after *alatam.*

Page 128, line 7, *diplazoideis* should be *diplazioideis.*

Page 129, line 5, *scolopendroid* should be *scolopendrioid.*

Page 129, line 19, *diplazoideis* should be *diplazioideis.*

Page 131, line 20, *transeuentibus* should be *transeuntibus.*

Page 131, line 24, *Extast* should be *Exstat.*

Page 135, line 14, **Antropium** should be **Antrophyum.**

Page 136, line 3, *Rhizomata* should be *Rhizomate.*

Page 136, line 25, *obtusis* should be *obtuses.*

Page 137, line 23, *New Calodonia* should be *New Caledonia.*

Page 138, line 9, *basicopicia* should be *basicopica.*

Page 139, line 17, *membraneceis* should be *membranaceis.*

Page 139, fourth line from bottom, *they* should be *their.*

Page 146, line 2, semicolon after *crasso* should be a comma.

Page 146, line 12, *aborvatis* should be *obovatis.*

Page 147, line 1, SCHIZAEACEÆ should be OSMUNDACEÆ.

Page 148, line 3, *Kge.* should be *Kze.*

Page 151, in list of illustrations, ***Polypodium caloplebium*** should be ***Polypodium calophlebium.***

Page 164, line 6 from bottom, for ***trunctailoba*** read ***trunctiloba*** Pres. Epim. (1849) 77.

Page 177, after line 9, insert:

"C. splendens (J. Sm.) This splendid species seems to be common in the islands.

"Luzon, Rizal Province, Morong (*Loher*).

"Mindanao, Davao (*Copeland*); Surigao (348 *Bolster*); Lake Lanao (503 Mrs. *Clemens*)."

Page 188, line 9, for **Aspidium melanorachis** read *Aspidium melanorachis.*

The second paragraph should read: "The form which I have identified in Bull. Herb. Boiss. VI. 1906, 1001, with *D. Smithianum* (Bak.) Diels is distinguished by the shorter sori, which touch somewhere before the sinus, by the rough scaly rachis armed like the costæ with sharp prickles, and by the more truncate lobes. I have this later from Ceylon." etc.

Add to the footnote for *D. dolichosorum* Copel. the reference p. 163 this volume.

63494——5

Page 206, *Memionitis prolifera* Retz. should be *Hemionitis prolifera* Retz.

Page 207, under **Dryopteris stenobasis** C. Chr. after *pinnæ*, insert "attenuated gradually to the base."

Page 208, under **Dryopteris Luerssenii** (Harringt.) C. Chr. *impleatibus* should be *implentibus*.

Page 260, last line, for *underscribed* read *undescribed*.

Page 264, line 8 from bottom, erase *stamen*.

Page 284, line 11 from bottom, for Mears read Mearns.

Pages 357, 371 and 400, for *Koordersiodendron pinnatum* Engl. read *Koordersiodendron pinnatum* (Blco.) Merr.

Page 372, **Artocarpus cumingii** Trec. should be **Artocarpus cumingiana** Trec.

The following errata occurred in the make-up at the printing office after the last page proof had been read:

Page 251, line 11, for *expidition* read *expedition*.

Page 255, line 2 from bottom, for *Mindor* read *Mindoro*.

Page 256, line 11, for *Stepf* read *Stapf*.

Page 265, lines 12 and 13 from bottom, erase from "longis" to "obovate" inclusive.

Page 278, line 14 from bottom, erase entire line except first word and insert **fastuosus** (Muell. Arg.) F.-Vill.

Page 290, under **Schefflera foetida**, line 3, for 4-rariter, read 4, rariter.

Page 294, under **Vaccinium hutchinsonii**, line 4, for pause read pauce.

Page 295, line 14, for axiliary read axillary.

Page 296, last line, for botanists read botanist.

Throughout the paper entitled ORCHIDACEÆ HALCONENSES, wherever the citation Ames orchidaceæ, fasc. 2 (1907) *ined.* occurs, read the date 1908.

Page 311, first line, for *orchirds* read *orchids*.

Page 313, fourth line from bottom, **Merillii** should be **Merrillii**.

Page 318, line 8, for Bonpl. 3: 222 (1885) read Otia bot. Hamb. 54 (1878).

Page 320, line 14 from bottom, (1882) should be (1822).

Page 322, line 6, for *O. aporaphylla* read *O. aporophylla*.

Page 326, lines 8 and 16, for *Calanthe triplicatis* read *Calanthe triplicata*.

Page 326, line 9, for *Orchis triplicatis* read *Orchis triplicata*.

Page 327, line 11 from bottom, for n. s. 4 read ser. 3, 4.

Page 328, line 2, for **alagensis** read **alagense**.

Page 334, line 14, for *speciments* read *specimens*.

Page 335, line 11 from bottom, for var. **angustun** read var. **angustum**.

THE PHILIPPINE

JOURNAL OF SCIENCE

SECTION C. BOTANY

VOL. II — JANUARY, 1907 — No. 1

THE COMPARATIVE ECOLOGY OF SAN RAMON[1] POLYPODIACEÆ.

By EDWIN BINGHAM COPELAND.
(From the Bureau of Education, Manila.)

INTRODUCTION.

Contemporary biological science is working toward two ends, namely:

(*a*) The reduction of the processes taking place in living things to understood reactions of chemistry and changes of physics; this is Physiology.

(*b*) Assuming the existence of living things, the causal interpretation of the forms they assume and the elucidation of their relationships; this is Bionomics, sometimes called "Evolution."

The application of principles established in the study of physiology to the problem of bionomics is called Ecology. Ecology must work most directly and clearly toward its ultimate aims as a part of bionomics—that is, the interpretation of form and structure and the elucidation of genetic affinities—if its systematic aspect is duly and strongly emphasized. That is, ecological work will contribute more evidently and directly to the progress of bionomics if its subject is primarily a homogeneous group of organisms, of whatever rank, than if the theme in the foreground is some geographical unit or some factor of the environment. Such a piece of

[1] San Ramon farm, Mindanao.

50146

work, for instance, was Brenner's "Klima und Blatt bei der Gattung Quercus." [2]

When geographical can be added to genetic unity, then the influence of local differences in environment, in the selection of adaptive peculiarities and the evolution of species should be especially clear and instructive. Such a combination of unities is usually impossible in a work of any scope because of lack of material. However, the homogeneous family of *Polypodiaceæ* has presented to me, within the radius of an easy day's walk in the wilderness back of San Ramon, Mindanao, material sufficiently ample to allow me to draw a variety of conclusions. While directly engaged in collecting on one visit, and on excursions at irregular times as official work on the physiology of the coconut permitted on a second one, I collected in this neighborhood 186 species in this family. I do not know that any other family anywhere in the world will yield such an amount of material; certainly no other will yield material so diverse in form and adapted to such varying local conditions.

In this family are xerophytic, trophophytic, mesophytic and almost hydrophytic plants. It includes terrestrial and epiphytic herbs, vines, and subarboreous species; with a remarkable number of peculiar adaptations to special conditions, such as the humus-cups of *Thayeria* and the ant-chambers of *Lecanopteris*.

Such remarkable structures as the two just mentioned have ever been favorite subjects of study; but of greater real importance than the study of these curiosities, interesting because of their rareness, is the causal interpretation of more common phenomena, such as length of stipe; its articulation; the presence or absence of indusia; the permanence of root hairs; and ciliate or serrate margins. For the explanation of each of these characteristics this family, as developed at San Ramon, furnishes concurrent evidence in two or more genera. I believe that the application of rich material to such everyday, but too often ignored problems as these will be regarded as a real service. The range of forms and habitats could have been widened by including some of the neighboring orders, taking in the aquatic *Ceratopteris,* the climbing leaves of *Lygodium,* the arboreous *Cyatheaceæ,* and the peculiarly xerophytic, sometimes rootless *Hymenophyllaceæ,* but the sacrifice of genetic homogeneity is too great a price to repay me for extending to this wider field.

The subjects in this paper are taken up in the following order:

I. The origin and geographical affinity of the San Ramon fern flora.
II. Local physiography and classification by environment.
III. Adaptations to common environment and to special conditions.
IV. Systematic application of the results.

[2] *Flora* (1902), **90**, 114–160.

I. THE ORIGIN AND GEOGRAPHICAL AFFINITY OF THE SAN RAMON FERN FLORA.

List of species with geographical range.

Name.	General distribution.	Greatest northward range from San Ramon.
Didymochlæna truncatula J. Sm.	Pantropic	Luzon.
Cyclopeltis presliana (J. Sm.) Berkeley	Malaya	Do.
Polystichum nudum Copel.	Local	
P. aristatum (Sw.) Presl	Africa to Polynesia	Japan.
Mesochlæna polycarpa R. Br.	Malaya	
Nephrodium 1712	Local	
Nephrodium 1649	do	
Nephrodium 1588	do	
Nephrodium 1714	do	
Nephrodium 1713	do	
Nephrodium erubescens (Wall.) Diels	India, Malaya	
N. syrmaticum (Willd.) Baker	India, Malacca	Luzon.
N. sparsum Don.	India, Malaya	China.
N. intermedium (Bl.) Baker	India, Borneo	Japan.
N. blumei (Moore) Hk.	Malaya	Luzon.
N. setigerum (Bl.) Baker	India to Polynesia	Japan.
N. Foxii Copel. ined.		Luzon.
N. immersum Hooker	Assam to New Caledonia	Do.
N. (Eunephrodium) truncatum Pr.	India to Polynesia	Do.
N. canescens (Bl.) f. nephrodiiformis Christ	Malaya	Do.
N. canescens (Bl.)	do	Do.
N. diversilobum Presl	Celebes	Do.
N. asymmetricum (Fée) Christ		
Nephrodium 1647	Local	
N. lineatum (Bl.) Presl	Malaya	Do.
N. urophyllum (Wall.) Bedd.	India to Australia	Do.
N. aridum (Don) J. Sm.	India to Polynesia	Do.
N. cucullatum (Bl.) Bak.	do	Do.
N. ferox Moore	Malaya	Do.
N. pteroides (Retz.) J. Sm.	India to Polynesia	China.
N. cyatheoides (Kaulf.) Presl	Sumatra, Hawaii	
N. procurrens (Mett.) Baker	India, Malaya	
Nephrodium 1685	Local	
Nephrodium 1605, 1677	do	
Nephrodium 1571	do	
N. Bordenii Christ ined.		Luzon.
Meniscium triphyllum Swtz. (?)	India, Java, Borneo	Formosa.
Aspidium difforme Bl.	Malaya	Luzon.
A. decurrens Presl	India to Polynesia	Do.
A. angulatum J. Sm.	Java, Borneo, New Guinea	
A. leuzeanum (Presl) Kze.	India to Polynesia	China.
A. leuzeanum var?	Local	
Psomiocarpa apiifolia Presl		Luzon.
Stenosemia aurita Presl	Malaya, Polynesia	Leyte.
S. pinnata Copel.	Local	
Leptochilus hydrophyllus Copel.	do	
L. heteroclitus (Pr.) Christensen	India, Melanesia	Luzon.
L. latifolius (Meyen) Christensen		Do.
Dipteris conjugata (Kaulf.) Reinw.	Malaya, Polynesia	Formosa.
Arthropteris obliterata J. Sm.	Africa to Polynesia	

List of species with geographical range—Continued.

Name.	General distribution.	Greatest northward range from San Ramon.
Nephrolepis cordifolia Presl	Pantropic	Japan.
N. hirsutula Presl	do	Luzon.
N. laurifolia Christ	Celebes	
Oleandra colubrina v. nitida Copel.	Mindanao, type in Luzon	
O. neriiformis Cav.	Pantropic ?	Do.
Humata heterophylla Desv.	Malaya, Polynesia	Do.
H. gaimardiana (Gaud.) J. Sm.	Burma, Polynesia	Do.
H. parvula (Wall.) Mett.	Malaya	
Davallia brevipes Copel.	Local	
D. denticulata (Burm.) Mett.	Malaya	China.
D. pallida Mett.	Borneo, Samoa	
D. solida Sw.	Malaya, Polynesia	Luzon.
D. decurrens Hk.		Bohol.
Microlepia hirsuta (J. Sm.) Presl		Luzon.
M. pinnata (Cav.) J. Sm.	Malaya, Polynesia	Do.
M. strigosa (Thunb.) Presl	India to Polynesia	Japan.
M. Speluncæ (L.) Moore	Pantropic	China.
Odontosoria retusa (Cav.)	New Caledonia, Celebes	Luzon.
Dennstaedtia Smithii (Hk.) Moore	Java (?)	Formosa.
D. Williamsi Copel.	Local	
D. cuneata (Hooker) Moore	Batjan	Luzon.
D. erythrorachis (Christ) Diels	Celebes	
Lindsaya hymenophylloides Bl.	Java to New Caledonia	Do.
L. Merrilli Copel.		Do.
L. Havicei Copel.		Mindanao.
L. scandens Hook.	Malay Peninsula	Luzon.
L. gracilis Bl.	Malaya	Do.
L. blumeana (Hook.) Kuhn	Java, Celebes	Leyte.
L. decomposita Willd.	India to Polynesia	Luzon.
L. pulchella (Hook) Mett.	Polynesia	Do.
Coniogramme fraxinea (Don) Diels	India to Polynesia	Japan.
C. serrulata (Blume) Fée	Java	
Loxogramme conferta Copel.		Mindanao.
L. iridifolia (Christ)	Celebes	
L. involuta (Don) Presl	India to Polynesia	Luzon.
Syngramma alismaefolia (Presl) J. Sm.	Singapore, Borneo	Do.
Callipteris cordifolia (Bl.)	Africa to Polynesia	Leyte.
C. prolifera Bory	Africa, Polynesia	Luzon.
C. esculenta (Retz.) J. Sm.	India, Polynesia	Formosa.
Diplazium pallidum Bl.	Java, Queensland	Luzon.
D. sp. (near tomentosum Bl.) 1667	Local	
D. bulbiferum Brack.		Do.
D. tenerum Presl	Malacca, Polynesia?	Leyte.
D. Williamsi Copel.		Luzon.
D. dolichosorum Copel.	Local	
D. sorsogonense Presl	India, Malacca	Do.
D. meyenianum Presl		Do.
D. fructuosum Copel.	Local	
D. polypodioides Bl.	India, Malaya, Australia	Do.
Asplenium musaefolium Mett.	Malacca to Polynesia	Do.
A. Phyllitidis Don.	India to New Guinea	Samar, Negros.
A. squamulatum Bl.	Malaya	Luzon.
A. epiphyticum Copel.		Mindanao.

List of species with geographical range—Continued.

Name.	General distribution.	Greatest northward range from San Ramon.
A. subnormale Copel.		Luzon.
A. resectum Smith	Africa to Polynesia	Japan.
A. vulcanicum Bl.	Ceylon, Malaya	
A. tenerum Forst.	India to Polynesia	Luzon.
A. macrophyllum Swtz.	do	Hongkong.
A. caudatum Forst.	Pantropic	Luzon.
A. cuneatum Lam.	do	Hongkong.
A. affine Sw.	Ceylon to Polynesia	Luzon.
A. Belangeri Kze.	Sumatra to Amboyna	
A. scandens J. Sm.	Borneo to Fiji	Leyte.
Athyrium silvaticum (Bl.) Milde	Java, Celebes	Luzon.
Scolopendrium pinnatum J. Sm.		Do.
S. schizocarpum Copel.	Local	
Stenochlæna sp.		
S. subtrifoliata Copel.	Local	
Blechnum egregium Copel.		Do.
Adiantum philippense L.	Pantropic	Hongkong.
A. diaphanum Bl.	Java to New Zealand	Do.
A. mindanaoense Copel.	Local	
Schizostege calocarpa Copel.	do	
S. pachysora Copel.	do	
Onychium tenue Christ	Java, New Guinea	Luzon.
Pteris opaca J. Sm.	Celebes?	Samar or Cebu.
P. longifolia L.	Cosmopolitan	Loo-Choo.
P. melanocaulon Fée		Luzon.
P. ensiformis Burman	India to Polynesia	Loo-Choo.
P. quadriaurita setigera Hk.	India, Malaya	Hongkong.
P. pluricaudata Copel.	Local	
P. excelsa Gaud.	India, Hawaii	Luzon.
P. tripartita Swtz.	Africa to Polynesia	Do.
Pteridium aquilinum (L.) Kuhn	Cosmopolitan	Siberia.
Monogramma trichoidea J. Sm.	Borneo	Luzon.
Vittaria alternans Copel.	Local	
V. falcata Kze. (?)	Malaya	Do.
V. minor Fée	Borneo	"Philippines."
Antrophyum latifolium Bl.	India, Java	
A. plantagineum (Cav.) Kaulf.	India to Polynesia	Luzon
A. semicostatum Bl.	Ceylon to Polynesia	Do.
A. reticulatum Kaulf.	Madagascar to Polynesia	Do.
Hymenolepis spicata (L. f.) Presl	do	China.
Taenitis blechnoides Swtz.	Ceylon to Polynesia	Luzon.
Niphobolus varius Kaulf.	Java, Mariannes	China.
N. adnascens (Sw.) Kaulf.	Africa to Polynesia	Luzon.
N. Lingua (Thunb.) J. Sm.	India, Malaya	Japan.
N. nummulariæfolius (Sw.) J. Sm.	India to Celebes	Luzon.
Prosaptia contigua Presl	Ceylon to Polynesia	Do.
P. cryptocarpa Copel.		Mindanao.
Polypodium caespitosum Mett.	Java	Luzon.
P. cucullatum Nees	Ceylon to Polynesia	Do.
P. gracillimum Copel.		Do.
P. pediculatum Baker	Borneo	
P. celebicum Bl.	Sumatra to Celebes	
P. obliquatum Bl.	India, Malaya	Do.

List of species with geographical range—Continued.

Name.	General distribution.	Greatest northward range from San Ramon.
P. papillosum Bl.	Malaya	Luzon.
P. (Goniophlebium) subauriculatum Bl.	India to Polynesia	Do.
P. Beddomei Baker	Malacca, Burma	Do.
P. (Phymatodes) accedens Bl.	Malaya, Polynesia	Do.
P. revolutum (J. Sm.) Christens.	Malaya, New Caledonia	Do.
P. punctatum (L.) Christ	Africa to Polynesia	Do.
P. Zippelii Bl.	Java, India	Do.
P. musaefolium Bl. ?	Malaya, New Guinea	
P. dolichopterum Copel.	Local	
Polypodium incurvatum Bl.	Malaya	
P. commutatum Bl.	Java	Do.
P. Phymatodes L.	Africa to Polynesia	Loo-Choo.
P. nigrescens Bl.	India to Polynesia	Luzon.
P. Schneideri Christ	Celebes	
P. palmatum Bl.	Malaya	Do.
P. angustatum Bl.	Java, Celebes	
P. albido-squamatum Bl.	Malaya, New Guinea	Do.
P. (Selliguea) caudiforme Bl.	Malaya, Polynesia	
P. macrophyllum Mett.	China to New Guinea	Do.
P. (Myrmecophila) sinuosum Wall.	Malaya, New Hebrides	Do.
P. (Drynariopsis) heracleum Kze.	Java, Celebes	Do.
Lecanopteris pumila Bl.	Malaya	
Photinopteris speciosa (Bl.) Presl	do	Do.
Dryostachyum pilosum J. Sm.	do	Do.
Thayeria Cornucopia Copel.	Local	
Drynaria quercifolia (L.) J. Sm.	India to Polynesia	China.
D. rigidula (Sw.)	do	Luzon.
Achrostichum aureum L.	Pantropic	Formosa.
Lomagramma pteroides J. Sm. (?)	Celebes (?)	Luzon.
Cheiropleuria bicuspis Presl	Java to New Guinea	Formosa.

According to their distribution, as now known, these ferns may be classified as follows:

Species common to the Malayan region and to China, Japan or Formosa	31
Malayan species reaching Luzon	77
Malayan species passing Mindanao but not reaching Luzon	8
Malayan species not passing Mindanao	23
Philippine species not confined to Mindanao	14
Species confined to Mindanao, not to San Ramon	4
Local species	27
Total	184

The local species and those confined to Mindanao are valueless in a statistical study of the affinities of the flora, and, in the same connection, the species confined to the Philippines may best be regarded as endemic and left out of account. There are then left 139 species known to occur outside this Archipelago.

Every one of these 139 species is Malayan, 31 of them being known in China, Japan or Formosa also. Nearly half of these 31 are known only in southern China, including Hongkong. Their migration northward into that region was almost certainly independent of that into the Philippines. Of the few remaining species, a few stop at Formosa and a few others reach as far as Japan. Of the entire 139 species, only a single one, *Pteridium aquilinum*, the most cosmopolitan of ferns, has such a distribution that it could as well be regarded as an immigrant from the north as from the south. Since it is impossible that many other ferns, and improbable that any others have come to Mindanao from north of Luzon, it is likewise unlikely that *Pteridium* is an immigrant from that side, especially since we have here the lanuginose form usual in the Tropics.

If, as is commonly done, we picture the migrations of plants as waves rolling on to a land from all sides on which it has points of biological contact, we should expect, on geographical grounds, to find such waves to have rolled on to the Philippines from Malaya on the south and Formosa on the north. The strength of these waves must depend on the ease of migration and on the "pressure" of the flora back of them. The character of the waves must be determined by the partial presssure of the different constituents of the parent flora.[3] The ease of migration depends upon the distance and the natural obstacles. Whatever their means of travel, ferns readily cross bodies of water.[4]

From its greater proximity and its incomparable wealth of ferns, it was easy to expect that Malaya would furnish by far the larger part of the San Ramon fern flora. Still, I did not at all anticipate that the entire flora would prove Malayan, and I know of no other case of any very rich flora which, aside from endemic species, is derived entirely from immigration from one side. Nor does the endemic element really constitute an exception here. *Polystichum nudum* belongs to the cosmopolitan "*aculeatum*" group. *Loxogramme conferta* is a very isolated species but in a Malayan genus. Two species are placed in *Schizostege*, a genus hitherto known only from Hawaii. *Schizostege* seems to be a generalized type, intermediate between *Cheilanthes* and *Pteris*, and its discontinuous distribution may be another evidence of antiquity, the Hawaiian fern flora being strongly Austro-malayan. The other local ferns all belong to groups characteristic of the islands to the south. This is no less true of the 14 species peculiar to the Philippines, of which only *Psomiocarpa apiifolia* is without very close Malayan cousins.

The non-polypodiaceous fern flora yields altogether harmonious evidence. A notable example is *Dicksonia chrysotricha* Hassk., the genus not being known hitherto from the Philippines.

The conclusion derived from the study of the San Ramon collections is applicable to the whole of Mindanao. I have at present fully 100

[3] Of course the effectiveness of migration depends also on the power of the migrants to live, with or without modification, in the new environment.

[4] Treub: Notice sur la nouvelle flore de Krakatoa. *Ann. Jard. Buitenz.* (1888) 7. Schimper: Plant Geography, Engl. Translation (1903), 80.

Polypodiaceæ from various parts of the island, which are not known about San Ramon. If these be considered each by itself, there is not one which can be regarded as a probable immigrant from the north.

To return to the wave figure, we see that no wave from north of Luzon has brought any ferns to Mindanao. Of the many waves from Malaya, some have reached only to Mindanao, a few to the Visayan Islands, the greatest force of them to Luzon, and a few still farther. It is not irrelevant to point out here that the Malayan element is little short of absolute, even in Luzon. One hundred and twenty-two, or 80 per cent, of the San Ramon ferns with the outside distribution of which I am at all acquainted, occur in Luzon. Beside sharing such a proportion as this of the ferns of the southern Philippines, Luzon has many characteristic Malayan species still unknown elsewhere in the Philippines. Beside these, it has a few Himalayan species. Since the accepted distribution of some species such as *Nephrodium flaccidum*, *Meniscium cuspidatum*, etc., not yet found in China, is from the Himalayas through Malaya to Luzon, and many species have this range and China in addition, it is not unlikely that plants now known only from the Himalayas and Luzon, or even from these and China, have at some time been continuously distributed through Malaya. Therefore, the existence of these species is not in itself strong evidence of a direct migration between Luzon and continental Asia.

There are a few ferns known only from Luzon and the regions north of it: *Polystichum varium* Pr., *P. deltodon* Diels, *Nephrodium erythrosorum* Hook., *Athyrium anisopteron* Christ, and *Plagiogyria stenoptera* Diels. Of these, *Polystichum varium* is as likely to have originated in one part of its range as in another. The *Nephrodium* is as likely to have migrated northward, because the greatest development of its genus and section (*Lastraea*), is Malayan. *Plagiogyria* is most probably a northward migrant from Luzon because the Philippines have about as many species in the genus as all the rest of the world. Finally, *Woodwardia radicans* has been regarded as a northern element of this flora but our plant is intermediate between the Japanese and Formosan form (*W. orientalis* Swtz.) and the widespread typical form found in Java; therefore, it is strongly probable that the Japanese plant was derived from Luzon. The boreal element in the fern-flora even of Luzon is, then, exceedingly insignificant.[5]

I have spoken of the Mindanao fern flora as immigrant from Malaya. The greater part of my evidence proves only a close affinity, but not the direction of migration. That this has been northward for the ferns as a whole is indicated by the high proportion of endemic species (showing the Philippines to be not truly continental islands, while floras are

[5] This is by no means so true for the whole of the flora of Luzon. Some families better represented to the north, such as Compositae, show relatively large numbers of immigrants here.

ultimately of continental origin), the greater aggregate fern wealth of Malaya as compared with the Philippines, the affinity of the Malayan fern flora with that of Polynesia, Australia, India, and even Africa (showing its great age), the scarcity of mammals in the Philippines and the geological evidence that Mindanao is younger than the highlands of Borneo. However, Mindanao is, for its area, richer in ferns than any other large island in the world and it can not be doubted that, in its measure, emigration southward has occurred and is occurring.

The comparative share of Borneo and Celebes in the composition of the Mindanao flora is a most interesting question, but one the present discussion of which may furnish the future with little of more final value than the evidence of present ignorance of the flora of all three islands. This much, at least, we can establish, that the connection with Celebes and presumably through it with other lands to the south and east, has been much more intimate than most writers in this field have imagined. San Ramon is at the extreme southwest of Mindanao, from which a chain of islands dots the sea at intervals of at most a few miles, quite across to Borneo. Several of these islands are sufficiently elevated to bear an abundance of tree-ferns. Migration to or from Borneo looks exceedingly easy.

Of the San Ramon ferns, *Vittaria minor* and *Polypodium pediculatum* are already known only from Borneo, or Borneo and the Philippines. From their further range, *Nephrodium intermedium, Lindsaya scandens, Syngramma alismæfolia, Monogramma trichoidea* and *Polypodium Beddomei* may be regarded as added evidence of connection with Borneo. Not one of the genera notable for its occurrence or development in Borneo, such as *Matonia, Syngramma, Dipteris* and *Taenitis,* is at all notable at San Ramon.

Valid evidence of floristic affinity with Celebes is furnished by *Nephrodium diversilobum* (even if it is not specifically distinct from *N. canescens*), *Nephrolepis laurifolia, Dennstaedtia erythrorachis, Pteris opaca* and *Polypodium Schneideri,* known only from Celebes and the Philippines. In a few cases, as in that of the conspicuous *Polypodium heracleum, Athyrium silvaticum,* and *Loxogramme iridifolia,* Celebes connects Mindanao with Java. A relationship to the flora of the remote east and south is shown by *Lindsaya pulchella* and *Odontosoria retusa.* More significant than mere community of species is the sharing of larger or more isolated groups. *Lomagramma* is probably a good instance in this connection, though its confusion with a *Leptochilus* has made its actual range dubious. *Thayeria* is found in Luzon, Mindanao and New Guinea.

Judging, then, from our present knowledge of the distribution of Malayan ferns, there is a closer affinity between the fern flora of San Ramon and that of Celebes than between the former and that of Borneo, in spite of the fact that the geographical connection with Borneo is [illegible] Mindanao as a whole naturally shows a stronger [illegible] affinities. The most conspicuous example is [illegible] with one species on Mount Apo, one in Celebes and [illegible] Sabah.

[illegible] GEOGRAPHY

San Ramon is situated [illegible] miles from Zamboanga. The coastal plain [illegible] meters [illegible] at least [illegible] meters in altitude [illegible] The [illegible] of this plain and the mountains are covered with virgin forest, while [illegible] cultivation. The streams, the largest of which is the San River, are [illegible] the plain [illegible] flood-plain a few meters wide [illegible] kilometers into the mountains. Above this the river [illegible] a cañon, in most places quite without any level [illegible] the walls of which are in [illegible] places too steep to be [illegible]. At the [illegible] of this cañon, at an altitude of 200 meters, the river forks, the two smaller [illegible] through practically impassable gorges. The ridge between these is 600 meters high at its [illegible]. This and the [illegible] ridges and the mountains nearer the coast are remnants of a plateau, which [illegible] kilometers farther inland, at an altitude of about 800 meters, is still well preserved, the creeks being hardly [illegible] meters below its level. From the back of this plateau, altitude 800 meters, Mount Balatoc rises abruptly to 1200 meters. The summit is flat, but small.

As I find it convenient to classify the vegetation, we have here:

The strand, including all places where the plants are subject to the deposition at any time of salt from the atmosphere.

The [illegible], only a few spots, the only [illegible] in which is [illegible].

The [illegible], including

a. The low jungle, low places near the coast, wet throughout the dry season.

b. The parang, drier grasslands without large trees, whether as a result of the dryness of the soil or because the trees have been removed.

The high forest, in Brandis' sense, a forest containing large trees, in which there is a pronounced dry season. This includes part of the coastal plain and the lower mountains and ridges.

The rain forest, a forest in which the dry season is broken by showers and practically no plants are deciduous. This includes the high plateau and the slopes of Mount Balaine.

The mossy forest, stunted woods with an exuberance of epiphytes. This is rather feebly developed on the main summit.

Like other classifications, this one is artificial; the several divisions [illegible], they are not absolutely coordinate, and they are not [illegible], but it is convenient, applicable to the field, and more approximately correlated with the ferns' physiology than is any other classification that preserves any semblance of geographical cohesion. From the standpoint of the fern, environments would be classified almost solely with regard to the physiological dryness of the air. In a general way, a classification can be made along this line, as this one is; but in detail it is not geographically feasible, because of great and very local differences. As an illustration of these; the Negritos of Bataan dig holes more than a meter deep and hardly 30 centimeters across, to gather a wild yam; these holes are common in the parang but harbor a fern vegetation characteristic of rocky gorges in the high forest.

Compared with the moisture, the temperature plays a very minor rôle in the control of the distribution of these ferns, and probably none at all in the modification of their structures, except indirectly as it influences their transpiration. Compared with the moisture of the air, that of the ground is likewise an insignificant factor both in distribution and structure. This is more true of plants in general than many ecologists seem to realize, but it is, of course, especially so of a group like the ferns, a large proportion of which display their independence of ground-moisture by being epiphytes; moreover, poor water-absorbing or water-conducting organs increase a plant's relative dependence on atmospheric moisture; ferns have ill-developed conducting systems, as compared with flowering plants. Tree ferns grow only where the air is most moist, in cañons, or forested pockets and cirque-like basins, and the few scandent *Polypodiaceæ*, in other than the wettest places, are strongly xerophytic in structure.

The moisture of the atmosphere and that of the soil are of course intimately related; if the ground is wet, air standing over it becomes so, and if the air in a given place has a very drying action on plants it will have the same effect on the soil. However, the influence of the moisture of the soil and of that of the air can sometimes very readily be distinguished; for instance, in the cases of plants of different height, growing together on the same ground, it is wholly dependent on the atmosphere whether or not the taller plant must differ from the shorter. There is no environment so damp but that on the whole the taller ferns have more xerophytic leaf-structures than those the leaves of which are close to the ground; but the limit of height is very different in different environments in which the structures of low-growing plants are essentially alike. The soil in the low jungle never dries; but there are times when the air does so, sufficiently and for a long enough period of time to make it an impossible habitat for any of the giant ferns of the rain forest or of any ferns of like structure and stature.

In the following table are listed the ferns of the several habitats the structure of which I was able to study with fresh material; the thickness of frond and of outer wall of each epidermis, the presence or absence of chlorophyll in the epidermis, and of a chlorophyll-less hypodermis, the number of stomata per square millimeter, and the average length and width of a stoma are given.

STRAND.

Name.	Thickness of frond.	Thickness of upper epidermal wall.	Thickness of lower epidermal wall.	Chlorophyll in epidermis.	Hypodermis.	Stomata per square millimeter.	Average length and width of stomata.
	mm	μ	μ				μ
Davallia denticulata	0.27	2.5	4.5	—	+	116	39 x 30
D. solida	.21	3	4	—	+	140	40 x 29
Niphobolus adnascens	1.49	10	10	—	—	28	51 x 39
Polypodium sinuosum	.57	5	7	—	—	28	47 x 45
Asplenium macrophyllum	.22	9	6	—	—	60	37 x 36
A. musaefolium	.71	5	7.5	—	+	26	56 x 49
Average	.58	5.75	6.5	16⅔%	83⅓%	36.3	45 x 36.5

SALT MARSH.

Name.	Thickness of frond.	Thickness of upper epidermal wall.	Thickness of lower epidermal wall.	Chlorophyll in epidermis.	Hypodermis.	Stomata per square millimeter.	Average length and width of stomata.
Achrostichum aureum	0.39	5	3.5	—	+	220	47 x 34

LOW JUNGLE.

Name.	Thickness of frond.	Thickness of upper epidermal wall.	Thickness of lower epidermal wall.	Chlorophyll in epidermis.	Hypodermis.	Stomata per square millimeter.	Average length and width of stomata.
Nephrodium procurrens	0.10	3	2.5	+	—	107	36 x 19
Nephrodium (1677)	.065	2	1	+	—	110	30 x 16
Callipteris prolifera	.14	1.5	1	+	—	60	37 x 25
Callipteris esculenta	.19	2	1	+	—	450	31 x 20
Polypodium Schneideri	.12	3	2	+	—	68	46 x 30
Average	12.3	2.3	1.5	90%	0	159	31.2 x 22

PARANG.

Name.	Thickness of frond.	Thickness of upper epidermal wall.	Thickness of lower epidermal wall.	Chlorophyll in epidermis.	Hypodermis.	Stomata per square millimeter.	Average length and width of stomata.
Nephrodium aridum	0.10	5	2.5	Trace.	—	200	32 x 17
N. cucullatum	.14	6.5	2	Trace.	—	360	36 x 19
N. pteroides	.10	3	1.5	+	—	250	27 x 18
Nephrolepis hirsutula	.23	3	3	—	—	182	45 x 26
Pteris longifolia	.18	3.5	2	+	—	94	51 x 32
P. ensiformis	.12	2.5	2.5	+	—	56	47 x 23
Average	.145	3.92	2.25	66⅔%	0	180.3	39.7 x 22.5
Average for plain	.185	3.18	1.91	77.3%	0	170.8	35.8 x 22.3

HIGH FOREST.

A. Gorges and Other Moister Places.

Name.	Thickness of frond.	Thickness of upper epidermal wall.	Thickness of lower epidermal wall.	Chlorophyll in epidermis.	Hypodermis.	Stomata per square millimeter.	Average length and width of stomata.
Nephrodium (1685)	0.07	4	2	+	—	120	40 x 19
N. syrmaticum	.08	2.5	1	+	—	400	37 x 19
N. setigerum	.12	3	1	+	—	210	33 x 26
N. Foxii	.10	2	1	+	—	200	36 x 21
N. immersum	.08	1	1	+	—	150	28 x 19
N. ferox	.13	4	3	Trace.	—	100	38 x 22
N. Bordenii (1688)	.09	3	1	+	—	225	36 x 21
Aspidium angulatum	.11	4	1	+	—	36	44 x 28
A. leuzeanum	.12	2	1	+	—	60	43 x 26
Leptochilus heteroclitus	.165	1.5	1.5	+	—	34	51 x 25
Microlepia Speluncæ	.19	4	3	+	—	92	60 x 30

HIGH FOREST—Continued.

A. Gorges and Other Moister Places—Continued.

Name.	Thickness of frond.	Thickness of upper epidermal wall.	Thickness of lower epidermal wall.	Chlorophyll in epidermis.	Hypodermis.	Stomata per square millimeter.	Average length and width of stomata.
	mm.	μ	μ				μ
Odontosoria retusa	0.175	1.5	3.5	+	—	36	65 x 38
Dennstaedtia cuneata	.13	4	3.5	Trace.	—	235	41 x 21
Diplazium tenerum	.24	2	1	+	—	88	58 x 43
Diplazium Williamsi	.15	2	1	+	—	64	43 x 28
D. meyenianum	.065	1.5	1	+	—	128	41 x 29
D. polypodioides	.10	1	1	+	—	300	25 x 29
Asplenium subnormale	.22	1	1	+	—	28	65 x 34
Scolopendrium pinnatum	.34	2.5	2	+	—	24	58 x 34
Adiantum philippense	.08	1	1	+	—	80	33 x 29
Pteris tripartita	.13	2	1	+	—	148	29 x 25
Polypodium macrophyllum	.30	1.5	1.5	+	—	26	58 x 31
Average	.145	2.32	1.56	95.5%	0	126.5	43.5 x 28.1

B. Terrestrial in Drier Woods.

(1) DIMORPHOUS ROSETTE-FORMERS.

Name.	Thickness of frond.	Thickness of upper epidermal wall.	Thickness of lower epidermal wall.	Chlorophyll in epidermis.	Hypodermis.	Stomata per square millimeter.	Average length and width of stomata.
Psomiocarpa apiifolia	0.11	4	1.5	+	—	38	58 x 25
Stenosemia aurita	.13	1.5	1	+	—	50	46 x 32
S. pinnata	.21	1.5	1	+	—	40	53 x 36
Leptochilus latifolius	.21	3	1	+	—	52	45 x 30
Average	.165	2.5	1.1	100%	0		

(2) WITHOUT PRONOUNCED ROSETTE-HABIT

Name.	Thickness of frond.	Thickness of upper epidermal wall.	Thickness of lower epidermal wall.	Chlorophyll in epidermis.	Hypodermis.	Stomata per square millimeter.	Average length and width of stomata.
Cyclopeltis presliana			2	+	—	24	60 x 37
			2	Trace.	—	44	64 x 34
Nephrodium diversilobum			2	+	—	140	41 x 19
Aspidium difforme			1	+	—	56	61 x 36
A. decurrens			1.5	+	--	44	43 x 31
Microlepia pinnata			4	+	—	28	72 x 51
Lindsaya gracilis			3.5	+	—	42	45 x 32
L. decomposita			1.5	+	—	80	48 x 34
Syngramma alismæfolia			3	+	—	24	77 x 57
Callipteris cordifolia			1	+	—	13	87 x 51
Diplazium pallidum			2	+	—	56	58 x 31
D. (near tomentosum)			2	+	—	22	72 x 41
D. bulbiferum			1	+	—	28	69 x 39
Onychium tenue		2.5	2.5	Trace.	—	100	36 x 28
Pteris opaca			3.5	—	—	128	52 x 36
P. melanocaulon			2	Trace.	—	204	52 x 27
P. quadriaurita			1	+	—	160	37 x 21
P. pluricaudata				+	—	60	47 x 20
Tænitis blechnoides				+	—	16	57 x 47
Average	.232	3.48	1.96	89.1%	0	63	55.4 x 34.3

HIGH FOREST—Continued.

C. EPIPHYTIC.

Name.	Thickness of frond.	Thickness of upper epidermal wall.	Thickness of lower epidermal wall.	Chlorophyll in epidermis.	Hypodermis.	Stomata per square millimeter.	Average length and width of stomata.
	mm.	μ	μ				μ
Nephrolepis laurifolia	0.28	6.5	7	—	—	90	40 x 26
Humata heterophylla	.28	7	7	—	—	4	58 x 36
H. gaimardiana	.22	6.5	1.5	—	+	95	26 x 18
H. parvula	.26	3.5	4.5	—	+	50	45 x 33
Davallia brevipes		3	4.5	—	—	26	49 x 34
D. pallida	.17	2.5	3	+	—	56	44 x 33
D. decurrens	.09	2.5	2.5	+	—	102	46 x 27
Lindsaya Merrilli	.065	2	1	+	—	44	44 x 33
Loxogramme conferta	.70	7	3.5	+	—	36	52 x 32
L. iridifolia	.42	3.5	2.5	+	—	8	82 x 55
L. involuta		4.5	6	+	—	22	57 x 52
Asplenium squamulatum	.60	10	10	—	Trace.	20	62 x 50
A. epiphyticum	.14	2	1	+	—	28	51 x 43
A. vulcanicum	.17	6	6	—	—	32	54 x 28
A. tenerum	.28	7.5	6	+	—	24	57 x 40
A. caudatum	.15	2	1.5	+	—	44	58 x 32
A. cuneatum	.11	4	4	+	—	{ 60	40 x 34
						{ 90	42 x 25
Stenochlæna sp.	.27	5	2.5	+	—	16	68 x 37
Vittaria falcata	.45	4.5	3	+	—.	7	79 x 41
Antrophyum semicostatum	1.06	7	6		—	13	93 x 87
Hymenolepis spicata	.60	8	4	—	+	32	51 x 40
Niphobolus varius	.87	7	8	—	+	25	60 x 40
N. Lingua	.82	5.5	16	—	+	172	38 x 28
N. nummulariæfolius	.86	4	8	Trace.	+	64	38 x 25
Polypodium revolutum	.56	8.5	8.5	—	+	26	68 x 34
P. incurvatum	• .23+	8	13	—	+	60	46 x 31
P. commutatum	.21	2	2	+	—	76	38 x 34
P. Phymatodes	.37	9.5	9.5	—	+	36	47 x 39
P. nigrescens	.13	1	1	+	—	52	51 x 31
P. albido-squamatum	.29	3	7	—	+	60	39 x 31
P. heracleum	.20	5	6	—	+	52	42 x 34
Lecanopteris pumila	.17	3.5	2	—	—	38	45 x 36
Photinopteris speciosa		6.5	7	—	+	60	52 x 43
Dryostachyum pilosum	.23	4	4	—	+	72	39 x 30
Drynaria quercifolia	.27	8	7	—	+	72	44 x 31
D. rigidula	.17	5	5	—	—(?)	96	44 x 37
Average	.3725	5.08	5.24	45.95%	39.19%	50.9	50.6 x 34.9
Average for high forest	.2682	3.89	3.83	72.56%	17.68%	74.6	50.1 x 32.4

RAIN FOREST.

A. TERRESTRIAL.

Name.	Thickness of frond.	Thickness of upper epidermal wall.	Thickness of lower epidermal wall.	Chlorophyll in epidermis.	Hypodermis.	Stomata per square millimeter.	Average length and width of stomata.
	mm.	μ	μ				μ
Didymochlæna truncatula	0.22	5	4	+	—	36	58 x 33
Polystichum nudum	.14	2	2	—	—	62	45 x 34
Nephrodium sparsum	.32	2	1.5	+	—	68	66 x 37
N. (1649)	.27	2.5	1.5	+	—	84	44 x 30
N. intermedium	.09	1	1	+	—	80	40 x 23
N. canescens nephrodiiformis	.065	1.5	1	+	—	185	38 x 21
N. canescens	.10	2.5	1	+	—	180	37 x 22
N. cyatheoides	.11	3	2.5	+	—	235	35 x 20
Meniscium triphyllum	.29	4.5	2	+	—	52	51 x 26
Aspidium leuzeanum. var.	.10	5	3	+	—	210	35 x 23
Microlepia strigosa	.10	1.5	1.5	+	—	165	39 x 19
Dennstaedtia Williamsi	.12	2	2	Trace.	—	260	42 x 24
Lindsaya blumeana	.12	3	1	+	—	42	49 x 25
Coniogramme fraxinea	.22	1	1	+	—	80	58 x 34
C. serrulata	.40	4	1	+ (?)	—	40	59 x 30
Diplazium dolichosorum	.29	3	1	+	—	80	56 x 33
D. fructuosum	.20	3	1	+	—	400	31 x 23
Asplenium resectum	.12	3	1.5	+	—	16	69 x 35
Athyrium silvaticum	.15	3	1.5	+	—	100	36 x 19
Blechnum egregium	.20	3.5	3	+	—	86	54 x 29
Adiantum diaphanum	.06	1	1	+	—	30	34 x 30
A. mindanaoense	.09	2	1.5	+	—	55	34 x 35
Schizostege calocarpa	.35	2.5	2	+	—	33	88 x 43
S. pachysora	.50	1	0.5	+	—	21	75 x 43
Pteris excelsa	.24	4	3	+	—	110	50 x 23
Antrophyum latifolium	.64	2.5	2.5	+	—	11	108 x 54
Polypodium dolichopterum	.21	1	1	+	—	15	93 x 51
Cheiropleuria bicuspis	.36	5	3.5	+	—	14	68 x 50
Average	.217	2.68	1.75	94.64%	0%	83.6	53.2 x 31

B. EPIPHYTIC.

Name.	Thickness of frond.	Thickness of upper epidermal wall.	Thickness of lower epidermal wall.	Chlorophyll in epidermis.	Hypodermis.	Stomata per square millimeter.	Average length and width of stomata.
Nephrodium (1712)	0.17	5	4	+	—	82	35 x 33
Arthropteris obliterata	.14	2	2	+	—	40	53 x 28
Nephrolepis cordifolia	.14	3	3	—	—	54	44 x 32
Microlepia hirsuta	.05	2	3	+	—	44	41 x 26
Lindsaya scandens	.14	2.5	1.5	+	—	29	53 x 31
Asplenium Phyllitidis	.53	5	12	—	+	28	64 x 36
A. Belangeri	.27	5.5	6	+	—	32	54 x 37
A. affine	.23	7	4	+	—	50	50 x 37
A. scandens	.20	2.5	1	+	—	40	42 x 21
Scolopendrium schizocarpum	.60	3	2	+	—	32	68 x 53
Stenochlæna subtrifoliata	.38	7	1.5	+	—	26	66 x 53
Monogramma trichoidea		2	2	+	—	10	50 x 25
Vittaria alternans	.50	4.5	3	Trace.	—	36	68 x 40
V. minor	.75	13	11	—	—	20	63 x 46

RAIN FOREST—Continued.

B. EPIPHYTIC—Continued.

Name.	Thickness of frond.	Thickness of upper epidermal wall.	Thickness of lower epidermal wall.	Chlorophyll in epidermis.	Hypodermis.	Stomata per square millimeter.	Average length and width of stomata.
	mm.	μ	μ				μ
Antrophyum plantagineum	0.39	11	9	+	—	5	102 x 48
A. reticulatum	.44	9	11	+	—	10	83 x 37
Prosaptia contigua	.22	5.5	2.5	+	—	46	63 x 54
P. cryptocarpa	.25	5	1	+	—	48	54 x 49
Polypodium cæspitosum	.77	2.5	2.5	+	—	88	47 x 46
P. pediculatum	.25	2	1.5	+	—	33	62 x 46
P. papillosum	.11	1.5	1	+	—	30	50 x 30
P. subauriculatum	.15	1	1	+	—	64	40 x 31
P. accedens	.47	4	4	+	—	23	46 x 34
P. punctatum	.40	3.5	3	+	—	29	42 x 31
P. Zippelii	.22	2	2	+	—	16	52 x 33
P. musæfolium (?)	.32	2	2	+	—	28	63 x 35
P. palmatum	.25	5	3	—	—	42	59 x 32
P. angustatum	.27	3.5	4	—	—	46	43 x 31
P. caudiforme	.32	4	5	—	+	60	50 x 36
Thayeria Cornucopia	.23	3	3.5	—	+	100	32 x 28
Lomagramma pteroides	.23	4	3	+	—	56	63 x 29
Average for epiphytes	.313	4.3	3.7	75.8%	9.68%	40.2	54.9 x 35.8
Average for rain forest	.2666	3.52	2.78	83.22%	5.08%	46.4	54.1 x 33.7

MOSSY FOREST.

A. TERRESTRIAL.

Name.	Thickness of frond.	Thickness of upper epidermal wall.	Thickness of lower epidermal wall.	Chlorophyll in epidermis.	Hypodermis.	Stomata per square millimeter.	Average length and width of stomata.
Dipteris conjugata	0.15	5	3	+	—	300	31 x 23
Oleandra colubrina (?)	.24	5	3	—	+	232	32 x 28
Dennstaedtia erythrorachis	.23	4	2	+	—	56	63 x 35
Average	.207	4.67	2.67	66⅔%	33⅓%	196	42 x 28.7

B. EPIPHYTIC.

Name.	Thickness of frond.	Thickness of upper epidermal wall.	Thickness of lower epidermal wall.	Chlorophyll in epidermis.	Hypodermis.	Stomata per square millimeter.	Average length and width of stomata.
Lindsaya Havicei	0.10	2.5	1	+	—	55	52 x 33
L. hymenophylloides	.13	1	1	+	—	85	66 x 40
L. pulchella	.19	1.5	1	+	—	22	66 x 34
Polypodium cucullatum	.17	5	5	+	—	28	53 x 45
P. gracillimum	.10	3	3	+	—	20	54 x 36
P. celebicum	.46	2	1.5	+	—	168	40 x 38
P. obliquatum	.26	5	2	+	—	55	50 x 38
Average for epiphytes	.21	2.86	2.07	100%	0%	54.7	54.4 x 37.7
Average for mossy forest	.209	3.4	2.25	90%	10%	113.5	50.7 x 35
Average for terrestrial	.195	2.95	1.82	90.34%	2.27%	105.2	49.1 x 29.5
Average for epiphytes	.3498	4.65	4.48	59.88%	27.78%	48.3	52.2 x 35.7
Average for all	.2663	3.77	3.09	75.74%	14.50%	77.9	50.6 x 32.5

A great many ferns range over more than one of these formations, both because they are not finely adapted to exact conditions—as, indeed, it is especially natural, in a mountainous country [a] that they should not be—and because, as already stated, the formations are not in most cases sharply bounded. In the table each plant is ascribed to the formation where the specimens represented in the table were collected; this was also the formation of which, on the whole, the plant seemed to me most characteristic.

THE STRAND.

The characteristic strand fern of the Philippines is *Niphobolus adnascens,* which is found on the tree trunks over almost every beach in the Islands. Associated with it on the coconut boles and trunks at San Ramon are a small *Davallia solida* and a remarkably dwarfed *D. denticulata.* A diminutive form of *Psilotum nudum* grows with them. A few miles from San Ramon, the mountains, bringing the high forest with them, come close to the sea and here *Polypodium sinuosum, Asplenium macrophyllum, A musaefolium,* and occasionally *Drynaria quercifolia,* are found on branches over the water where the reflected light and occasional spray must combine to make the habitat arid; these might all have been classified with the flora of the high forest, but the comparative want of ferns on the strand makes all but the last more conspicuous here. *Nephrolepis hirsutula* grows near enough to the beach at San Ramon to be a possible recipient of salt spray in hard winds. It is normally terrestrial but sometimes grows in a manner in which it has little if any connection with the ground. Next to *Niphobolus, Nephrolepis* is our most xerophytic genus, to judge by the difficulty encountered in drying the plants. With this occasional exception, every fern found on the San Ramon strand is epiphytic. The explanation of this habit is simple. The majority of the epiphytes in every formation—at any rate below the mossy forest—are vigorous xerophytes, in contrast on the whole, to the terrestrial ferns. On the strand, the epiphytic habitat is little if at all more arid than the terrestrial; it is more so with regard to the light, very little more so with regard to the wind, less so with regard to the salt. While the strand is altogether too arid to permit many terrestrial ferns even to become adapted, epiphytes can be at home on it with a comparatively slight modification of their usual life and structure. Structures enabling a plant to endure intense insolation make strong illumination a necessity; no other plants receive as much light as strand epiphytes. On rocky shores of the Gulf of Davao a terrestrial fern, *Cheilanthes Boltoni,* grows just above high tide; its genus is characteristic of arid America.

Of course, epiphytes are far from equally xerophytic, and only those which are the most so can endure strand conditions. These epiphytes

[a] Copeland: Variation of some California plants. *Bot. Gaz.* (1905). 38; 413.

all have thick leaves and this fact is the more conspicuous if other species in their genera are used for comparison, rather than if other formations in their entirety are brought into consideration. Thus the two *Davalliæ* on the strand average 0.24 millimeter in thickness; *D. pallida* and *D. decurrens*, of the high forest, average only 0.13 millimeter. If *Antrophyum* were represented here it would raise the average. *A. obtusum* is a strand fern along Illana Bay.[7] Whether thick or comparatively thin, all strand ferns here are coriaceous in texture; all except the *Niphobolus* are glabrous, and the latter is notably glabrous in its genus; the woolly covering of some of our *Niphobolus* species would be fatal where salt spray blows. All have a waxy-shining epidermis. All except one have a hypodermis lacking chlorophyll; this serves to mitigate the insolation more often than it operates as a reservoir of available water.

THE SALT MARSH.

The habitat of *Achrostichum* is a part of what Schimper calls the *Nipa* formation, which Whitford, to show how far it is from being a unit, calls the *Nipa-Acanthus* formation. *Achrostichum* never grows with luxuriant *Nipa*, but does with *Acanthus* and in marshes still fresher than those which are characterized by this plant, its landward boundary being the extreme limit of saltiness at any tide. Since *Acanthus* has a rather narrower landward range and is not pantropic, as is *Achrostichum*, this fresh-brackish marsh, where *Nipa* does not grow, might well be named after *Achrostichum*.

Achrostichum is structurally so perfectly adapted to its part arid, part watery habitat that, in spite of its occasional salty substratum and its exposure to the sun and at times strong winds, the fronds reach to a height of 1 meter to 1.8 meters, or more. The pinnæ are thick, coriaceous and entire, with thick epidermal walls and hyaline epidermis, underlain above by two layers of irregular hyaline hypodermis and below by one incomplete one. There are two layers of compact, palisade-like parenchyma. So far, the structure is xerophytic, which seems to be an adaptation to the insolation rather than to any saltiness of the substratum, since the hypodermis is rather thick-walled. But a large, stout fern with very vigorous neighbors must not have xerophytic structures which interfere with active photosynthesis. The stomata of *Achrostichum* are very numerous and rather large, occupying 35 per cent of the whole nether surface, a greater proportion than in any other fern of our flora. The air which surrounds this fern, though sometimes drying, is more often humid and a wet nether surface would long remain so. As a protection against the closure of the stomata by water, the margin, 0.33 millimeter broad, is thin, cartilaginous, hyaline, and deflexed; and the young fruiting surface is effectively covered by large paraphyses with branched, oily

[7] *Drymoglossum piloselloides*, a thick-leaved relative of *Niphobolus*, is a strand-epiphyte in some places in the Philippines.

heads (fig. 24). The roots of *Achrostichum* are fleshy, with the cortex composed mostly of intercellular spaces. An occasional root, without structure which is evidently different, grows upward out of the ground.

SAVANNA-WOOD.

This formation includes very nearly the whole of the coastal plain. Here, at and near the sea level, the dry season is longest and driest and the hills give no protection against the winds which, though that far from as strong as they would be farther from the equator, suffice constantly to remove the moister air next the ground and vegetation. More than one-third of this plain has been made drier artificially by clearing for cultivation, and there are areas of many hectares which seem naturally to be in grass. The most conspicuous features of this formation are the moderate size of the trees and their very open stand; still, in the back part of the plain are several rather close groves of some enormous *Myrtaceous* species, believed to be a *Eucalyptus*.[8] In spite of the lofty trees and close stand this portion is bionomically, unmistakably a part of the savanna-wood and not of the high forest. With regard to the ferns, the notable feature of this formation is the absolute absence of any characteristic epiphyte. *Asplenium musaefolium*, *Drynaria quercifolia*, and a few other species of the high forest are occasionally to be found, but no single one can be regarded as at home here. The commonest of these, *D. quercifolia*, is almost always deciduous in the savanna-wood, being quite without green leaves during the driest months, but in the high forest it is very rarely found in this condition.

I made hygrometric determinations, on the strand and about 300 meters inland, three times daily, for a long enough period to get a very good idea of the comparative atmospheric dryness of the two places during the dry season. These are published in detail as an appendix to my paper on the coconut,[9] and a summary will suffice here. The figures state the relative humidity.

Relative humidity.

NOVEMBER 15–30.	6 a. m.	12 noon.	4 p. m.
Strand	88.8	79.9	83.3
Interior	91.5	81.2	87.6
DECEMBER.	7.30 a. m.	11.30 a. m.	4 p. m.
Strand	84.27	79.2	80.52
Interior	85.00	79.28	81.85
JANUARY.			
Strand	84.74	75.65	80.28
Interior	84.13	74.65	80.60

[8] A. Gray: *Bot. Wilkes Explor. Exped.* (1854), 554. Caldera, near which this tree was found, is 4 miles from San Ramon, and the collection was doubtless from these groves. Maiden has confirmed the generic determination. *Proc. U. S. Nat. Mus.*, Wash. (1903), **26**, 691.

[9] *This Journal* (1906), **1**: 146.

The rains ceased in November. These results indicate—what indeed was reasonably to be anticipated, that during the rains and while their influence lasts, the air is moister away from the strand, but that with the advance of the dry season the savanna district becomes even drier than the strand. Occasional determinations showed this to be the case at all times except shortly after the rare showers throughout the succeeding months up to May.

As compared with the strand, the savanna-wood is less suited for epiphytic vegetation, because:

(a) During the driest season its air is even drier than that of the strand, thus necessitating most perfect structures in the ferns for the restriction of the loss of water.

(b) The illumination is not in general as intense as it well may be for plants with exceedingly xerophytic structure.

(c) Epiphytes here have no such supply of concentrated mineral food, as the strand plants have in the salt spray, which latter will allow an extremely limited transpiration to be sufficient to satisfy their needs.

The conditions are less severe for terrestrial than for epiphytic vegetation, but they still limit the development of ferns to a condition which is insignificant as compared with that found in the higher formations. While the atmospheric dryness stamps its character on all the vegetation of the coastal plain and amply justifies treating it as a single formation, there are great inequalities in the soil-moisture. Immediately back of the strand there are in several places considerable areas where the ground is wet throughout the severest droughts. In the wet season the air here must at all times be almost saturated. The two dominant genera in these low jungles, *Nephrodium* and *Callipteris*, are alike in venation, and so strikingly alike in aspect that sterile *C. esculenta* is constantly confused with *Nephrodium* by collectors. I have not yet been able to correlate the resemblance with the common environment. Several of these ferns have special devices to preserve the dryness of the reproductive structures; these will be taken up later.

The drier parts of the savanna have ferns with smaller fronds and a distinctly more xerophytic structure, as appears for instance, in the thicker epidermal walls. *Nephrodium* and *Pteris* are the dominant genera. Of *Pteris*, *P. opaca*, *P. melanocaulon* and *P. quadriaurita* might almost as appropriately be included here as in the high forest; the two former are very local near the boundary between the two formations. The very large average number of stomata is a result of the large representation of *Nephrodium* and may or may not be correlated with the environment.

THE HIGH FOREST.

The high forest occupies the larger part of the land area within reach of San Ramon and has yielded me more ferns than any other formation. It includes all the forest in which epiphytes are a characteristic part of the vegetation and in which there is an evident difference between the rainy and the dry season. The lower boundary is approximately the back line of the coastal plain. It is marked in most places by large-leaved epiphytic aroids or scandent ferns and sometimes by luxuriant *Calamus*. The high forest gets its name from the many high trees, but in the size of these there is the utmost diversity. Of Whitford's formations [10] I construe it as including the *Anisoptera-Strombosia*, the *Dipterocarpus-Shorea*, and the *Shorea-Plectronia*. I have not adopted these names because so fine a stratification does not appear justified by my material, and because I do not know how far they, being descriptive, are applicable—that is, whether or not the trees the names of which they bear, are dominant, or representative, or represented, here. On the other hand, the classification and nomenclature which I am using are generally applicable though not exclusive.[11]

There are, indeed, a number of societies, some very distinctly localized and locally characteristic of different altitudes, but the narrower strata as entities are without other than the weakest structural or floristic characters.

From the bionomic standpoint, nearly all the ferns of the high forest fall readily into one or another of three not very unequal divisions, reasonably definable by the environment, by the plants' structures and systematically. These are terrestrial plants of humid places, terrestrial plants of dry places, and epiphytes. In each of these divisions in the high forest many ferns have wide distribution, but no fern of one division is common in another and the dubious ground between the moisture-loving and the drought-enduring terrestrial species has but a very limited representation.

The hygrophilous geophytes grow along the river and its branches, in narrow gulches usually without running water, and near the upper boundary of the high forest, in the sheltered cirque-like basins which are the characteristic abode of tree-ferns and of *Diplazium Williamsi*. *Diplazium tenerum* occurs along the river, never high enough to escape floods, and *Nephrodium Foxii* is exceedingly common in the same places. *N. philippinense*, *Aspidium irriguum*, *Leptochilus inconstans* and *Polypodium rivulare* have the same restricted range along creeks in the mountains near Manila, as *Odontosoria* does throughout the Islands. Except the *Aspidium* (and, as compared with most *Aspidia*, it is so) these are all small, or finely dissected, or both. The *Aspidium* is usually found with its fronds torn and broken by freshets. *N. ferox* and *Odontosoria retusa* are strictly confined to stream banks, but high enough up and with long enough stipes to escape floods. *Leptochilus*

[10] *This Journal* (1906), 1, 384.

[11] Beside the formations represented at San Ramon, we have in the Philippines, prairies in northern Luzon, and a montane brush above the mossy forest on Apo. The savanna-wood can be construed broadly to include cogonal, parang, and pine lands.

heteroclitus, L. hydrophyllus and *Polypodium macrophyllum* grow on and between rocks, in gulches usually without running water. The rockiness of this habitat makes it, during the dry season, approach the character of the dry woods.

The narrow belt of low ground along the river and the bottoms of the moist tributary depression of course permit a decidedly luxuriant growth as compared with the savanna-wood or the drier high forest. *Nephrodium setigerum, N. immersum, Aspidium leuzeanum, Microlepia Speluncæ, Diplazium meyenianum, D. polypodioides* and *Odontosoria retusa* have ample fronds, usually more than 1 meter high; while 2 meters is a commoner height for *N. ferox, Dennstaedtia cuneata, D. Smithii* and *Pteris tripartita.* In the wettest part of the savanna-wood only *Polypodium Schneideri*, and, among the plants of the drier high forest, only *Pteris opaca*, reach a height of 1 meter. Stature is as much an index of the character of the environment as is structure, sometimes a much better one, because great size may require some xerophytic structures in plants having moist habitats.

Thickness of frond is a xerophytic character, as exhibited by many epiphytes of *coriaceous texture,* but it is also often an hygrophytic character, exhibited by many moisture-loving plants with *fleshy* but not at all coriaceous fronds. One of these is *Scolopendrium pinnatum*, the mesophyll of which is composed of large, roundish cells with copious intercellular spaces. Its veins are sometimes 2 millimeters apart. Its stipe and rhizome are also fleshy. *Polypodium macrophyllum* as well has an open mesophyll, with intercellular spaces, even in the uppermost layer. Except for these two, the thickness of which is itself an hydrophytic character, the average thickness of these leaves (0.122 millimeter) is slightly less than that in the low part of the savanna-wood.

Floristically, this division is characterized by the development of *Nephrodium* § *Lastraea*, and of *Diplazium*.

The more xerophytic terrestrial part of the high forest flora includes several elements which have nothing in common except the comparative physiological dryness of the environment. *Pteris opaca* and *P. melanocaulon* grow only over the river but in sunny places almost without soil. *P. opaca* is especially xerophytic in structure because of its size; it has a waxy upper surface, no chlorophyll in the thick-walled epidermis, sharply differentiated palisade-like mesophyll, and veins only 0.19 millimeter apart, but, like *Achrostichum*, it is adapted to life in damp air also, having a likewise well-developed spongy parenchyma and numerous large, mobile stomata. Sheer banks above creeks, sunny and often devoid of other vegetation, are the habitat of *Onychium*, here and throughout the Islands.

On small breaks in very steep shaded ground, as under the brow of ridges, are found *Cyclopeltis presliana* in the lower woods and *Microlepia pinnata* at altitudes above 400 meters. *Lindsaya decomposita* and a

Kaulfussia are occasional neighbors of the latter. Scattered through the forest at lower altitudes are *Aspidium difforme*, *A. decurrens*, *Diplazium bulbiferum* and *Pteris quadriaurita*, beside the strongly dimorphous species, *Psomiocarpa apiifolia*, *Stenosemia aurita*, *S. Pinnata*, and *Leptochilus latifolius*, and the subdimorphous *Nephrodium diversilobum*; and *Lindsaya gracilis*, *Syngramma alismæfolia*, *Diplazium pallidum*, *D. tomentosum* (?) and *Pteris pluricaudata*, near the rain forest. The last is intermediate, sometimes growing with *Diplazium Williamsi*.

The dimorphism mentioned above is, as far as it goes, a tropophytic adaptation. *Asplenium subnormale* and the little *Ophioglossum* growing on creek edges are more completely tropophytic, usually disappearing during the dry season. Unless *Adiantum philippense* is one, I found no specialized terrestrial tropophytes in the savanna-wood at San Ramon. Elsewhere in the Islands, *Nothochlæna densa*, *Cheilanthes tenuifolia* and *Helminthostachys* are parang plants active during the wet season only.

The best defined society in this group of plants is that of the narrow ridges in the upper part of the high forest. These have a very limited flora, in which *Callipteris cordifolia* and especially *Taenitis* and *Polystichum aristatum* are characteristic. *Syngramma* and *Lindsaya gracilis* are the only other ferns likely to be found here.

The characteristic feature modifying all epiphytic vegetation is the limited and uncertain water supply. The structure and life of the plant are so profoundly altered in adaptation to this general feature, that details of the environment, especially those in regard to the moisture present in the latter, are able to exert much less influence on either the structure or the distribution of epiphytes than on those of terrestrial ferns. Therefore, although many of the epiphytic ferns are confined to the cañons and only a few have not been found in them, a division of the epiphytes along the lines adopted in treating the terrestrial species would encounter too many doubtful cases and would be supported by altogether too little difference in structure to be justified. For the same reason, the epiphytic vegetation of the high forest is more closely related to that higher up the mountain than is the terrestrial.

Epiphytic plants are those growing essentially without contact with the soil. They are so called because they usually grow upon other plants, but some flourish indifferently either in such a situation or on rocks, and a few grow as a rule, or it may be entirely, on cliffs or bowlders. *Davallia pallida* I found but once and then on stone. Such species might technically be called lithophytes or petrophytes, but such a distinction would express no important biological difference. Species often or usually on rocks are *Nephrolepis laurifolia*, *Asplenium vulcanicum*, *Antrophyum semicostatum* (sometimes even terrestrial), *Hymenolepis spicata*, *Niphobolus nummulariæfolius*, *Polypodium revolutum*, *P. nigrescens*, *P. commutatum*, *P. albido-squamatum*, *Photinopteris speciosa*, and the *Davallia* already cited. As rock-dwellers, these are practically confined to the main river bed, because rocks nowhere else receive sufficient illumination; but some are scattered through the forest on the trunks of trees. *Polypodium albido-squamatum, Photinopteris,* and *Drynaria rigidula* are not uncommon terrestrial plants, on arid, treeless mountain sides in Benguet; and on the edges of cliffs in forests, various normally epiphytic plants are found on the ground. These cases illustrate what everyone knows, that the epiphytic habit has been assumed by plants because of the more ample light; they are a caution against any disposition too much to accredit the abundance of epiphytes along rivers to the moisture of the air.

Another group of species finds adequate illumination only in the tops of the upper-story trees, the spreading branches of which are veritable gardens. This tree-top vegetation is of necessity most imperfectly known in a forest in which the timber has never been cut. Ferns mixed with mosses, orchids, etc., which are encountered in such windfalls as may be present are *Oleandra neriiformis, Humata heterophylla, H. gaimardiana, H. parvula, Davallia brevipes, Polypodium incurvatum, Lecanopteris pumila, Drynaria quercifolia,* and *D. rigidula.* Along the river, but never elsewhere, the tree-top species of *Humata* come down to the crowns and trunks of the smaller trees. *Niphobolus Lingua* and *Dryostachyum pilosum* grow with them and should probably be added to this very imperfect list of tree-top species.

Growing on tree trunks at the lower border of the high forest are *Lindsaya Merrilli* and *Asplenium epiphyticum.* Just as thickness of frond, usually a xerophytic character, is also under opposite conditions at times an appropriate adaptation to a very wet environment, so thinness of frond although usually a character of moisture-loving plants, is characteristic as well of one highly developed type of xerophytes.[12] This type, familiar in the cases of mosses, many lichens, and the *Hymenophyllaceæ,* endures dryness, not by devices to prevent the loss of water, but by becoming dry with impunity. *Lindsaya Merrilli* is a plant of this type. Growing in what, during a part of the year, is the driest environment to which any *Lindsaya* of our flora is exposed, it has still the thinnest leaves in its genus. Where it grows, on the trunks of small trees in the densest part of these lower woods, the illumination during the wet months is quite inadequate for any plant with structures which would allow it to retain any activity through the dry season. *Asplenium epiphyticum* is less specialized, in the same direction, and does not grow quite to the boundary of the savanna-wood. Somewhat above it but together with it, are *Asplenium caudatum* and two or three other *Asplenium*

[12] A xerophyte is a plant adapted to enduring great dryness; one which, as compared with other plants, can endure a desiccating environment *at any time.*

A tropophyte is a plant adapted to a periodically very drying environment. *Bull. Bureau of Education.* Manila (1906), 24.

species related to it and to *A. hirtum*, all having the same ecological character. All our *Loxogramme* species dry up and shrivel during prolonged drought but do not lose their fronds.

The epiphytes of the high forest include no really tropophytic element, since the plants just mentioned, always ready for drought and able to recover instantly from it, are rather to be considered as xerophytes. It has already been noted that *Drynaria* is a potential tropophyte, which, in the high forest, usually retains its foliage. The strong representation of *Asplenium, Loxogramme, Humata* and *Davallia*, and *Phymatodes* and its derivatives, *Niphobolus, Drynaria, Dryostachyum, Lecanopteris* and *Photinopteris*, and the absence of *Eu-Polypodium*, are floristically notable among the high forest epiphytes.

THE RAIN FOREST.

The rain forest is our most constantly humid region. The deep cañons in the upper part of the high forest belt preserve a moist atmosphere when the neighboring uplands have become dry, and in these cañons the flora merges, formally and floristically, into that of the rainy forest. In the rain forest the air is everywhere moist, and so few species find anything like a dry habitat that all the terrestrial plants are best grouped together. The dry, *Taenitis*- and *Polystichum*-marked ridges of the upper high forest flatten out and the ridges that rise in turn from the rain forest, although at times arid because of their exposure, are floristically most distinct. The characteristic ferns along these upper ridges are *Cheiropleuria* and *Dipteris*. There are, of course, places where ridges continue across the rain forest, but, where I have visited them, the vegetation plainly showed that they were more constantly humid than the ridges above or below.

As compared with the high forest the rain forest above San Ramon is a very limited area. It is very inaccessible, absolutely trailless and uninhabited and I was certainly the first white man to visit it, and then I was able to enter it but three times, for a day each; when this is considered and the number of species which I collected is understood, its real wealth of ferns will begin to be appreciated. I have little doubt that its actual wealth in species is greater than that of the high forest. Of course aside from trees, ferns are here the dominant and characteristic vegetation. The luxuriance of this fern vegetation in the number of individuals and in the size thereof, is beyond the comprehension of anyone who has not seen it. A large majority of the terrestrial species reach to a height of above 1 meter; several are commonly above 2 meters; *Aspidium lenzeanum* var. and *Diplazium fructuosum* are some 4 meters in height, stipe included; and an enormous *Dennstaedtia* has a tuft of spreading fronds reaching more than 7 meters.

Small streams everywhere thread this wilderness and furnish the most available paths. On their banks are *Adiantum diaphanum* and *Polypodium dolichopterum*. Our other smaller species, *Nephrodium sparsum*, *N. canescens*, *Lindsaya blumeana* with its finely dissected fronds, *Asplenium resectum* and *Adiantum mindanaoense*, find sufficient illumination on the *steeper* hillsides. *Antrophyum latifolium* grows on sheer walls over cataract-carved pools.

Antrophyum latifolium and *Adiantum diaphanum* are preëminent in structural adaptation to the very moist habitat. *A. diaphanum* is almost our thinnest fern, 0.06 millimeter or less in thickness, with a single layer of mesophyll in part of the frond and the upper and nether epidermis in immediate contact elsewhere. The outer walls are scarcely 1 μ thick. The epidermal cells are very rich in chlorophyll, very wavy in surface view and irregular in section, with outgrowths half as deep as the cells, which make the surface, especially the nether one, unwettable. These water-repellant projections are aided by fine, dark hairs, 0.5 millimeter long, scattered over the nether surface but only near the basiscopic margin above. The stomata are in the general level of the epidermis, protected against wetting by the projections, which, however, as they do not form any closed wall, do not seriously interfere with the circulation of the air. Acroscopic and outer margins are shallowly lobed by sharp incisions. The filiform rachis is underlain by the pinnæ. The frond will shed water in any direction, but the water usually runs off at the apex. In this forest a low plant is likely to receive a nightly sprinkling from those above it, although the sky be clear. The margins of the pinnæ (or pinnules, of bipinnate individuals) are almost parallel; the frond presents a practically unbroken surface to the light, almost without waste by overlapping. The sori are at the bottom of fine sinuses in the middle of the lobes, certainly the driest spots on the nether surface, and the indusia are beset with long hairs. The filifom stipe allows the frond to be agitated by any breath of breeze.

The quasi-epiphytic habitat of *Antrophyum latifolium* probably subjects it to rare and very brief desiccation. Some of its structures, the mass of felty roots, for instance, may be correlated with this danger; but its general structure is as truly hygrophilous as that of *Adiantum diaphanum*, yet two ferns could hardly be more unlike. Instead of being delicate, the *Antrophyum* has a frond 0.64 millimeter thick, which in its best development is orbicular, with a diameter of 17 centimeters, a glabrous surface, and entire margin. Even in these it is a hydrophyte. In its thickness it is fleshy, rather than coriaceous, as compared with its relatives, its smooth, subwaxy surface will hold practically no water and the entire margin is a sharp edge. Its epidermal walls are 2.5 μ in thickness. The kind of adaptation which this is appears from a comparison with its local congeners; *A. semicostatum* has the walls 7 μ and 6 μ thick, while the average thickness of those of *A. plantagineum* and *A. reticulatum* is 10 μ. *A. latifolium* is without the spicular idioblasts which reinforce the epidermis of most *Vittarieæ*. The stomata are flush with the surface and show a remarkable development of the knife-like ridge of entrance which is characteristic of floating plants.[13] The large, respiratory chamber underlies guard cells and hyaline subsidiary cells. The stomata are enormous, 108 μ by 54 μ, sometimes even 114 μ long, with the rift correspondingly large. The apex of the pendent fronds is abruptly acuminate—that is, caudate. The stipes are fleshy, 7 millimeters in diameter near the base. The sori are superficial and protected by numerous paraphyses with oleaginous heads, 100 μ long, 70 μ broad.

In spite of such examples as are given by these two ferns, the terrestrial vegetation of the rain forest as a whole reflects its environment in stature only, not in structure. It has already been pointed out that great size is made possible by moist air, and that under no actual atmospheric conditions can a very large fern be as hydrophytic in structure as a small one. This will explain the thicker epidermal walls possessed by such ferns than are encountered in those of the more hydrophytic plants of the high forest. However, the greater average thickness of frond is largely due to the considerable number of decidedly fleshy species, the largest fronds (*Aspidium*, *Dennstaedtia*) being thinner than the average. The only completely dimorphous species is the single pronounced xerophyte, *Cheiropleuria*. *Blechnum egregium* and *Nephrodium canescens* are sub-dimorphous.

Floristically, the rain forest is well marked by the presence of *Didymochlæna*, *Coniogramme* and *Schizostege*.

The size of epiphytes is in general fixed by their position under not very elastic limits. Not merely is a large size unsuitable to plants the water supply of which is limited and uncertain, but, still more, the necessity of maintaining the plant's attachment to its support usually makes much weight or much area of exposure to the wind perilous. Therefore, the epiphytes of the rain forest, unable to respond to the moister environment by a much greater stature of frond, as is the case with the terrestrial ferns, differ from the epiphytes of the high forest in being distinctly less xerophytic in structure. In spite of the presence of some fleshy species as *Scolopendrium schizocarpum*, the fronds of the rain forest are on the average less thick. They also have thinner epidermal walls and in harmony with the much denser vegetation on and near the ground, the nether epidermis is considerably thinner than the upper, as it is in terrestrial plants everywhere. Seventy-six per cent of the species have chlorophyll in the epidermis, as against 46 per cent of the high forest species; while less than 10 per cent, as against 39 per cent in the high forest, have any hyaline layer beneath the upper epidermis.

[13] Haberlandt: *Flora* (1887), **70**: 97. Copeland: *Ann. of Bot.* (1902), **16**: 349.

The large proportion of scandent species should be noted, as evidence that it is on mechanical grounds, rather than from danger of desiccation, that the rain forest epiphytes are not conspicuous for huge fronds. Scandent plants increase the opportunity for a firm attachment at least as rapidly as they do the number of fronds. Numerous species climb in a moderate way, like *Lindsaya Merrilli* and *Asplenium epiphyticum*, both notable climbers in their high forest neighborhood; among these moderate climbers in the rain forest are *Nephrodium 1712*, *Arthropteris obliterata*, *Lindsaya scandens*, *Asplenium scandens*, *Scolopendrium*, *Polypodium papillosum*, etc. *Stenochlaena*, *Lomagramma* and *Thayeria* are most luxuriant climbers but the two former retain a limited ground connection for a long time or even permanently.

All the peculiar ecological types found among the high forest epiphytes recur in the rain forest. *Polypodium papillosum* retains its fronds on the rare occasions when they dry and curl up, in spite of the fact that its stipe is jointed. This is probably true of *Microlepia hirsuta*, also, but I have never seen it dry. There are also epiphytes in the rain forest which are not coriaceous, do not normally dry out, and which have not articulate stipes. Among these, *Polypodium pediculatum* is notable because of its genus. Plants of this character do not occur in the high forest. In the rain and mossy forests they are never very large and are found where mossy trunks insure the constancy of the otherwise rarely interrupted humidity.

A most interesting inhabitant of these mossy trunks is *Monogramma trichoidea*, in form perhaps the most hygrophytic of all ferns. Its fine, scaly stems are woven into a network, along which are the very numerous "fronds," making a closer and finer tuft than mosses often do. The single fronds are pendent, 3 to 7 centimeters long, and diamond-shaped in section. The longer diameter, 0.33 millimeter, is morphologically horizontal; the shorter, 0.25 millimeter, vertical. The very slender, fibro-vascular bundle is axial, surrounded on all sides by 1 to 2 layers of mesophyll. The four surfaces are exactly alike, stomata scattered over all, the epidermal cells rich in chlorophyll, with occasional idioblasts 0.35 millimeter long, the outer half of the lumen of which is obliterated. In northern Negros, where the rain forest descends to near the sea level, I have found the shorter and slightly stouter *M. dareæcarpa*.

The tree-top vegetation is still richer and less known than in the high forest. Together with several of the species already listed from the high forest, I have found in rain forest windfalls *Vittaria alternans*, *V. minor*, *Polypodium* (*Goniophlebium*) *subauriculatum*, *P. Beddomei*, *P.* (*Selliguea*) *caudiforme*, and a sterile *Elaphoglossum*. With these are a great *Nepenthes*, *Melastomataceæ* and *Gesneriaceæ*. The trees most in evidence are *Quercus*, with huge *Agathis* on the higher ridges.

Floristically it is to be noted that *Eu-Polypodium* and its offshoot

Prosaptia are, together with *Phymatodes*, strongly represented here. With the exception of *P. cæspitosum* and *Thayeria*, which are epiphytes where *Cheiropleuria* is terrestrial, the epiphytic vegetation is, like the terrestrial, well scattered through the forest.

The rain forest as a whole, as compared with the high forest, has two notable features not yet mentioned in treating their divisions. The first is the larger proportion of epiphytes, 53 per cent in comparison with 45 per cent. The reasons for this are evident. The more luxuriant vegetation puts a greater premium on the better illumination of the epiphytes and the greater and more constant humidity removes part of the difficulty of the epiphytic habit. The other notable feature of the rain forest flora is the strong representation of isolated, new, or supposedly rare species and genera and of small genera. This peculiarity is doubtless in part merely apparent, due to our greater ignorance of the rain forest in comparison with our knowledge of the high forest in other places; but in larger part it is real and explicable. The rain forest, just as is locally the case at San Ramon, in all this part of the world occupies a small area as compared with the high forest; it is therefore less likely to develop large genera adapted to its conditions. More important than this, the rain forest is in small areas, often very isolated, and the uplands are the oldest and longest isolated situations of every part of the globe. Under such conditions, isolated types of vegetation are bound to occur. The most notable illustrations are *Schizostege* and *Thayeria*. Others, aside from species elsewhere unknown, are *Antrophyum latifolium*, *Cheiropleuria*, *Microlepia hirsuta*, *Monogramma* and *Lomagramma*.

THE MOSSY FOREST.

The mossy forest is the one in which the extreme humidity permits such a luxuriance of epiphytes that they stunt the growth of the trees. Such a condition is most readily reached where strong light acts together with moisture to favor the epiphytes,—that is, on ridges and peaks—and in such places the exposure to wind is another powerful factor in dwarfing the trees. However, at an altitude of near 6,000 feet on Apo a sheltered plateau exists, with a moist and sufficient soil. The trees in this place are not notably dwarfed in height—they are not high-growing species—but the development of their branches and foliage is very weak. The most abundant tree here, as is usual elsewhere on ridges in the mossy forest, is *Leptospermum amboinense*. The ground in this situation is cold; it was somewhat below 15° C. at the place where I camped in April and October. This coldness, and the mantle of epiphytes on the stems and even on the leaves, check the development of the trees, even though no high winds aid in bringing about this condition. The mossy forest may or may not be elfin-wood, and elfin-wood may or may not be mossy.

The mossy forest is ill-developed on mount Balabac. Its area comprises but a few hectares and the number of ferns is insignificant compared with that on Apo. *Quercus* is common: less abundant on or about the summit are *Podocarpus*, *Dacrydium* and *Phyllocladus*. Conspicuous companions of the terrestrial *Polypodiaceæ* are *Dicksonia chrysotricha*, *Gleichenia dolosa*, and a *Lycopodium* of the *cernuum* group which spreads or climbs for many meters. I reached this summit but once and doubtless missed some characteristic ferns, but the whole number can not be large. Beside the species listed there are a number of rain forest species in this situation. *Hymenophyllaceæ* are very abundant and comprise many species. *Tmesipteris* grows on the trunks of tree ferns.

The startling feature of the mossy forest fern vegetation is that, in sharp distinction to the condition in all lower formations, the terrestrial species are structurally much more conspicuously xerophytic than are the epiphytic species. This reversal has been brought about from both sides. The terrestrial species have on the average thinner fronds than the rain forest species, because none are at all fleshy: and the epidermal walls are very much thicker than in any other terrestrial group (except *Achrostichum*). One of these, *Oleandra*, has an hyaline epidermis and hypodermis, resembling no other of our terrestrial ferns except *Achrostichum*. The xerophytic character of these ferns can be ascribed to the coldness of the ground and to their size. Excepting where they are especially exposed, all are 1 to 2 meters in height, which height, on the mountain top, gives them an exposure to the wind greater than that possessed by the ridge-dwellers, *Cheiropleuria* and *Taenitis*, in the lower forests.

The epiphytes, on the other hand, are much less xerophytic in structure here than in any other formation. For this also there are two evident reasons. These epiphytes are small plants, indeed with two exceptions they are very small and delicate, and their environment is moister, even if the atmosphere is less constantly moist, than that of the epiphytes anywhere else, for the thick mantle of moss and other vegetation on the tree trunks, where fogs are of almost daily occurrence, furnishes a supply of water such as is nowhere else available to epiphytes. Many species in the mossy forest on Apo are epiphytic or terrestrial without evident preference. It may be that some of these small ferns are of the biological type of the *Hymenophyllaceæ*, but I have never seen them dry.

But two genera *Polypodium* and *Lindsaya* are represented in the typical epiphytes of this summit. *Eu-Polypodium* is the dominant genus in this formation everywhere and *Lindsaya* is always well represented. Three of the five *Eu-Polypodia* in this situation namely, *P. cucullatum*, *P. gracillimum*, and *P. macrum*, have erect rhizomes and nonarticulate stipes. From this group (of *P. cucullatum*), and in the same environment, the genus *Acrosorus* of Apo and of summits in Celebes and Samoa, has been derived.

III. STRUCTURAL ADAPTATIONS.

The following exposition of the adaptive nature of fern structures is arranged according to the anatomy and physiology of the plants, instead of according to outside conditions, for two reasons. One of these is that every botanist is more familiar with the outlines of plant anatomy and physiology than he is with the environment of any tropical plant and he therefore would always stand on unfamiliar ground if the other arrangement were adopted; the other reason is that each part of the plant constitutes an environment for the rest of the plant—I have already shown that largeness of frond usually demands thickness of epidermis—and that it would not be easy to fit correlations into a classification based on outside conditions.

THE VEGETATIVE FROND.

Size.—It has already been shown that large fronds are characteristic of habitats having a moist air, and that, on the other hand, large fronds must by virtue of their size be more or less xerophytic in their finer structure. Very large fronds must have stout stipes and rhizomes which are well anchored. The caudex of the huge rain forest variety of *Aspidium leuzeanum* is 10 centimeters in diameter. All our *Dennstaedtias* except *D. erythrorachis* have very stout, prostrate rhizomes. The enormous fronds of *Angiopteris* and *Marattia* spring from a globose caudex which often is 30 centimeters in diameter. Epiphytes have comparatively small fronds, the few exceptions being supported in an exceptional manner, namely, *Asplenium musaefolium, Polypodium heracleum*, and *Platycerium*, by massive nests which, in large specimens, completely invest the supporting branch or trunk, as is sometimes the case with the stout rhizomes of *Polypodium musaefolium, Drynaria quercifolia* and *Thayeria.*

The margins of large fronds are always reinforced to give a protection against tearing. This protection may be by marginal anastomoses of the veins, as is the case in *Syngramma, Callipteris cordifolia, Asplenium musaefolium* and *A. Phyllitidis;* by more copious anastomoses in a great many ferns, such as *Drynaria* and its relatives; by walls merely thicker near the margin, as in *Cyclopeltis;* or by a more or less broad and rigid cartilaginous border, as is the case in *Hymenolepis, Polypodium affine, P. heracleum, Dryostachyum, Thayeria, Photinopteris* and *Achrostichum.* When the margin is deflexed, a very common occurrence, it is less likely to tear. If the frond is lobed or incised, the sinuses are the places needing reinforcement. They are reinforced by the venation of *Goniopteris* and *Callipteris* and by a broader border of cartilage in *Polypodium affine* and other species. The special reinforcement of the sinuses can serve only as a protection against tearing, but the reinforcement of the margin as a whole is equally a protection against gnawing animals. It is probably of use in this way to many smaller ferns. Cartilaginous borders, if sharp or deflexed, also help to keep the nether surface dry. Their greatest development in Philippine ferns is in *Elaphoglossum.*

Removal of water.—The ready removal of water from the frond is insured and facilitated in a variety of ways. One of these is by a smooth, even, waxy, unwettable surface, as in *Asplenium musaefolium* (the cuticle of which is firm enough to be stripped off), *A. Phyllitidis,* and *Pteris opaca.*

Caudate tips are a very familiar structure serving this end and, of course, acuminate tips in general are more common and less conspicuous structures of the same kind. Among conspicuously caudate tips are those of *Oleandra colubrina, Coniogramme fraxinea, C. serrulata,* 2 centimeters long, *Asplenium musaefolium, A. Phyllitidis, Antrophyum latifolium, A. semicostatum* and *Polypodium papillosum.* Long-acuminate, rather than caudate, are those of *Dipteris, Nephrolepis laurifolia, Asplenium vulcanicum, A. caudatum, A. affine* (the pinnæ), *Scolopendrium schizocarpum, Stenochlaena, Blechnum egregium, Antrophyum plantagineum, Prosaptia, Polypodium subauriculatum, P. Zippelii, P. incurvatum, P. Phymatodes* and all its relatives, *Drynaria* and all its relatives, *Lomagramma* and *Cheiropleuria.* Very nearly all of these are epiphytes.

The removal of water from an erect fern is brought about in the same way by an attenuate base, like that of *Dipteris, Syngramma, Meniscium, Polypodium sinuosum* and *Cheiropleuria.* Essentially like these are plants such as *Odontosoria* with cuneate, erect pinnules. Pinnæ drawn down at the base, instead of attached horizontally will drain in the same way; illustrations are *Lindsaya Havicei* and *Asplenium caudatum* and other species. The reduction of the basiscopic half of the pinna has the same effect, the part of the lamina which is removed being that portion which could not readily drain down the rachis; this modification is begun in *Asplenium vulcanicum;* carried farther in *A. tenerum* and its relatives and in *Polystichum;* farther still in *Asplenium resectum;* and completed in our dimidiate *Lindsayas* and *Adiantum.*

If detracted pinnæ are carried farther they become decurrent, forming a wing on the rachis and stipe which serves at once as a drain for water and mechanically, aside from any value it may have in increasing the leaf-area. *Microlepia pinnata* and *Callipteris esculenta* show slightly decurrent pinnæ. Winged rachises and stipes are exceedingly common; they are illustrated by *Aspidium decurrens, Stenosemia pinnata* (upper part), *Davallia decurrens, Diplazium Williamsi, Schizostege pachysora, Polypodium dolichopterum, P. affine, P. Schneideri,* and *Lecanopteris.* A broad wing is sometimes convex upward on both sides, effecting a depression along the axis, as in *P. papillosum, P. Schneideri* and *P. heracleum,* the pinnæ of *Pteris longifolia* and the whole frond of *Polypodium caudiforme.* The fine divisions of *Onychium* are concave above.

Many ferns have very narrow, erect wings, continuous along the rachis, or along the rachis and stipe, opening laterally at the insertions of the pinnæ. Notable illustrations of this are *Davallia decurrens* (Fig. 1), *Diplazium polypodioides, Asplenium subnormale, A. resectum, A. tenerum, A. Belangeri, A. macrophyllum*, etc. Merely channeled are the stipes of *Stenosemia, Humata parvula* and *Lindsaya decomposita;* the rachis of *Nephrolepis cordifolia, Stenochlaena subtrifoliata, Blechnum egregium* and *Polypodium angustatum;* the rachis and stipe of *Davallia pallida, Microlepia strigosa, Diplazium pallidum, D. tomentosum* (?), *D. dolichosorum*, etc., *Athyrium silvaticum, Pteris ensiformis, P. pluricaudata*, etc., *Taenitis, Polypodium palmatum*, etc.; the branches of the rachis of *Odontosoria;* the rachis and costæ of *Schizostege calocarpa;* the costæ of *Oleandra neriiformis* and *Polypodium revolutum;* and the veins of *Nephrodium sparsum* and other species. The costa of *Polypodium* 1741 (*P. musaefolium* ?) is triangular in section, a flat side raised well above the level of the frond, while the third angle stands out sharply below it.

An effect similar to that of the winged stipe is produced by the auricles of the pinnæ of some species standing close to or against the rachis and each underlaying the base of the succeeding pinna; *Nephrolepis cordifolia* is an illustration.

The convexity of the major areolæ of *Polypodium affine* and *P. heracleum* can conduct water along the main veins and costæ, just as a wing convex as a whole does, and the rows of close-set papillæ on the segments of *P. papillosum* must operate in the same way.

Irrespective of the ease or difficulty of becoming wet (as a matter of fact, for a reason which I shall presently develop, cut, incised and dissected fronds shed water), finely dissected fronds dry readily by evaporation. There are two reasons for this; the limited single surfaces, preventing the holding of much water, and the ready agitation brought about by any movement of air. San Ramon ferns with very fine, ultimate divisions are *Nephrodium setigerum, Psomiocarpa apiifolia, Lindsaya hymenophylloides, L. blumeana, Diplazium meyenianum, Asplenium Belangeri, A. scandens, Onychium, Monogramma* (whole frond) and *Polypodium gracillimum*.

Dryness of nether surface.—Aside from these general adaptations to promote the facile escape of water from the frond, there are various other devices which prevent the passing of water to its nether surface. A very simple structure of this kind is the convexity of the ultimate divisions of the frond, such as that of the segments of *Blechnum egregium*, the pinnæ of *Nephrodium canescens*, and the ultimate pinnules of *Nephrodium setigerum, Dennstaedtia cuneata* and *Diplazium polypodioides*. To reach the nether surface of any of these structures water would have to run uphill from the margin.

This is equally the case when the margin of an otherwise essentially plane frond or division is deflexed, as in *Nephrodium* 1712, *N. intermedium*, *Oleandra nitida*, *Microlepia pinnata*, *Dennstaedtia Williamsi*, *Callipteris cordifolia*, *Antrophyum plantagineum*, *A. reticulatum*, *Polypodium punctatum*, *P.* 1741, *P. incurvatum*, *P. palmatum*, *Photinopteris* and *Achrostichum*. In the majority of these plants the margin is sharp as well as deflexed, but that of *P. incurvatum* is rather thicker than the rest of the frond, as a result of increase in the sclerenchyma. A few species, namely, *Loxogramme conferta*, *Scolopendrium schizocarpum*, *Antrophyum latifolium*, *Polypodium cæspitosum*, *P. dolichopterum* and *Thayeria*, have a sharp margin which is not deflexed. The sharpness alone must prevent a *drop* of water from running to the nether surface. If the entire surface is wet, water need perhaps not move in drops, but might move in a film around even a rather sharp edge; but so long as the nether surface is not wet, or is imperfectly so, the surface tension of a drop would cause it to become spherical on an edge as sharp as the ones under discussion, and it would therefore fall off.

Over-fullness of the margin causes an effect like that produced by convexity of frond. Such margins are wavy or crisped, alternately raised and deflexed. Water will, of course, run to the margin where it is lowest and only to this point; these are the places from which it would have to run upward if it were to wet the nether surface. Examples are *Polypodium Schneideri* and *P. macrophyllum*. Such fronds can hardly be torn because the extra length of their margins allows them merely to straighten if the fronds are bent toward the other side.

A ciliate margin is also, as a rule, an obstacle to the passage of water; for, if the hairs are not wet, a drop must pass over their ends from which it will inevitably fall off, but if they are wet, they usually furnish an opportunity for water to run down far enough to fall instead of allowing it to pass to the nether surface. As a matter of fact, neither the hairs nor the cuticles of plants in general are very readily wet. Local ferns with ciliate margin are *Nephrodium procurrens*, *N.* 1685, *Oleandra colubrina nitida*, *Polypodium celebicum* and *Dryostachyum pilosum*. Several Philippine species of *Elaphoglossum* are remarkable in this respect.

If we suppose a frond to be horizontal, then the possibility of a drop passing from the upper to the nether surface depends on the area of contact which it can preserve with the frond in rounding the margin. If the frond is in some other position, the area of contact is still a very important factor. Unless this area is sufficient to allow the drop to flatten into a broad enough oval markedly to reduce its relative surface over what it would be were the drop a sphere, and thus to overcome the force with which gravity can act to remove the drop from the leaf, it will inevitably assume a spherical form and fall. Of course, if the contact is sufficiently reduced, the drop will become spherical independently of gravity and it will fall from any leaf which is not absolutely horizontal and quiet; this is the case in respect to any leaf the surface of which is unwettable.

We have just seen that the surface tension of a drop must prevent its passing around a sharp edge; now, if drops run to a toothed margin, they must in part run over the sinuses. If the sinus were a curve with a radius equal to or greater than that of the drop, or if the radius were not very much less, the drop would have an area of contact greater than it would have in passing an entire margin, and so it would pass more readily. Such sinuses are found in *Dipteris,* but they are so placed that water, to reach the majority of them, must run uphill. I have found such sinuses in no other fern. If a drop runs to a sharp sinus, or to one much narrower than the diameter of the drop, then the sides of the sinus will hold it up and make it run outward until the sinus widens or the ends of the teeth are reached. If the sinus widens sufficiently to allow the drop to run through, then even though the frond is quiet enough so as not to cause it to be shaken off when its contact is limited, and although it might return to the body of the frond while still touching both sides of the sinus, without its having to run uphill, nevertheless it is more than likely to leave the frond because of its impetus in falling through the sinus; this condition is easily demonstrated. If a drop runs out *onto* a tooth, it loses its opportunity to pass to the nether surface by diminishing its possible contact, just as it does if it runs on a caudate tip. In general water must run to the teeth, rather than over a sinus, because its surface tension prevents its starting over an edge if it can run along it. Thus a toothed or cut margin in all parts of its periphery is provided with a water-removing structure such as caudate leaves have at the apex. I have demonstrated the inability of water to pass around a narrowly or sharply cut margin by experiments on various ferns.

Among ferns with serrate margins are *Nephrodium syrmaticum, Davallia decurrens, Diplazium pallidum, D.* 1667, *D. polypodioides, Asplenium vulcanicum, A. macrophyllum, Athyrium silvaticum, Blechnum egregium,* sterile parts of *Pteris ensifolia* (a serrate fertile margin of *Pteris* is impossible), and the fertile frond of *Drynaria rigidula.* Beside being serrate (rather obtusely), *Davallia decurrens* has the segments so close together that a drop will run on the upper surface from one to another. In practice I have watched a drop run across eight segments without a particle passing through between them. Ferns with more deeply incised margin are *Polystichum, Nephrodium* 1712, *N. intermedium, N. Bordenii, Aspidium difforme, Davallia solida, Microlepia hirsuta, Lindsaya Havicei, Diplazium Williamsi, D. dolichosorum, D. meyenianum, D. fructuosum, Asplenium caudatum, A. cuneatum, Adiantum diaphanum, A. mindanaoense, Schizostege calocarpa* and *S. pachysora.* A very much larger proportion of terrestrial than of epiphytic ferns have such margins; as epiphytes by virtue of their position

devices which further obstruct the sinuses, is in itself a proof that cut margins are devices to prevent the passing of water to the nether surfaces.

A mass of phanerogamic evidence for my thesis was accumulated by Anheisser,[15] who showed that in a great number of plants of various families, serrate or biserrate margins and the restriction of stomata to the nether surface were concurrent phenomena.

The epidermis.—Beside the excretion of a cuticle sufficiently waxy to be more or less unwettable, many ferns have outer walls the convexity of which is a strong factor in the same direction. These convex walls are sometimes confined to the nether surface, sometimes they are more convex in that situation. San Ramon ferns the epidermal walls of which are sufficiently convex to be difficult to wet are *Stenosemia aurita*, *Davallia pallida*, *Microlepia hirsuta*, *Odontosoria retusa*, *Dennstaedtia Williamsi*, *D. cuneata*, *Asplenium subnormale*, *Adiantum philippense*, *A. diaphanum*, *A. mindanaoense*, *Antrophyum reticulatum* and *Polypodium affine*. Such an adaptation is naturally to be found chiefly in ferns growing in places where they are likely often to be wet; as a matter of fact, it is entirely confined to ferns growing in such places.

It has already been remarked that half of the depth of the cell of *Adiantum diaphanum* is made of the projections. Between walls which are convex to this extent, and trichomes, no line can be drawn. Dry trichomes, unicellular or pluricellular, occur over the mesophyll of *Nephrodium setigerum*, *N. immersum*, *N. canescens*, *N. aridum*, *N. cucullatum*, *N. pteroides*, *N.* 1677, *N. Bordenii*, *Nephrolepis hirsutula*, *Microlepia strigosa*, *M. Speluncæ*, *Adiantum diaphanum*, *Niphobolus* and *Photinopteris*. Very many species have them on the veins, the reason for this restriction being the mechanical one that their bases can be firmly anchored in the more solid walls to be found there. A better protection against wetting than is produced by dry hairs is furnished by glandular ones such as are to be found on *Nephrodium* 1712, *N. setigerum*, *N. Foxii*, *N. immersum* (few), *N. canescens*, *N. aridum*, *N. cucullatum*, *N. pteroides* (few), *N.* 1685, *N. Bordenii*, and "*Mesochlaena*." These are of characteristic size, form and color in each species.

The thickness of the outer walls of the epidermis has already been tabulated, which tabulation clearly brings out their relation to the general environment and the differences between epiphytes and terrestrial plants. These measurements are of the outer walls as they are apparently normally developed over parenchyma. However, many species have thin spots or pits around the periphery of each cell. These cells have their lateral walls wavy throughout their whole depth, or straight in the inner part but wavy next to the outer wall. The thin spots in the outer wall are in the excurrent angles or lobes. Ferns in which such pits are well developed are *Didymochlaena*, *Dipteris*, *Humata heterophylla*, *H.*

[15] Anheisser, R.: Ueber die aruncoide Blattspreite. *Flora* (1900), **87**: 64.

gaimardiana, *H. parvula* (fig. 3), *Davallia brevipes* (fig. 4), *D. solida*, *Asplenium vulcanicum*, *A. macrophyllum*, *A. cuneatum*, *A. Belangeri*, *Pteris ensiformis*, *Monogramma*, *Antrophyum latifolium*, *Polypodium cucullatum* and *P. Phymatodes*. In *Humata gaimardiana* and *Davallia brevipes* these pores reach a depth of three-fourths to four-fifths of the thickness of the wall. In leaving these lobes of thick walls thin, the plant makes the most economical mechanical use of its plastic material, for the lateral walls effectively reinforce the outer ones at the edges. Straight, lateral walls act less effectively in the same way and a few ferns have the outer walls thinner around the even margin, as is the case in *Nephrodium 1712*, or they are pitted around the margin, as in *Asplenium squamulatum* (fig. 5). On the other hand, *Scolopendrium pinnatum*, the epidermal cells of which are rather large in proportion to the thickness of the outer and lateral walls, has the outer wall reinforced centripetally for a short distance from each of the obscure entrant lobes (fig. 6). The same is true of some of the entrant angles in *Athyrium silvaticum*, whereas *Asplenium scandens* has the outer wall thinned in the excurrent lobes and reinforced from the entrant curves.

The "spicular" idioblasts of the *Vittarieæ* (figs. 7, 8) have long been familiar objects. They are wanting in *Antrophyum latifolium* and *Vittaria minor*, the former growing in sheltered places and being too broad to need any longitudinal reinforcement, and the latter being very small and stout.

It has just been pointed out that wavy, lateral walls are a reinforcement of the outer wall, contributing greatly to the rigidity and strength of the epidermal framework as a whole. On other than mechanical grounds, most terrestrial ferns have the outer walls much thicker above than below. In evident correlation with this fact, these ferns almost invariably have wavy walls for the nether epidermis alone, or have the latter decidedly more wavy than those above. *Diplazium* furnishes many good illustrations. Wavy, lateral walls, by reinforcing the outer walls or in extreme cases, dividing the cell into comparatively small lobes or parts, make larger epidermal cells possible than would otherwise be tolerated from mechanical reasons. The relation of size of cell and waviness of wall to the stomatal movement will be discussed presently. The relation of the thickness of outer wall to size of cell needs no argument; it is well illustrated within one plant by *Odontosoria retusa*, which, though a terrestrial fern along creeks, has the outer wall 1.5 μ thick above, and 3 μ to 4 μ thick below, the cells above being remarkably narrow.

Stomata.—Stomata occur on the upper surfaces of the fertile fronds of *Achrostichum* and *Cheiropleuria* and on all four faces of those of *Monogramma;* otherwise they are entirely confined to the nether surfaces of all our ferns. Here they may be equally distributed or they may be in streaks, or to a limited measure, in groups. They occur only over the parenchyma. As the table shows, the number varies from 7 to 400 per square millimeter. As a general rule the number and size vary in opposite directions.

In most ferns the outer walls of the guard cells are in the same plane as the outer walls of the epidermal cells. The guard cells are usually not very different from the epidermal cells in depth, but sometimes they are decidedly shallower. They are slightly elevated above the level of the surface in *Nephrodium 1712, Microlepia Speluncæ, Odontosoria retusa* (by one-half their depth), *Loxogramme conferta, Diplazium dolichosorum, Asplenium subnormale, Blechnum egregium, Pteris opaca,* etc., *Polypodium gracillimum, P. pediculatum, P. revolutum* and *P. affine.* They are in effect immersed by outgrowths from the epidermal cells of *Adiantum philippense* and *A. diaphanum,* fill the bottom of pits 10 μ deep in *Dipteris,* and occupy the middle of deeper pits in all our species of *Niphobolus.*

In the majority of *Polypodiaceæ,* no ridge of exit can be detected. The exceptions at San Ramon are *Nephrodium sparsum, Oleandra nitida, Humata heterophylla, H. gaimardiana. H. parvula, Davallia solida, Lindsaya gracilis, L. blumeana,* etc., *Coniogramme fraxinea, Syngramma, Diplazium pallidum, D. meyenianum, Asplenium Phyllitidis, A. tenerum, A. macrophyllum, A. caudatum, A. cuneatum, A. affine, A. Belangeri, Scolopendrium schizocarpum, Stenochlæna. Pteris tripartita, Hymenolepis, Niphobolus, Prosaptia cryptocarpa, Polypodium cæspitosum, P. celebicum, P. obliquatum, P. 1741, P. dolichopterum, P. incurvatum, P. Phymatodes, P. palmatum, P. angustatum, P. albido-squamatum, P. caudiforme, P, macrophyllum, Drynaria rigidula, D. quercifolia, Thayeria, Dryostachyum* and *Lomagramma.* In most of these the ridge of exit is comparatively undeveloped; but in a few such as *P. celebicum* and *P. Phymatodes* the two ridges are about equal. Species with an especially strongly developed ridge of entry are *Meniscium triphyllum, Davallia pallida, Microlepia pinnata, M. Speluncæ, Odontosoria retusa, Dennstædtia Williamsi, D. erythrorachis, Callipteris esculenta, Diplazium tenerum, Asplenium subnormale, A. resectum, Stenochlæna subtrifoliata, Pteris tripartita, Antrophyum latifolium, A. semicostatum, A. plantagineum, A. reticulatum, Polypodium affine, P. macrophyllum* and *Cheiropleuria bicuspis.* In nearly all of these this ridge is prominent and aimed obliquely outward, as in *Dennstædtia punctilobula* [10]; but in *Diplazium tenerum* and in *Antrophyum* it is plane with the surface and in section remarkably like shears in appearance. In *Stenochlæna subtrifoliata* the ridge of entrance incloses a vestibule of considerable size.

[10] Copeland: Mechanism of Stomata. *Ann. of Bot.* (1902) **16:** 349. *Pl. 13. Figs. 40, 41.*

In my paper on the mechanism of stomata, just cited, I showed that there was diversity even among our few *Polypodiaceæ* in the eastern United States; the range in the Philippines naturally is much wider in all respects. The turgor of the guard cells has an exceedingly wide range. In *Pteris opaca*, it is the same as in the epidermal cells, or (using a familar ellipsis) 0.3 to 0.4 normal potassium nitrate. In *Leptochilus heterophyllus*, it is 0.5 normal potassium nitrate in the guard cells, but only 0.4 normal in the epidermis. In *Loxogramme iridifolia*, plasmolysis of the guard cells begins in 0.5 normal solution. In *Prosaptia cryptocarpa* and *Polypodium dolichopterum*, the turgor is 0.7 normal, that of the epidermis of the latter being the same; and in *Niphobolus varius* (other *Niphobolus* species are not very different) the pore closes in a normal solution of potassium nitrate, but plasmolysis is doubtful even in this concentration. In *Niphobolus* all neighboring cells of the stoma plasmolyze before the pore appreciably begins to close.

Probably a majority of the ferns here and elsewhere in the world are similar to *Dennstædtia punctilobula* in their stomatal mechanism although in most of them the development of the ridge of entrance is less excessive. Representative of these are *Nephrodium sparsum*, *Leptochilus heterophyllus*, *Pteris opaca* and *Polypodium dolichopterum*. The stomata of this type are considerably longer than they are broad, and are struck by anticlinal walls only near the ends; usually, in fact, only near one end, because of the mode of origin of the stoma. Measurements of the stomata of *N. sparsum* are:

	Open.	In $\frac{0.5}{N}$ KNO_3
Width of stoma.......... microns....	36.5	35
Width of ridge of entrance.......... do........	3	0
Width of guard cell.......... do........	16	17.5
Width of pore.......... do........	4	0

The stoma of *Polypodium dolichopterum* does not quite close. Measurements on it are:

	Open.	In $\frac{0.5}{N}$ KNO_3	In $\frac{0.7}{N}$ KNO_3
Width of stoma.......... microns....	46	45	45
Width of pore.......... do........	5	0.5	0.5

The stoma of *Pteris opaca* widens 1 μ while the pore opens 3 μ.

Callipteris cordifolia is not very far from this type, opening chiefly by a swinging outward of the ridge of entrance; but its width does not change. Measurements on it are:

	Open.	In $\frac{0.5}{N}$ KNO_3
Length .... microns....	81	82
Width of stoma.......... do........	35	35
Width of "pore".......... do........	12	5
Width of ridge of entrance.......... do........	6	0.5

*Some of the stomata close completely.

The stomatal movement of *Niphobolus adnascens* is like that of *Callipteris*. In the (very moderate) thickening of the walls it suggests the *Amaryllis* type, but it is raised above and overlies the subsidiary cell sufficiently to cause the latter to hold the inner half, so that the outer half alone can move freely outward and backward. Its closure is effected by the ridge of entrance. The stoma of *Nephrodium syrmaticum* seems to be quite of the *Amaryllis* type; its dorsal walls are perpendicular to the surface.

Another large number of fern stomata are of the mechanical type of *Medeola*, or of that first described modification of it called the type of *Mnium*. These are broad stomata, with shallow guard cells the dorsal walls of which are rigid. The rigidity of the dorsal wall may be due to its own heaviness, as in *Oleandra colubrina nitida*, to local folds or thickenings, like those of *Medeola*, but less pronounced, as is the case with *Scolopendrium schizocarpum, Vittaria minor, Polypodium cæspitosum* and *P. obliquatum*; it may be due to its being struck only dorsally by the two anticlinal walls, as in *Hymenolepis* and *P. sinuosum*, or to being struck irregularly by more numerous walls, as in *Davallia denticulata, Asplenium musæfolium, A. macrophyllum, Adiantum mindanaoense, Vittaria minor, Prosaptia cryptocarpa*, and others. Most ferns with these stomata have them slightly longer than broad, but exceptions are not infrequent. In the fertile frond of *Loxogramme conferta* they vary from round stomata to those which are 59 μ long by 74 μ broad, and in *Asplenium macrophyllum* from some which are longer than they are broad to examples which are 24 μ long by 33 μ broad. The movement of these stomata is illustrated by *Niphobolus varius* and *Prosaptia cryptocarpa*. Measurements on the former are:

		Open.	In $\frac{1}{N}$ KNO_3
Width of stoma	microns....	34	34
Width of ridge of entrance	do........	3	0
Width of ridge of exit	do........	8	5

Measurements on *P. cryptocarpa* are:

		Open.	In $\frac{0.5}{N}$ KNO_3	In $\frac{1}{N}$ KNO_3
Length	microns....	45	......	45
Width of stoma	do........	45	......	45
Width of pore	do........	5. 5	3. 5	2[a]

In some of these stomata, closure is still less complete. Thus, in *Loxogramme conferta* the open pore is 14 μ wide, and closes only to 8 μ. That of *L. iridifolia* remains open 4 μ after three days in darkness, but closes to 0.5 μ if plasmolyzed. The stoma of *Scolopendrium schizocarpum* is 68 μ long by 53 μ broad when closed; each guard cell is then 26.5 μ broad, or very nearly that, but only 10 μ in depth. Some species have still shallower guard cells. In *Drynaria quercifolia* and *Photinopteris*, but not in *Dryostachyum*, the lumen in median cross-sections is

[a] Plasmolyzed.

exceedingly shallow, namely, 3 μ deep in a stoma 44 μ by 31 μ, but it is thick-walled. Whatever motility these cells may have is like that possessed by *Osmunda*,[17] the activity being restricted to the thin-walled ends, which open the pore by an increase in depth. Finally, the stoma of *Dipteris* is almost exactly like that of the *Coniferæ* and *Allium*, with a very oblique dorsal wall and large overlying twin subsidiary cells.

It is obvious in regarding the occurrence of the different mechanical types that those the movement of which involves a change of outline can operate well only where the rest of the epidermis is not too rigid; while those the movement of which involves no change of outline can operate in spite of rigid neighboring walls, and indeed are often protected by such walls from interference with them by the neighboring tissues. These stomata (types of *Medeola* and *Mnium*) regulate the openness of the pores by changes in the depth of the guard cells, and these changes are effective because the stomata are broad. The firm structure of epiphytes therefore puts a premium on broad stomata. The ratio of average length to average width of the stomata of all San Ramon epiphytes is 1 to 0.67; of terrestrial species, 1 to 0.47.

Even apparently differentiated subsidiary cells practically never occur with the stomata of the *Medeola* and *Mnium* types, which are independent of the contiguous cells. However, in stomata the movement of which involves any accommodation on the part of the adjacent cells, as is the case in stomata of the types of *Amaryllis, Dennstædtia* and the *Coniferæ*, specialized subsidiary cells are or are not necessary, according to the nature of the general epidermal cells. If the latter are very large, as in *Diplazium*, and without too rigid walls, a specialization of subsidiary cells seen in surface view or in section is unnecessary and does not occur. Subsidiary cells are without one or the other of the two properties of the majority of epidermal cells which interfere with the movements of the guard cells; that is, either the subsidiary cells contain less chlorophyll than the epidermal cells, or they have less rigid walls, or both. They contain less chlorophyll in *Aspidium angustatum* and *Lindsaya pulchella* and none at all in any *Antrophyum*, in all species of which genus other epidermal cells contain a certain quantity. It has already been pointed out that wavy, anticlinal walls increase the rigidity of the epidermis. An epidermis wholly wavy or wavy next the outer wall (as in *Antrophyum*) has subsidiary cells with plane walls in *Cyclopeltis, Nephrodium syrmaticum, Aspidium angulatum, Arthropteris, Pteris ensiformis* (not always), *Antrophyum latifolium*, etc., and *Photinopteris* (not always). *Humata heterophylla* has the anticlinal walls which strike the backs of the guard cells very thin, while elsewhere they are thick. *Monograma* has conspicuously broad, subsidiary cells with the narrow stomata. *Dipteris* has small, angular epidermal cells rich

[17] Copeland, l. c. 347, figs. 31 to 33.

in chlorophyll, but very large subsidiary cells occupying half of the entire area.

Assimilating tissue.—The specialization of the epidermal cells of ferns is what their environment demands. In terrestrial species, with very few exceptions, they are not extremely differentiated from the parenchyma, but that this difference between the majority of ferns and the majority of spermaphytes is an adaptive, not a primitive character on the part of the ferns, is amply proved by the exceptions. Some species in every tribe of *Polypodiaceæ* represented at San Ramon are without chlorophyll in the epidermis. The spicular cells of the *Vittarieæ* have already been mentioned, and in four tribes, *Davallieæ, Asplenieæ, Polypodieæ* and *Achrosticheæ,* are species which have carried protective specialization deeper than the epidermis, having a specialized hypodermis. In the majority of ferns it is more correct to describe the epidermis as specialized in other directions than for protection, than to call it undifferentiated. In very numerous ferns it is, indeed, a highly specialized photosynthetic tissue which is not infrequently more specialized than any part of the parenchyma.

Ferns with epidermal cells conspicuously deep and rich in chlorophyll are *Cyclopeltis, Nephrodium immersum, N. diversilobum, N. pteroides, Aspidium decurrens, A. angulatum, Stenosemia aurita, Leptochilus latifolius, Diplazium pallidum, D. 1667, D. bulbiferum, D. tenerum, D. Williamsi, Asplenium subnormale, A. resectum, A. scandens, Schizostege calocarpa* and *Pteris quadriaurita.* The inner ends of deep epidermal cells are out of contact with one another, leaving intercellular spaces, in *Nephrodium 1712, N. canescens, Meniscium, Microlepia pinnata, Lindsaya hymenophylloides, L. pulchella, Adiantum mindanaoense, Polypodium obliquatum* and *P. dolichopterum.* In several of these the chlorophyll is concentrated in the inner end. Instead of one, there are several inward projections, making the cells breeches-like in section in *Nephrodium procurrens, N. Foxii, Stenosemia pinnata, Polypodium cæspitosum* and *P. cucullatum;* the subepidermal layer has this character in *Humata heterophylla.* In *Adiantum diaphanum,* an especially large share in the photosynthesis falls to the epidermis, the upper and nether epidermis being in direct contact in a considerable part of the frond. It will be noticed that the ones which have been mentioned in this connection are nearly all terrestrial species. But some very large terrestrial species are like many epiphytes in the more or less complete suppression of the chlorophyll in the epidermis, this being the case in *Nephrodium ferox, N. cyatheoides,* and the huge variety of *Aspidium leuzeanum.*

It is the mesophyll rather than the epidermis, which shows less specialization in the ferns than in the seed-plants. In the parenchyma the differentiation is especially backward. A completely and typical developed palisade parenchyma does not occur, but layers which are like it in compactness and more or less approaching it in form and arrangement of the cells are found in *Nephrodium cucullatum, Humata gaimardiana, H. parvula, Pteris opaca, P. longifolia, Hymenolepis, Niphobolus, Polypodium albido-squamatum* and *Achrostichum*. Except the first, these have the underlying parenchyma very loose and open.

The mesophyll is more compact and green above, without however being palisade-like, in *Didymochlaena* (6),* *Cyclopeltis, Odontosoria, Dennstaedtia erythrorachis* (7–8), *Diplazium* 1667 (4), *D. bulbiferum D. tenerum, D. dolichosorum* (9), *Asplenium Belangeri* (4–5), *Blechnum egregium* (6), *Pteris melanocaulon* and *Taenitis* (8); more open below but about as green, in *Nephrodium immersum* (5); with chlorophyll even in the sclerenchyma, *N. pteroides* (3–6), *Meniscium* (6–7), *Aspidium difforme* (7), *Dipteris* (8–9), *Pteris excelsa* (7–8), *Polypodium Zippelii* (8), and *P. Phymatodes* (8).

In very many more species there is no evident differentiation. According to my cards, the parenchyma is undifferentiated and compact in *Polystichum, Nephrodium syrmaticum* (5), *N. intermedium* (3), *N. setigerum* (5), *N. canescens* (2), and its var. *nephrodiiformis* (2–3), *N. diversilobum* (2–3), *N. 1685* (3), *N. Bordenii* (2–3), *Nephrolepis cordifolia* (5), *N. hirsutula* (6–7), *N. laurifolia* (7), *Oleandra, Davallia decurrens* (5), *Microlepia strigosa* (4), *Dennstaedtia Williamsi* (4–6), *D. cuneata* (6–7), *Lindsaya gracilis* (2), *Callipteris esculenta, Diplazium meyenianum* (2–4), *D. fructuosum* (6), *D. polypodioides* (4–5), *Asplenium caudatum* (4–5), *A. cuneatum, Adiantum mindanaoense* (2), *Schizostege pachysora* (8), *Onychium* (7–8), *Pteris pluricaudata* (3–4), *Vittaria falcata* (10), *Polypodium papillosum* (3–4), *P. angustatum* (6–7), *P. heracleum* (4–5), *Lecanopteris* (5), *Drynaria rigidula* (5), *Dryostachyum* (4–5), and *Lomagramma* (6). It is subcompact in *Nephrodium procurrens* (3), *N. 1677* (2–3), *Asplenium Phyllitidis* (10–13), *A. epiphyticum* (4), and *Vittaria alternans* (8). It is quite open throughout in *Nephrodium 1712* (3), *Aspidium angulatum* (3–4), *Stenosemia aurita* (3), *Arthropteris* (4), *Lindsaya hymenophylloides* (2), *L. blumeana* (2), *L. pulchella* (3), *Loxogramme conferta* (10), *L. iridifolia* (8), *Syngramma* (8), *Diplazium pallidum* (3), *D. Williamsi* (2–4), *Asplenium resectum* (3), *A. affine* (5), *Athyrium silvaticum* (4), *Stenochlaena sp.* (6), *Adiantum philippense* (2), *Schizostege calocarpa* (4), *Antrophyum latifolium* (6), *Polypodium cucullatum* (2), and *P. Schneideri* (4). It is stellate, throughout or in part, in *Nephrodium sparsum* (4–5), *Microlepia pinnata* (7), *Coniogramme fraxinea* (6–7), *Loxogramme involuta, Callipteris cordifolia* (10) (the last three below only), *Vittaria minor* (14) except two layers above, *Prosaptia cryptocarpa, Polypodium caespitosum* (8) (fig. 14), *P. celebicum* (5), and *P. obliquatum* (6).

* Figures in parentheses state the number of layers of parenchyma.

According to its necessity, a hyaline hypodermis has been differentiated in one or more genera of every tribe, as shown in the table (pages 12 to 16). This is usually found only beneath the upper epidermis, but underlies the nether as well in *Davallia solida, Asplenium musaefolium, Polypodium incurvatum, P. albido-squamatum* and *P. caudiforme,* and incompletely in *Dryostachyum* and *Achrostichum.* This tissue is found only in xerophytes and the notion has sometime had vogue that its function is to act as a water store. That this is not in general the case, I have pointed out elsewhere [18] among the ferns the walls of the hypodermis are almost invariably so thick that any change in size or form, which is necessary if they are to give up any water, is quite impossible. Thus, in *Polypodium albido-squamatum* the hypodermal walls are 12 μ thick, almost obliterating the lumen; *Humata gaimardiana* has two layers of hyaline cells with walls 8 to 10 μ thick, and *H. parvula* has two layers with walls 12 μ thick. Species with thin hypodermal walls and some with walls thick enough to seem rigid if plane, have the walls wavy or angular (fig. 15), as seen from the surface and therefore not collapsible under vertical pressure; or there are thickened intruding folds of the walls (fig. 16), such as brace the stomata of *Medeola* and other plants. Such walls are found in all species of *Phymatodes,* and its offshoots, *Drynaria,* etc., which have any differentiated hypodermis, and in *Achrostichum.* They are also found in the uppermost parenchyma layer of *Davallia pallida, Dennstaedtia Williamsi* and *Polypodium subauriculatum.* Beside a hyaline hypodermis with thick walls with the *Phymatodes* contour, *P. sinuosum* has the uppermost layer of green mesophyll and, in less measure, the next two or three layers, provided with heavily thickened lines (fig. 17), perpendicular to the surface, to prevent collapse. Giesenhagen [19] reports the same structures in *Niphobolus stigmosus, N. Gardneri* and other species, and cites Poirault as authority for their occurrence in some other species of *Polypodium.* The rays of the stellate parenchyma of *Humata parvula* usually have thickened walls and the fine, close veinlets of *Drynaria* and its relatives, *Polypodium heracleum, Dryostachum* and *Thayeria* are all connected with the epidermis by bands of sclerenchyma, inhibiting even an incipient collapse.

On the other hand, there are a very few species provided with an evidently available store of water. Thus, *Polypodium caudiforme,* with two layers of noncollapsible cells under the upper epidermis, has one layer of collapsible cells next the nether one. The walls of the green parenchyma of *Loxogramme iridifolia, Antrophyum reticulatum,* and *Polypodium accedens* are somewhat collapsible with loss of water, but not greatly so. In this direction again it is *Niphobolus,* of all our ferns,

[18] *This Journal* (1906) 1: 25.

[19] Giesenhagen: *Schwendener Festschrift* (1899), 6, 8, 17, 18, *Pl. I. Farngattung Niphobolus* (1901), 67–79.

in which specialization has gone farthest.[20] All the anticlinal walls of *N. varius* are freely collapsible. *N. adnascens* has hypodermal walls 4 μ thick, but still, as they are placed, subject to a slight folding; while the anticlinal walls of the parenchyma are accordion-like (fig. 18). Sections of a leaf, 0.5 millimeter thick (weather very dry), widened to 0.83 millimeter as soon as cut, and to 1.49 millimeter in water; and the walls were still pleated. *N. nummularifolius* has a single, uppermost layer with rigid walls, the remainder, hyaline and green, being collapsible. Sections of a frond, 0.56 millimeter thick, widened to 0.67 millimeter when cut and to 0.86 millimeter when wet. *N. lingua* has 2 to 3 layers of hyaline cells, of which only the inmost can collapse at all, and this much less readily than can the parenchyma.

Hydathodes.—*Niphobolus* is likewise the only genus having trichome-hydathodes.[21] These hairs are different in form,[22] those of each of our species being characteristic, and our most xerophytic species, *N. adnascens* being glabrescent; but they are all alike in insertion, each hair growing in a pit which is practically filled by the basal cell of the trichome. This basal cell is alive, with considerable evident contents. When the leaf is damp, this contents fills the cell; it can then absorb water from the cells borne on it whether they are dead or alive, and give water to the cells within. Judging by the high turgor in *Niphobolus* leaves, this movement must be fairly active. When the outside of the leaf becomes dry, the outer cells of these trichomes lose their water and promptly draw on the basal cell. If the connection were maintained, the basal cell would then supply itself from the interior of the leaf. But this does not happen because its protoplasm instead of keeping in connection with the cells within and without, shrinks away from its wall and contracts into a lump touching but one end of the cell. A dead air space, or approximate vacuum in the basal cell then protects the interior of the leaf from evaporation. The protoplasm of the basal cell collapses instead of maintaining its turgidity, because it loses water outward without faster than it can get it from within, this condition must be due both to the very high turgor of the mesophyll and to the unequal permeability of the end-walls of the basal cell, their outer end being pitted.[23]

Very many ferns have the vein-tips hyaline, and, as a rule, the clear spots are hydathodes, clear because of the absence of air-containing spaces. Such hydathodes are found, among other ferns, on *Meniscium*, *Arthropteris*, *Nephrolepis*, *Asplenium vulcanicum*, *A. tenerum*, *A. Belangeri*, *A. scandens*, *Hymenolepis*, *Niphobolus Lingua*, *Polypodium Zippelii*, *P.*

[20] Giesenhagen, l. c.

[21] Giesenhagen: *Farngattung Niphobolus*. P. 44, and under each species.

[22] Using the word hydathode to include water-absorbing structures.

[23] Giesenhagen: *Schwendener Festschrift* (1899), p. 5.

1741, *P. affine*, *P. Phymatodes*, *P. palmatum*, *P. albido-squamatum*, *P. macrophyllum*, *P. heracleum*, *Dryostachyum* and *Photinopteris*. White incrustations of lime are regularly found on these hydathodes on some species of *Nephrolepis*, *P. albido-squamatum*, and young fronds of *Dryostachyum* and *Photinopteris* and occasionally on various other ferns. There are other hyaline vein-tips, as in *Asplenium subnormale*, which are not active hydathodes, though perhaps potential ones.

Venation.—Except as it is modified by correlation with other structural peculiarities, such as the fineness of dissection of the frond, the venation in general is decidedly closer in species of arid than in those of humid habitats. Thus, among plants of arid places, the distance between veinlets is:

Of *Nephrodium cucullatum*, 0.35 millimeter; *N. aridum*, 0.4 millimeter; *Pteris opaca*, 0.10 millimeter; *P. longifolia*, 0.5 millimeter; *P. melanocaulon*, 0.5 millimeter; among plants of moister places; of *N. ferox*, 0.75 millimeter; *N. Foxii*, 0.5 millimeter; *N. syrmaticum*, 0.7 millimeter; *N. pteroides*, 0.7 millimeter; *Lindsaya scandens*, 0.7 millimeter; *Syngramma*, 1.5 millimeters; *Diplazium pallidum*, 1 millimeter; *Asplenium caudatum*, 1 millimeter; *Scolopendrium pinnatum*, 2 millimeters; *Stenochlæna subtrifoliata*, 2 millimeters; *Pteris pluricaudata*, 0.7 millimeter; *Pteris excelsa*, 1 millimeter.

Anastomosis of the veins makes the venation closer in effect, and as a general proposition, with many exceptions, ferns with anastomosing veins are more xerophytic in habitat than those with free veins. As illustrations on the largest scale, *Goniopteris* and *Callipteris* are dominant in the savanna-wood where *Lastræa* and *Diplazium* do not occur, and *Phymatodes* and its offshoots have fifteen species in the high forest, but *Eu-Polypodium* is unrepresented. The frequent correlation between large size and ampleness of frond and reticulate venation is too obvious to need elaboration.

Articulate stipe.—The articulation of the stipe to the rhizome, and of the pinnæ or segments to the stipe, facilitate the reduction or removal of the leaf surface whenever it is necessary. It is thus an adaptation to life where plants must sometimes endure a more or less prolonged want of water. Like other adaptive characters, but in greater measure than many, because it involves a deeper specialization, it has a taxonomic value, as species, genera, and even larger groups have developed in constant adaptation to certain conditions. Under the conditions at San Ramon (and under tropical conditions in general), then, the characteristically epiphytic groups have articulate stipes; the characteristically terrestrial ones, nonarticulate stipes. *Davallieæ* and *Polypodieæ* are typically epiphytic tribes with articulate stipes; *Asplenieæ*, *Aspidieæ* and *Pterideæ*, typically terrestrial tribes, without articulations.

It is the exceptions to this general rule for the tribes which put this interpretation of articulations beyond any question. Among the *Aspidieæ*, we have at San Ramon a single epiphyte, *Nephrodium 1712*, and it is our only species in the tribe with articulate stipes. Among the *Davallieæ*, the constantly epiphytic genera are *Humata* and *Davallia*; *Arthropteris* and *Oleandra* are biologically epiphytes, though the former is probably terrestrial in origin, and one *Oleandra* maintains its ground-connection. *Nephrolepis* is likewise epiphytic in fact, or in its exposure and independence of ground water. Except *Nephrolepis* the pinnæ of which are articulate, these all have articulate stipes. The terrestrial genera are *Microlepia, Odontosoria* and *Dennstaedtia;* they are without articulate stipes with the exception of the single epiphytic species, *Microlepia hirsuta.* This argument is equally valid, whether *M. hirsuta, M. Speluncæ* and *M. pinnata* are regarded as congeneric or in three related genera. *Lindsaya* is not specialized as an epiphytic genus; three San Ramon species are terrestrial; *L. Merrilli,* and very likely its near relatives; *L. hymenophylloides* and *L. Havicei* are ecologically like the *Hymenophyllaceæ; L. pulchella, L. hymenophylloides* and *L. Havicei* grow in a habitat where even *Polypodium* is not usually articulate; and *L. scandens* grows in the very moist rain forest.

Among our *Asplenieæ* a single genus exists which when mature is always epiphytic in exposure; namely, *Stenochlæna.* Its pinnæ under these circumstances are articulate, but young plants, near the ground and growing from it, are without articulations. *Asplenium* can not be regarded as a specialized epiphytic genus as its many epiphytic species are altogether too diverse in their adaptations, indicating that they have assumed this habit separately; some are sclerophyllous xerophytes, as the *Neottiopteris* group; others, such as the *A. caudatum* group, seem to become dry without great injury, and a few, such as *A. Belangeri* are rain forest species, just as is our epiphytic *Scolopendrium, S. schizocarpum.*

Our *Pterideæ* include no epiphytes and no plants with structural articulations. However, there are species of *Adiantum,* notably *A. opacum* of Palawan, the pinnules of which are deciduous in an emergency. The *Vittarieæ* as a group are nonarticulate epiphytes. To endure this condition they have thick, rolling leaves with very heavy epidermal walls and very few stomata. *Loxogramme* is ecologically like them.

The *Polypodieæ,* with the exception, perhaps, of *Taenitis,* are a very natural tribe in which the axis of evolution has been in the air under standard epiphytic conditions. From this axis are many offshoots, of which one remote one may be *Taenitis* which is terrestrial and nonarticulate. Again articulate stipes have been lost by the *Drynaria* group, the humus-collecting habit of which demands permanent fronds, but which still fit their dry environment by being able to shed their pinnæ or segments. *Dryostachyum* shows advances toward the loss of the articulation; *D. splendens* of Mindanao having a joint evident to the eye, but without function (that is, a vestigial structure), while the otherwise identical Luzon fern has not even an apparent joint. Again in the very moist rain forest and mossy forest, where the rhizome and roots are imbedded in enough other vegetation to be situated as if they were in the soil itself, articulations cease to be necessary, and their disuse and eventual disappearance can be observed, usually in ferns with small fronds, in many different groups of species: In the *P. cucullatum* group and in *Acrosorus;* in *P. pediculatum* and *P. macrum,* with many articulate relatives; in *P. caespitosum* and some other species of *"Grammitis;"* and in the Panay species, *P. Yoderi* [24] but not in *P. tenuisectum.* On the other hand; a Luzon *Phymatodes, P. Proteus,* growing among rocks on arid ground, has articulate pinnæ as well as stipes. *Loxogramme* is probably an offshoot of *Eu-Polypodium.* which has developed a decidedly xerophytic structure with or after the loss of articulations.

Of the San Ramon *Achrosticheæ,* two are terrestrial and nonarticulate, the other is high-scandent with articulate pinnæ.

RHIZOME.

The stems of *Polypodiaceæ* are moderately modified in adaptation to a wide range of conditions—more modified and more variously so than one might imagine from text-book comparisons with *Equisetum* and the *Lycopodineæ.* The most primitive form of fern stem is probably a short, erect one such as is observed in *Aspidium, Diplazium* and *Pteris.* This may be subterranean, or barely superficial, or, in damp and darker places, may rise into the air, as is the case in most of the large *Diplazia,* in *Callipteris esculenta,* and notably in the huge variety of *Aspidium leuzeanum. Blechnum Fraseri,* of Luzon [25] and New Zealand, is remarkable among Philippine *Polypodiaceæ* for its tall and graceful stem. On trunks deeply covered with vegetation, some small ferns have stems standing out radially, with a dense, apical tuft of small fronds. Among these are *Vittaria minor,* and all the *Polypodia* with nonarticulate stipes. A large number of ferns lift the fronds above competition with their terrestrial neighbors by assuming the scandent habit. Such are *Nephrodium 1712,* all scandent species of *Leptochilus, Arthropteris, Nephrolepis volubilis,* all scandent *Lindsayæ* except *L. gracilis, Asplenium epiphyticum, Lomagramma,* and *Stenochlaena.* The majority of these maintain their connection with the ground, but are still, if we classify all ferns as either terrestrial or epiphytic, rather to be regarded as belonging to the latter class, because of their exposure.

Numerous other ferns keep to the ground or to their original aërial support, but remove their leaves from competition with one another by a creeping habit and by bearing them at considerable intervals. Neither the geotropism of the stem, determining whether it shall be prostrate

[24] *This Journal* (1906). **1**: Suppl. 161.

[25] Christ: *Bull. Herb. Boissier* (1898). **7**, 149. *Pl.* 6. It is usually more slender than this figure shows.

or erect, nor its symmetry, radial or bilateral, seems to be a very deep-seated or firmly fixed character, for both change in many instances within universally recognized generic or subgeneric limits, as in "*Goniopteris*" and *Eu-Polypodium*. The correlation between length of stem and length of stipe is too obvious to need any discussion; ferns with scandent or wide-creeping rhizomes have short stipes, while those with erect rhizomes have the tufted stipes long enough to separate the fronds.

Fleshy rhizomes serving as water-reservoirs are found in *Drynaria* and its relatives, most notably in *Polypodium heracleum*,[26] and less developed in *Photinopteris* and *Polypodium affine*. The rhizome and stipe of *Scolopendrium pinnatum* are fleshy, as is the stipe of *Antrophyum latifolium*.

All rhizomes are protected against loss of water at the apex, and many throughout their length, by scales which vary in form, size, and texture. Exceedingly harsh paleæ are found on *Dipteris* and *Dennstaedtia Williamsi*, two ferns with notably stout rhizomes. It is very probable that these are protective against animals, such as deer and hogs which are very numerous, but which never, so far as I have observed, touch these species. Similar scales protect the fleshy crowns of various *Cyatheaceæ*. The muricate stems of *Stenochlaena*, and muricate stipes of *Dennstaedtia erythrorachis*, *Diplazium polypodioides* and other species, as well as of *Athyrium silvaticum* probably have the same function. Dead bases of stipes must provide other rhizomes with an unpalatable mantle, but most fern stems are too hard to need protection of this kind.

Many stems contain chlorophyll when exposed to the light. It is regularly present in those of *Polypodium accedens*, *P. dolichopterum*, *P. commutatum* and *P. Schneideri*.

Light correlations.—The correlation between length of rhizome and length of stipe has just been mentioned. A similar correlation exists between length of one or the other of these and the development of the lowest pinnæ. Deltoid fronds—that is, fronds with elongate lowest pinnæ—would seriously interfere with each other's light if they were not borne on wide-creeping rhizomes, as is the case in *Davallia* and *Humata*; or on very long ascending stipes, as is true of most species of *Aspidium*, *Nephrodium sparsum*, *N. intermedium*, *Dennstaedtia erythrorachis*, *Adiantum mindanaoense*, *Schizostege pachysora*, *Pteris excelsa* and *P. pluricaudata*; or on comparatively short, but more horizontal stipes, as in *Leptochilus latifolius* and *Psomiocarpa apiifolia*. Fronds with short stipes, unless these are very remote, usually have the pinnæ reduced toward the base; illustrations with creeping rhizomes are *Nephrodium aridium*, *Arthropteris*, *Lindsaya hymenophylloides*, *L. Havicei*, *Polypodium celebicum*, *P. obliquatum*, *Prosaptia*, *Drynaria* (normal

[26] Goebel: *Pflanzenbiologische Schilderungen* (1889), 1, 202. However, Professor Goebel is in error when he cites the *Hymenophyllaceæ*, because they have no store of water, as being quickly killed by dry air.

fronds) and *Dryostachyum pilosum.* Under the same conditions, entire fronds are narrowed below, as in *Scolopendrium schizocarpum.* Among tufted fronds, such forms are the rule, as in *Diplazium, Blechnum, Polypodium cucullatum,* etc., and, as an example with entire fronds, *P. cæspitosum* may be cited. All the lower pinnæ are sometimes equally and extremely reduced, the largest ones being immediately above these; such fronds are physiologically like those with long stipes and large, lowest pinnæ; illustrations are *Nephrodium cucullatum, N. 1685,* and *N. Bordenii.*

Fronds with broad bases which are so placed as not to overlap, might lose considerable light between the stipes, but in general this space is utilized. A long, broad wing on the stipe sometimes extends the assimilating area, as may be observed in *Aspidium decurrens, Leptochilus latifolius* and *Polypodium dolichopterum.* In many ferns the lowest pinnæ are flexed forward so that they practically fill the space between the frond bases; this is the case with *Nephrodium procurrens, N. diversilobum, N. cucullatum* (lowest functional pinnæ), *N. 1685, N. Bordenii, Humata gaimardiana, Asplenium caudatum, Polypodium Schneideri* and *Achrostichum. Asplenium subnormale* has the stipes sufficiently erect to bring the large lowest pinnæ near together and the rachis is bent strongly outward just above these, the upper parts of the fronds being divergent and nearly horizontal. Similarly, the close-set fronds of *Cheiropleuria* are curved outward and downward above the often cuneate base, so that most of the frond slopes downward toward the apices. In many cases such as *Nephrodium canescens, N. 1677, Diplazium dolichosorum, D. 1667,* and *Athyrium silvaticum,* the lower pinnæ are deflexed into the space otherwise lost between the frond bases. Deltoid fronds usually reach the same end by a strong basiscopic development of the lowest pinnæ; this is true in most species of *Aspidium, Humata, Davallia, Cheilanthes, Pteris* and other genera.

Since half the margin of a frond is longer than its axis, most pinnæ, being narrowed toward their apices, lose considerable space between their distal ends. This form economizes the conduction of water and food and is mechanically good because it is compact, but it involves a waste of light, which is saved by *Nephrodium diversilobum* and *N. Bordenii,* which broaden toward almost truncate apices. *N. 1677* has its pinnæ somewhat narrow near the base, where they bear large, foliose auricles.

The pinnæ, as well as the fronds of ombrophilous plants, as is to be expected, are in general fitted together so as to utilize all possible light consistent with the disposible surface. Notably perfect mosaics are presented by *Davallia solida, Dennstaedtia Williamsi* and all fronds with trapezoidal or "lunulate" pinnæ or pinnules, such as are presented by *Polystichum amabile, Didymochlaena, Lindsaya scandens, L. pulchella, Asplenium resectum,* and *Adiantum.* The pinnæ of *Lindsaya pulchella* would overlap wastefully were they not set at such an angle that they act like a grating.

ROOT.

Perhaps the most interesting specialization of the roots of ferns, but one which I have seen mentioned nowhere else, is the massing of very numerous roots, all densely covered by a felt of long, brownish, persistent root-hairs which form a structure for the storage of water. Appropriately to their function, these masses of hairy roots are commonly found on ferns growing on naked rocks or tree-trunks, but never on ferns with abundant soil, nor on trunks laden with moss. Persistent root-hairs as organs of attachment are very common among ferns and other plants, and it is doubtless through roots clinging by such means that these water-stores have been evolved. Obviously, too, the deepest roots in every mass of this kind still fasten the plant to its support, but that more than the deepest layer in a mass, which is sometimes more than 2 centimeters thick, can serve in this way is of course impossible. There is every gradation from these thick pads down to those so thin they may serve for attachment alone, as is the apparent case with *Polypodium macrophyllum*. San Ramon ferns with a sufficient mass of felty roots so that they must store water are *Nephrodium Foxii* (on rocks), *Davallia pallida*, *Loxogramme conferta* (few), *L. iridifolia*, *Asplenium tenerum*, *A. Belangeri*, *Antrophyum latifolium*, *A. reticulatum* (very thick pad), *Niphobolus nummulariæfolius* (few), *Polypodium accedens* (few), *P. 1741* (few), and *P. nigrescens*, beside all humus-collecting species. Fuzzy roots are found on *Antrophyum plantagineum*, *Polypodium Zippelii*, *P. angustatum*, *P. albido-squamatum*, *P. caudiforme* and *Dryostachyum pilosum*, growing on submossy trunks or subnaked rocks, the hairiness of the roots of *P. angustatum* being evidently dependent on the nakedness of their substratum. In contrast with the preceding, the following epiphytes on *mossy* trunks have fine, naked roots: *Humata parvula*, *Davallia brevipes*, *Microlepis ciliata*, *Lindsaya Havicei*, *L. pulchella*, *Prosaptia contigua*, *Polypodium celebicum* and *P. palmatum*.

Asplenium epiphyticum has roots of two kinds; those of unlimited length, positively geotropic, forming a jacket around the stem, diarch, flanked by sclerenchyma, unbranched, with hairs along the sheltered side; and roots 2 to 3 centimeters long, slightly negatively geotropic, freely branched, closely appressed to the support, clinging by copious hairs, of similar structure to the preceding but with more sclerenchyma. These are the clinging roots. The former, under favorable conditions, will reach the ground and then branch. *A. scandens* likewise has roots of two kinds. In various scandent ferns stems are massed, and hold water as these two *Asplenia* do by means of the stems and mantles of roots, and in a few cases, as in *Lindsaya Merrilli*, persistent, decurrent leaf bases or stipes are useful in the same way.

The bracing "roots" of *Nephrolepis* are very familiar objects. Those of several species of *Diplazium* are very stiff and somewhat spreading above the ground. It has already been mentioned that *Achrostichum aureum* sometimes has pneumathode-roots which are not greatly differentiated.

HUMUS COLLECTORS.

The amount of study which has already been devoted to two of the most extraordinary specializations of ferns, those for collecting humus and for association with ants, spare me the necessity of entering into the details of either. Of humus-collectors, we have at San Ramon the nest-builders, *Asplenium musaefolium*, *A. Phyllitidis* and *Drynaria rigidula; Polypodium punctatum*, which makes brackets of leaf-bases interlaid and overlaid with humus and detritus which are sometimes 15 centimeters broad and almost as deep, but which does not normally form round nests; *P. heracleum* and *Drynaria quercifolia*, which, in their best development, form spiral brackets, the supporting leaves being in a single series, but imbricate; and *Thayeria*, which makes a most perfect, independent receptacle with each leaf. Other Philippine humus-collectors are *Dryostachyum splendens* in Mindanao and Luzon, and *"Polypodium" meyenianum* in Luzon. This character of *D. splendens* is not generally recognized and my determination might be in error, but it is based on a comparison with a plant of the type number of Cuming's collection.

Thayeria is so remarkable and recent a discovery that I take the liberty of repeating a part of the description, from this JOURNAL (Volume I Suppl. (1906) page 165). "Fronde solitaria in ramo laterale rhizomatis endogena, cornucopiæforme; ramo in fundo cornucopiæ in radiculas multas dissipato." "In its humus-collecting structures *Thayeria* is wholly unlike any other known plant, the specialization having gone beyond the frond to the rhizome. Each leaf is a unit, a complete receptacle, wholly out of contact with the main rhizome. It is the most perfect of the humus-collecting organs developed in its group, the material collected being inclosed on all sides and protected against desiccation with a throughness not attained even by *Asplenium Nidus*. The specialization of the branch end as a root bearer in the bottom of the cornucopia is a very novel feature."

MYRMECOPHILY.

Our two remarkable myrmecophilous ferns, *Polypodium sinuosum* and *Lecanopteris*, have recently been thoroughly studied by Yapp,[27] in whose paper the previous literature is summarized. With regard to the anatomy, there is nothing essential to add; but with regard to the significance of the bizarre form and structure of these and other myrmecophilous plants of this region, Yapp followed Treub and Goebel in a puzzling oversight of the service rendered the plant by the ants, which insects furnish their hosts with mineral food.

[27] *Ann. of Bot.* (1902) **16**, 185.

Our myrmecophilous plants are, without exception, epiphytes. As such, they are exposed to dearth of water and dearth of mineral food. When they protect themselves against injury by the former by using devices to reduce the transpiration, they aggravate the latter difficulty. Epiphytes have many ways of overcoming their difficulty of obtaining mineral food, such as the maintenance of remote ground-connections; parasitism; complete exhaustion of their own dead parts; coöperation in the accumulation of an aërial "soil," in the mossy forest and in tree-top gardens at lesser altitudes; special humus-collecting structures, such as have just been described; the insectivorous habit, in *Nepenthes,* and the attraction of insects for the sake of the débris they bring, or for their excreta or their carcasses, as is the case with the plants now under discussion. The plants waste none of their parts to support the ants, offering them only a tolerably moist shelter, and this is very evidently a sufficient inducement for the ants to seek them, for I have never found a healthy individual of one of these plants without its tenants. The latter are not specialized in adaptation to their specific hosts, for the same ants inhabit the chambers of different plants; for instance I have found one kind in *Polypodium sinuosum, Myrmecodia,* and *Hydnophytum* all in a single tree. Although ants have not the reputation of being untidy housekeepers, the chambers which they occupy are never really clean. The plant can of itself effect the quick removal of liquid ejecta; if can get rid of sölid ones, only as they are dissolved. I have found a fungus in an apparently healthy *Polypodium sinuosum,* growing in the lining of the chamber and at first imagined that it might be analogous in function to mycorhiza, but it is not always present and it was probably merely accidental. Both of these ferns are without other roots than such as are necessary for their firm attachment and they habitually grow on bare branches, without any mass of other epiphytes; therefore, they would be in especial straits for mineral food if it were not for their tenant ants. Nevertheless they are conspicuous for the very ready falling off of their leaves, conclusive evidence that they are not in practice obliged to husband their ash-constituents. The facts that *Polypodium sinuosum* can live after its chamber is plugged (Goebel), and that *Hydnophytum* and *Myrmecoidea* can grow and develop their chambers without the presence of ants (Treub), do not prove that the ants are useless to the plants any more than the power of *Drosera* to live under favorable conditions without insects is a demonstration that the plant is not insectivorous. Of the two ferns, *Lecanopteris* is the more highly developed in myrmecophily, not only in grosser, conspicuous characters, but also in the perfection of its chamber, the walls of which, as described and figured by Yapp, are made up of pockets, which are doubly serviceable as collectors of possible food, and as increasing the absorbing area.

The doctrine that these stems are enlarged as water-reservoirs, and chambered and the reservoir-tissue removed because they are too fleshy, has a fit companion in that other which interprets the leaves of *Dischidia* as protectors of the roots, but does not tell us what purpose roots serve in such a place.[28] As a matter of fact, these plants are also myrmecophilous, the leaves furnishing shelter for ants, and the ants furnishing food which the roots absorb. *Dischidia* is rarely without ants and rarely without a considerable amount of débris about the roots inside each leaf brought by them. There are other *Asclepiadaceæ*, epiphytic without evident structural modifications, the roots of which are invariably in aërial ants' nests.

In all these cases it is likely enough that the plant derives some organic as well as mineral food from its tenants.

REPRODUCTIVE STRUCTURES.

The principles underlying the adaptations of the reproductive structures of fern's (sporophytes) are very simple. The sporangia must be protected during their development against injury by desiccation or otherwise; the mature spores must dry thoroughly enoughly to be easily and well scattered; and the drying of the spores must not involve too great a desiccation of the frond, for an insignificant number of Philippine ferns suffer an annual loss of their leaves. The structures found in ferns are a compromise between these rather antagonistic principles.

Ferns almost always protect their sporangia, at the same time that they avoid interference with the illumination of the assimilating organs, by restricting the former to the nether surface; our physiological exceptions are *Psomiocarpa* and *Stenosemia,* the vegetative and reproductive fronds of which are distinct, and *Lecanopteris,* which may not be entirely dependent on photosynthesis for its organic food.

For the sake of facile nutrition and to preserve the normal exercise of its functions by the nether epidermis, the sporangia of practically all ferns, the vegetative and reproductive fronds (or pinnæ) of which are alike, are collected into sori. Most ferns protect these sori by means of indusia. At San Ramon, 60 per cent of all *Polypodiaceæ* have indusia, the remaining 40 per cent including 13 members of the old genus *Achrostichum* and a number formerly put into *Gymnogramme,* beside all those with well-defined nude sori. In the indusiate list are included the *Pterideæ* (not including *"Gymnogramme"*), they having, bionomically, indusia as truly as any ferns do. Any full discussion of the forms and origin of indusia would be superflous here, in view of the attention they have received as most important structures in taxonomy, but it is pertinent to the subject of this work to point out that their structure fits the local demands upon it. Thus, it is leathery in the two strongly

[28] Scott and Sargant, in *Ann. of Bot.* (1893), 7: 243, suggest that the roots are to absorb water, those in the inverted pitchers of *D. rafflesiana* condensing the water transpired by the interior of the leaves.

xerophytic genera, *Davallia* and *Humata*, but not in their mesophytic relatives, *Microlepia* and *Leucostegia*, and in *Asplenium* it is the xerophytic section, with entire fronds, which has by far the firmest indusia.

A heavy coating of hairs protects the sori as well as the stomata against undue loss of water in *Niphobolus lingua* and various of its congeners. While the function of paraphyses is in general to protect against water rather than desiccation, there are some ferns the paraphyses of which cover the sporangia so thoroughly that they must serve in their time in both ways. Among these are *Achrostichum*, *Lomagramma*, *Cheiropleuria*, various *Vittarieæ*, *Hymenolepis*, and *Polypodium subauriculatum;* and, most conspicuously of all Philippine ferns, *Polypodium lineare* Thunb., of Luzon.

The protection of the sorus by the folding backward of the margin of the frond is familiar to all in the "indusia" of most *Pterideæ*. The same effect is reached very thoroughly by two of our species of *Polypodium*—*P. cucullatum* and *P. gracillimum*—which have one half of each pinna wholly or partly folded backward against the other half, covering the single sorus. This, or a convexity approaching the same effect, characterizes Presl's genus *Calymmodon*. In *Acrosorus*, the folding is complete and permanent, the edge being grown fast and the sorus opening toward the apex.

Numerous ferns protect their young sori by more or less completely sinking them below the level of the frond's surface. According to the extent of the immersion and the thickness of the frond, the spots occupied by the sori may or may not be prominent on the upper surface of the frond. When they make moderately convex spots it strengthens the frond mechanically, so that the fertile part of the frond of *Nephrodium Foxii*, for instance, retains its form for some time after the sterile part begins to wilt. Ferns with indusiate sori moderately immersed are *Didymochlæna*, *Nephrodium Foxii*, *N. immersum*, *Microlepia hirsuta*, *Humata immersa*, and *Davallia pallida*. In *Asplenium Phyllitidis* (fig. 19) and its immediate relatives, they are sunken approximately half the depth of the frond, opening obliquely, and the part of the frond outside them merging into the indusium. In *Scolopendrium pinnatum* (fig. 20) the double sori are immersed, the entire broad depression being covered by the indusia. Non-indusiate sori shallowly immersed are found in some species of *Antrophyum*, *Loxogramme conferta*, *Taenitis*, *Polypodium pediculatum*, *P. Phymatodes*, *P. palmatum*, *P. angustatum*, *P. sinuosum*, *P. heracleum*, *Drynaria rigidula* and *Lecanopteris*. *Polypodium revolutum* and *P. cæspitosum* have them deeply immersed in fleshy fronds. This goes farther in *P. celebicum*, and reaches an extreme in the less fleshy fronds of *P. obliquatum* (fig. 21), which has the cavity deepened by a crater-like rim. In this species and its immediate relatives (*Cryptosorus* Fée) the cavity is closed when the sorus is very young, but opens later. In *Prosaptia*, the cover has become permanent, and the sorus opens toward the margin of the frond. In *Monogramma* and *Vittaria* (fig. 22), the sori are in deep slits, the effect being as in *Asplenium Phyllitidis*, but the protection of the more open slits is perfected by capitate paraphyses. In *Polypodium incurvatum*, and more prominently in *P. subauriculatum*, *P. nigrescens*, *P. Schneideri* and *P. papillosum*, the sori are "immersed" for several times the thickness of the frond, forming very prominent projections from the upper surface.

The structures which serve to prevent the desiccation of young sori serve also, without exception, to make their exposure to liquid water impossible and there are a considerable number of ways in which they are adapted to perform this latter function well. In other cases, structures at first clearly protective are done away with or changed in such a way as to make the mature sporangia as exposed as possible. Thus, in a large part of our *Nephrodia* and in many of their relatives, the indusia partly or completely disappear as the sporangia mature. The segments of *Polypodium cucullatum* and *P. gracillimum* flatten out, as do, in varying measure, the reflexed margins of the *Pterideæ*. The indusia of the *Asplenieæ* curl or bend outward to permit the drying and scattering of the spores. In *Asplenium scandens*, and without doubt in many other species, the indusia are motile, bending outward when dry, but closely appressed when wet. This movement deserves careful study, both as to its commonness and its mechanism. I have noticed it to exist, but in a less pronounced manner, in *Onychium*.

The indusia are beset with hairs, which I interpret as water-repellant structures, in *Nephrodium procurrens*, *N. aridum*, *N. cucullatum* (few), *N. 1677*, *N. Bordenii*, (decidedly hispid), *Microlepia strigosa* (long basal hairs), and *Adiantum diaphanum;* and glandular-hairy or glandular-ciliate in *Nephrodium 1712*, *N. setigerum* (with fugacious indusia), *N. 1685*, *Aspidium angulatum* (fig. 23), and *Oleandra colubrina nitida*.

It has already been stated that paraphyses are in general water-repellant structures, in adaptation to which function they are provided with oily heads. Among the San Ramon ferns provided with these are *Aspidium leuzeanum*, *Oleandra neriiformis*, *Microlepia pinnata*, *Dennstaedtia Williamsi*, *Vittaria*, *Anthrophyum*, *Taenitis*, *Hymenolepis*, *Polypodium subauriculatum*, *Lomagramma*, *Achrostichum* and *Cheiropleuria*. The paraphyses are in part a substitute for indusia and often occur on ferns such as the *Achrostichea*, which could not have indusia; but they are not rarely present in indusiate sori. They are notably developed on *Lomagramma* and *Achrostichum* (fig. 24), the brown color of the fruiting surface of the latter being due to them, while the sporangia are green. The branched form, like the oiliness, is evidence that they are specialized for protection against water rather than against desiccation.

Hairs on the end of the sporangia have the same effect. They are found in a number of species of *Nephrodium*, such as *N. setigerum* (glandular), *N. diversilobum* (fig. 25), but not in *N. canescens*, and *Meniscium*.

These hairs on the ends of all mature sporangia make the whole sorus incapable of being wet.

Spores of ferns are in general not readily wet, because of their waxy and often rough or reticulate surfaces. They are rough, for instance, in *Nephrodium procurrens, N. 1677, N. Bordenii, Aspidium angulatum* and *Asplenium resectum;* granular in *N. setigerum,* and reticulate in N. *1685.* Their resistence to wetting not merely facilitates their dispersal, but also insures them against germination under too temporarily favorable conditions.

The immersed sori of *Prosaptia contigua* and several species of *Polypodium* are very effectively protected against any danger of wetting by a few long brown hairs standing across the mouth of the pit.

Very numerous ferns provide, in a variety of ways, that the dryness necessary for the dispersal of the spores shall involve the least possible danger of desiccation of the vegetative frond. One very simple means to this end is the location of the sori on the margin, or even on teeth.

The marginal or apical position of the sori has been assumed independently by the plants in many different groups of ferns. Mindanao illustrations are the *Hymenophyllaceæ, Dicksonieæ, Psomiocarpa,* etc. (in bionomic effect), *Nephrolepis acutifolia,* many *Davalliæ* and *Humatæ, Dennstaedtia, Odontosoria, Lindsaya,* the *Pterideæ, Prosaptia, Acrosorus, Lecanopteris,* and *Lomagramma.* As the primary purpose of this position of the sori is to insure the dryness of the sporangia and spores, it is characteristic of plants growing in the most moisture-laden atmosphere; as in the rain and mossy forest where the *Hymenophyllaceæ. Dicksonia, Dennstaedtia, Lindsaya* and *Lomagramma* are examples, or along creeks, in the case *Odontosoria*; they also occur on some vigorous xerophytes such as *Davallia, Humata, Nephrolepis acutifolia,* and *Lecanopteris.* The relation between the atmospheric moisture-conditions and the position of the sori is well illustrated by ferns other than those with sori actually on the margin. *Diplazium meyenianum* is much more constantly restricted to moist hollows than is its occasional companion, *D. polypodioides,* the former having flat segments, with long sori reaching the margin, while the latter has short, costal sori, protected against liquid water by the concavity of their surface of the segment. *Asplenium Phyllitidis* of the rain-forest has long sori reaching nearly or quite to the margin, while *A. musaefolium* of the high forest and strand has them short and costal. It is common in *Nephrodium* § *Goniopteris,* and the *Lastræa* species of similar form, for the sori to be more nearly costal at the ends of the segments than at the base; *N. 1685* is a good example.

The tooth position is obviously drier than the merely marginal. The *Lindsayæ* growing in the moistest places are deeply cut—even finely dissected in *L. blumeana,* and in *L. capillacea* Christ, of the mossy forest of Luzon. As a rule, these sori lack just enough of being marginal to be out of danger of liquid water, the end of the segment projecting slightly farther than that of the indusium, and often being toothed—with one tooth in *L. Merrilli,* with three in *L. capillacea.* Yapp suggests that the peculiarly placed sori of *Lecanopteris* will let the spores escape only when there is wind enough to be likely to scatter them into such places as the plant normally occupies, which are in the crowns of lofty trees.

It is probable that a considerable majority of all ferns have the fructification developed toward the apex rather than toward the base of the frond, obviously favoring the greater dryness of the fertile region. There are all grades of specialization in this respect, from that in which the preference of the sori for the distal end is doubtful, or not emphasized, as is the case in many species of *Nephrodium, Polypodium,* etc., through those in which the restriction is clear and constant, as in *Asplenium musaefolium,* and those in which the fertile region is moderately restricted in its development in area, as *Niphobolus adnascens, N. varius, Polypodium accedens,* and *P. angustatum,* or otherwise modified in form, as in some species of *Nephrolepis,* or in structure, as in *Onychium* and *Achrostichum,* to those with the most completely metamorphosed fertile region, as *Dryostachum, Photinopteris,* and *Hymenolepis.* In these extreme cases, the fern seems to gain the most of the advantages of dimorphism, with decided economy of material, and still more in the conduction of food to the fertile part.

The adequate dryness of the mature reproductive structures, without jeopardy to the proper performance of the vegetative functions, is accomplished in many ferns by a specialization of entire fronds for one or the other end. In many ferns there is little or no specialization other than a difference in the length of the stipes, those of the fertile frond being the longer. Among the ferns the dimorphism of which does not involve a great reduction of the assimilating area are *Nephrodium diversilobum, Syngramma, Pteris ensifolia, P. pluricaudata, Taenitis, Niphobolus adnascens, N. Lingua, Polypodium palmatum* and *P. sinuosum.* All of these have the fertile fronds with the longer stipes. In *Pteris* there is a difference in the margin, and in *P. ensifolia* the fertile frond is less compound than the sterile. The fertile fronds of *Polypodium sinuosum* yield to drought and fall off before the sterile. Other ferns with a moderate reduction of the assimilating area of the fertile frond are *Humata heterophylla, H. parvula, Loxogramme conferta, Niphobolus nummulariæfolius, Polypodium incurvatum* and *Drynaria rigidula.* These also either have longer stipes of the fertile frond or else, as in *N. nummulariæfolius* and *L. conferta,* the fertile frond is itself elongate, while the sterile frond is more or less round. The stipe of the fertile frond of *Humata parvula* is 5 to 8 centimeters high, that of the sterile frond 1 to 1.5 centimeters. *Drynaria rigidula* has the pinnæ of the fertile frond more serrate than those of the sterile and much more readily deciduous. *Humata heterophylla* has the sterile frond entire, but the fertile one toothed or lobed. The significance of these differences in maintaining the dryness of the fruiting surface has been made clear by a preceding discussion of the value of such margins.

As a general rule, among these moderately dimorphous ferns, the stomata are, area for area, more numerous on the fertile frond than on the sterile, as the following table illustrates, the numbers being the stomata per square millimeter.

	Sterile.	Fertile.
Nephrodium diversilobum	140	200
Aspidium angulatum	36	180
Loxogramme conferta	8	28
Niphobolus nummulariafolius	64	100
Polypodium accedens	23	52[a]

The stomata of the fertile and those of the sterile frond of *Cheiropleuria* differ in the series of divisions by which they are formed, the latter being unlike those which I have found in any other fern (figs. 26, 27.)

There still remain a few ferns in which the differentiation has gone so far that the assimilating, but not spore-bearing, surface of the fertile frond has practically been obliterated. These are *Leptochilus* and *Cheiropleuria,* still with some expansion of green lamina, the nether surface of which is completely covered, at least at maturity, with sporangia; *Blechnum egregium,* the fertile pinnæ of which are expanded at the base only; and *Psomiocarpa, Stenosemia, Stenochlæna,* and *Lomagramma,* whose fertile fronds are almost completely without assimilating surface. Of these, *Stenochlæna* and *Lomagramma* are scandent, all the others except *Blechnum egregium* having the stipes of the fertile fronds notably long. The two scandent genera have the pinnæ articulate to the rachis, and the pinnæ of the fertile fronds of both are much more caducous than those of the sterile. This, with the further fact that only plants of a very considerable age are fertile, makes fertile fronds of both hard to find. The fructification of *Lomagramma* originates on the nether surface but becomes lateral, as-exposed as possible, by the curling of the frond (fig. 28).

The fertile frond of *Blechnum egregium* as well, seems to be both rare and transitory, and on all these ferns they are to be found only in season.

Leptochilus (most species), *Psomiocarpa* and *Stenosemia* are terrestrial plants characteristic of the border between high forest and savanna-wood. Their close neighbors are *Nephrodium diversilobum* and *Aspidium angulatum.* To endure the dryness of the dry months, these plants have their fronds close to the ground; such species as *Psomiocarpa apiifolia, Leptochilus latifolius,* and their Luzon associate, *Hemionitis arifolia,*[29] are often real rosette-formers. If their spores were matured at this time,

[a] Fertile apical region.

[29] Whitford, l. c., 399. "All but geophilous during the dry season."

they would need no especial devices to insure their dryness, but they would have only a very remote prospect of germinating, and when the spores are formed, when they do have a chance to germinate, they would be very unlikely to become dry enough to scatter if borne in the position of the vegetative frond.

To summarize: dimorphism, whether merely begun, or highly developed, whether a character of whole fronds or of their parts, has in all cases the object of permitting the proper dryness of the mature sporangia without an improper desiccation of the vegetative structures. This is done by merely raising the reproductive structures farther above the substratum; or (rarely) by special structural devices, such as notched margins; by a restriction of the assimilating surface of the reproductive frond or region, so that it may be sacrificed in emergency; or by a more complete elimination of the vegetative structures in constitutionally ephemeral fertile fronds.

IV. TAXONOMY.

In discussing the San Ramon ferns from the systematic side, I shall hold myself chiefly, but not absolutely, to the local material, and in using this I shall be contented with pointing out some characters observed in groups which locally are notably well represented, and in suggesting briefly the probable genetic affinities of these ferns.

The past decade has been that of the greatest advance in systematic pteridology, because it has seen the general and surely the final abandonment of the idea that any single structure is of equal importance in the natural classification of all groups of ferns, or is even in every case of any value at all. The indusia are very useful in the proper characterization of many genera and tribes, but are not always of certain specific value in *Nephrodium,* or *Aspidium.* The shape of the sorus is sometimes of generic or even tribal value as a diagnostic character, but is variable in some species—and individuals—of *Aspidium, Athyrium,* and *Phymatodes.* In a single species of *Leptochilus, L. lanceolatus* Fée, the sporangia may be in distinct sori, or may cover the whole nether surface. The veins are sometimes free, sometimes anastomosing, on single individuals of *Schizoloma fuligineum* and *Polypodium californicum. Dryostachyum splendens,* descended from ferns with articulate stipes, has sometimes evident vestigial articulations, sometimes apparently none; yet each of these characters, presence of indusia, shape of sorus, venation, and articulation of stipe, has sometime been held to be a fit basis for the initial or general classification of all *Polypodiaceæ.*

Since all of these characters, and all other real characters are, phylogenetically (not, so far as we know, ontogenetically), adaptations to the environment and since the family is an old one and environments not only change, but are seldom found to be sharply differentiated, and the dissemination of spores is unceasing, it would be very wonderful if there were a single character by which all ferns could be divided into great and natural groups, or if any character were of equal value in all groups. The fixedness of any character depends partly on grounds we understand, or certainly will understand with sufficient study along familiar lines—that is, on the relation of the character in question to the environment of the plant or group in question—and in part, or ground we can name, but do not yet know how to investigate; that is, the different heredity-strength among variations (including mutations), and characters already called hereditary.

While the applicability of no character is universal, neither are there uniformly fixed limits to the value of any character. The margin of the frond is unstable in many or most genera, yet entire margins characterize the simple fronds of *Vittarieæ,* and genera in various other tribes, and a simple and entire *Lastræa,* or *Athyrium,* or *Microlepia,* or *Schizostege* is unknown. Even geographical characters are useful. The whole of the character of not a few genera, as *Prosaptia, Niphobolus,* and *Achrostichum,* is intelligible when, and only when, the habitat is included and recognized as the dominant character of all. More broadly geographical characters are of value too, for no plant has progeny in places inaccessible to its reproductive structures. The relative antiquity of groups, as definable by their present characters, is important evidence in judging their relationships. If a species or genus is confined to one locality or one part of the world, it is probably not very ancient. If it has a very wide and continuous distribution, its age can not be less than sufficient to permit such a dissemination. Our oriental *Prosaptia, Acrosorus, Loxogramme, "Schellolepis," Niphobolus, Drynaria, Dryostachyum, Thayeria* and *Lecanopteris,* and the American *Lepicystis, Campyloneuron* and *Phlebodium* must all be younger groups than the cosmopolitan *Polypodium,* ancestor and cousin at once of them all. A group with wide and discontinuous distribution must be ancient enough to have become widely distributed, and to have died out in the intermediate territory; it may not be older than a group with equally wide continuous distribution, but its minimum probable age is greater.

I have tried to become so well acquainted with the San Ramon ferns that I might know each species and larger group as the sum of its characters and not by any single character; so that I might fairly judge in each of the stability and diagnostic value of each character, and recognize the more elusive peculiarities, as well as those which lend themselves readily to description. Success in such an attempt is at best a matter of degree, but the probable degree is very much greater in the field, especially in the field with some laboratory equipment, than it is in the herbarium.

I have also tried to decide what might fairly be regarded as a primitive form of Polypodiaceous fern. Such a fern must be a very generalized type, not highly specialized in adaptation to any conditions; it should be, or have been, world-wide in distribution; it should be found, in its essentials in the various tribes, excepting as some tribes may probably be derived from others, rather than directly from the primitive form; and, in the more ancient tribes, it should be possible to outline the development of the more recent and more specialized genera from those most like the generalized primitive form.

I believe that this primitive fern, through whatever stages it may have been evolved, from which all *Polypodiaceæ* have been derived, was a terrestrial plant of humid woods, with a short, stout rhizome, with ample, compound or decompound leaves, the fertile and sterile not differentiated, with nonarticulate stipe, and with small, distinct, more or less round sori. I shall show that such ferns as these meet every demand laid down in the preceding paragraph. Nearly or exactly this primitive fern exists now in the genus *Nephrodium,* more particularly in the subgenus *Lastræa.* As is true of all generalized types, it is impossible by any character, or any practicable combination of characters, to diagnose *Lastræa* as a natural group, retaining all species which as a matter of highly probable genetic affinity should be included and excluding all plants the genetic affinity of which is very remote. *Lastræa* merges into *Goniopteris* through species with a single pair of irregularly anastomosing veinlets; or else, as at least in part is probably the case, we include in *Lastræa* a considerable number of species descended through *Goniopteris* but with free veins; in either case the natural separation of the two groups is not feasible. The line between *Lastræa* and *Pleocnemia* is but little less vague.

The indusium of *Lastræa* is utterly inconstant. The lines between *Phegopteris* and *Dryopteris* and between *Goniopteris* and *Cyclosorus* appear to me to be purely artificial; nor is the shape of the indusium, when present, invariable. It is not rarely peltate in *Nephrodium immersum,* just as it is sometimes reniform in *Aspidium angulatum,* and in the plant known as *Mesochlaena* it is diplazioid in form. Again the sorus is elongate, in those immediate relatives of *Nephrodium urophyllum* sometimes called *Meniscium.* And even the nonarticulate stipe is not a constant character, for my No. *1712* is unmistakably a *Lastræa,* with as evidently articulate a stipe as that of any other scandent fern. While the fronds are characteristically compound, there are exceptions, and there are species, both in *Lastræa* and *Goniopteris,* with the fronds subdimorphous.

As *Nephrodium* is altogether the most generalized and indefinable genus of ferns, its general characters—compound, ample, thin fronds; nonarticulate stipe; short, stocky rhizome, and round sori—can be accepted as those of the most generalized, and therefore primitive *Polypodiaceæ.* *Lastræa* is also thoroughly cosmopolitan.

Goniopteris is more specialized, having a relatively stable frond-form, almost always simply pinnate, and firmer texture. Glandular trichomes, occasional in *Lastraa,* are common in *Goniopteris,* and hair-like trichomes are found in some species on frond, indusium, and even on the sporangia. The group is also characterized on the whole by numerous small stomata. *Mesochlaena* is a *Goniopteris* in form, texture and even in the characteristic pubescence, and, *in a genus notable for the instability of indusium characters* should hardly be separated by the indusium alone. As a matter of fact, the more fundamental peculiarity is the elongation of the sorus, which is very moderate, but sufficient on mechanical grounds to effect the division of the indusium as in *Diplazium.* The recognition of *Meniscium* as a genus characterized by elongate sori is generally and properly abandoned. Though a very large and pantropic group, *Goniopteris* has a narrower range than *Lastræa.*

Both *Polystichum* and (through *Pleocnemia?*) *Aspidium* are so intimately related to *Lastræa* that the proper assignment of species is sometimes difficult and able botanists are not wanting who would still include all in one great genus. *Polystichum* is probably older [30] than *Aspidium,* being cosmopolitan, while *Aspidium* is tropical, and rather nearer in character to *Lastræa,* but the indusium characters of *Polystichum* are hardly as unstable as those of *Aspidium.* In the latter the reticulate venation is directly correlated with the simplification of the frond. The indusium furnishes valid characters for most species of *Aspidium,* and, like any other character, is to some extent a guide to affinities; but the distinction of *Sagenia, Tectaria* and *Arcypteris,* genera to be recognized by this character alone, breaks up some certainly natural groups. I am not ready to judge the real naturalness of *Pleocnemia,* as a genus.

Judging by its distribution, *Didymochlaena* must be a rather old genus. Its aspect suggests that it may be descended through *Polystichum. Cyclopeltis,* a likewise homogeneous genus in the tropics of both hemispheres, is generally recognized as an offshoot of *Polystichum.*

In the group of small genera split out of *Achrostichum* and included by Diels in *Aspidieæ,* dimorphism has so obscured most other characters that it is not easy to be sure of their real affinity. However, the group as a whole is almost certainly unnatural, for I can not imagine our two so-called *Polybotryæ* to be congeneric, or that both are intimately related to *Stenosemia* or *Leptochilus.* In view of their conspicuous morphological *and* geographical isolation, both of the former ought clearly to be restored to generic rank, as *Egenolfia* and *Psomiocarpa.* The former may, with a high measure of probability, be regarded as derived from *Polystichum,* in which genus *P. auriculatum* is the most similar species in this part of the world; but a direct *Lastræa* ancestry is not impossible and it sometimes approaches *Leptochilus.* Neither the loss of the indusium, nor a considerable dimorphism is any novelty in *Polystichum.*

Psomiocarpa is possibly more doubtful, but I do not believe that

[30] That is, as a group with its present characters; all plants may be assumed to have an equally long ancestry.

it can have originated elsewhere than in *Lastræa*. The fertile frond is too reduced to offer any clue, but the sterile is altogether Lastræa-like. As it is, so far as known, strictly endemic in the Philippines it is quite reasonable to look for its ancestry among the most similar Philippine ferns.

With the same measure of probability, *Stenosemia* may be regarded as an offshoot of *Aspidium*, not, in spite of the sparingly anastomosing veinlets of *Pleocnemia*, but of *Euaspidium*. *A. Griffithii* is a species, in the most similar *Aspidium* group, with "sori" anastomosing along the veins. Indusia, if present in the ancestral forms, would inevitably have been lost in the reduction of the fertile frond. *Leptochilus* must also be derived from *Aspidium*. *L. lanceolatus* is still occasionally found with the sporangia confined to the main veins (as in *Loxogramme*); and *L. latifolius* usually has the "sori," at least until they are old, anastomosing along the veins instead of covering the frond, being in this respect like its possible relative *Hemionitis*. The instability of the frond-form of many species of *Leptochilus* is a familiar phenomenon.

If the *Woodsieæ* of the *Natürlichen Pflanzenfamilien* are a homogeneous group, it has a common origin with the *Aspidieæ*, the most primitive representatives of the two groups being much closer together in all respects than either is to the highly specialized members of its own tribe. In fact, they are so alike that it is impossible to call either the more primitive, and I have ascribed that place to *Lastræa* only on geographical grounds, and because it is now a great and conspicuous group. The primitive member of *Woodsieæ* is *Acrophorus*. The generalized character of this fern is shown by its history. Hooker, who calls it *Davallia* (§ *Leucostegia*), remarks:[31] "Blume arranges it in *Aspidium*, and expresses no doubt as to the propriety of so doing. Presl makes a distinct genus of it, and places it between *Cystopteris* and *Leucostegia*. Judging from his figure, I do not see how it differs from *Davallia*, but he says 'hocce genus *Cystopteridi* valde affine est, differt soris in venulis apicalibus,' and under *Leucostegia* he says '*Acrophoro* affinissimum est' " Diels[32] says of it "Habituell an *Diacalpe* erinnernd; durch die gleichseitige Entwickelung der Segmente sowohl wie das Indusium von den Davallieen zu unterscheiden," the indusium being "breit eiförmig, am Grunde angewachsen, sonst frei."

I have no doubt that all these authors were describing the same species, and that, so far as insufficient material is ever a justification, each was justified in his view as to the affinity of the plant. When I first found the plant, hitherto unknown in the Philippines, I ascribed it to *Lastræa* as unhesitatingly as Blume had done; but it is indeed strikingly like

[31] Species Filicum 1: 157.

[32] Nat. Pflanzenfam. I, 4: 164.

Diacalpe in aspect and likewise suggestive in the same way of several species of *Lastræa,* and, more remotely, of *Monachosorum.* I have a single specimen, the majority of the indusia of which are reniform and fixed by the sinus, as in *Nephrodium,* but which has certain of them fixed by the base, more or less broad, and a few, unequal-sided, exactly like those of *Athyrium.* I should not know where else to look for the origin of *Diacalpe,* if not in *Acrophorus.*

On the internal evidence of the *Davallieæ, Microlepia* is certainly to be regarded as the central genus, to which most other genera are evidently related; and I am strongly inclined to believe that it is also the most primitive, although a possible relation between *Leucostegia* and *Acrophorus* has just been mentioned. In habitat and aspect, and in all essential characters except the indusium, *Microlepia* agrees with some of the most primitive *Aspidieæ* (*Lastraæ*). *M. strigosa* and *M. rhomboidea* are strikingly Polystichum-like in aspect, but I do not regard this as due to affinity. Of the other genera of the *Davallieæ, Wibelia* is sometimes included in *Microlepia,* and is certainly near it. Of all our ferns it is the most constantly unstable in form. A fern of my Mount Apo collection stands so exactly on the line between *Microlepia* and *Dennstaedtia* that it might be included in one practically as easily as in the other. Because the sori are not always quite marginal, and by the structure of the receptacle, which I do not regard as really diagnostic, I described it as *Microlepia, M. dennstaedtioides;* but Christ was rather disposed to call it *Dennstaedtia. Dennstaedtia* is still too clearly a natural group to lose its generic identity because it is intimately related to its parent.

In *Microlepia,* as in *Lastræa,* there is at San Ramon a solitary epiphytic species, *M. hirsuta,*[33] the stipe of which is appropriately articulate. While it is hardly probable that this species is an ancestor of our epiphytic *Davalliaeæ,* it shows how easily they may have originated in *Microlepia. Davallia* is the nearest epiphytic genus, and there is also in Mindanao a *Davallia, D. wagneriana,* the lowest pinnæ of which are not enlarged, but it differs from *M. hirsuta* in several important respects.

Unless, which is hardly probable, *Leucostegia* had an independent origin without the *Davallieæ,* it may best be regarded as a near derivative of *Microlepia,* though *Davallia pallida* approaches *Leucostegia* in the large free part of the indusium. *Humata* is probably derived from *Leucostegia,* the strong resemblance to *Davallia* being in adaptation to their common environment. However, there are other points of possible contact between all these genera, and real affinities are still somewhat a matter of guesswork. Even the natural generic limits are not certain.

[33] This argument loses none of its force if the species be removed from *Microlepia,* as a separate genus; its affinity to *Microlepia* is unmistakable, whatever name it bears.

Microlepia is undefinable because generalized. *Leucostegia* may also be primitive, or it is possibly heterogeneous but it is as natural a group as is made by combining it with *Humata*, and its union with *Davallia* seems to me still less proper.

My reasons for believing that the nearest affinity of *Oleandra* is to *Humata* have already been published.[34] These do not constitute good proof, but they are the best evidence we have as to the affinities of *Oleandra;* but *Oleandra* would appear from its distribution to be the older group.

The mutual affinity of the other genera treated as *Davalliea* is still more dubious. *Arthropteris* seems to me to be very near *Lastræa*, in which group its first species was described. Our scandent *Lastræa* (No. 1712) shares with *Arthropteris* the articulate stipe and the terminally placed sori. In spite of the very striking resemblance between their fronds, comparing, for instance, *Arthropteris ramosa* with *Nephrolepis Lauterbachii* or *Nephrolepis cordifolia*, the affinity of these two genera is by no means above doubt, and if one is descended from the other, it is not a proven fact that *Arthropteris*, in spite of its apparently much closer affinity to *Lastræa*, is the parent, for *Nephrolepis* is shown by its distribution, and still more by its conspicuous morphological isolation, and by the diversity of its fructification, to be a very ancient genus.

Nephrolepis acutifolia is like *Schizoloma* in two conspicuous characters, the articulate pinnæ and the unbroken marginal sori, but this is probably only a coincidence. The latter genus is an unmistakable relative of *Lindsaya*, and, less intimately, of *Odontosoria*, but the common ancestry of the group is doubtful. Of the three, *Odontosoria* seems the nearer to *Microlepia*. The group is certainly terrestrial in origin.

Monachosorum is, as Diels says, "habituell an *Davallia* erinnerndes," but the suggestion of *Leucostegia* is stronger; and this is due almost exclusively to their common share in the aspect of *Acrophorus* and various species of *Lastræa*; that is, *Monachosorum* is more like the generalized ferns than like the more highly developed ferns of any tribe, and its assignment to any tribe, by our present knowledge, is purely arbitrary.

The most primitive genus of the *Asplenieæ* is unquestionably *Athyrium*. It is a generalized group, sharing, on the one hand, the characters of *Diplazium* and *Asplenium*, and merging into both, and on the other, being indistinguishable from *Lastræa*. *Athyrium cyclosorum* Rupr., of Asia and western North America, usually regarded as a form of *A. filix-fæmina*, is Lastræoid in its indusia. I have recently described a new

[34] Polypodiaceæ of the Philippines, *Govt. Lab. Publ.* (1905) **28:** 48. "The resemblance to * * * the simple species of *Humata*—the creeping, scaly rhizome, the articulate stipe, the free, forked, closely parallel veins, the shape, attachment, and texture of the indusium, and its opening obliquely toward the apex of the frond—all these can not well be construed otherwise than as evidences of real affinity."

species as *Athyrium*, *A. hyalostegium*, so *Lastræa*-like in character that I should not have hesitated to accept a former name in either group. *Aspidium Fauriei* Christ is likewise rather an *Athyrium*, but on the border. *Athyrium* meets all the demands previously laid down of a fern to be regarded as primitive.

The constant restriction of the sorus to one side of its vein is the most obvious character distinguishing *Asplenium* from *Athyrium*, while the equal and long development of both sides of the vein, and the consequent breaking apart of the halves of the indusium, characterize *Diplazium*. The larger part of *Athyrium* is nearer to *Diplazium*, and, as the genera are usually construed, there is probably more than one point of contact; that is, the line between them is not quite natural and can not be made sharp. Most *Athyria* share the stout habit and rather harsh, dark paleæ of *Diplazium*. *Diplazium* as a natural group is also characterized by the exceedingly deep and irregular cells of the nether epidermis (figs. 29, 30). *Athyrium silvaticum* shares this character also, and is very much nearer to every species of *Diplazium* than it is to the primitive *Athyria*. Yet, if a line is to be drawn between the genera, the indusium must be the diagnostic character, leaving *Diplazium*, what *Athyrium* can not be made, a clear-cut, definable, and within itself a natural group; but, even so, I have some apparently undescribed plants not positively referable to one or the other.

Callipteris is an offshoot of *Diplazium*, still so close that its generic separation is a matter of taste. *Diplaziopsis*, of Christensen, long known as *Allantodia*, is a near derivative of *Callipteris*, confined to this part of the world.

Blechnum is an old group. The distribution of the genus, and of some species attests this, as do its general morphological isolation and the varied specializations of different species. A very close connection with any other forms is no longer to be expected. The merging-point between *Eublechnum* and *Lomaria* must be older than one subgenus and is likely to be as old as either, and is therefore most likely to give a clue to the affinities of both. *Blechnum egregium* is in this position. It is a stout but not very large fern, with stout, erect caudex; rigid, black, partly subaërial roots; harsh, black paleæ, and very deep epidermal cells of the upper surface of the frond, those of the nether surface being very irregular (fig. 31). These are all notable characters, and their combination in this *Blechnum*, and in *Diplazium*, and in *Athyrium silvaticum* is unmistakable proof of affinity.

The nearest affinity of *Asplenium* is to the more primitive part of *Athyrium*. Within itself, it is as diversified in form as might be expected of a nonprimitive genus growing in every land in all save extreme habitats. But, unlike our other large genera, it is clearly definable, and has not given rise to a large number of other groups conveniently distinguished as genera. The reasons for this are that it is not, as *Lastraa* is, primitive, and that it is not so particularly fitted to any set of widespread conditions—as *Polypodium* is to the epiphytic habitat—that it can become dominant under them and then become further specialized under peculiar subordinate conditions.

Of our fern genera, only two, *Scolopendrium* and *Stenochlaena*, can be treated at all positively as derived from *Asplenium*. The case of the former is already well known. If my *S. pinnatum* is what it is called, which I do not doubt, the genus *Triphlebia* owes its origin to an error, for its sori originate exactly as in *S. vulgare*.

Asplenium epiphyticum is a fern described from material collected on the Gulf of Davao. It is common along the lower border of the high forest at San Ramon, and I have now specimens from Surigao. Its resemblance to some other scandent *Asplenia* with simple fronds was strong enough to demand some care in the diagnosis; yet Christ says it is unmistakably *Stenochlaena*, and, indeed, its vegetative structures, root, stem, and leaf, are apparently identical with those of occasional immature forms of *S. aculeata* (Blume) Kunze.[35] Knowing a fern as well as I do this one, and never having seen any indication that it is other than *Asplenium* in its fruit, or has any other structures that as an *Asplenium* it might not have, I can only believe it to be an *Asplenium*; but neither can the affinity to *Stenochlaena* be mistaken. In my opinion we have in *S. aculeata* a very striking and perfect example of the repetition in the development of the individual of the race-history of *Stenochlaena*, proving beyond any doubt that it is an offshoot of *Asplenium*. The remarkably complete preservation of the race-history in this case is because the forms which are gone through are themselves well adapted to the environment.

It is not impossible that *Coniogramme* is also an offshoot of *Asplenium*, but the evidence on which any particular ancestry might be ascribed to this genus and to *Syngramma* is still inadequate.

The *Pterideæ* are so poorly represented in the Philippines that a discussion of the affinities of most of the group is not called for here. I have already pointed out the interesting position of *Schizostege* as a probable ancestor of *Pteris* and *Cheilanthes*. The genus was first described by Hillebrand from a very rare Hawaiian plant. Baker reduced it to *Cheilanthes*, and Christ, with rather better reason, transferred it to *Pteris*. My Mindanao plants agree in every essential character with the Hawaiian, but in aspect incline toward *Cheilanthes*, rather than toward *Pteris*. In the irregular and imperfect marginal anastomoses, *Schizostege* stands directly between *Cheilanthes* and *Pteris*. Its antiquity is attested by its discontinuous distribution. The resemblance of *S. Lydgatei* to *P. quadriaurita* suggests that the latter may be the most primitive form of *Pteris*, and this idea receives some support from the

[35] Underwood: The Genus Stenochlaena. *Bull. Torr. Bot. Club.* (1906), **33**: 40.

wide distribution of this species and the great number of clearly derived forms, with simple and with reticulate veins.

The suggestion that *Hypolepis tenuifolia* is really a *Lastræa*, or near it, is not new,[36] but even if this be its nature it may not throw any light on the origin of the real *Pterideæ*, in which group this species is at best a rather foreign element.

Nowhere else in the world do the *Polypodieæ* reach a development comparable either in number of species or in diversity, as is shown by the number of genera, to that attained in this Archipelago. They therefore present an especially advantageous field for local study. Within this great tribe, it seems too clear for question that *Eupolypodium* is the parent, the center from which the other groups have been derived, but *Eupolypodium* itself, whatever its antiquity, is not in its characters a primitive group, being so specialized in adaptation to a habitat different from that of primitive ferns that it and its progeny have, the world over, become the predominant epiphytic ferns. Articulate stipes and comparative simplicity of frond are conspicuous adaptations to this habitat, and are characteristic of this tribe, but these same characters are, as has been seen, assumed by epiphytic plants of diverse origin. and the articulation is readily lost under conditions that render it superfluous. These facts make the tracing of the true origin of the tribe a difficult, perhaps an impossible, task.

The articulation of the stipe may be of utility to a fern of cold countries, such as *Polypodium vulgare*, permitting the ready casting of the leaves. Such a fern would readily adopt the epiphytic habit, under favorable conditions, and might have given rise to this tribe, but we have no evidence of the development of articulate stipes by a terrestrial fern, and know that it has not occurred in most terrestrial ferns, even in cold regions. Therefore, it seems more reasonable to believe that *P. vulgare* has retained the articulations of epiphytic ancestors, being probably a plant the terrestrial habit of which, has permitted it to remain where it now lives, dating its origin from a time when the climate of the same region was favorable to epiphytic vegetation. Or, more probably, *Polypodium* may have been derived from primitive ferns which evolved the present characters of the tribe in adaptation to epiphytic life. In this case, the ancestry was probably in *Lastræa*, considering the instability of the indusium of this subgenus, and the known fact that it can develop articulations when an epiphyte. In the preceding section of this paper it was shown that in various minor groups within *Polypodium* there are species the stipes of which are inarticulate. It is not possible by any usable definition to exclude these species of *Eupolypodium* from *Lastræa* and any one of them can readily be imagined to be a connecting link; but there is little reason to

[36] Polypodiaceæ of the Philippines. *Govt. Lab. Publ.* (1905), **28:** 95.

believe this of one rather than of another, and all of them known to me impress me as being of almost certain Polypodioid derivation. *P. Merritti*, of Mindoro, is superficially one of the most Lastræa-like. It has a short rhizome, inarticulate stipes, a pinnate frond with obscurely toothed pinnæ, and glandular pubescence, but the glandular trichomes are very unlike those of *Lastræa*, and the plant is distinctively Polypodioid in the paleæ and in the attachment of the pinnæ.

In my opinion, *Polypodium*, in the broad sense in which the genus is usually construed, is more cumbrous than is justified by its unquestionable naturalness. No group of whatever rank is justified except by naturalness (or by ignorance). The tribe is natural and among its various subdivisions can be seen every measure of consanguinity. A genus should be convenient as well as natural, and it ought not, within itself, to be more diverse than can well be avoided. Our *Polypodia* fall into four natural, readily recognizable, and definable groups: *Polypodium* in the narrower sense, *Goniophlebium*, *Phymatodes*, and *Drynariopsis* (*P. heracleum*); to these *Myrmecophila*, but not *Selliguea*, should perhaps be added. However, the scope of this paper does not include the settling of generic nomenclature where taste alone is at stake, and individual initiative in such a matter may well await the joint action of pteridologists. I therefore follow Diels and Christensen, so far as they agree.

Within *Eupolypodium* are several groups which in their full development are well characterized by the form of the frond, a character which, even if they did not intergrade, would but weakly justify their elevation to generic rank. Yet they are natural groups. The most probable central group is that with the fronds pinnate or nearly so, and superficial sori, representatives of which are *P. minutum* and *P. macrum*. From these, a series of forms can be found leading to Blume's *Ctenopteris*, *P. celebicum* and *P. obliquatum*, with very deeply immersed sori. From this latter group, as has already been shown, *Prosaptia* has been derived. *Polypodium papillosum* is a rather aberrant species, suggesting *Goniophlebium* in some respects, but not so much so that I would wish to combine them, nor would I treat it as a distinct genus (*Thylacopteris* Kze.).

Polypodium trichomanoides is intermediate between *P. macrum* and the *Calymmodon* group, the best known representative of which is *P. cucullatum*. From this group, as has already also been shown, *Acrosorus* has been derived.

Through such forms as *P. solidum* and its variety *denticulata*, or through species like *P. loherianum* and *P. pleiosoroides*, with irregular margins, the *P. macrum* group is connected with the large section with entire fronds, called *Grammitis* by Blume, who recognized, however, that the sorus was not always elongate. When ferns in any group have developed a considerable area of uncut lamina, whether by the shortening of the sinuses between lobes, or otherwise, there has been a tendency toward an anastomosis of the veins. The illustrations of this are very numerous, as in the evolution of *Goniopteris* and *Aspidium* from *Lastræa;* of *Callipteris* from *Diplazium;* of the *Nidus* group in *Asplenium;* of *Pteris biaurita* from *P. quadriaurita;* of *Goniophlebium* and *Phymatodes* from *Eupolypodium.* In *Prosaptia,* the least divided species, with broadest segments, *P. Toppingi,* has usually some anastomosing veins. From the *Grammitis* section has been derived *Loxogramme,* characterized by the anastomosing veins of the usually ample but simple and entire fronds, and also by the total loss of articulate stipes. These two characters together, in a group certainly independent of *Phymatodes,* amply warrant the recognition of *Loxogramme* as a distinct genus. Its most Grammitis-like species is *L. parallela;* at the other extreme, it contains one species with dimorphous fronds, *L. conferta.*

It is more than possible that Blume was correct in treating *Antrophyum* and *Loxogramme* as relatives; I know of no other likely origin for the *Vittarieæ,* and the resemblance of these two genera is certainly strong. If this be the origin of the tribe, *Vittaria* is an offshoot of *Antrophyum* and *Monogramma* is, in any case, derived through *Pleurogramme* from *Vittaria.*

Goniophlebium[37] and *Phymatodes* are independent in origin from all the preceding offshoots of *Polypodium* and nearly or quite independent of each other. The former is a small, compact, uniform, therefore very natural group, confined to this part of the world, the union of which with the American so-called *Goniophlebia* is quite unwarranted.

Phymatodes,[38] too, is a natural group, but very far from compact and uniform, being internally diverse and the parent of numerous highly specialized groups commonly treated as genera. In view of their unstable venation and evidently broad affinities, it seems probable that the oldest representatives of *Phymatodes* are small ferns with simple fronds, the fertile and sterile alike, and without specialized fertile region (*Craspedaria,* in part). Such fronds are found on young plants of various *Phymatodes* species, more complicated at maturity. *Grammitis* is the probable source of this group. Within the *Craspedaria* group, specialization has taken place in the dimorphism of the fronds of most species, and in the nearly related species, *P. accedens,* the sori are restricted to the specialized apex. I believe that both *Hymenolepis* and *Niphobolus* have been derived from some ancestor not very unlike *P.*

[37] J. Smith's *Schellolepis* is the original *Goniophlebium* of Blume, *P. cuspidatum* being the first species mentioned, and the whole subgenus being Malayan. *Flora Javae* (1828), II, 132.

[38] If raised to generic rank, it must probably be called *Selliguea. Pleopeltis* would be an older name; but it (*P. angustum*) has certainly been distinct from *Phymatodes* much longer than *Selliguea* has.

accedens. Drymoglossum and *Niphobolus* are unmistakably related, and *Elaphoglossum* is probably parallel in genesis to *Niphobolus,* with a reasonably near common ancestor. *Elaphoglossum* is no small genus in the Philippines.

From the *Craspedaria* group are descended the species of *Phymatodes* with more ample, entire fronds, among which several natural groups are more easy to recognize than to define. From one of these, perhaps from some such fern as *P. triquetrum,* the *Selliguea* group is derived. More than one of these groups with entire fronds has descendants with the fronds deeply pinnatifid, as in *P. phymatodes.* One of the most natural of the latter groups with the fronds almost pinnate, is exceedingly thin in texture, and inhabits water courses and very moist banks. Among its species are *P. insigne* and *P. dolichopterum. P. ellipticum* is an outgrowth of this group, and is, therefore, not intimately related to the other species called *Selliguea.* Another group of the pinnatifid species includes *P. incurvatum* and *P. palmatum.* The former is decidedly dimorphous, and strongly suggests the probable ancestry of *Christiopteris. Cheiropleuria,* too, has possibly its source here, but I strongly suspect that both it and *Taenitis* are Aspidioid in origin. Another group in *Phymatodes* is *Myrmecophila* (*Aspidopodium*), the commonest species of which is *P. sinuosum.* This group has given rise to *Lecanopteris.*

Finally, the most diversified and highly specialized outgrowth of *Phymatodes* is the *Drynaria* group of genera. Within the usually accepted limits of *Phymatodes, P. musaefolium* and *P.* (*Drynariopsis*) *heracleum* lead to this group, from the large group of species with ample, entire fronds. From some such plant as *P. heracleum, Drynaria* has evolved by the restriction of the humus-collecting work to distinct fronds. From a similar ancestor, *Aglaomorpha* and *Dryostachyum* have been evolved by the restriction of the fertile region to the apex of the frond and the fusion of the sori. From these, *Thayeria,* the most highly specialized of the group, has probably been derived. The *Drynaria* group is a very natural one, characterized by its habit, by the fleshy rhizome, and its scales, by the frond form (exception, *D. rigidula*) and venation, and by the remarkable cutting-off of the segments. *Photinopteris* is a probable relative of *Dryostachyum,* the affinity being indicated by the location of the fertile region, absence of definite sori, and the glands at the bases of the pinnæ; young fronds of *D. pilosum* have rows of lime dots.

The *Phymatodes* group, as a whole, is xerophytic and characterized by the presence of a specialized hypodermis beneath the upper epidermis. The cells of this tissue are irregular in surface view, very often more so than the epidermis; sometimes one, sometimes the other has the thicker walls. Under conditions that render the hypodermis unnecessary as a protective structure, it loses its ecological character, but usually maintains to a recognizable extent the characteristic form of the cells; in such cases, intercellular spaces are often formed by incurrent folds of the walls. In my notes such a structure is mentioned in *P. accedens, P. revolutum, P. Zippelii, P. dolichopterum, P. 1741* (*musaefolium?*), *P. incurvatum, P. commutatum* (fig. 32), *P. Phymatodes, P. Schneideri, P. palmatum, P. angustatum, P. albido-squamatum, P.* (*Selliguea*) *caudiforme* (fig. 33), *P. sinuosum* (fig. 34), *Lecanopteris* (fig. 35), *P. heracleum, Drynaria rigidula, D. quercifolium, Dryostachyum pilosum* (fig. 36), *Thayeria* (fig. 37), and *Photinopteris.* It also occurs in *P. subauriculatum*; but is wanting in *Taenitis* and *Cheiropleuria,* and in *P. angustum,* the type species of *Pleopeltis.* This is one reason why, if I were raising the subgenera of *Polypodium* to generic rank, I would keep *Selliguea,* including *Phymatodes,* distinct from *Pleopeltis.*

ILLUSTRATIONS.

PLATE I.

FIG. 1. Davallia decurrens. Transverse section of rachis of pinna. X 20.
2. Stenosemia pinnata. Sinus seen from above, with hairs. X 20.
3. Humata parvula. Transverse section, nether epidermis. X 320.
4. Davallia brevipes. Same. X 320.
5. Asplenium squamulatum. Same. X 320.
6. Scolopendrium pinnatum. Cell of upper epidermis, surface view. X 320.
7. Vittaria falcata (?). Spicular cell, surface view. X 175.
8. Same, in transverse section. X 175.
9. Polypodium revolutum. Stoma. X 320.
10. Niphobolus nummulariæfolius. Stoma. X 320.
11. N. varius. Stoma, transverse section. X 320.
12. Same. Surface view; limit of exterior chamber shown by dotted lines. X 320.
13. Stenochlaena subtrifoliata. Stoma. X 320.
14. Polypodium caespitosum. Transverse section of frond. X 75.
16. P. accedens. Surface view; upper epidermis dotted, hypodermis in solid lines. X 175.
17. P. sinuosum. Tangential ("surface") section of upmost green layer X 320.

PLATE II.

FIG. 15. Drynaria quercifolia. Surface view; upper epidermis dotted, hypodermis in solid lines. X 320.
18. Niphobolus adnascens. Mesophyll cell, in *water*. X 175.
19. Asplenium Phyllitidis. Section across two sori. X 20.
20. Scolopendrium schizocarpum. Section across double sorus. X 20.
21. Polypodium obliquatum. Section of sorus. X 41.
22. Vittaria alternans. Transverse section of frond, showing paraphyses, but without sporangia. X 20.
23. Aspidium angulatum. Indusium, with depressed center. X 27.
24. Achrostichum aureum. Paraphysis head. X 175.
25. Nephrodium diversilobum. Sporangia. X 75.
26. Cheiropleuria bicuspis. Stoma of fertile frond. X 175.
27. Same. Stoma of sterile frond. X 175.
28. Lomagramma pteroides (?). Half of transverse section of frond, sporangia wanting. X 27.
29. Diplazium bulbiferum. Epidermis of nether surface. X 175.

PLATE III.

FIG. 30. **Diplazium** pallidum. Epidermis of nether surface. X 175.
31. **Blechnum egregium.** Same. X 320.
32. **Polypodium affine.** Surface view; upper epidermis dotted, hypodermis in solid lines. X 320.
33. **P. caudiforme.** Upper epidermis in solid lines, hypodermis dotted. X 75.
34. **P. sinuosum.** Upper epidermis in solid lines, hypodermis dotted. X 175.
35. **Lecanopteris** pumila. Upper epidermis dotted, hypodermis in solid lines. X 175.
36. **Dryostachyum** pilosum. Upper epidermis in solid lines, hypodermis dotted. X 320.
37. **Thayeria Cornucopia.** Upper epidermis dotted, hypodermis in solid lines. X 75.

PLATE IV.

Genealogical tree of Polypodiaceæ.

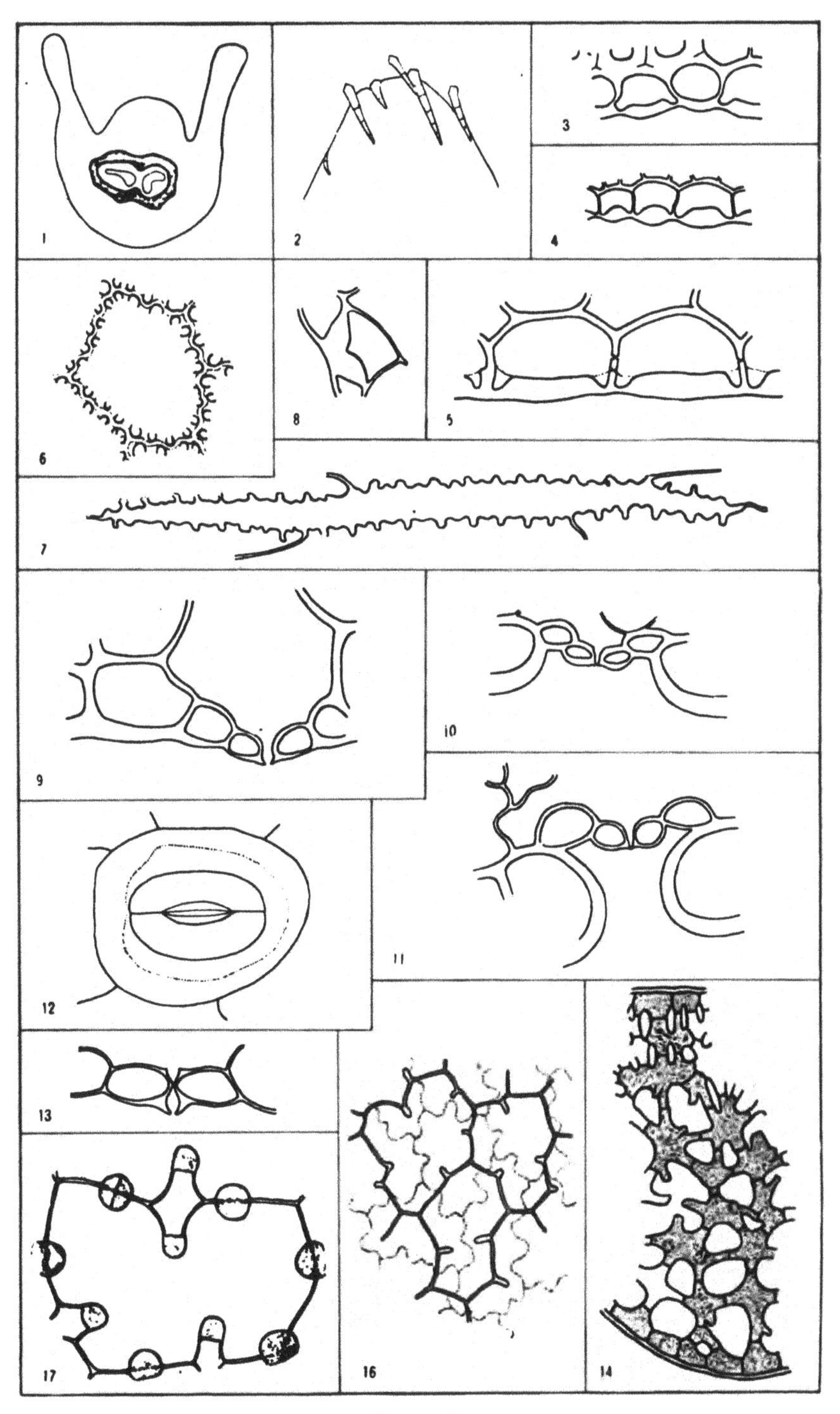

PLATE I.

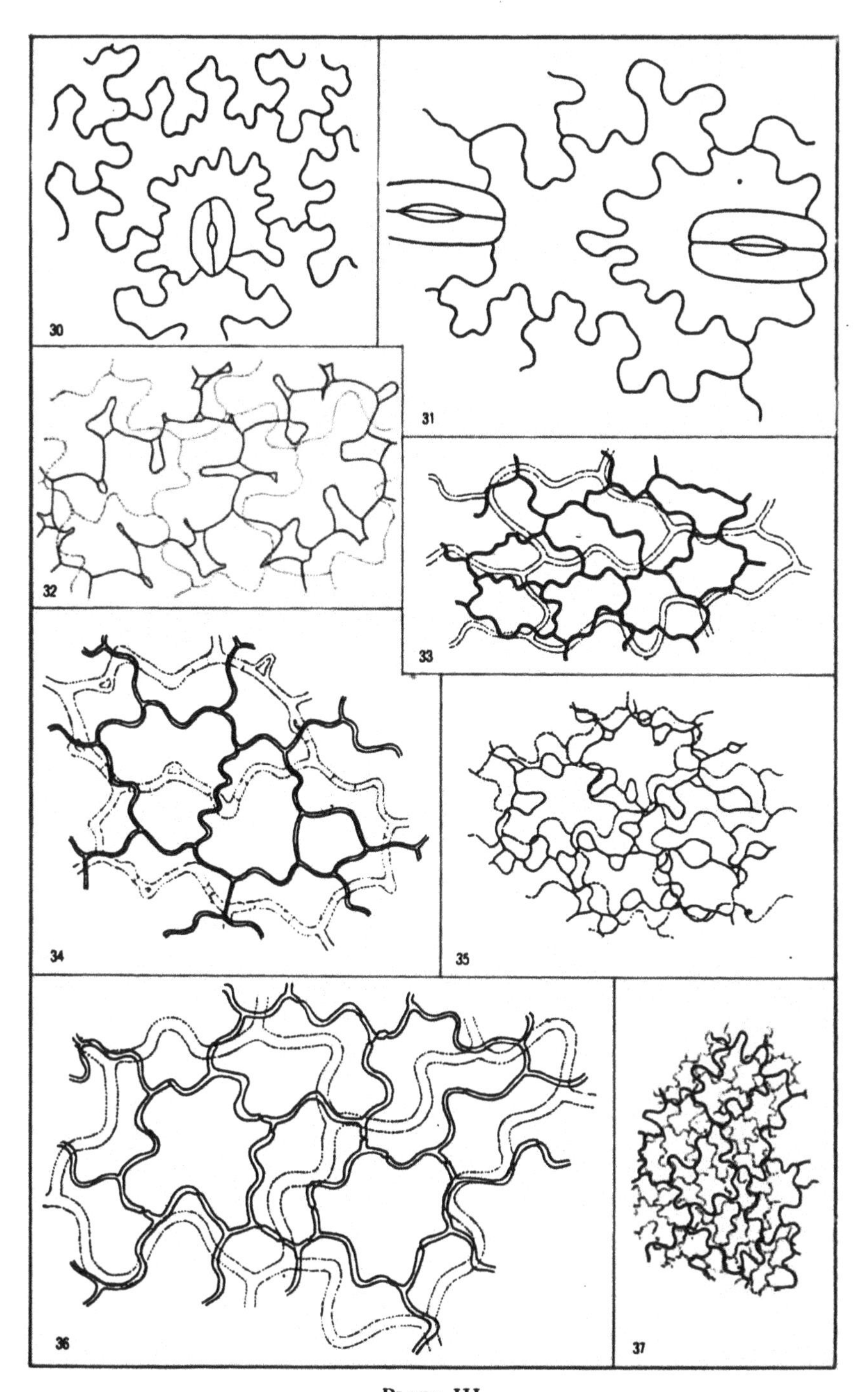

PLATE III.

THE PHILIPPINE

JOURNAL OF SCIENCE

C. BOTANY

VOL. II APRIL, 1907 No. 2

CYPERACEÆ OF THE PHILIPPINES: A LIST OF THE SPECIES IN THE KEW HERBARIUM.

By C. B. CLARKE.[1]
(*Kew, England.*)

The present list contains only the species in the Kew Herbarium and not quite all of these. I have added some of the more important synonomy relating to the Philippines and adjacent lands. The short notes on the genera and species are not given as sufficient diagnoses, but are intended to be useful to assist collectors in the field.

The present list is not therefore complete in any respect for Philippine *Cyperaceæ,* which perhaps is of minor importance when so many species and localities are being constantly added. The list will have the merit that each species stands on plants examined, though the various species may not invariably have been determined correctly.

1. KYLLINGA Rottb.

In all the Philippine species of this genus, the spikelet is 1-nutted.

* *Rhizome creeping.*

1. **Kyllinga monocephala** Rottb. Descr. et Ic. (1773) 13. *t. 4. f. 4.* syn. quibusdam excl.

Keel of the flowering glume crested, scarious, full of oil glands.

Hook. f. Fl. Brit. Ind. **6**: 588; Hemsl. Bot. Challenger Voy. **2**: 85; Merrill, Philip. Journ. Sci. **1** (1906) Suppl. 30. *Kyllinga Mindorensis* Steud. Cyp. 67;

[1] It is with the deepest regret that this Bureau has learned that Mr. Clarke died at Kew, England, on August 15, 1906.—E. D. M.

Miq. Fl. Nederl. Ind. **3**: 292. *K. monocephala* var. *Mindorensis* Boeck. in Linnæa **35** (1867–68) 428; Vidal, Phaner. Cuming. Philip. 155; Rev. Pl. Vasc. Filip. 284. (*K. Mindorensis* is no variety of *K. monocephala;* it is the typical state of the species.)

PHILIPPINES, (*Callery*); (*Llanos*); (*Moseley*); (1558 *Cuming*, hb. Kew). LUZON, (*Kastalsky*); (*Leclanches*); (1860 *Wichura*); (3981 *Vidal*); (86 pro parte, 3160, 5103 *Merrill*); (6026 *Leiberg*); (457 *Topping*); (3283 *Ahern's collector*). MINDANAO, (249 *DeVore & Hoover*); (548 *Copeland*).

Common in the warm regions of the Old World, as in East Asia and Polynesia. From South America I have seen two examples only, which I suppose to be "allata."

2. **Kyllinga brevifolia** Rottb. Descr. et Ic. (1773) 13. *t. 4 f. 3.*

Keel of the flowering glume not crested, green, minutely scabrous or with a few scattered hairs.

Hook. f. Fl. Brit. Ind. **6**: 588. *Kyllinga triceps* Blanco Fl. Filip. 34, *non* Rottb; Steud. Cyp. 72. *K. rigidula* Steud. Cyp. 71, quoad pl. Philippensem; Miq. Fl. Nederl. Ind. **3**: 294. *K. caespitosa* var. *robusta* Boeck. in Linnæa **35** (1867–68) 413, *minime* Nees; Vidal Phaner. Cuming. Philip. 155; Rev. Pl. Vasc. Filip. 283.

PHILIPPINES, (552 *Cuming*). LUZON, (772, 1656 *Loher*); (86A *Merrill*). BASILAN, (43 *DeVore & Hoover*). MINDANAO, (101 *Clemens*); (585 *Copeland*).

The warmer parts of both hemispheres; the commonest species of the genus.

3. **Kyllinga intermedia** R. Br. Prodr. (1810) 219.

Slender, with small lateral globose or depressed heads; keel of the flowering glume green, smooth, not crested.

Benth. Fl. Austr. **7**: 521; Hemsl. in Journ. Linn. Soc. **36** (1903) 223. *Kyllinga oligostachya* Boeck, in Linnæa **35** (1867–68) 407. *Kyllinga monocephala* Seem. Fl. Viti. 318, pro majore parte, *non* Rottb. *K. brevifolia* Boeck. ! ms. in A. Dietrich No. 717.

LUZON, Benguet, (4371, 4421, 4695 *Merrill*); (6494 *Elmer*): Lepanto, (4618, 4624 *Merrill*).

Frequent in north and east Australia. I also refer to this species plants collected in Formosa and Fiji. It is hardly more than a variety of *Kyllinga brevifolia* Rottb., as indeed *Boeckeler* sometimes esteemed it. The distribution in Formosa and Fiji might therefore be disputed.

4. **Kyllinga pungens** Link. Hort. Berol. **1** (1827) 326.

Rhizome stout, conspicuously squamose; leaves and bracts short; keel of flowering glume not crested.

Dyer Fl. Trop. Afr. **8**: 269, 277. *K. obtusata* Presl Rel. Haenk. **1** (1828) 183; Boeck. in Linnæa **35** (1867–68) 418. *K. bifolia* Miq. Fl. Nederl. Ind. **3**: 293. *Cyperus aphyllus* Hassk. Cat. Hort. Bogor. (1844) 24.

LUZON, (*Loher*); (4649 *Merrill*); (6495 *Elmer*).

Abundant in tropical America, frequent in tropical Africa. I have, in all, four examples from Malaya, none from India.

** *Tufted.*

5. **Kyllinga cylindrica** Nees ! in Wight Contrib. (1834) 91.

Stems tufted, rootstock hardly any; heads 3–1, the central one usually cylindric.

Hook. f. Fl. Brit. Ind. **6**: 588; Hemsl. in Journ. Linn. Soc. **36** (1903) 223.
LUZON, Benguet, (6500 *Elmer*). MINDANAO, Lake Lanao, (102 *Clemens*).
Frequent in warm regions of the Old World, Africa, Asia to tropical Australia. The representative species in the New World is *Kyllinga odorata* Vahl, which may be treated as a geographical variety.

SPECIES EXCLUDED FROM KYLLINGA.

KYLLINGA ALBESCENS Steud., based on 1418 *Cuming*, is *Lipocarpha orgentea* R. Br.

2. PYCREUS Beauv.

In all the Philippine species of this genus the nut is reticulate, not zonate.

1. **Pycreus sanguinolentus** Nees in Linnæa **9** (1834) 283.

Stem decumbent, clothed by leaf-sheaths one-third the way up; heads of spikelets simply umbelled.

Hook. f. Fl. Brit. Ind. **6**: 590; Hemsl. in Journ. Linn. Soc. **36** (1903) 206. *Cyperus sanguinolentus* Vahl Enum. **2**: 351; Miq. in Ann. Mus. Lugd. Bat. **2**: 140. *C. Eragrostis* Vahl Enum. **2**: 322 ? (syn. Retz. excl.); Moritzi ! Verz. Zoll. Pfl. 96; Benth. Fl. Hongk. 385; Hemsl. Bot. Challenger Voy. **2**: 85. *C. areolatus* R. Br. Prodr. 216. *C. atratus* Steud. ! in Zoll. Verz. Ind. Archip. heft. 2, 62; Miq. Fl. Nederl. Ind. **3**: 259.
LUZON, Benguet, (4706 *Merrill*).
Old World, from the Black Sea to Australia, and from tropical Africa to Manchuria; a common plant.

2. **Pycreus pulvinatus** Nees in Linnæa **9** (1834) 283.

A slender annual, heads (short spikes) umbelled, pale; glumes mucronate.

Hemsl. in Journ. Linn. Soc. **36** (1903) 206. *Pycreus nitens* Nees in Nova Acta Nat. Cur. **19** (1843) Suppl. **1**: 53; Hook. f. Fl. Brit. Ind. **6**: 591. *Cyperus pumilus* L. Amoen. Acad. **4** (1788) 302. *Cyperus nitens* Retz. Obs. **5** (1789) 13 ?; Miq. in Ann. Mus. Lugd. Bat **2**: 140 ?; Vidal ! Rev. Pl. Vasc. Filip. 283; Phaner. Cuming. Philip. 155. *C. gymnoleptus* Steud. ! Cyp. 3; Miq. Fl. Nederl. Ind. **3**: 255.
PHILIPPINES, (559 *Cuming*); (198 *Chamisso*). LUZON, (3404 *Ahern's collector*).
Old World, from tropical Africa to China and north Australia, a common plant. Once received from Martinique. Received in five collections from Florida (named erroneously *P. leucolepis*).

3. **Pycreus polystachyus** Beauv. Fl. d' Owar. **2** (1807) 48. *t. 86. f. 2.*

Moderately stout, heads (short spikes) in a dense umbel or compound head; glumes muticous.

Hook. f. Fl. Brit. Ind. **6**: 592; Hemsl. Journ. Linn. Soc. **36** (1903) 205. *Cyperus polystachyus* R. Br. Prodr. 214; Benth. Fl. Hongk. 385; Miq. Fl. Nederl. Ind. **3**: 258, Suppl. 260, *non* Rottb. *C. odoratus* Linn. ! Sp. Pl. ed. 1, 46, ed. 2, 68 et herb. propr., pro majore parte. *C. teretifructus* Steud. ! in Zoll. Verz. Ind. Archip. heft. 2, 62, et Cyp. 3.

PHILIPPINES, (196 *Chamisso*); (1870 *Wichura*). LUZON, (876 *Jagor*); (792 *Loher*); PANAY, (98 *Copeland*).

In all warm countries, especially near the sea, one of the most generally distributed and abundant species of *Cyperaceæ*.

Sir *J. D. Hooker* (in Trimen Fl. Ceylon **5**: 20) says that this species may have a 3-fid style. Unfortunately the sheet from Ceylon, 800 *Thwaites*, contains *Pycreus polystachyus* Beauv., mixed with *Cyperus Zollingeri* Steud., which latter was dissected by Hook. f., and found with 3-fid styles.

Var. β **laxiflorus** Benth. ! Fl. Austral. **7**: 261.

Spikes (and umbels) opened out, the spikelets often quite solitary.

Hook. f. Fl. Brit. Ind. **6**: 592. *Cyperus paniculatus* Rottb. Descr. et Ic. 40; Hook. et Arn. Bot. Beechy Voy. 99.

LUZON, (877 *Jagor*); Manila, (*Wichura*).

In all warm countries, especially near the sea; a common plant.

4. **Pycreus sulcinux** C. B. Clarke in Hook. f. Fl. Brit. Ind. **6**: 593.

Slender; umbel and spikes very loose; spikelets elongate; nut unsymmetric, subconcave on the margin next the axis.

Cyperus sulcinux C. B. Clarke in Journ. Linn. Soc. **21** (1884) 56.

PHILIPPINES (*Moseley*). LUZON, Benguet, (4289 *Merrill*); (6579 *Elmer*). MINDANAO, Lake Lanao, (44 *Clemens*):

Scattered throughout the warmer regions of the Old World. Frequent in India.

5. **Pycreus globosus** Reich. Fl. Germ. Excurs. (1830–32) 140.[10]

Umbel simple or reduced to one head; spikelets clustered, usually pale or greenish.

Hemsl. in Journ. Linn. Soc. **36** (1903) 203. *Pycreus capillaris* Nees in Linnæa **9** (1834) 283; Hook. f. Fl. Brit. Ind. **6**: 591. *Cyperus globosus* Allioni Fl. Pedem. Auctuar. 49. *C. vulgaris* Kunth Enum. **2**: 4; Benth. Fl. Hongk. 385; Miq. Fl. Nederl. Ind. **3**: 256. *C. mucronatus* Moritzi ! Verz. Zoll. Pfl. 95. *C. jungendus* Steud. ! in Zoll. Verz. Ind. Archip. heft 2, 63, et Cyp. 3. *C. trachirrhachis* Steud. ! ll. cc. 62, 3. *C. flavescens* Benth. ! Fl. Austral. **7**: 259 (excl. syn.); Miq. in Ann. Mus. Lugd. Bat. **2**: 140, 211.

PHILIPPINES, (875 *Jagor*).

Old World, from Spain to Japan and north Australia, a common plant. Rare in tropical Africa.

Var. β **Nilagiricus** Clarke in Hook f. Fl. Brit. Ind. **6**: 592.

Spikelets narrower, often chestnut-colored or black.

Hemsl. in Journ. Linn. Soc. **36** (1903) 204. *Cyperus Nilagiricus* Steud. ! Cyp. 2. *C. Junghuhnii* Miq. Fl. Nederl. Ind. **3**: 260.

LUZON, Benguet, (791 *Loher*); (6483 *Elmer*); Lepanto, (4630 *Merrill*).

Warmer parts of the Old World, especially in the mountains; a common plant. Frequent in Africa, Japan, India and China.

PHILIPPINES SPECIES OF CYPERUS WITH 2-FID STYLES; NOT IN THE KEW HERBARIUM.

CYPERUS ALBUS Presl Rel. Haenk. **1**: 174. "Stylus bifidus," "Nux oblonga," "Spiculæ dense fasciculatae."

LUZON, fide Presl.

CYPERUS LUZONENSIS Presl Rel. Haenk. **1**: 174.

"Stylus 2-fidus," "Nux elliptica."

LUZON, fide Presl.

An example of Llanos, marked "*C. luzonensis* Presl," is not Presl's plant, as the styles are all 3-fid; it is *Mariscus microcephalus* Presl below.

3. JUNCELLUS C. B. Clarke.

1. **Juncellus pygmaeus** C. B. Clarke in Hook. f. Fl. Brit. Ind. **6** (1893) 596.

Style 2-fid.

Hemsl. in Journ. Linn. Soc. **36** (1903) 207. *Cyperus pygmaeus* Rottb. Descr. et Icon. (1773) 20. *t. 14. f. 4, 5;* Miq. Fl. Nederl. Ind. **3**: 261.

LUZON, (769 *Loher*).

Old World, from Africa to Korea and north Australia; a common weed.

In the genus *Cyperus*, we have in many species the primordial nut equally trigonous; in the majority of species the nut is more or less flattened on the anterior face (the back, of authors) so that the nut is approximately plano-convex in section, the style remaining always 3-fid. In *Juncellus* the "dorsal" flattening appears to have been carried so far that the style is 2-fid; in two species of *Juncellus*, however, the style is 3-fid or 2-fid indifferently. There is thus no difference in structure, and no definite line to be drawn between *Cyperus* and *Juncellus*. The only reason for retaining *Juncellus* as a genus is convenience, even after it has been taken out, *Cyperus* being inconveniently large.

4. CYPERUS Linn.

Style 3-fid. Spikelets persistent. Glumes deciduous.

Sect. 1. PYCNOSTACHEÆ. Spikelets digitate or clustered; not spicate.

1. **Cyperus tenellus** Linn. f. Suppl. (1781) 103.

Stem 5 to 15 cm. long; with one lateral cluster of 1 to 4 spikelets; glumes obtuse.

Benth. ! Fl. Austr. **7**: 265 (syn. dub. *C. modestulo* Steud. excl.).

LUZON, Benguet, Pauai, (4740 *Merrill*), in a cold swamp at about 2,200 m.

Cape of Good Hope; in Australia rare. No. 4740 *Merrill* is a very small plant; it may possibly represent a new species near *C. tenellus*.

2. **Cyperus leucocephalus** Retz. Obs. 5 (1789) 11.

Stem with one dense globose head, 8 to 14 mm. in diameter.

Hook. f. Fl. Brit. Ind. 6: 602; Vidal, Rev. Pl. Vasc. Filip. 283; Phaner. Cuming Philip. 155. *Cyperus Sorostachys* Boeck. in Linnæa 35 (1867–68) 588. *Sorostachys Kyllingioides* Steud. Cyp. 71; Miq. Fl. Nederl. Ind. 3: 296.

PHILIPPINES, (1417 *Cuming*).

Old World, in warm countries, from Senegal to Cochin-China and north Australia; also in Brazil, but less frequent.

3. **Cyperus uncinatus** Poiret in Lam. Encycl. 7 (1806) 247.

Annual, 5 to 15 cm. high; heads umbelled; glumes with a hooked mucro.

Hemsl. in Journ. Linn. Soc. 36 (1903) 219. *Cyperus cuspidatus* H. B. K. Nov. Gen. et Spec. 1 (1815) 204; Vidal, Rev. Pl. Vasc. Filip. 282; Phaner. Cuming Philip. 155. *C. solutus* Steud. Cyp. 14; Miq. Fl. Nederl. Ind. 3: 263.

PHILIPPINES, (676 *Cuming*). LUZON, (713 *Loher*). MINDANAO, Lake Lanao, (*Clemens*).

In the warmer parts of both hemispheres; a common species.

4. **Cyperus difformis** Linn. Amoen. Acad. 4 (1759) 302.

Annual, 30 to 65 cm. high; spikelets very small, closely agglomerated, umbel compound (or simple).

Miq. Fl. Nederl. Ind. 3: 369; Hook f. Fl. Brit. Ind. 6: 599; Vidal Rev. Pl. Vasc. Filip. 283; Phaner. Cuming. Philip. 155; Hemsl. in Journ. Linn. Soc. 36 (1903) 210. *Cyperus subrotundus* Llanos Fragm. Pl. Filip. 14, fide Naves et Villar.

LUZON, (549 *Cuming*); (726 *Loher*); (3650 *Merrill*); (5680, 6298 *Elmer*).

Warmer Europe, Africa, warmer Asia, Australia, Polynesia; an abundant species. From America I have but two examples, one from New Mexico and one from Michoacan (Mexico).

5. **Cyperus Haspan** Linn. Sp. Pl. ed. 1, 45, ed. 2, 66 partim.

Root in the second year creeping; plant 3 to 6 dm. high, scantily leaved; umbel usually compound.

Hook. f. Fl. Brit. Ind. 6: 600; Miq. Fl. Nederl. Ind. 3: 267; Hemsl. in Journ. Linn. Soc. 36 (1903) 213.

LUZON, (724, 725 *Loher*); (54 *Merrill*); (5845 *Elmer*); MINDANAO, (1999 *Wichura*); (*Clemens*); (582 *Copeland*).

Tropical and subtropical countries throughout the World; an abundant species. The type specimen in hb. Linn. propr., marked by him *C. Haspan*, is not this species, while among the pieces of *C. Haspan* scattered through his Herbarium, Linnæus has named no one "*C. Haspan.*"

6. **Cyperus flavidus** Retz. Obs. 15 (1789) 13.

An erect, short-lived annual, becoming, in three months, yellow or blackish.

Hook. f. Fl. Brit. Ind. 6: 600.

PHILIPPINES, (*Llanos*). LUZON, (718 *Loher*); (3657 *Merrill*).

Tropical and subtropical Asia and Africa; abundant. Also in North Australia. This is the common small *Cyperus* of dibbled rice. It rapidly withers as the rice field dries. It is difficult to draw a line between it and *Cyperus Haspan*.

7. **Cyperus diffusus** Vahl Enum. 2 (1806) 321.

Large, leaves and bracts 1 to 2 cm. broad; umbel very compound, of heads of few spikelets (solitary pedicelled spikelets often added); points of glumes spreading in fruit.

Miq. Fl. Nederl. Ind. 3: 264; Hook. f. Fl. Brit. Ind. 6: 603; Hemsl. in Journ. Linn. Soc. 36 (1903) 211; Merrill in Philip. Journ. Sci. 1 (1906) Suppl. 1: 30. *Cyperus elegans* Swartz Obs. Bot. 30; Kunth, Enum. 2: 28. *C. longifolius* Decaisne in Nouv. Ann. Mus. Paris 3 (1834) 359; Miq. Fl. Nederl. Ind. 3: 265. *C. moestus* Kunth Enum. 2: 31; Miq. Fl. Ind. Bat. 3: 265. *C. scirpoides* Presl, Rel. Haenk. 1: 178, non Vahl. *C. Sorzogonensis* Presl l. c. 174.

PHILIPPINES, (*Micholitz*). LUZON, (*Wichura*); (463, 465 *Topping*).

Tropical and subtropical, southeast Asia and America, a sylvan (not rice field) species.

8. **Cyperus pubesquama** Steud.! in Zoll. Verz. Ind. Archip. heft 2 (1854) 62.

Glumes mostly puberulous, their points (in fruit) closely, rigidly appressed to the spikelet; otherwise as in *C. diffusus* Vahl.

Miq. Fl. Nederl. Ind. 3: 266; Hook. f. Fl. Brit. Ind. 6: 604. *Cyperus Lagorensis* Steud. ! Cyp. 36; Miq. Fl. Nederl. Ind. 3: 275. *C. Calacaryensis* Steud. ! Cyp. 34; Miq. l. c. 275. *C. diffusus* Kunth, Enum. 2: 30; Vidal ! Rev. Pl. Vasc. Filip. 283; Phaner. Cuming. Philip. 155, non Vahl.

PHILIPPINES, (445, 533, 534 *Cuming*). LUZON (5054 *Merrill*). MINDANAO, (280 *DeVore et Hoover*).

Frequent in India and through Malaya to New Guinea.

This species is closely allied to, as well as very like, *C. diffusus* Vahl, but is specifically distinguishable.

(8a., CYPERUS BANCANUS Miq. Fl. Nederl. Ind. Suppl. (1860) 260, 599; C. B. Clarke in Journ. Linn. Soc. 34 (1898) 27; Merrill in Philip. Journ. Sci. 1 (1906) Suppl. 1: 30.

LUZON, (2581 *Meyer*); (291 *Copeland*), fide *Merrill*.

Malayan Peninsula and Archipelago, Tonkin; frequent.

This plant is likely to occur in the Philippines, though I have no specimen before me. *Merrill* places a ? after his determination.)

Sect. 2. CHORISTACHEÆ. Spikelets spicate.

9. **Cyperus iria** Linn. Sp. Pl. ed. 1, 45; ed. 2, 67, tab. Rheede citat. excl.

Medium large annual, umbel often compound; glumes obovate, ultimately hardly imbricate.

Miq. Fl. Nederl. Ind. 3: 269; Hook. f. Fl. Brit. Ind. 6: 606; Vidal, Rev. Pl. Vasc. Filip. 283; Phaner. Cuming. Philip. 155; Hemsl. in Journ. Linn. Soc. 36 (1903) 213. *Cyperus Nuttalii* Llanos ! Fragm. Pl. Filip. 14.

PHILIPPINES, (*Llanos*); (563 *Cuming*); (3982 *Vidal*); (874 *Jagor*). LUZON, (731, 732 *Loher*); (200 *Chamisso*); (2307, 3651 *Merrill*); (5706 *Elmer*). MINDANAO, (1998 *Wichura*); (602 *Copeland*).

A weed in rice fields of the Old World (examples from the Santee Canal in Florida were introduced from the Old World, fide *Chapman*).

Var. β **paniciformis** Clarke in Hook. f. Fl. Brit. Ind. 6: 607.

Spikelets 2 to 4-nutted; racemes nearly linear.

Cyperus paniciformis Franch. et Savat. Pl. Japon. 2: 103, 537.

LUZON, (1867 *Wichura*).

India, China, Japan and Malaya.

10. **Cyperus compressus** Linn. Sp. Pl. ed. 1, 46, ed. 2, 68.

A green annual, umbel simple; spikelets much flattened; glumes acutely boat-shaped, mucronate.

Miq. Fl. Nederl. Ind. **3**: 263 et Suppl. 260, 599; Hook. f. Fl. Brit. Ind. **6**: 605; Vidal, Rev. Pl. Vasc. Filip. 282; Phaner. Cuming. Philip. 155; Hemsl. Journ. Linn. Soc. **36** (1903) 210. *Cyperus humilis* Llanos Fragm. Pl. Filip. 13, fide Naves.

PHILIPPINES, (546 *Cuming*); (1960 *Vidal*). LUZON, (870 *Jagor*); (715 *Loher*); (199 *Chamisso*); (183 *Meyen*); (1130 *Merrill*).

Very common in the warmer parts of both hemispheres.

11. **Cyperus distans** Linn. f. Suppl. (1781) 103.

Umbel open, compound; spikelets spicate, linear; glumes small, very remote.

Hook. f. Fl. Brit. Ind. **6**: 607; Vidal Rev. Pl. Vasc. Filip. 283; Phaner. Cuming. Philip. 155; Hemsl. in Journ. Linn. Soc. **36** (1903) 211. *Cyperus elatus* Presl. ! in Oken Isis **21** (1828) 271; Steud. in Zoll. Verz. Ind. Archip. heft 2, 63; Miq. Fl. Nederl. Ind. **3**: 284; Boeck. ! in Flora **62** (1879) 551, non Linn.

PHILIPPINES, (*Llanos*); (444 *Cuming*); (*Moseley*). LUZON, (733, 734 *Loher*); (6302 *Elmer*); (1133 *Merrill*).

In warm countries almost throughout the World. An abundant species.

12. **Cyperus eleusinoides** Kunth ! Enum. **2** (1837) 39.

Umbel compound, rays long, unequal; spikelets densely spicate; glumes scarious at the tips, hardly distant.

Miq. Fl. Nederl. Ind. **3**: 270; Benth. Fl. Austral. **7**: 277; Hook. f. Fl. Brit. Ind. **6**: 608; Hemsl. in Journ. Linn. Soc. **36** (1903) 212. *Cyperus xanthopus* Steud. in Flora **25** (1842) 595; Cyp. 36 (pl. Japon excl.)

LUZON, (728, 729 *Loher*).

Abyssinia and the Dead Sea to Queensland.

13. **Cyperus Malaccensis** Lam. Ill. **1** (1791) 146.

Stem stout, almost 3-winged at the top; leaves few, short; umbel compound, rather dense; spikelets spicate, linear; glumes obtuse, concave with incurved margins.

Miq. Fl. Nederl. Ind. **3**: 279; Hook. f. Fl. Brit. Ind. **6**: 608; Hemsl. in Journ. Linn. Soc. **36** (1903) 214; Merrill, Philip. Journ. Sci. **1** (1906) Suppl. **1**: 30. *Cyperus odoratus* Linn. ! Sp. Pl. ed. 1, 46; ed. 2, 48; et hb. propr. pro parte. *C. scoparius* Decaisne in Nouv. Ann. Mus. Paris, **3** (1834) 359; Miq. Fl. Nederl. Ind. **3**: 279. *C. spaniphyllus* Steud. ! in Zoll. Verz. Ind. Archip. heft 2, 62; Miq. l. c. 267. *C. difformis* Blanco ! Fl. Filip. 32, non Linn.

PHILIPPINES, (*Llanos*). LUZON, (*Loher*); (376, 4247 *Merrill*). MINDANAO, (1337 *Copeland*).

From the Persian Gulf to the Philippines and north Australia.

14. **Cyperus pilosus** Vahl Enum. **2** (1806) 354.

Spikes 1 to 2 in. long, the rhachis minutely hairy; umbel simple or compound; spikes often subdigitate.

Hook. f. Fl. Brit. Ind. **6**: 609; Vidal, Rev. Pl. Vasc. Filip. 283; Phaner. Cuming. Philip. 155; Hemsl. in Journ. Linn. Soc. **36** (1903) 215; Merrill, in Philip. Journ. Sci. **1** (1906) Suppl. **1**: 30. *Cyperus venustus* Moritzi ! Verz. Zoll. Pfl. 96, non R. Br. *C. piptolepis* Steud. in Zoll. Verz. Ind. Archip. heft 2, 63; Cyp. 40; Miq. ! Fl. Nederl. Ind. **3**: 279.

PHILIPPINES, (535 *Cuming*). LUZON, (867 *Jagor*); (1865 *Wichura*); (2308 *Merrill*). MINDANAO, (142 *Clemens*); (827 *Copeland*).

Rare in tropical Africa; abundant in India, extending to Japan and Queensland.

15. **Cyperus Zollingeri** Steud. in Zoll. Verz. Ind. Archip. heft 2, (1854) 62.

Umbel compound or simple; spikelets spicate, long linear, yellowish.

Miq. Fl. Nederl. Ind. **3**: 264; Hook. f. Fl. Brit. Ind. **6**: 613; Hemsl. in Journ. Linn. Soc. **36** (1903) 219. *Cyperus rotundus* Presl. ! Rel. Haenk. **1**: 175; Miq. Fl. Nederl. Ind. Suppl. 260, 600.

LUZON, (*Haenke*); (*Meyen*); (6473 *Elmer*).

Tropical Africa, southeastern Asia to Bouru and Queensland; a plentiful species.

16. **Cyperus rotundus** Linn. Sp. Pl. ed. 2, 67 partim, nec. Linn. hb. propr.

Stem thickened at base; spikelets flattened, reddish or pale, not yellow.

Miq. Fl. Nederl. Ind. **3**: 274; Hook. f. Fl. Brit. Ind. **6**: 615; Vidal, Rev. Pl. Vasc. Filip. 282; Phaner. Cuming. Philip. 155; Hemsl. in Journ. Linn. Soc. **36** (1903) 216. *Cyperus hexastachyus* Rottb. Descr. et Ic. 28. *t. 14. f. 2;* Decaisne in Nouv. Ann. Mus. Paris, **3** (1834) 358. *C. Hydra* Michx. Fl. Bor. Am. **1**: 27; Presl, Rel. Haenk. **1**: 175. *C. bulboso-stolonifer* Miq. ! Fl. Nederl. Ind. Suppl. 260, 559; Kurz in Tidschr. Nederl. Ind. 27 (1864) 222, *non* Steud. *C. laevissimus* Steud. ! Cyp. 32. *C. curvatus* Llanos ! Fragm. Pl. Filip. 15.

PHILIPPINES, (557, 715 *Cuming*); (1961 *Vidal*). LUZON, (1869 a, b, *Wichura*); (2786 *Merrill*). SAMAR, (944 *Jagor*). PANAY, (97 *Copeland*).

In all warm countries; one of the worst pests in cultivated lands.

There is in Linnæus's Herbarium only *one* example marked by his hand as *Cyperus rotundus;* it is therefore the "type" of the species. It may be *Mariscus Thunbergii* Schrader; it certainly is not our *Cyperus rotundus* Linn.

17. **Cyperus stolonifer** Retz. Obs. **4** (1786) 10.

Spikelets nearly terete (very obscurely compressed); glumes concave, obtuse; otherwise as in *Cyperus rotundus.*

Miq. Fl. Nederl. Ind. **3**: 265; Hook. f. Fl. Brit. Ind. **6**: 615; Hemsl. in Journ. Linn. Soc. **36** (1903) 217. *Cyperus bulboso-stolonifer* Steud. ! in Zoll. Verz. Ind. Archip. heft 2, 62; Cyp. 18; Miq. Fl. Nederl. Ind. **3**: 266.

LUZON, (717 *Loher*); Manila, (*Chamisso*).

Southeastern Asia, extending from Mauritius to Formosa and north Australia; a frequent, hardly common, species.

18. **Cyperus radiatus** Vahl. Enum. **2** (1806) 369.

A large erect annual; spikes cylindric, subdigitate; rhacheola of the spikelet with oblong persistent wings.

Miq. Fl. Nederl. Ind. **3**: 277; Hook. f. Fl. Brit. Ind. **6**: 617; Vidal Rev. Pl. Vasc. Filip. 283; Phaner. Cuming. Philip. 155; Hemsl. Journ. Linn. Soc. **36** (1903) 216. *Cyperus involucratus* Poiret in Lam. Encycl. **7** (1806) 253;

Decaisne in Nouv. Ann. Mus. Paris, **3** (1834) 360. *C. verticillatus* Roxb. Fl. Ind. **1**: 206 (ed. Wallich, p. 209); Miq. Fl. Nederl. Ind. **3**: 276. *C. macrosciadion* Steud. Cyp. 37; Miq. l. c. 277. *C. longifolius* Decaisne ! in Nouv Ann. Mus. Paris, **3** (1834) 359.

PHILIPPINES, (537 *Cuming*). LUZON, Manila, (*Wichura*); (38 *Merrill*); (727 *Loher*); (4 *Fenix*).

In warm countries throughout the World; an abundant plant.

19. **Cyperus exaltatus** Retz. Obs. **5** (1789) 11.

Spikes cylindric, peduncled, otherwise as *C. radiatus* Vahl.

Miq. Fl. Nederl. Ind. **3**: 276; Hook. f. Fl. Brit. Ind. **6**: 617; Hemsl. in Journ. Linn. Soc. **36** (1903) 212. *Cyperus venustus* R. Br. Prodr. 217; Miq. Fl. Nederl. Ind. **3**: 280. *C. altus* Nees in Linnæa **9** (1834) 285; Miq. l. c. 276. *C. elatus* Hassk. in Flora **45** (1862) 191, non Linn.

PHILIPPINES, (*Llanos*).

Warm parts of the Old World; abundant. From America I have only two or three examples.

This plant, much confused with *C. elatus* Linn., has the anthers ecristate.

20. **Cyperus digitatus** Roxb. Hort. Beng. 1814 (1813) 81.

A large erect annual; spikes cylindric; rhacheola of the spikelet with yellow, early soluble, lanceolate wings.

Hook. f. Fl. Brit. Ind. **6**: 618; Hemsl. in Journ. Linn. Soc. **36** (1903) 211. *Cyperus strigosus* Llanos Fragm. Pl. Filip. 16, non Linn. *C. auricomus* Vidal ms, non Sieber.

PHILIPPINES, (*Llanos*); (3985 *Vidal*); (714 *Loher*); (2317 *Merrill*). MINDANAO, (1997 *Wichura*).

Warm parts of Asia, Polynesia and America; rare in Africa, where its place is taken by the closely allied representative, *C. auricomus* Sieber.

21. **Cyperus elatus** Linn. Amoen. Acad. 4: 301; Sp. Pl. ed. 2, 67 et hb. propr.

A tall erect annual; spikes cylindric; anthers narrowly oblong with a lanceolate crest one third their own length.

Hook. f. Fl. Brit. Ind. **6**: 618. *Cyperus racemosus* Miq. Fl. Nederl. Ind. **3**: 270, 278 (syn. excl.) ? of Retz. *C. bispicatus* Steud. ! in Zoll. Verz. Ind. Archip. heft 2, 62; Miq. l. c. 285. *C. exaltatus* hb. Vidal, ! non Retz.

PHILIPPINES (3986 *Vidal*); (592 *Jagor*).

Southeastern Asia, from Madras to Cochin-China and Timor; a rare plant.

PHILIPPINE SPECIES OF CYPERUS, WITH THE STYLE 3-FID (OR UNKNOWN) NOT IN THE KEW HERBARIUM.

CYPERUS CAESPITOSUS Llanos Fragm. Pl. Filip. 14, non Poiret. *C. dehiscens*, Naves Nov. App. 303, non Nees.

PHILIPPINES, fide *Llanos*.

CYPERUS ANABAPTISTUS Steud. Cyp. 37. *C. Cumingii* Steud. olim in litteris, nec Cyp. 25.

LUZON; hb. *Cuming* fide *Steudel*.

CYPERUS MINUTIFLORUS Presl, Rel. Haenk. 1: 351, non Nees. *C. micranthus* Presl, l. c. 178. *C. breviflorus* Dietr. Sp. Pl. 2: 316. *C. multiflorus* Kunth, Enum. 2: 562 (err. typogr.)

LUZON, fide *Presl.*

CYPERUS PHILIPPENSIS Presl Rel. Haenk. 1: 174.

LUZON, fide *Presl.*

CYPERUS SPICATUS Presl, Rel. Haenk. 1: 173.

LUZON et MEXICO, fide *Presl.*

According to *Fenzl*[1] this species must be near *Cyperus radiatus* Vahl, and *C. elatus* Linn.

5. MARISCUS Vahl.

As *Cyperus,* but spikelets in fruit deciduous. Style 3-fid.

Sect. I. Stem at base apparently thickened by the inflated scariose-colorate basal leaf sheaths.

1. **Mariscus Merrillii** C. B. Clarke, sp. nova.

Culmo 4 ad 7 cm. longo, 1-cephalo; foliis culmo longioribus, basi inflatis, scarioso-coloratis; capite 10 ad 12 mm. in diam., globoso, spiculis densissimo; spiculis maturitis oblongo-linearibus, 4 ad 8-floris, sub-2-nucigeris; glumis lanceolato-elongatis, striatis; nuce anguste oblongo, sublineari, trigono, nigro; stylo 3-fido.

LUZON, Province of Cavite, Maragondong (4170 *Merrill*).

Species *M. Dregeano* Kunth, affinis et similis, ab spiculas nuces angustas diversa. Spiculae iis *M. flabelliformis* Kunth, magis simile videntur.

Sect. II. Leaf sheaths herbaceous; spikelets ripening 1 to 3 nuts.

2. **Mariscus cyperinus** Vahl Enum. 2 (1806) 377.

Umbel simple, rays short; spikes cylindric, dense, with suberect 2-flowered spikelets.

Hook. f. Fl. Brit. Ind. 6: 621; Hemsl. in Journ. Linn. Soc. 36 (1903) 220; Merrill in Philip. Journ. Sci. 1 (1906) Suppl. 1: 30. *Mariscus umbellatus* Moritzi ! Verz. Zoll. Pfl. 98, vix Vahl. *M. Sundaicus* Miq. ! Fl. Nederl. Ind. 3: 289. *Kyllingia cyperina* Retz. Obs. 6: 21. *Cyperus Manilensis* Boeck. ! in Engler Jahrb. 5 (1884) 501.

PHILIPPINES, (*Moseley*). LUZON, Manila, (1871 *Wichura*); (766 *Jagor*); (780, 781, 789, 790 *Loher*); (3311 *Ahern's collector*); (464, 456 *Topping*); (485 (in part) *Whitford*); (6675 *Elmer*).

Southeastern Asia, from Ceylon to Petropaulovski and Otaheiti, frequent.

3. **Mariscus tenuifolius** Nees in Mart. Fl. Brasil. 2[1] (1843) 46.

Stolons slender; umbel simple, small, spikes loose; spikelets slender, usually maturing two nuts.

Hook. f. Fl. Brit. Ind. 6: 622; Schrader ms. *Cyperus umbellatus* var. *laxata* C. B. Clarke in Journ. Linn. Soc. 21 (1884) 201.

LUZON, Arayat, (782 *Loher*). MINDANAO, (617 *Copeland*).

India from Behar to Ceylon and to Malacca; a very rare plant in herbaria.

[1] *Denkschr. Acad. Wissen. Wien.* 8 (1855) 47.

4. **Mariscus Sieberianus** Nees in Linnæa **9** (1834) 286.

Umbel simple, spikes cylindric, peduncled; spikelets in fruit spreading or decurved, maturing one or two nuts.

Hook. f. Fl. Brit. Ind. **6**: 622; Hemsl. in Journ. Linn. Soc. **36** (1903) 221. *Mariscus umbellatus* Vahl. Enum. **2**: 376 pro parte; Miq. Fl. Nederl. Ind. **3**: 288. *M. cyperinus* Presl ! in Oken Isis **21** (1828) 270, non Vahl. *Cyperus umbellatus* Miq. in Ann. Mus. Lugd. Bat. **2**: 142, non Roxb.

LUZON, (783, 784, 785 *Loher*); (6472 *Elmer*).

In all warm countries; very common.

5. **Mariscus Philippensis** Steud. Cyp. (1855) 66, char. emend.

Spikes exactly cylindric, very dense with small spikelets.

Miq. Fl. Nederl. Ind. **3**: 290; Merr. in Philip. Journ. Sci. **1** (1906) Suppl. **1**: 30. *Mariscus umbellatus* Presl, Rel. Haenk. **1**: 181. *Cyperus cylindrostachys* Boeck. ! Linnæa **36** (1869–70) 383, partim; Vidal, Rev. Pl. Vasc. Filip. 283; Phaner. Cuming Philip. 155.

LUZON, (568 *Cuming*) (No. 1422 in hb. Hooker); (1863, 1866 *Wichura*); (3277 *Merrill*).

Philippines and Hainan.

This is hardly more that a geographic form of the universally distributed *Mariscus Sieberianus.*

Sect. III. Leaf sheaths herbaceous; spikelets ripening 3 or more (sometimes 15) nuts.

6. **Mariscus microcephalus** Presl ! Rel. Haenk. **1** (1830) 182.

Large and with large compound umbels; spikelets linear, often 1 cm. long, brown.

Miq. Fl. Nederl. Ind. **3**: 290; Hook. f. Fl. Brit. Ind. **6**: 624; Hemsl. in Journ. Linn. Soc. **36** (1903) 221; Merrill, Philip. Journ. Sci. **1** (1906) Suppl. **1**: 30. *Cyperus dilutus* Vahl. Enum. **2**: 357; Miq. l. c. 285; Vidal, Rev. Pl. Vasc. Filip. 283; Phaner. Cuming. Philip. 155. *C. Haenkeanus* Kunth, Enum. **2**: 93. *C. septatus* Steud. ! in Zoll. Verz. Ind. Archip. heft 2, 62; Cyp. 46; Miq. Fl. Nederl. Ind. **3**: 284, et Suppl. 260. *C. cuadriflorus* (i. e. *quadriflorus*) Llanos, Fragm. Pl. Filip. 18. *C. microcephalus* Naves in Blanco Fl. Filip. Nov. App. 304, non R. Br. *C. Grabowskianus* Boeck. in Engl. Jahrb. **5** (1884) 502. *C. rufus* Nees in Linnæa **9** (1834) 285.

LUZON, (538 *Cuming*); (869 *Jagor*); (788 *Loher*); (*Meyen*); (3983 *Vidal*); (1566 *Merrill*); (5531 *Elmer*). CULION, (470 *Merrill*). PANAY, (1656 *Cuming*). MINDANAO, (2000 *Wichura*); (581 *Copeland*).

Usually very large, but as in many species of *Cyperaceæ*, small examples occur. The "type" example of *Presl*, is very young, the heads being quite small; hence his inappropriate specific name.

Southeastern Asia; common. Also in Bourbon, Mauritius.

7. **Mariscus albescens** Gaudich. in Freycinet Voy. (1826) 415.

A strong plant, leaves almost spongy with transverse lines; umbel compound; spikelets turgid, ripening 3 to 6 nuts.

Hook. f. Fl. Brit. Ind. **6**: 623; Hemsl. in Journ. Linn. Soc. **36** (1903) 220; Merrill in Philip. Journ. Sci. **1** (1906) Suppl. **1**: 30. *Cyperus stuppeus* Forst. f.

Prodr. (1786) 89. *C. pennatus* Lam. Ill. 1 (1791) 144; Decaisne in Nouv. Ann. Mus. Paris, 3 (1834) 359; Miq. Fl. Nederl. Ind. 3: 281; Vidal, Rev. Pl. Vasc. Filip. 283; Phaner. Cuming. Philip. 155. *C. holciflorus* Presl, Rel. Haenk. 1: 171; Miq. Fl. Nederl. Ind. 3: 282. *C. firmus* Presl, Rel. Haenk. 1: 171, fide Boeck. *C. anomalus* Steud. Cyp. 37; Miq. l. c. 279. *C. imbricatus* Llanos, Fragm. Pl. Filip. 17, fide Naves. *C. ovatus* Llanos ! Fragm. Pl. Filip. 15. *C. nitidulus* Vidal ! Rev. Pl. Vasc. Filip. 283.

PHILIPPINES, (*Llanos*); (436, 1636 *Cuming*); (*Moseley*). LUZON, (786, 787 *Loher*); (4254 *Merrill*); (592, 1305 *Whitford*). CULION, (587 *Merrill*).

Southeastern Asia and Polynesia; common. In tropical Africa, rare. The one specimen seen by me, from Valparaiso, may have been an herbarium mixture.

8. **Mariscus flabelliformis** H. B. K. Nov. Gen. et Sp. 1 (1815) 215; Dyer Fl. Trop. Afr. 8: 397; Merrill, Philip. Journ. Sci. 1 (1906) Suppl. 1: 30.

LUZON, (405 *Whitford*). PANAY, (96 *Copeland.*)

Common in tropical America. I have referred to this Kunthian species various examples from tropical Africa, Java and Polynesia.

6. TORULINIUM Desv.

Spikelets linear, when mature breaking up into several 1-nutted pieces.

1. **Torulinium confertum** Desv. in Hamilt. Prodr. Ind. Occid. (1825) 15.

A large plant with large compound umbels.

Hemsl. in Journ. Linn. Soc. 36 (1903) 222. *Torulinium ferox* Kunth, Enum. 2: 90, in citat. *Cyperus odoratus* Linn. Sp. Pl. ed. 1, 46, partim; Boeck. ! in Linnæa, 36 (1869–70) 407. *C. ferax* L. C. Rich. ! in Act. Soc. Hist. Nat. Paris, 1 (1792) 106; Vidal, Rev. Pl. Vasc. Filip. 283; Phaner. Cuming. Philip. 155. *C. ferox* Vahl. Enum. 2: 357 (saltem pro parte); Decaisne in Nouv. Ann. Mus. Paris, 3 (1834) 359. *C. Haenkei* Presl, Rel. Haenk. 1: 172. *C. cephalophorus* Presl. Rel. Haenk. 1: 170. *C. calopterus* Miq. Fl. Nederl. Id. 3: 282 et Suppl. 260. *C. holophyllus* Miq. l. c. 283. *Mariscus ferax* Hook. f. Fl. Brit. Ind. 6: 624.

PHILIPPINES, (868 *Jagor*). LUZON, (536 *Cuming*); (5532 *Elmer*).

In all warm countries; very abundant in America.

For this plant there are 79 published names; but this number does not include several subspecies, often regarded as varieties.

7. ELEOCHARIS R. Br.

Culm with but one spikelet, leafless. Hypogynous bristles present.

Subgenus I. LIMNOCHLOA. Stoloniferous, somewhat robust. Glumes rather rigid, not (or obscurely) keeled.

1. **Eleocharis equisetina** Presl, Rel. Haenk. 1 (1828) 195.

Dried stems apparently septate.

Miq. Fl. Nederl. Ind. 3: 302; Hook. f. Fl. Brit. Ind. 6: 626. *Eleocharis esculenta* Viellard ! in Ann. Sc. Nat. IV. 16 (1862) 37. *Heleocharis esculenta* F. Muell. Fragm. Phyt. Austral. 8: 239, in citat. *H. equisetina* Naves, Nov. App. 306. *H. plantaginea* Vidal, Rev. Pl. Vasc. Filip. 284; Phaner. Cuming. Philip. 156.

LUZON, (1255 *Cuming*).

From Madagascar to the Philippines and New Caledonia; a rare species.

2. **Eleocharis variegata** Presl ! in Oken Isis **21** (1828) 269.

Dried stems nearly terete, not transversely septate.

Mauritius, Madagascar.

Var. β **laxiflora** Hook. f. Fl. Brit. Ind. **6**: 626; Hemsl. in Journ. Linn. Soc. **36** (1903) 229. *Scirpus laxiflorus* Thwaites, Enum. Pl. Zeyl. 435.
LUZON, (738, 739 *Loher*).
India and Malaya, frequent; also in Polynesia and Central America.

Subgenus 2. ELEOGENUS. Style 2-fid; annuals.

3. **Eleocharis ochreata** Nees in Linnæa **9** (1834) 294.

Leaf-sheaths with a large scarious margin which easily rubs away.

LUZON, (742 *Loher.*)
Scattered in the warm parts of the Old World; abundant in the New World.

4. **Eleocharis atropurpurea** Kunth, Enum. **2** (1837) 151.

Margins of the leaf-sheaths herbaceous; hypogynous bristles white, shining.

Hook. f. Fl. Brit. Ind. **6**: 627; Hemsl. in Journ. Linn. Soc. **36** (1903) 226.
LUZON, Manila, (*Barthe*) (in herb. Paris).
Tropical and warm countries throughout the World.

5. **Eleocharis capitata** R. Br. Prodr. (1810) 225.

Margins of the leaf-sheaths herbaceous; hypogynous bristles pale-reddish, or red-brown.

Presl, Rel. Haenk. **1**: 196; Decaisne in Nouv. Ann. Mus. Paris, **3** (1834) 361; Miq. Fl. Nederl.-Ind. **3**: 299 et Suppl. 261; Hook. f. Fl. Brit. Ind. **6**: 627; Hemsl. in Journ. Linn. Soc. **36** (1903) 227. *Eleocharis atropurpurea* Presl, Rel. Haenk. **1**: 106, non Kunth.
LUZON, (*Llanos*); (55 *Merrill*).
In tropical and warm countries; common in both the Old and New World.

Subgenus 3. EUELEOCHARIS. Style 3-fid; annuals.

6. **Eleocharis chaetaria** Roem. et Sch. Syst. **2** (1817) 154, Mant. 90, 540.

Nut conspicuously cancellate.

Moritzi Verz. Zoll. Pfl. 96; Hook. f. Fl. Brit. Ind. **6**: 629. *Eleocharis setacea* R. Br. Prodr. 224, in adn., neque *E. setacea* R. Br. Prodr. 225. *Chaetocyperus setaceus* Nees in Linnæa **9** (1834) 289; Miq. Fl. Nederl. Ind. **3**: 298 et Suppl. 261.
LUZON, Benguet, (741 *Loher*).
In southeastern Asia and tropical America, abundant; rare in tropical Africa.
The plant is usually a small annual; but sometimes has a slender rhizome 2.5 to 5 cm. long, and is (at least) biennial.

7. **Eleocharis afflata** Steud. ! in Zoll. Verz. Ind. Archip. heft 2, (1854) 62 et Cyp. 76.

Nut smooth, hypogynous bristles 6, overlapping the nut.

Miq. Fl. Nederl. Ind. **3**: 279; Hook. f. Fl. Brit. Ind. **6**: 629; Hemsl. in Journ. Linn. Soc. **36** (1903) 226. *Eleocharis subprolifera* Steud. ! in Zoll. Verz. Ind.

Archip. heft 2, (1854) 62 et Cyp. 80; Miq. Fl. Nederl. Ind. **3**: 300. *E. pellucida* Presl, Rel. Haenk. 1 (1828) 196, e descript. dubie huc allata. *Heleocharis pellucida* et *H. afflata* Naves in Blanco, Fl. Filip. Nov. App. (1883) 307.

LUZON, *Benguet* (740 *Loher*); (5751, 6299 *Elmer*); *Lepanto* (4621 *Merrill*).

Abundant in southeastern Asia, from India to Japan.

8. **Eleocharis microcarpa** Torrey in Ann. Lyceum New York, **3** (1836) 312.

Nut smooth, hypogynous bristles wanting.

Heleocharis Schweinfurthiana Boeck. ! in Flora, **62** (1879) 562.

LUZON, (5193 *Loher*).

Abundant in warmer and tropical America; rare in tropical Africa.

No. 5193 *Loher* may be taken to be a depauperated state of the common *Eleocharis afflata* Steud. In *Eleocharis*, as in most large genera of *Cyperaceæ*, the comparison of the Old World species against those of the New World has been done, as yet, imperfectly.

8. FIMBRISTYLIS Vahl.

Glumes many, placed spirally, or in the section *Abildgaardia* the lower ones subdistichous. Hypogynous bristles none. Style-base swollen, with a constriction or articulation between it and the nut, deciduous, leaving no button on the nut.

Sect. I. ELEOCHAROIDES Benth.

Culm with one spikelet (rarely one to three).

1. **Fimbristylis tetragona** R. Br. Prodr. (1810) 226.

Leaves hardly any; nut cylindric-oblong, straw colored.

Hook. f. Fl. Brit. Ind. **6**: 631; Hemsl. in Journ. Linn. Soc. **36** (1903) 246. *Fimbristylis abjiciens* Steud. in Zoll. Verz. Archip. Ind. heft 2, 62; Miq. Fl. Nederl. Ind. **3**: 316. *Scirpus tetragonus* Poiret in Lam. Encycl. Suppl. **5**: 98, neque in Lam. Encycl. **6**: 767.

LUZON, Manila, (749 *Loher*).

From India to the Philippines and New South Wales; frequent.

2. **Fimbristylis acuminata** Vahl, Enum. **2** (1806) 285.

Leaves hardly any; nut obovoid, transversely ridged.

Miq. Fl. Nederl. Ind. **3**: 314, var. β excl.; Hook. f. Fl. Brit. Ind. **6**: 631; Hemsl. in Journ. Linn. Soc. **36** (1903) 230.

LUZON, (748 *Loher*); (3390 *Ahern's collector*).

From North India to the Philippines and to New South Wales; frequent.

3. **Fimbristylis setacea** Benth. ! in Hook. Lond. Journ. Bot. **2** (1843) 239.

Stem filiform, spikelet slender; nuts minute, leaves sometimes as long as the stems.

Fimbristylis acuminata β *minor* Miq. Fl. Nederl. Ind. **3**: 314; Vidal, Rev. Pl. Vasc. Filip. 284. *F. bursifolia* Vidal, l. c. et Phaner. Cuming. Philip. 156. *Abildgaardia brevifolia* Steud. ! Cyp. 72, cf. Rolfe in Journ. Bot. **24** (1886) 59, in nota. *Isolepis cochleata* Steud. ! Cyp. 100.

LUZON, (675 *Cuming*); Manila, (*Barthe*).

From Burma to the Philippines and to Queensland; rare.

(*Fimbristylis nutans* Vahl, Enum. **2**: 285, is a species closely allied to *F. acuminata* Vahl, above, and is said to occur in the Philippines by *Vidal*. I expect that it does occur, but the two examples on which *Vidal* relies, viz, Nos. 1413 *Cuming* and 1975 *Vidal*, are at Kew and are *Fimbristylis schoenoides* Vahl, below. I have seen no *Fimbristylis nutans* from the Philippines.)

4. **Fimbristylis polytrichoides** Vahl, Enum. **2** (1806) 248 (*polythricoides*).

Nut obovoid, brown-black, smooth, slightly scaly on the shoulders.

Miq. Fl. Nederl. Ind. **3**: 315, syn. excl.; Hook. f. Fl. Brit. Ind. **6**: 632; Hemsl. in Journ. Linn. Soc. **36**: (1903) 241. *Fimbristylis albescens* Steud. in Zoll. Verz. Ind. Archip. heft 2, 61; Miq. Fl. Nederl. Ind. **3**: 316. *F. juncea* Boeck. ! in Linnæa **37** (1871) 4, non Roem. et Sch. *Scirpus polytrichoides* Retz. Obs. **4**: 11. *Abildgaardia Javanica* Steud. ! in Zoll. Verz. Ind. Archip. heft 2, 63; Miq. Fl. Nederl. Ind. **3**: 297, non Nees.

PHILIPPINES, (129 *Chamisso*). LUZON, (753, 754, 755 *Loher*); (4249 *Merrill*).

India to Japan and to north Australia, frequent. Rare in east tropical Africa.

Sect. II. DICHELOSTYLIS Benth.

Style 2-fid. Stem generally with more than one spikelet. Lower glumes spirally imbricated.

Series A. *Spikelets all solitary.*

5. **Fimbristylis schoenoides** Vahl, Enum. **2** (1806) 286.

Stem with 3–1 spikelets; nut obovoid, smooth, minutely reticulate.

Miq. Fl. Nederl. Ind. **3**: 315; Hook. f. Fl. Brit. Ind. **6**: 634; Hemsl. in Journ. Linn. Soc. **36** (1903) 243. *Fimbristylis bispicata* Nees ! in Linnæa **9** (1834) 290; Miq. Fl. Nederl. Ind. **3**: 317. *F. nutans* Vidal ! Rev. Pl. Vasc. Filip. 284; Phaner. Cuming. Philip. 156, non Vahl. *Scirpus schoenoides* Retz. Obs. **5**: 14. *Abildgaardia nervosa* Presl, Rel. Haenk. **1**: 180.

LUZON, (1413 *Cuming*); (1975, 3980 *Vidal*); (750, 751, 752, 1975 *Loher*).

In India very common, extending to the Philippines and Queensland. Recorded by *Britton* in Florida, where occur several Old World plants.

6. **Fimbristylis subbispicata** Nees in Nova Acta Nat. Cur. **19** (1843) Suppl. **1**: 75.

Stem with 2–1 large cylindric spikelets; otherwise as *F. schoenoides* Vahl.

Hook. f. Fl. Brit. Ind. **6**: 634; Hemsl. in Journ. Linn. Soc. **36** (1903) 245.

From India to China and Japan; frequent.

Var. β **caesia** Miq. ! Fl. Nederl. Ind. **3**: 315.

With many leaves; spikelets one or two to a culm, hardly different from those of *F. schoenoides.*

PHILIPPINES, (747 *Loher*).

Java.

This species should perhaps be reduced to *Fimbristylis schoenoides* Vahl, as several species of the genus occur in leafless and leafy forms. However this may be, No. 747 *Loher* appears to match exactly the authentic example of *F. caesia* Miq.

7. **Fimbristylis dipsacea** Benth. in Benth. et Hook. f. Gen. Pl. **3** (1883) 1049.

Umbel nearly simple with (often) 12 spikelets; nut oblong-cylindric, slightly curved, microscopically marked with wavy transverse lines.

Hook. f. Fl. Brit. Ind. **6:** 635; Hemsl. in Journ. Linn. Soc. **36** (1903) 235. *Scirpus dipsaceus* Rottb. Descr. et Ic. 56. *t. 12. f. 1.* *Echinolytrum dipsaceum* Desv. in Journ. Bot. **1** (Paris 1808) 21. *t. 1.* *Isolepis dipsacea* Roem. et Sch. Syst. **2:** 119; Miq. Fl. Nederl. Ind. **3:** 309.

LUZON, Manila, (1855 *Wichura*); Laguna, (5105 *Merrill*).

Tropical Africa and India, common, extending thence to Amurland, but not yet received from the Malayan Archipelago.

In this species the very young ovary is often (not always) ornamented with prominent clavate glands, as shown by *Desveaux*, which often disappear in the ripe fruit; a character considered by *Desveaux* to be of generic value.

8. **Fimbristylis dichotoma** Vahl, Enum. **2** (1806) 287.

Spikelets numerous, oblong, somewhat angular by reason of the keeled glumes; nut obovoid, transversely trabeculate between the longitudinal ribs.

Miq. Fl. Nederl. Ind. **3:** 319; Hook. f. Fl. Brit. Ind. **6:** 635; Hemsl. in Journ. Linn. Soc. **36** (1903) 232. *Scirpus dichotomus* Linn. Sp. Pl. ed. 2, 74 et herb. propr.

PHILIPPINES, (*Llanos*). LUZON, Manila, (1342 *Loher*).

Throughout the warmer parts of the Old World; an abundant species.

This species is only distinguishable from *Fimbristylis diphylla* Vahl, by characters of trifling importance; examples marked by eminent cyperologists *F. dichotoma* Vahl, are, not very rarely, marked by other eminent cyperologists *F. diphylla*.

9. **Fimbristylis diphylla** Vahl ! Enum. **2** (1806) 289.

Spikelets oblong, terete, the glumes less keeled than in *F. dichotoma* Vahl, the nuts indistinguishable from those of the latter.

Hook. f. Fl. Brit. Ind. **6:** 636; Hemsl. in Journ. Linn. Soc. **36** (1903) 233; Vidal, Rev. Pl. Vasc. Filip. 284; Phaner. Cuming. Philip. 156; Merrill in Philip. Journ. Sci. **1** (1906) Suppl. **1:** 30. *Fimbristylis juncifolia* Presl, Rel. Haenk. **1:** 190. *F. communis* Kunth, Enum. **2:** 234 (syn quibusdam excl.); Miq. Fl. Nederl. Ind. **3:** 323. *F. ambigua* Steud. ! in Zoll. Verz. Ind. Archip. heft 2, 61; Miq. l. c. 323. *F. Philippica* Steud. ! Cyp. 116; Miq. l. c. 324. *F. circinnata* Steud. ! Cyp. 116; Miq. l. c. 324. *F. Nukahiwensis* Steud. ! Cyp. 117. *F. calocarpa* Steud. ! Cyp. 117; Miq. Fl. Nederl. Ind. **3:** 325. *F. squarrosa* Miq. ! l. c. 319, saltem pro majore parte, non Vahl. *F. polymorpha* Boeck. ! in Vidensk. Meddel. Kjob. (1869) 141, 158. *Scirpus diphyllus* Retz. Obs. **5:** 15.

PHILIPPINES, (*Meyen*). LUZON, (1854 *Wichura*); (558 *Cuming*); (763, 764, 765 *Loher*); (1141, 4252, 4632 *Merrill*); (70 *McGregor*); (6300, 6470 *Elmer*); (6135 *Leiberg*). MINDANAO (2001 *Wichura*); (545 *Copeland*); (*Clemens*).

In all warm countries; perhaps the most widespread and abundant of *Cyperaceæ*.

The species is here taken in rather a narrow sense, and has about 150 published names. If the species is limited as in *Kunth* it would have about

200 published names. The synonyms cited (taken largely from *Steudel's* herbarium) may prevent trouble in looking vainly for his Malayan species in the Philippines.

Taking the species in the restricted sense, it is most variable, sometimes filiform, sometimes robust, normally glabrous but frequently very hairy, the culm often carrying one spikelet only, sometimes 200. The commonest Malayan form of the species has the nut somewhat scaly or scabrous on the shoulders, the stem and leaves not rarely hairy (Nos. 756, 757, 758 *Loher*).

10. **Fimbristylis aestivalis** Vahl. Enum. **2** (1806) 288.

Small, tufted, umbel compound or decompound; spikelets small; nut obovoid, straw colored, not transversely barred.

Hook. f. Fl. Brit. Ind. **6**: 637; Hemsl. in Journ. Linn. Soc. **36** (1903) 230. *Fimbristylis dichotoma* Presl ! Rel. Haenk. **1**: 191, non Vahl. *F. squarrosa* Steud. ! in Zoll. Verz. Ind. Archip. heft 2, 61. *F. tricholepis* Miq. Fl. Nederl. Ind. **3**: 319, et Suppl. 262. *Scirpus aestivalis* Retz. Obs. **4**: 12.

PHILIPPINES, (*Llanos*); (*Haenke*). LUZON, Manila, (*Wichura*); (3174 *Didrichsen*); (746 *Loher*); Benguet, (743, 744, 745 *Loher*); (6070 *Elmer*).

India to Amurland and Australia; common.

Var. *β* **macrostachya** Benth. Fl. Austral. **7**: 310.

Stronger, with larger spikelets; nut obscurely ribbed longitudinally.

LEYTE, (1008 *Jagor*).

North Australia.

This variety tends toward *Fimbristylis dichotoma* Vahl.

11. **Fimbristylis podocarpa** Nees ! in Wight. Contrib. (1834) 98.

Nut obovoid, on a conspicuous gynophore, subcancellate but hardly transversely barred, otherwise as *F. diphylla* Vahl.

Nees in Linnæa **9** (1834) 290, et in Hook. et Arn. Beechy Voy. 225, partim; Hook. f. Fl. Brit. Ind. **6**: 638; Hemsl. in Journ. Linn. Soc. **36** (1903) 241. *Fimbristylis communis* (forma) Kunth, Enum. **2**: 234. *F. polymorpha* (forma) Boeck. in Linnæa **37** (1871) 14.

LUZON, (766 *Loher*).

India, frequent; extending through Malaya to the Marianne Islands. Also from Brazil, there are two examples which are referred to *F. podocarpa*, but which are possibly depauperated *F. spadicea* Vahl.

12. **Fimbristylis ferruginea** Vahl, Enum. **2** (1806) 291.

Umbel often simple; glumes minutely pubescent on the shoulder; nut obovoid, smooth.

Decaisne in Nouv. Ann. Mus. Paris, **3** (1834) 362; Hook. f. Fl. Brit. Ind. **6**: 638; Vidal, Rev. Pl. Vasc. Filip. 284; Phaner. Cuming Philip. 156; Hemsl. in Journ. Linn. Soc. **36** (1903) 235; Merrill, Philip. Journ. Sci. **1** (1906) Suppl. **1**: 30. *Fimbristylis cyrtophylla* Miq. Fl. Nederl. Ind. **3**: 325. *F. trispicata* Steud. Cyp. 107; Miq. l. c. 317. *Scirpus ferrugineus* Linn. ! Sp. Pl. ed. 1, 50, ed. 2, 74, et hb. propr. partim.

LUZON, (127 *Chamisso*); (1341 *Loher*); (1396 *Cuming*); (1304 *Whitford*). CULION, (546 *Merrill*).

The warmer parts of the whole World, very common especially near the sea.

This species is very generally named correctly in herbaria, for the reason that with a lens, and without dissecting the spikelets, the characteristic gray pubescence of the shoulders of the glumes can be seen.

Series B. *Spikelets many, solitary, some paired or digitate.*

13. **Fimbristylis rigidula** Nees in Wight Contrib. (1834) 99.

Rhizome short, creeping; spikelets, some solitary, pedicelled, some paired.

Hook. f. Fl. Brit. Ind. **6**: 640; Hemsl. in Journ. Linn. Soc. **36** (1903) 242. *Fimbristylis communis* Kunth, Enum. **2**: 235 partim. *F. ferruginea* Vidal ! Rev. Pl. Vasc. Filip. 284 partim. (i. e., No. 1396 partim).

PHILIPPINES, (1396 *Cuming*), in herb. Mus. Brit.

India and China, frequent.

Out of 130 species of *Fimbristylis*, only two have a creeping rhizome.

14. **Fimbristylis spathacea** Roth. Nov. Pl. Sp. (1821) 24.

Leaves rigid, umbel compound, usually close; spikelets some solitary, pedicelled, some digitate or clustered; style 2-fid.

Hook. f. Fl. Brit. Ind. **6**: 640; Hemsl. in Journ. Linn. Soc. **36** (1903) 244. *Fimbristylis glomerata* Nees in Linnæa **9** (1834) 290; Boeck. in Linnæa **37** (1871) 47, partim. *F. rigida* Kunth ! Enum. **2**: 246; Moritzi ! Verz. Zoll. Pfl. 97; Steud. ! in Zoll. Verz. Ind. Archip. heft 2, 61; Miq. Fl. Nederl. Ind. **3**: 327. *F. ciliolata* Steud. ! in Zoll. Verz. Ind. Archip. heft 2, 61; Miq. l. c. 317. *F. laevissima* Steud. ! Miq. l. c. 324. *Scirpus glomeratus* Retz. Obs. **4**: 11. *Isolepis Haenkei* Presl, Rel. Haenk. **1**: 187 partim.

LUZON, (*Martens*); (1857 bis *Wichura*); (1344 *Loher*); (5683 *Elmer*).

Tropical Asia, America and the Mascarene Islands; a frequent plant.

Boeckeler regards this species as a 2-stigma form of *F. obtusifolia* Kunth and *F. cymosa* R. Br.

Sect. III. TRICHELOSTYLIS.

Style 3-fid. Stem generally with more than one spikelet. Lower glumes spirally imbricate.

Series A. *Spikelets all solitary.*

15. **Fimbristylis tenera** Roem. et Sch. Syst. Mant. **2** (1824) 57.

Leaves many short; stems slender, with often 3 to 7 spikelets; glumes often minutely ciliate-pubescent; nut sub-tubercular.

Hook. f. Fl. Brit. Ind. **6**: 642, varr. incl.; Dyer Fl. Trop. Afr. **8**: 412, 420. *Fimbristylis oxylepis* Steud. ! Cyp. 110. *F. firmula* Boeck. in Flora **42** (1859) 69, partim, non Steud.

LUZON, (760 *Loher*).

In tropical Africa rare; in India frequent; in Malaya rare.

The Philippine example has minutely pubescent glumes, and is altogether more like the typical African examples than it is to the Indian; the "varieties" can be maintained only as "forms."

16. **Fimbristylis miliacea** Vahl, Enum. 2 (1806) 287.

Lower sheaths without leaves, the upper with or without leaves; spikelets numerous, small, globose; nut transversely lineolate.

Miq. Fl. Nederl. Ind. **3**: 321; Vidal, Rev. Pl. Vasc. Filip. 284; Phaner. Cuming. Philip. 156; Hook. f. Fl. Brit. Ind. **6**: 644; Hemsl. in Journ. Linn. Soc. **36** (1903) 239; Merrill in Philip. Journ. Sci. **1** (1906) Suppl. **1**: 30. *Fimbristylis flaccidula* Steud. ! in Zoll. Verz. Ind. Archip. heft 2, 61. *F. flaccida* Steud. ! Cyp. 113; Miq. Fl. Nederl. Ind. **3**: 321. *Scirpus miliaceus* Thunb. Fl. Japon. 37, non Linn. hb. *Isolepis miliacea* Link. Hort. Berol. **2**: 316; Presl, Rel. Haenk. **1**: 188, var. β excl.

PHILIPPINES, (*Llanos*). LUZON, (564 *Cuming*); (865 *Jagor*); (759 *Loher*); (*Meyen*); (85 *Gaudichaud*); (130 *Chamisso*); (1858 *Wichura*); (450 *Topping*); (3, 2316 *Merrill*). PANAY, (99 *Copeland*).

In southeastern Asia and Oceania very common. In tropical Africa and America scattered.

17. **Fimbristylis quinquangularis** Kunth. Enum. 2 (1837) 229.

Lower leaves longer than the upper, but stems sometimes leafless; spikelets oblong: otherwise as *F. miliacea*.

Miq. Fl. Nederl. Ind. **3**: 321; Hook. f. Brit. Ind. **6**: 644; Hemsl. in Journ. Linn. Soc. **36** (1903) 242. *Scirpus miliaceus* Linn. ! hb. propr; an Sp. Pl. 75 partim ?

LUZON, (1340 *Loher*). PANAY, (100 *Copeland*).

In India very common; extending to the Liu Kiu and the Marianne Islands.

This species is difficult to separate from *Fimbristylis miliacea* Vahl, the nut being very similar. The stems are sometimes marked 5-angular at the top.

18. **Fimbristylis globulosa** Kunth. Enum. 2 (1837) 231.

Uppermost sheath without a leaf: spikelets cuboid-ellipsoid larger than those of *F. miliacea*: otherwise much as that species.

Miq. Fl. Nederl. Ind. **3**: 322; Hook. f. Fl. Brit. Ind. **6**: 644; Hemsl. in Journ. Linn. Soc. **36** (1903) 237. *Fimbristylis efoliata* Steud. ! in Zoll. Verz. Ind. Archip. heft 2, 61; Miq. Fl. Nederl. Ind. **3**: 318 et Suppl. 261. *Scirpus globulosus* Retz. Obs. **6**: 19.

PHILIPPINES, (*Loher*). LUZON, (3652 *Merrill*); (5582, 5705 *Elmer*).

In India very common, extending through Malaya to the Marianne Islands.

19. **Fimbristylis complanata** Link Hort. Berol. 1 (1827) 292.

Bracts usually 2, shorter than the umbel, linear-ligulate with an abrupt triangular tip: otherwise as *F. quinquangularis* Kunth.

Miq. Fl. Nederl. Ind. **3**: 320; Hook. f. Fl. Brit. Ind. **6**: 646; Hemsl. in Journ. Linn. Soc. **36** (1903) 231, varr. incl. *Fimbristylis autumalis* Boeck. in Vidensk. Meddel. Kjob. (1868–69) 141 var. γ; Vidal ! Rev. Pl. Vasc. Filip. 284; Phan. Cuming. Philip. 156, non Roem. et Sch. *Fimbristylis anceps* Steud. ! in Zoll. Verz. Ind. Archip. heft 2, 61, cf. Boeck. in Flora **42** (1859) 68. *F. amblyphylla* Steud. ! Cyp. 116; Miq. Fl. Nederl. Ind. **3**: 324. *Scirpus complanatus* Retz. Obs. **5**: 14. *Cyperus complanatus* Willd. Sp. Pl. **1**: 270. *Isolepis Willdenowii* Roem. et Sch. Syst. **2**: 120; Mant. 69; Presl ! Rel. Haenk. **1**: 189. *Isolepis complanata* Roem. et Sch. Syst. **2**: 119; Mant. 68, 533; Decaisne ! in Nouv. Ann. Mus. Paris, **3** (1834) 360.

PHILIPPINES, (*Llanos*); (530 *Cuming*); (46 hb. *Presl*). LUZON, (871, 872 *Jagor*); (762 *Loher*).

Very common in southeastern Asia; common in all warm countries.

Scirpus autumnalis Linn. Mant. 180, is a much more slender plant, the top of the stem not at all flattened, which is abundant in America. There occur, especially from Australia, forms or varieties of *Fimbristylis complanata* Link, which are difficult to separate from the American *F. autumnalis* Roem. et Sch. However, whether these are to be referred to *F. autumnalis* or not, I have received no similar forms from the Philippines; the Philippine *Fimbristylis complanata* is a robust plant with the stem conspicuously flattened at the top, often almost 2-winged.

Series B. *Spikelets in clusters.*

20. **Fimbristylis cymosa** R. Br. Prodr. (1810) 228.

Clusters few, sometimes only one head; nut chestnut black.

Decaisne in Nouv. Ann. Mus. Paris, **3** (1834) 361; Miq. Fl. Nederl. Ind. **3**: 328; Hemsl. in Journ. Linn. Soc. **36** (1903) 232. *F. multifolia* Boeck. ! in Linnæa **38** (1874) 397.

PHILIPPINES, "Toubonia" (1433 *Cuming*) in hb. Kew (a false number).

Common in Malaya, Oceania and Australia, not extending to India.

There is some doubt about the example above cited, not however concerning the species. I do not know an island named "Toubonia," but the ticket is written up distinctly by *Bentham*, and it is highly probable that the species in indigenous in the Philippines.

21. **Fimbristylis junciformis** Kunth ! Enum. **2** (1837) 239.

Clusters of spikelets very numerous; nut straw-yellow.

Miq. Fl. Nederl. Ind. **3**: 327; Hook. f. Fl. Brit. Ind. **6**: 647. *F. brevifolia* Presl ! Rel. Haenk. **1**: 192, non R. Br. *F. brachyphylla* Presl ! Rel. Haenk. **1**: 351, non Schultes. *F. Haenkei* Dietr. Sp. Pl. **2**: 161. *F. falcata* Kunth, Enum. **2**: 239; Miq. Fl. Nederl. Ind. **3**: 326. *Scirpus junciformis* Retz. Obs. **6**: 19.

PHILIPPINES, (*Presl*). LUZON, Manila, (*Callery*).

In India, common.

Two bundles of *Haenke's* collection, which *Presl* supposed were collected in Monterey and adjacent localities in California, were collected in the Philippines.

Sect. IV. ABILDGAARDIA.

Lower glumes of the spikelet two-ranked, the uppermost spiral. Style 3-fid.

22. **Fimbristylis monostachya** Hassk. Pl. Jav. Rar. (1848) 61.

Stem with one, rarely a second, spikelet; nut almost stalked, more or less tubercled.

Hook. f. Fl. Brit. Ind. **6**: 649; Hemsl. in Journ. Linn. Soc. **36** (1903) 240. *Cyperus monostachyus* Linn. ! Mant. 180 et hb. propr. *Abildgaardia monostachya* Vahl, Enum. **2**: 296; Moritzi, Verz. Zoll. Pfl. 95; Miq. Fl. Nederl. Ind. **3**: 297. *A. compressa* Presl ! Rel. Haenk. **1**: 179; Miq. Fl. Nederl. Ind. **3**: 297.

LUZON, (*Meyen*); (*Haenke*); (1859 *Wichura*); (761 *Loher*); (114 *Merrill*). MINDANAO, (145 *DeVore and Hoover*); (398 *Copeland*).

Common in all warm countries, very common in India in turf.

23. **Fimbristylis fusca** Benth. in Benth. et Hook. Gen. Pl. **3** (1883) 1048.

Stem with about 10 spikelets; nut obovoid, verrucose.

Hook. f. Fl. Brit. Ind. **6**: 649; Hemsl. in Journ. Linn. Soc. **36** (1903) 236. *Fimbristylis Kamphoveneri* Boeck. ! in Engl. Jahrb. **5** (1884) 505. *Gussonea pauciflora* Brongn. in Duperry Voy. Coquille, **2** (1829) 171. *t. 34. B.* *Abildgaardia pauciflora* Kunth, Enum. **2**: 249; Miq. Fl. Nederl. Ind. **3**: 298. *Rhynchospora ? anomala* Steud. ! in Zoll. Verz. Ind. Archip. heft 2, 61; Miq. Fl. Nederl. Ind. **3**: 337. *Isolepis longispica* Steud. ! Cyp. 104.
LUZON, (1343 *Loher*); (2785 *Merrill*).
From India to the Philippines and Bouru; frequent.

24. **Fimbristylis Actinoschoenus** Hook. f. Fl. Brit. Ind. **6** (1893) 650.

Spikelets subcapitate, 3–1-flowered; glumes almost aristate; nut obovoid, smooth.

Hemsl. in Journ. Linn. Soc. **36** (1903) 230. *Arthrostylis Thouarsii* Kunth, Enum. **2**: 284; Miq. Fl. Nederl. Ind. **3**: 335. *A. Chinensis* Benth. Fl. Hongk. 397. *Actinoschoenus filiformis* Benth. in Hook. f. Ic. Pl. **14**: 33. *t. 1346.*
CULION, (553 *Merrill*).
Madagascar; Ceylon; Malayan Peninsula to the Liu Kiu Islands.
The three geographic forms of this, from Madagascar, Ceylon and Malaya, differ somewhat, and have been esteemed distinct species.

9. BULBOSTYLIS Kunth.

1. **Bulbostylis barbata** Kunth, Enum. **2** (1837) 208, 205.

Stem with one head of spikelets.

Hook. f. Fl. Brit. Ind. **6**: 651; Hemsl. in Journ. Linn. Soc. **36** (1903) 247; Merrill in Philip. Journ. Sci. **1** (1906) Suppl. **1**: 31. *Scirpus barbatus* Rottb. Descr. et Ic. 52. *t. 17 f. 4.*; Vidal Rev. Pl. Vasc. Filip. 284; Phaner. Cuming. Philip. 156. *S. capillaris* Linn. Sp. Pl. ed. 2, 73, partim. *Isolepis barbata* R. Br. Prodr. 222; Presl Rel. Haenk. **1**: 187; Decaisne in Nouv. Ann. Mus. Paris, **3** (1834) 360. *I. involucellata* Steud. ! in Zoll. Verz. Ind. Archip. heft 2, 62; Miq. Fl. Nederl. Ind. **3**: 311. *I. Cumingii* Steud. ! Cyp. 101; Miq. l. c. 310. *Fimbristylis barbata* Benth. ! Fl. Austral. **7**: 32.
PHILIPPINES, (*Llanos*); (*Haenke*, hb. Presl No. 56); (1508 *Cuming*). LUZON, Manila, (1857 *Wichura*); (204 *Chamisso*); (1337, 1338 *Loher*); (6478 *Elmer*); (410 *Whitford*); (321, 1121 *Merrill*).
Warm parts of the Old World; abundant in India. There are also two collections from Florida and two from Georgia, North America.

2. **Bulbostylis capillaris** Kunth, Enum. **2** (1837) 205, 212.

Umbel simple or compound, rarely reduced to a single spikelet.

In America, abundant.

Var. ε **trifida** Hook. f. Fl. Brit. Ind. **6**: 652.

Nut smooth, not, or obscurely transversely undulate.

Hemsl. in Journ. Linn. Soc. **36** (1903) 248. *Bulbostylis trifida* Kunth, Enum. **2**: 213. *Scirpus capillaris* Linn. Mant. 321 et hb. propr. *Isolepis capillaris* Ledeb. Fl. Ross. **4**: 257; Miq. Fl. Nederl. Ind. **3**: 311. *I. trichokolea* Steud. ! in

Zoll. Verz. Ind. Archip. heft 2, 62; Miq. l. c. 308. *Fimbristylis capillacea* Steud. ! l. c. 61; Miq. l. c. 320.

LUZON, (*Loher*), (4291, 4487, 4708 *Merrill*).

Warm parts of the Old World; abundant.

10. SCIRPUS Linn.

1. **Scirpus supinus** Linn. Sp. Pl. ed. 2, 73, neque hb. propr.

Stem with one head of few spikelets; hypogynous bristles usually 0; style usually 3-fid; nut strongly transverse undulate.

Hook. f. Fl. Brit. Ind. **6**: 655; Hemsl. in Journ. Linn. Soc. **36** (1903) 254. *Isolepis supina* R. Br. Prodr. 221; Miq. Fl. Nederl. Ind. **3**: 309. *I. juncoides* Miq. l. c. 312, e descript.

LUZON, (1346 *Loher*); (3654 *Merrill*).

From France and South Africa to the Philippines and South Australia; very common.

Var. β **uninodis** Hook. f. Fl. Brit. Ind. **6**: 656.

Head of spikelets loose, often with one or two short rays.

Isolepis uninodis Delile Fl. Egypt. 8. *t. 6. f. 1;* Miq. Fl. Nederl. Ind. **3**: 311, partim. *I. ambigua* Steud. ! in Zoll. Verz. Ind. Archip. heft 2, 62, nec Steud. Cyp. 91. *I. oryzetorum* Steud. Cyp. 97. *Eleocharis tristachyos* Moritzi ! Verz. Zoll. Pfl. 97.

LUZON, (796 *Loher*).

From Senegambia to Queensland, frequent.

2. **Scirpus erectus** Poir. in Lam. Encycl. **6** (1804) 761.

Hypogynous bristles 5 to 6; style 2-fid; nut very obscurely transverse-undulate; otherwise as *Scirpus supinus* Linn.

Hook. f. Fl. Brit. Ind. **6**: 656; Hemsl. in Journ. Linn. Soc. **36** (1903) 248. *Scirpus debilis* Pursh, Fl. Amer. Sept. **1** (1814) 55. *S. juncoides* Roxb. Hort. Beng. 1814 (1813) 81; Steud. in Miq. Fl. Nederl. Ind. **3**: 303 et Suppl. 261. *S. Luzonensis* Presl ! Rel. Haenk. **1**: 193; Decaisne in Nouv. Ann. Mus. Paris **3** (1834) 361; Miq. l. c. 304. *S. Timorensis* Kunth ! Enum. **2**: 162; Miq. l. c. 305.

LUZON, (*Meyen*); (*Haenke*); (1348, 1349 *Loher*); (6301 *Elmer*).

Abundant in India, China, Japan; common in the eastern United States, also in Mascarenia, Asia Minor.

The spikelets in this species are terete, and it is easily distinguishable from *Scirpus supinus* Linn., the spikelets of the latter being many-angled.

3. **Scirpus inundatus** Poir. in Lam. Encycl. Suppl. **5** (1817) 303; Spreng. Syst. **1** (1825) 207; Benth. Fl. Austral. **7** (1878) 329. *Isolepis inundata* R. Br. Prodr. (1810) 222.

MINDANAO, Mount Apo (298 *DeVore & Hoover*); (1046, 1435 *Copeland*).

Australia, New Zealand, Norfolk Island, and north Borneo.

4. **Scirpus articulatus** Linn. Sp. Pl. ed. 2, 70 et hb. propr.

Stem leafles, terete, with one dense lateral head of large spikelets.

Hook. f. Fl. Brit. Ind. **6**: 656.

PHILIPPINES, (*Moseley*). LUZON, (3979 *Vidal*); (800, 801 *Loher*).

Abundant in the warmer parts of the Old World.

5. **Scirpus mucronatus** Linn. Sp. Pl. ed. 1. 50; ed. 2, 73, pro majore parte.

Stem nearly leafless, triquetrous upward, with one dense lateral head of large spikelets.

Decaisne in Nouv. Ann. Mus. Paris, **3** (1834) 361; Miq. Fl. Nederl. Ind. **3**: 304; Hook. f. Fl. Brit. Ind. **6**: 657; Hemsl. in Journ. Linn. Soc. **36** (1903) 252. *Scirpus acutus* Presl ! Rel. Haenk. **1**: 192. *S. Preslii* Dietr. Sp. Pl. **2**: 175; Miq. Fl. Nederl. Ind. **3**: 305. *S. Javanus* Nees in Wight Contrib. 112; Moritzi, Verz. Zoll. Pfl. 97; Steud. in Zoll. Verz. Ind. Archip. heft 2, 62; Miq. Fl. Nederl. Ind. **3**: 305. *S. Sundanus* Miq. l. c. 304.

LUZON, (*Mertens*); (3978 *Vidal*); (799 *Loher*); (5948 *Elmer*); (84, 2261, 4342 *Merrill*).

Warmer parts of the Old World, abundant in southeastern Asia; frequent in Oceania; rare in Africa.

Britton reports this plant from Pennsylvania, and *Coulter* has sent examples from California, supposedly introduced.

6. **Scirpus triqueter** Linn. Mant. (1767) 29.

Leaves short; stem triquetrous; umbel lateral, thin; style 2-fid; hypogynous bristles 3 to 6, retrosely scabrous, not plumose.

Hook. f. Fl. Brit. Ind. **6**: 658; Hemsl. in Journ. Linn. Soc. **36** (1903) 255.
Europe, north Asia, Japan, common; also at the Cape of Good Hope.

Var. *β* **segregata** Hook. f. Fl. Brit. Ind. **6**: 658.

Spikelets mostly solitary; hypogenous bristles 3–2.

S. subulatus Prain in Journ. Asiat. Soc. Beng. **60**[2] (1892) 335, non Vahl.
LUZON, (802 *Loher*).
Coasts of the Bay of Bengal; also in New Guinea.

7. **Scirpus grossus** Linn. f. Suppl. (1781) 104, et Linn. hb. propr.

Very large, the umbel large, compound; hypogynous bristles 6, scabrous; style 3-fid.

Miq. Fl. Nederl. Ind. **3**: 307 et Suppl. 261; Hook. f. Fl. Brit. Ind. **6**: 659; Hemsl. in Journ. Linn. Soc. **36** (1903) 250. *Scirpus aemulans* Steud. ! in Zoll. Verz. Ind. Archip. heft **2**, 62. *S. maritimus* var. *β aemulans* Miq. Fl. Nederl. Ind. **3**: 306 (infauste).
India, common.

Var. *β* **Kysoor** Hook. f. Fl. Brit. Ind. **6**: 660.

Tubers edible; hypogynous setæ villous by reason of flaccid many-celled hairs.

Scirpus Kysoor Roxb. Hort. Beng. 1814 (1813) 6.
PHILIPPINES, (*Llanos*). LUZON, (878 *Loher*); (62 *Merrill*). MINDANAO (1336 *Copeland*).
Frequently cultivated in India. Many examples have been received from the Malayan Peninsula and Archipelago.

8. **Scirpus Ternatensis** Miq. Fl. Nederl. Ind. **3** (1855) 307.

Tall, with nodes and leaves in the upper half of the stem; umbel large, compound; style 2-fid.

Hemsl. in Journ. Linn. Soc. **36** (1903) 254. *Scirpus Chinensis* Munro in Seem. Voy. Herald (1857) 423; Hook. f. Fl. Brit. Ind. **6**: 662.

LUZON, (797 *Loher*); (5790, 6287 *Elmer*).

In India and China, common; also received from Tonking, Liu Kiu, Celebes and the Bonin Islands.

11. FUIRENA Rottb.

1. **Fuirena glomerata** Lam. Ill. **1** (1791) 150.

Annual, petals subquadrate, clawed, cordate at the base.

Decaisne in Nouv. Ann. Mus. Paris, **3** (1834) 360; Miq. Fl. Nederl. Ind. **3**: 326; Hook. f. Fl. Brit. Ind. **6**: 666; Hemsl. in Journ. Linn. Soc. **36** (1903) 256. *Fuirena Rottboellii* Nees ! in Wight Contrib. 94; Steud. in Zoll. Verz. Ind. Archip. heft 2, 61. *F. striata* Llanos Fragm. Pl. Filip. 21. *Scirpus ciliaris* Linn. Mant. 182 et hb. propr. !

PHILIPPINES, (*Llanos*). LUZON, (866 *Jagor*); (767 *Loher*); (96 *Gaudichaud*); (3653 *Merrill*); (5596 *Elmer*).

Warm parts of the Old World, very common; a frequent weed in rice lands.

2. **Fuirena umbellata** Rott. Descr. et Ic. (1773) 70. *t. 19*, i. e. *t. 18* altera, *f. 3*.

Perennial, petals obovate, narrowed at the base, hardly clawed.

Presl, Rel. Haenk. **1**: 186; Moritzi, Verz. Zoll. Pfl. 97; Steud. in Zoll. Verz. Ind. Archip. heft 2, 61; Vidal, Rev. Pl. Vasc. Filip. 284; Phaner. Cuming. Philip. 156; Hook. f. Fl. Brit. Ind. **6**: 666; Hemsl. in Journ. Linn. Soc. **36** (1903) 256. *Fuirena tereticulmis* Presl ! Rel. Haenk. **1**: 186. *F. pentagona* Schum. Guin. Pl. 42; Miq. Fl. Nederl. Ind. **3**: 329.

LUZON, (1254, 1834 *Cuming*); (768 *Loher*).

Warm countries of both hemispheres, very common.

12. RYNCHOSPORA Vahl.

Style 2-fid. Hypogynous bristles present.

* *Culm with one dense head of spikelets.*

1. **Rynchospora Wallichiana** Hook. f. Fl. Brit. Ind. **6** (1893) 668.

Nut laterally flattened.

Hemsl. in Journ. Linn. Soc. **36** (1903) 260. *Rhynchospora Haenkei* Presl, Rel. Haenk. **1** (1828) 199; Miq. Fl. Nederl. Ind. **3**: 336. *Rh. Wallichiana* Kunth, Enum. **2** (1837) 289; Moritzi, Verz. Zoll. Pfl. 98; Steud. in Zoll. Verz. Ind. Archip. heft 2, 61; Miq. Fl. Nederl. Ind. **3**: 262. *Mariscus umbellatus* var. *procerior* Steud. in Zoll. Verz. Ind. Archip. heft 2, 63.

LUZON, (794, 795 *Loher*); (6510 *Elmer*). SEMERARA, (4153 *Merrill*).

From India to Japan and Queensland, common; rare in tropical Africa.

2. **Rynchospora Wightiana** Hook. f. Fl. Brit. Ind. **6** (1893) 669.

Nut dorsally flattened.

Rhynchospora Wightiana Steud. Cyp. 148. *Rh. discolor* Steud. l. c. *Rh. longisetis* var. F. Muell. Fragm. Phyt. Austral. **9**: 75, in obs. *Haplostylis Wightiana* Nees ! in Nova Acta Nat. Cur. **19** (1843) Suppl. **1**: 101. *H. Meyenii* Nees ! ms. partim.

LUZON, (798 *Loher*).

Malabar Peninsula from Poona to Ceylon; also in Cochin-China.

There is an American plant (No. 2385 *Gardner*) collected in Piauhy, Brazil, which I can not distinguish specifically from *R. Wightiana*.

** *Spikelets copiously umbelled.*

3. **Rynchospora aurea** Vahl, Enum. 2 (1806) 229.

Style-branches 2, very short, almost wanting.

Hook. f. Fl. Brit. Ind. **6**: 670. *Rhynchospora aurea* R. Br. Prodr. 230; Presl, Rel. Haenk. **1**: 179; Miq. Fl. Nederl. Ind. **3**: 336 et Suppl. 262; Vidal, Rev. Pl. Vasc. Filip. 285; Phaner. Cuming. Philip. 156. *Rh. articulata* Spreng. Syst. **1**: 197; Moritzi. Verz. Zoll. Pfl. 98; Steud. in Zoll. Verz. Ind. Archip. heft **2**, 61; Miq. Fl. Nederl. Ind. **3**: 337. *Scirpus corymbosus* Linn. Amoen. Acad. **4**: 303.

PHILIPPINES, (15 hb. *Presl*). CEBU, (1763 *Cuming*). MINDORO, (807 *Merrill*).

In the warm parts of the World, common in the Old; very common in the New World.

4. **Rhynchospora glauca** Vahl, Enum. **2** (1806) 233.

Style-branches 2, long.

Hook. f. Fl. Brit. Ind. **6**: 671. *Rhynchospora laxa* R. Br. Prodr. 230; Miq. Fl. Nederl. Ind. **3**: 337. *Rh. ferruginea* Roem. et Sch. Syst. **2**: 85; Presl, Rel. Haenk. **1**: 199.

In all warm countries; frequent in India.

Var. β **Chinensis** Hook. f. Fl. Brit. Ind. **6**: 672.

Spikelets often ripening two nuts; hypogynous bristles often reaching the top of the beak of the nut.

Hemsl. in Journ. Linn. Soc. **36** (1903) 259. *Rhynchospora Chinensis* Boeck. in Linnæa **37** (1873) 586.

LUZON, Benguet, (5757 *Elmer*); Lepanto, (4623 *Merrill*).

From Madagascar to Japan and the Sandwich Islands, frequent.

13. SCHOENUS Linn.

1. **Schoenus apogon** Roem. et Sch. Syst. **2** (1817) 77. *S. imberbis* Poir. Encycl. Suppl. **2**: 251, neque (homonyma) p. 250, nec. R. Br. *S. laxiflorus* Steud. ! Cyp. 166. *S. Brownii* Hook. f. ! Handb. New Zeal. Fl. 298. *Chaetospora imberbis* R. Br. ! Prodr. 233. *C. tenuissima* Steud. ! Cyp. 162. *C. Japonica* Franch. et Savat. Pl. Japon. **2**: 122, 548, e descript. *C. umbellulifera* Boeck. ! in Flora, **65** (1882) 28. *Isolepis margaritifera* Nees in Ann. Nat. I, **6** (1849) 46. *Scirpus margaritifer* Boeck. ! in Linnæa **36** (1869–70) 697.

LUZON, (1347 *Loher*).

From Japan to New Zealand; common in Australia.

14. CLADIUM R. Br.

1. **Cladium distichum** sp. nova.

Glumis 12 ad 16, oblongis, specie distichis; paniculae laxae, ramis anfractuoso-flexuosis; nuce parva, sessili, obovoidea, papyracea, viridi-lutea, grosse laxe rugosa, rostro subnullo, i. e., brevi depresso-ovoideo, glabro.

Plant 1 m. high. Leaves very few, 2 to 7 cm. long, linear, rigid. Spikelets with 3 or 4 short ovate glumes at the base, the upper one perfecting a nut, succeeded by 10 or 12 oblong, much larger, remarkably distichous glumes. I detected in the ripe fruit no hypogynous setæ. The species is very like *Cladium undulatum* Thwaites, but the nut is smaller and the beak 0.

LUZON, Principe, (1124 *Merrill*).

A young plant, No. 758 *Merrill*, Palawan (Paragua) may also be referable to this species.

15. GAHNIA J. G. et R. Forster.

1. **Gahnia Javanica** Moritzi, Verz. Zoll. Pfl. (1845–46) 98.

Tall, scabrous; panicle long, compound, dense, of black 1- to 2-flowered spikelets.

Miq. Fl. Nederl. Ind. **3**: 340; Vidal, Rev. Pl. Vasc. Filip. 285; Hook. f. Fl. Brit. Ind. **6**: 676; Hemsl. in Journ. Linn. Soc. **36** (1903) 262. *Phakellanthus multiflorus* Steud. ! in Zoll. Verz. Ind. Archip. heft 2, 61. *Syzyganthus multiflorus* Steud. Cyp. 153. *Schoenus Hasskarlii* Steud. ! l. c. 166.

LUZON, Mount Banajao, (1965 *Vidal*). MINDANAO, Mount Apo, (288 *DeVore et Hoover*); (1038 *Copeland*).

From Yunnan and Penang to New Guinea and the Viti Islands; frequent.

16. REMIREA Aublet.

1. **Remirea maritima** Aubl. Pl. Guian. 1 (1775) *45. t. 16.*

Spikelets 1-flowered, in dense digitate spikes.

Vidal, Rev. Pl. Vasc. Filip. 285; Phaner. Cuming. Philip. 156; Hook. f. Fl. Brit. Ind. **6**: 677; Hemsl. in Journ. Linn. Soc. **36** (1903) 258. *Mariscus capitatus* Steud. ! in Zoll. Verz. Ind. Archip. heft 2, 63; Miq. Fl. Nederl. Ind. **3**: 288. *M. pungens* Steud. ! l. c. 60; Miq. l. c. 288. *M. maritimus* Miq. Fl. Nederl. Ind. Suppl. 600. *Cyperus Kegelianus* Steud. ! Cyp. 60. *Lipocarpha foliosa* Miq. ! Fl. Nederl. Ind. **3**: 332 et Suppl. 262.

PHILIPPINES, (867 *Cuming*). LUZON, Manila, (168 *Chamisso*). MINDORO, (881 *Merrill*). MINDANAO, (857 *Copeland*).

Tropical seacoasts of the Old and New Worlds.

17. SCLERIA Berg.

Series A. *Many of the spikelets (apparently) 2-sexual.*

1. **Scleria lithosperma** Swartz Prodr. (1788) 18.

Medium sized; panicle thin, lax; nut white, smooth.

Moritzi, Verz. Zoll. Pfl. 98; Steud. ! in Zoll. Verz. Ind. Archip. heft 2, 61; Miq. Fl. Nederl. Ind. **3**: 344; Vidal, Rev. Pl. Vasc. Filip. 285; Phaner. Cuming. Philip. 156; Hook. f. Fl. Brit. Ind. **6**: 65; Hemsl. in Journ. Linn. Soc. **36** (1903) 265. *Scleria glaucescens* Presl, Rel. Haenk. **1**: 202. *Scirpus lithospermus* Linn. Sp. Pl. ed. 1, 51. *Schoenus lithospermus* Linn. Sp. Pl. ed. 2, 65 pro parva parte. *Olyra orientalis* Lour. Fl. Cochinch. **2**: 674.

PHILIPPINES, (1817 *Cuming*). LUZON, (603 *Whitford*); (3176 *Merrill*); (6144 *Leiberg*).
Warm regions of the World; very common in Asia, Oceania and America; rare in Africa.

Var. β **Roxburghii** Thwaites ! Enum. Pl. Zeyl. 354.

Rather stouter, nut with transverse wrinkles, at least when young.

Hook. f. Fl. Brit. Ind. **6**: 685.
LUZON, (805, 806 *Loher*).
South India.

2. **Scleria corymbosa** Roxb. Hort. Beng. 1814 (1813) 103.

One to three meters high, panicle large, compound, dense.

Hook. f. Fl. Brit. Ind. **6**: 686. *Scleria androgyna* Nees in Linnæa **9** (1834) 303.
CULION, (656 *Merrill*).
India, frequent.

3. **Scleria Motleyi** C. B. Clarke in hb. Kew, ms. in Motley nn. 72, 74, 152.

Minutely loosely floccose all over; nut ellipsoid, with triangular top, white, with rusty hairs.

Borneo, frequently received.

Var. β **densi-spicata.**

Spikes compound, dense, of many spikelets; bracts linear, caudate, standing out from the spike.

LUZON, Pampanga, Mount Arayat, (803 *Loher*).

Series B. *Spikelets all 1-sexual.*

(*a*) *Plants medium sized or small; annual or with scarcely any rhizome.*

4. **Scleria tessellata** Willd. ! Sp. Pl. 4 (1805) 315, tab. Rumph. citat. excl.

Nut tessellated, often minutely hairy.

Hook. f. Fl. Brit. Ind. **6**: 686; Hemsl. in Journ. Linn. Soc. **36** (1903) 267. *Scleria propinqua* Steud. ! Cyp. 169; Miq. Fl. Nederl. Ind. **3**: 343.
PHILIPPINES, (*Loher*). LUZON, Benguet, (4370 *Merrill*).
From India to Japan and Queensland; common in India.

5. **Scleria annularis** Kunth, Enum. **2** (1837) 359.

Nut smooth, sometimes obscurely fenestrate; margin of the disc truncate.

Hook. f. Fl. Brit. Ind. **6**: 687; Hemsl. in Journ. Linn. Soc. **36** (1903) 263. *Hypopyrum annulare* Nees in Linnæa **9** (1834) 303.
LUZON, (807 *Loher*).
India and China (Ichang), rare.

6. **Scleria Zeylanica** Poiret in Lam. Encycl. **7** (1806) 3, exemplo Madagascarensi excluso.

Nut smooth; margin of the disc subentire, glandular, colored.

Hook. f. Fl. Brit. Ind. **6**: 687. *Scleria lateriflora* Boeck. ! in Linnæa **38** (1874) 455.

LUZON, (808 *Loher*); (3665, 4617 *Merrill*).

Ceylon to New Caledonia, scattered.

This species is very near *Scleria annularis* Kunth, but the inflorescence is looser and with divaricate and nodding branches.

(*b*) *Plants more robust, with woody horizontal rhizomes.*

* *Sheaths of the leaves 3-winged.*

7. **Scleria hebecarpa** Nees in Wight Contrib. (1834) 117.

Nut slightly verrucose, hairy; lobes of the disc lanceolate, appressed to the nut.

Hook. f. Fl. Brit. Ind. **6**: 689; Hemsl. in Journ. Linn. Soc. **36** (1903) 264. *Scleria scrobiculata* Moritzi ! Verz. Zoll. Pfl. 98, partim; Steud. ! in Zoll. Verz. Ind. Archip. heft 2, 61, non Nees. *S. Japonica* Steud. ! Cyp. 169. *S. Wichurai* Boeck. ! in Engl. Jahrb. **5** (1884) 510.

LUZON, Manila, (*Wichura*).

India, frequent, to Japan and the Viti Islands.

8. **Scleria Chinensis** Kunth, Enum. **2** (1837) 357.

Nut verrucose or tessellate, hairy; lobes of the disc short-ovate, rounded.

Hook. f. Fl. Brit. Ind. **6**: 690; Hemsl. in Journ. Linn. Soc. **36** (1903) 263; Merrill in Philip. Journ. Sci. **1** (1906) Suppl. **1**: 31. *Scleria scrobiculata* Moritzi ! Verz. Zoll. Pfl. 98 partim.

LUZON, (804 *Loher*); (3964, 3958 *Merrill*).

China to Singapore, Malaya and Queensland.

9. **Scleria oryzoides** Presl, Rel. Haenk. **1** (1828) 201.

Panicle-branches numerous, suberect, long, with many solitary spikelets; nut smooth, margin of the disc truncate.

Miq. Fl. Nederl. Ind. **3**: 342; Hook. f. Fl. Brit. Ind. **6**: 691.

LUZON, (*Haenke*); (144 *Merrill*).

Southern India to north Australia, frequent; also received twice from Mozambique.

10. **Scleria purpureo-vaginata** Boeck. ! in Engler Jahrb. **5** (1884) 513.

Nearly glabrous, leaves appearing in twos and threes, subopposite; nut white, tessellated; lobes of the disc short-ovate, obtuse.

Scleria sumatrensis Vidal ! Rev. Pl. Vasc. Filip. 285, non Retz.

LUZON, (1936 *Vidal*); (1852 bis *Wichura*); (34 *Whitford*). NEGROS, (95 *Copeland*). PALAWAN, (353 *Bermejos*).

Ceram Laut.

The falsely opposite or ternate leaves brings this plant apparently near the group of *Scleria Sumatrensis*. The present species has strongly 3-winged leaf-sheaths which separates it. As regards *S. Sumatrensis* Retz., an abundant species, it has a very tall disc sticking up and encircling the nut up to two-thirds its height, which makes the species an easy one to recognize.

** *Sheaths of the leaves usually triangular, scarcely winged. Leaves (chiefly) ternate or opposite.*

11. **Scleria scrobiculata** Nees in Wight Contrib. (1834) 117.

Panicle dull grey-purple; lobes of the disc short.

Moritzi ! Verz. Zoll. Pfl. 99 partim; Miq. Fl. Nederl. Ind. 3: 342; Hemsl. in Journ. Linn. Soc. 36 (1903) 266. *Scleria Timorensis* Nees ! in Linnaea 9 (1834) 302. *S. tessellata* Decaisne ! in Nouv. Ann. Mus. Paris. 3 (1834) 362, non Willd. *S. Neesiana* Hook. et Arn. Bot. Beechy Voy. 229 (non *S. Neesii* Kunth). *S. Wrightiana* Steud. Cyp. 173; Miq. Fl. Nederl. Ind. 3: 345. *S. [illegible]* Llanos Fragm. Pl. Filip. 108, non Carr. *S. Kepensis* K. Schum. ! in Engl. Jahrb. 15 (1891) 267.

PHILIPPINES, *Llanos*; (*Mearley*); (*King*). LUZON, (409, 530 *Loher*); (458 *Topping*); (3813 *Ahern's collector*); (24 *Fenwick*); (5534 *Elmer*). MINDORO, (1258 *Merrill*). MINDANAO, (591 *Copeland*).

From the Andaman Islands to Liu Kiu and New Guinea; scattered, not very common.

12. **Scleria multifoliata** Boeck. in Linnaea 38 (1874) 514.

Panicle red; lobes of the disc 1/3 to 2/3 the height of the nut.

Hook. f. Fl. Brit. Ind. 6: 693. *Scleria tessellata* Brongn. in Duperry Voy. Coquille, 2 (1829) 164. *Carex Amboinica* Rumph. Herb. Amb. 6: 20. t. 8. f. 1.

LUZON, Manila, (52 *Gaudichaud*); Bataan, (400 *Topping*); Benguet, (4534 *Merrill*).

Malayan Peninsula and Archipelago; frequent.

Scleria pubescens Zoll. ! Verz. Ind. Archip. heft 2. 61, which is not *S. pubescens* Steud., I consider a variety only of *S. multifoliata* Boeck.

13. **Scleria Sumatrensis** Retz. Obs. 5: 19. *t.* 2.

Disc tall, encircling the nut up to 3/4 its height.

Boeck. in Linnaea 38: 513; Hook. f. Fl. Brit. Ind. 6: 693. *S. purpurascens* Steud. Cyp. 169.

BASILAN (7 *De Vore & Hoover*).

Southern Bengal to Malaya.

16. **DIPLACRUM** R. Brown.

1. **Diplacrum caricinum** R. Br. Prodr. (1810) 241; Brongn. in Duperry. Voy. Coquille, 2 (1829) 160; Moritzi, Verz. Zoll. Pfl. 99; Steud. in Zoll. Verz. Ind. Archip. heft 2, 60; Miq. Fl. Nederl. Ind. 3: 345 et Suppl. 262; Hemsl. in Journ. Linn. Soc. 36 (1903) 267. *Scleria caricina* Benth. Fl. Austral. 7: 426; Hook. f. Fl. Brit. Ind. 6: 688.

LUZON, (736 *Loher*); (3626 *Merrill*).

Frequent in India, thence scattered to Hongkong, the Philippines and Queensland.

There is no possible line to be drawn between *Diplacrum* and *Scleria*. In *Diplacrum* the female flower—that is, spikelet—appears terminal; but so it does in *Scleria flaccida* very often.

19. CAREX Linn.

Subgenus I. VIGNEANDRA. Style 2-fid; spikelets male at top.

1. **Carex brunnea** Thunb. Fl. Jap. (1784) 38; C. B. Clarke in Journ. Linn. Soc. **37** (1904) 3, 5, cum syn.; Hemsl. in Journ. Linn. Soc. **36** (1903) 278; Merrill, Philip. Journ. Sci. **1** (1906) Suppl. **1**: 31.

LUZON, (1346 *Whitford*); (711 *Loher*); (3197, 3880, 4223, 4224, 4514, 4529, 4731, 4819 *Merrill*).

From Madagascar to Japan, Australia and the Sandwich Islands; common.

2. **Carex Graeffeana** Boeck. in Flora, **58** (1875) 123; C. B. Clarke in Journ. Linn. Soc. **37** (1904) 3, 5.

LUZON, Benguet, (690 *Loher*). MINDANAO, (1250 *Copeland*). Samoa.

Subgenus II. CARICINICA. Style 3-fid; stem with one spike.

3. **Carex capillacea** Boott, Carex, **1** (1858) 44. *t. 110;* C. B. Clarke in Journ. Linn. Soc. **37** (1904) 3, 7, cum syn.; Hemsl. l. c. **36** (1903) 278.

LUZON, Benguet, (705 *Loher*); (4732 *Merrill*).

Southeastern Asia, extending from Laristan to Sachalin and to New South Wales.

Subgenus III. CARICANDRA. Style 3-fid; spikes numerous, very many of them male at the top.

4. **Carex scaberrima** C. B. Clarke in Journ. Linn. Soc. **37** (1904) 4, 10. *Carex Bengalensis* var. *γ scaberrima* Boeck. in Linnæa, **40** (1876) 347; Vidal, Rev. Pl. Vasc. Filip. 285; Phaner. Cuming. Philip. 156.

LUZON, Albay, (936 *Cuming*); Benguet, (4796 *Merrill*).

Endemic; that is to say, *C. scaberrima* can not well be made a variety of *C. Bengalensis* Roxb., or of *C. cruciata* Wahl., unless a large series of admitted species of this subgenus are reduced to one.

5. **Carex fuirenoides** Gaudich. ! in Freycinet. Voy. (1826) 412; C. B. Clarke in Journ. Linn. Soc. **37** (1904) 6, 11, cum syn.

LUZON, (109 *Merrill*). CEBU, (1764 *Cuming*).

Queensland and the Marianne Islands.

6. **Carex Cumingii** Vidal ! Phaner. Cuming. Philip. (1885) 156; Rev. Pl. Vasc. Filip. 286, non Boott; C. B. Clarke in Journ. Linn. Soc. **37** (1904) 4, 11.

LUZON, (1408 *Cuming*); (704, 712 *Loher*); (6449 *Elmer*); (*Alberto*).

Endemic.

7. **Carex filicina** Nees in Wight. Contrib. (1834) 123; Hemsl. in Journ. Linn. Soc. **36** (1903) 285; C. B. Clarke in Journ. Linn. Soc. **37** (1904) 4, 11.

LUZON, (707 B *Loher*); (866 *Klemme*); (949 *Whitford*); (4513, 4743 *Merrill*).

Abundant in India; also in China.

Many admitted species of *Carex*, and among these is the next, *C. continua* C. B. Clarke, are very close to *C. filicina* Nees.

8. **Carex continua** C. B. Clarke in Hook. f. Fl. Brit. Ind. **6** (1894) 717; Hemsl. in Journ. Linn. Soc. **36** (1903) 281; C. B. Clarke in Journ. Linn. Soc. **37** (1904) 4, 11; Merrill, Philip. Journ. Sci. **1** (1906) Suppl. **1**: 31.

LUZON, (707, 708, 709, 710 *Loher*); (189, 1121, 1145 *Whitford*); (6985 *Elmer*); (3197 *Merrill*).

Widely distributed in North India; also in Yunnan.

9. **Carex rhizomatosa** Steud. in Zoll. Verz. Ind. Archip. heft 2, (1854) 60; C. B. Clarke in Journ. Linn. Soc. **37** (1904) 4, 12, cum syn.
NEGROS, (1795 *Cuming*).
Assam, Burma.

10. **Carex turrita** C. B. Clarke in Journ. Linn. Soc. **37** (1904) 4, 13.
LUZON, Benguet, (700 *Loher*).
Endemic in the Philippines.

11. **Carex baccans** Nees in Wight Contrib. (1834) 122; Hemsl. in Journ. Linn. Soc. **36** (1903) 274; C. B. Clarke, l. c. **37** (1904) 4, 14.
LUZON, (706, 1948 *Loher*); (4515, 4555, 4794 *Merrill*); (6270 *Elmer*).
Common in north India, extending to Java and Formosa.

12. **Carex Loheri** C. B. Clarke in Journ. Linn. Soc. **37** (1904) 4, 14.
LUZON, (701, 702, 703, 708 bis *Loher*); (4488 *Merrill*).
Endemic.

Subgenus IV. EUCAREX. Style 3-fid; terminal spike wholly male.

13. **Carex rhynchachaenium** C. B. Clarke in Govt. Lab. Publ. **35** (1905) 5; Merrill, Philip. Journ. Sci. **1** (1906) Suppl. **1**: 31.
LUZON, Bataan, (6983 *Elmer*).
Endemic.

14. **Carex subtransversa** C. B. Clarke, sp. nova.

Glabra, mediocris, foliis cum culmo aequilongis, 3 ad 4 mm. latis; spicis 3 ad 4, ima plus minus remota, summa mascula; bracteis inflorescentia vix brevioribus; utriculis (rostro incluso) 2.5 mm. longis, globoso-trigonis, 5 ad 8 nervatis, glabris, luride viridibus, in rostrum subito angustatis, rostro cum ⅔ parte utriculi aequilongo, fere lineari, paullo obliquo; stylo 3-fid.

LUZON, Benguet, Pauai, (4730 *Merrill*), 2,100 m. s. m.
C. transversæ Boott similis et affinis, differt utriculos multo minores, maturatos oblique erectos.

20. HYPOLYTRUM L. C. Rich.

Style 2-fid. Inflorescence paniculate-corymbose.

1. **Hypolytrum latifolium** L. C. Rich. in Pers. Syn. **1** (1805) 70.

Nut wrinkled, brown or chestnut, with a small conic whitened beak.

Miq. Fl. Nederl. Ind. **3**: 333; Hook. f. Fl. Brit. Ind. **6**: 678; Hemsl. in Journ. Linn. Soc. **36** (1903) 258; Merrill, Philip. Journ. Sci. **1** (1906) Suppl. **1**: 30. *Hypolytrum schoenoides* Moritzi ! Verz. Zoll. Pfl. 97; Steud. in Verz. Ind. Archip. heft 2, 61. *H. myrianthum* Miq. ! Fl. Nederl. Ind. **3**: 333. *Albikkia scirpoides* Presl, Rel. Haenk. **1**: 185. *t. 35;* vix Presl in Oken Isis. **21**: 269. *A. schoenoides* Presl l. c. *t. 34.*
LUZON, (2089 *Borden*).
In India frequent, extending to Formosa, Queensland and the Viti Islands.
These species of *Hypolytrum* are very close together; some botanists include Mascarene plants in *H. latifolium* L. C. Rich.

2. **Hypolytrum compactum** Nees ! in Linnæa, **9** (1834) 288; Nova Acta Nat. Cur. **19** (1843) Suppl. **1**: 73.

Panicle congested nearly into one mass; beak conic, flattened, whitened, nearly as long as the nut.

Miq. Fl. Nederl. Ind. **3**: 333; Merrill, Philip. Journ. Sci. **1** (1906) Suppl. **1**: 29.

LUZON, (*Meyen*); (776, 777, 778 *Loher*); (6011 *Leiberg*); (782, 2920 *Borden*); (2496 *Merrill*); (51 *Whitford*); (3977 *Vidal*).

Cochin-China and the Andaman Islands.

I identified two of the above plants with a species from the Aru Islands (as yet unpublished); but *Merrill* l. c. says the Philippine plants cited above are all one species, and I consider that he is correct.

3. **Hypolytrum viridinux** C. B. Clarke, ms. (1895) hb. Kew.

Spicis parvis; stylo 2-fido; nuce parva, matura viridi, reticulato-rugosa, rostro vix ullo albescente.

LUZON, Montalban, (775 *Loher*); Benguet, (6223 *Elmer*).

Borneo.

This species is allied to *H. latifolium* L. C. Rich., but has much smaller spikes and nuts. In the type (from Sandakan) the nuts are remarkably green.

4. **Hypolytrum Philippense** C. B. Clarke, ms. (1887) hb. Kew.

Panicula composita, spiculis parvulis, numerosis, brunneis; ceteroquin ut *H. latifolium* L. C. Rich.

LUZON, Isabela, (3987 *Vidal*); Union, (779 *Loher*); Laguna, (5126 *Merrill*); Rizal, (77 *Foxworthy*). PALAWAN, (744 *Merrill*).

Endemic.

This differs from *H. viridinux* in the very copious panicle; in the type (No. 3987 *Vidal*) there are about 600 spikes. I have not seen the fruit.

21. MAPANIA Aubl.

1. **Mapania humilis** F.-Vill. Nov. App. (1883) 309.

Leaves as though petioled, 30 to 45 mm. broad, flagellate at the tip.

Hook. f. Fl. Brit. Ind. **6**: 683. *Pandanophyllum humile* Hassk. ! in Tijdsch. Nat. Vereen. **10** (1843) 119; Miq. Fl. Nederl. Ind. **3**: 334. *Lepironia cuspidata* Miq. Fl. Nederl. Ind. Suppl. 603, fide auctoris. *L. humilis* Miq. Illustr. Fl. Arch. Ind. 61. *t. 23.*

LUZON, (774 *Loher*). MINDORO, (4030 *Merrill*).

Malayan Peninsula and Archipelago; frequent.

22. LIPOCARPHA R. Br.

1. **Lipocarpha argentea** R. Br. in Append. Tuckey Congo (1818) 459.

Spikes terete, points of the glumes appressed, almost incurved.

Miq. Fl. Nederl. Ind. **3**: 331; Vidal, Rev. Pl. Vasc. Filip. 285; Phaner. Cuming. Philip. 156; Hook. f. Fl. Brit. Ind. **6**: 667; Hemsl. in Journ. Linn. Soc. **36** (1903) 257. *Kyllinga albescens* Steud. ! Cyp. 68; Miq. Fl. Nederl. Ind. **3**: 294.

LUZON, Albay, (1418 *Cuming*); Benguet, (6484 *Elmer*).

Warm parts of the Old World; common.

2. **Lipocarpha microcephala** Kunth. Enum. **2** (1837) 268.

Spikelets squarrose, points of the glumes recurved.

Miq. Fl. Nederl. Ind. **3**: 331; Hook. f. Fl. Brit. Ind. **6**: 668; Hemsl. in Journ. Linn. Soc. **36** (1903) 257. *Ascolepis kyllingioides* Steud. in Zoll. Verz. Ind. Archip. heft 2, 61; Miq. Fl. Nederl. Ind. **3**: 313. ***Kyllinga squarrosa*** Steud. Cyp. 68; Miq. l. c. 294. *Isolepis squarrosa* Miq. ! in Ann. Mus. Lugd. Bat. **2** (1865–66) 211, non Roem. et Sch.

LUZON, (798 *Loher*); (3625 *Merrill*).

Japan to Australia, common; Singapore its extreme western habitat.

THE OCCURRENCE OF ANTIARIS IN THE PHILIPPINES.

By ELMER D. MERRILL.

(From the botanical section of the Biological Laboratory, Bureau of Science.)

ANTIARIS Lesch.

Antiaris toxicaria (Pers.) Leschen. in Ann. Mus. Paris, **16** (1810) 478. *t. 22;* Blume, Rumphia, **1** (1835) 56. *t. 22, 23;* Benn. Pl. Jav. Rar. (1838–1852) 52. *t. 13;* Miq. Fl. Ind. Bat. **1**[2] (1859) 291; Hook. f. Fl. Brit. Ind. **5** (1888) 537; F. Vill. Nov. App. (1883) 202.

MINDORO, Bulalacao (1551 *Bermejos*) August 27, 1906. V., *Salogón;* T., *Dalít.*

This interesting species was first called to our attention by the Honorable *Dean C. Worcester*, Secretary of the Interior of the Government of the Philippine Islands, who brought from Bulalacao early in the year 1906, a small quantity of a substance used by the natives in that vicinity for poisoning arrows, but without botanical material by which the species yielding the product could be identified. As Dr. *R. F. Bacon* of the Bureau of Science had undertaken the chemical examination of the different arrow poisons used by the natives of the Philippines, a native collector was sent to Bulalacao with instructions to secure a quantity of the poison as well as botanical specimens from the tree yielding the product. Although the material secured was without fruit or flowers, a careful examination of it leads me to conclude that it is identical with *Antiaris toxicaria*, while Dr. *Bacon* informs me that a chemical examination of the poison shows it to be identical with that yielded by this species.

Miquel[1] credits the species to the Philippines, citing *Camell* for authority for its occurrence here. That the species was known from the Philippines over two hundred years ago, and that from *Camell's* time up to the year 1906 this much discussed and well known plant had not been rediscovered in the Archipelago, is at least interesting.

F.-Villar includes the species in his Novissime Appendix, citing *Miquel* and *Camell* for authority for its occurrence in the Philippines, but stating that he had not seen specimens.

Camell[2] states the following regarding this plant, under his "*De quibusdam Arboribus Venenatis:*"

"1· *Ipo*, seu *Hypo* arbor est mediocris, folio parvo, & obscurè virenti, quae tam malignae, & nocivae est qualitatis, ut omne vivens umbra sua interimat, unde narrant in circuitu, & umbrae distinctu plurima ossium, mortuorum hominum, anamaliumve videri. Circumvicinas etiam plantas enecat, & aves insidentes interficere ferunt, si *Nucus* Vomicae *Igasur*, plantam non invenerint, qua reperta vita quidem donantur, & servantur, sed defluvium patiuntur plumarum. *Antonius Molero* mihi retulit, post iter per Sylvosam viam, passum

[1] *Fl. Ind. Bat.* **1**[2] (1859) 292.

[2] J. Ray: *Hist. Plant.* **3** (1704) App. 87.

fuisse defluvium capillorum unius lateris, an forsan ex hac arbore ? Hypo *lae Indi Camucones*, & *Sambales*, *Hispanis* infensissimi longis excipiunt arundineis perticis, sagittis intoxicandis deserviturum, irremediabile venenum, omnibus aliis alexipharmacis superius, praeterquam stercore humano propinato. *An* Argensolae *arbor comosa*, quam *Insulae Celebes* ferunt, cujus umbra occidentalis mortifera, orientalis Antidotum. *An* Machucae *Zeuca* ? qui addit: Sagittis lacte fructus arboris *Mansanillo*, illitis vulneratos, non emori, sed intumescere, & hebetissimos reddi. Num *Mansanillo* idem, ac *Mansanan* seu *Pomum portus Acapulco* ? quod referunt primum bene sapere, sed mox infernali ardore fauces, & interiora adurendo excruciare, quod si non perimit, saepius mortales accelerat accidentes: Asportatur in naves, ut mures, & glires eo comesto intereant."

This species has long been known to Europeans, and many of the early travelers in the Malayan Archipelago wrote fabulous accounts of the tree and its deadly properties. *Robert Brown*[3] has given an exhaustive historical account of it.

The distribution of true *Antiaris toxicaria* is somewhat doubtful, *Hooker* f., reducing to it *Antiaris innoxia* Blume and some other species, giving its distribution as from the Deccan Peninsula, Pegu to Martaban, Ceylon and the Malay Islands, stating that the north Australian *A. macrophylla* R. Br. may be the same. *Engler* gives the distribution as from Java to the Sunda Islands.

[3] Bennett, Brown and Horsefield: Plantæ Javanicæ Rariores (1838–52) pp. 53–63.

PHILIPPINE MYXOGASTRES.

By GEORGE MASSEE.

(From the Royal Botanic Gardens, Kew, England.)

In the material examined, it is not surprising, but on the contrary somewhat gratifying to announce that no new species have been discovered. The number of species dealt with is yet too small to justify any statement as to the predominance of any given group in the area investigated.

TUBULINA Pers.

Tubulina cylindrica Rost. Monog. (1875) 220; Mass. Monog. Myx. (1892) 39.

LUZON, Manila (*Copeland*) December, 1904, on dead bamboo.
Europe, North America, India, Ceylon and Australia.[1]

STEMONITIS Gled.

Stemonitis atra Mass. Monog. Myx. (1892) 78.

MINDANAO, without locality (*Copeland*) 1904.
New Zealand.

Stemonitis Bauerlinii Mass. Monog. Myx. (1892) 79.

LUZON, Manila (*Copeland*) December, 1904, on dead bamboo; Province of Benguet, Sablan (6216 *Elmer*) April, 1904, on prostrate logs; Province of Bataan, Mount Mariveles (4122, 4123, 4129, 4130 *Merrill*) January, 1904, on prostrate logs.
New Guinea.

LYCOGALA Mich.

Lycogala epidendrum Rost. Monog. (1875) 85; Mass. Monog. Myx. (1892) 121.

MINDANAO, Davao (981 *Copeland*) April, 1904, on prostrate logs.
Temperate and tropical regions of the World.

ARCYRIA Hill.

Arcyria punicea Rost. Monog. (1875) 268; Mass. Monog. Myx. (1892) 142.

LUZON, Province of Bataan, Lamao River, Mount Mariveles (4119, 4128 *Merrill*) January, 1904. MINDANAO, District of Zamboanga, San Ramon (729 *Copeland*) May, 1904, on prostrate logs.
Temperate and tropical regions of the World.

[1] Geographical distribution of species taken from Massee's Monograph of the Myxogastres (1892). (E. D. M.)

Arcyria serpula Mass. Monog. Myx. (1892) 164. *Hemiarcyria serpula* Rost. Monog. (1875) 267.

LUZON, Province of Bataan, Mount Mariveles (4126 *Merrill*) January, 1904, on prostrate logs.

Tropical and temperate regions of the World.

TRICHIA Haller.

Trichia Balfourii Mass. Monog. Myx. (1892) 186.

LUZON, Province of Bataan, Mount Mariveles (4124 *Merrill*) January, 1904, on prostrate logs.

Cape of Good Hope.

DIDYMIUM Schrad.

Didymium farinaceum Schrad.; Mass. Monog. Myx. (1892) 219.

MINDANAO, District of Davao, Todaya (1175 *Copeland*) April, on rotten abacá (*Musa textilis*); trail to Mount Apo (1174 *Copeland*) April, 1904, on rotten wood in forests at 1,300 m. alt.

Europe, United States, Australia and the Bonin Islands.

Didymium macrospermum Rost. Monog. (1875) 166; Mass. Monog. Myx. (1892) 228.

LUZON, Manila (4121 *Merrill*) January, 1904, on dead fruits and bracts of *Pandanus*. MINDANAO, District of Davao, Catalonan (921 *Copeland*) April, 1904, on various objects near rotten log.

Germany.

Didymium clavus Rost. Monog. (1875) 153; Mass. Monog. Myx. (1892) 230.

LUZON, Manila (5179 *Merrill*) October, 1905, on fallen petioles of *Poinciana regia*.

Europe, North America, Egypt and Ceylon.

Didymium Barteri Mass. Monog. Myx. (1892) 231.

LUZON, Province of Laguna, Los Baños (4118 *Merrill*) June, 1905, on aborted fruits of *Phytocrene*.

West Africa.

PHYSARUM Pers.

Physarum cinerum Rost. Monog. (1875) 102; Mass. Monog. Myx. (1892) 298.

LUZON, Province of Bataan, Lamao River (*Copeland*) January, 1904, on base of an orchid on dead tree.

Temperate and tropical regions of the World.

Physarum rubiginosum Rost. Monog. (1875) 104; Mass. Monog. Myx. (1892) 302.

MINDANAO, District of Zamboanga, San Ramon (*Copeland*) November, 1904, on *Musa textilis*.

Sweden, Finland, United States.

Physarum cerebrinum Mass. Monog. Myx. (1892) 306. *fig.* 275.

LUZON, Manila (4116 *Merrill*) July, 1905, on dead *Hibiscus*.

Java.

TILMADOCHE Rost.

Tilmadoche nutans Rost. Monog. (1875) 127; Mass. Monog. Myx. (1892) 327;

MINDANAO, District of Zamboanga, San Ramon (*Copeland*) December, 1904; District of Davao, Davao (854 *Copeland*) April, 1904, on dead wood.
Europe, North Africa, North America, Ceylon, India and Australia.

Tilmadoche oblonga Rost. Monog. App. (1876) 13; Mass. Monog. Myx. (1892) 334.

MINDANAO, District of Zamboanga, San Ramon (*Copeland*) November, 1904, on abacá (*Musa textilis*) waste.
United States.

Tilmadoche gyrocephala Rost. Monog. (1875) 131; Mass. Monog. Myx. (1892) 335. *Didymium gyrocephalum* Mont. in Ann. Sci. Nat. II. **8**: 362.

LUZON, Manila (4117 *Merrill*) July, 1905, on dead *Hibiscus*.
United States and South America.

FULIGO Rost.

Fuligo varians Rost. Monog. (1875) 134; Mass. Monog. Myx. (1892) 340.

LUZON, Province of Cavite, Maragondong (4191 *Merrill*) July, 1905, in forests.
Temperate and tropical regions of the World.

CIBOTIUM BARANETZ J. SM., AND RELATED FORMS.

By H. CHRIST.
(*Basel, Switzerland.*)

In treating the oriental forms of the genus *Cibotium*, Hooker and Baker, Synopsis Filicum, and C. Christensen, Index Filicum 183, have recognized the Sandwich Island species as distinct, reducing the others to one species, *Cibotium Barometz* (typographical error for *Baranetz*, see Kunze, Suppl. Schkuhr. **1**: 63, in note). Kunze, however, separated from this collective species two forms, *C. glaucescens* Kunze, l. c., *Tab. 31*, and *C. Cumingii* Kunze, l. c., 64. As a matter of fact the recognition of several subspecies is well justified, and from an examination of the material in my herbarium I recognize the following:

Cibotium Baranetz (Linn. *Polypodium*, errore calami *Barometz*) J. Sm. in Journ. Bot. **1**: 437.

Pinnulis fertilibus 2 cm. latis, usque ad costam incisis, segmentis lanceolato-falcatis, acutis, *1 cm. longis*, subintegris aut parce serratis, sinubus angustis separatis, rhachibus costisque floccoso-araneosis; soris utroque costulae latere *binis rarius ternis aut quaternis* basi segmenti insertis, minutis, non prominentibus, a costa aliquantulum remotis, subobliquis sed margini segmenti subparallelis, 1 mm. latis; valvis inaequalibus (i. e., externa majore, cupulam semiglobosam formante, interiore angustiore, rotundato-ovali) tenuiter coriaceis, flavo-brunneis nec pruinosis.

Hab, Assam, Makum forest, (*D. Brandis* 1397). Tonkin français, (*P. Bon* 3971). Annam, Prov. Quang Binh, (*P. Cadière* 22). Bujong, Malacca, (*Ridley* 9532). Hongkong, (*Faber*); Happy Valley, (*Warburg*). China, Futschau, (*Warburg*); Tunkin, (*Faber* 1904); Yunnan, Mengtze, (*Henry* 9418); Szemao, (*Henry* 11739). Formosa, (*Faurie* 687, 677); *Warburg;* Tamsui, (*Henry* 1385). Liu Kiu, (*Warburg.*)

The typical form, widely distributed in tropical Asia, on the one hand from North East India through South China to Formosa and Liu Kiu, and on the other hand to Malacca, Tonkin and Annam, usually acaulescent, but found by Henry in Yunnan with stems 8 feet high.

Cibotium Assamicum Hook. Sp. fil. **1**: 83. *Tab. 12 B.*

Pinnulis usque ad 23 mm. latis, usque ad costam incisis, segmentis fertilibus lineari-lanceolatis, acutis, falcatis, *12 mm. longis*, 2.5 mm. latis, manifeste dentatis, sinubus angustis separatis, rhachi subglabra, costis parce araneoso-floccosis: soris in apice dentium *prominentibus* in frondibus plene evolitis utroque costulae latere *quinis* senisve, ultra segmenti dimidiam partem occupantibus, ejusque apice solummodo soris exempto, vix 1 mm. latis, tenuiter coriaceis, obscure brunneis non pruinosis: valvis *inaequalibus* exteriore semigloboso, interiore angustiore, ovato.

Habitu Cyatheae magnae soris marginalibus praeditae.

Hab. Makum, Luckinpore (*C. B. Clarke* 37812). Tonkin français. (*P. Bon* 5408.)

The most distinct form.

Cibotium Sumatranum n. subsp.

Pinnulis fertilibus *minoribus*, 8 mm. latis, non usque ad costam incisis, ala 1 mm. lata utrinque superstite (sinubus obtusis *separatis*) 3 mm. longis, ligulato-obtusiusculis, denticulatis; soris utroque costulae latere *binis rarius ternis,* basi segmenti insertis, a costa remotis, *minutis* 0.5 ad 0.7 mm. latis; valvis *subaequalibus*, tenuiter coriaceis, rufo-fulvis, haud prominentibus, margini segmenti subparallelis, non pruinatis.

Hab. Sumatra, Linggalang Volcano. (*Henry Rouyer* 1905), Herb. Turic.

More delicate in all parts, segments small, blunt, sinus rounded, the plant glabrous or nearly so.

Cibotium Cumingii Kunze Suppl. Schkuhr. 1: 64.

Pinnulis 1.3 ad 1.5 cm. latis, fere ad costam incisis, segmentis confertis, sese tangentibus (sinubus paene nullis aut acutis) *ovato-trigonis, acutis,* 0.5 ad 0.7 cm. longis, decumbentibus, minute subserrulatis, costis costulisque dense rufo- aut albido-*pubescentibus;* soro utroque costulae latere solitario, costae adpresso, magno, 1.5 usque ad 2 mm. lato non prominente, transverse ovato, *valde obliquo* usque ad verticali quoad marginem segmenti et costulam; valvis subaequalibus, rigide coriaceis, dilute brunneis *saepeque coeruleo-pruinatis.*

Hab. Philippines, Luzon, Mount Alagut 1,900 m., (*Loher*), Province of Benguet, Baguio, (*Elmer* 6027, 6503, 6559).

Hooker[1] incorrectly unites *Cibotium Cumingii* Kze., with *C. glaucescens* Kze., l. c. 63. *Tab. 31,* as Kunze has well indicated the differences. *C. glaucescens* was based on a cultivated, and as the illustration shows, an immature specimen, of unknown origin, the sori scarcely showing, and is almost without doubt identical with *C. Baranetz*, in young stage of development, and should be reduced to that species.

[1] Sp. Fil. 1: 83.

PTERIDOPHYTA HALCONENSES: A LIST OF THE FERNS AND FERN-ALLIES COLLECTED BY ELMER D. MERRILL ON MOUNT HALCON, MINDORO.

By EDWIN BINGHAM COPELAND.
(*From the Bureau of Education, Manila.*)

Mount Halcon is probably third in height among Philippine mountains, being nearly 2,700 meters in altitude and, so far as known, surpassed only by Mount Apo and Mount Malindang, both in Mindanao. The trip on which the ferns enumerated below were collected was undoubtedly the first conquest, by white men, of the highest peak. It was undertaken by the order of and with the support of Major-General *Leonard Wood,* by Major *Edgar A. Mearns,* Surgeon, United States Army, accompanied by *Elmer D. Merrill,* botanist of the Bureau of Science, and *W. I. Hutchinson,* of the Forestry Bureau. All the botanical material was collected by Mr. *Merrill,* 759 numbers being secured, representing about 700 different species, of which 206 species and varieties were ferns and fern-allies.

Previous collections have been made since the American occupation of the Philippines by *Merrill* in 1903 and 1905 on the Baco River at the north base of Halcon, and by *R. C. McGregor* at the same place in 1905. In June, 1906, Mr. *M. L. Merritt* of the Forestry Bureau ascended the mountain to an altitude of about 2,100 meters, making a small but very interesting botanical collection. In 1895 *John Whitehead,* an English naturalist, made a small botanical collection on Mount Dulangan, a spur of Halcon reaching an altitude of 1,800 meters, and between the years 1836 and 1840 *Hugh Cuming,* also an Englishman, collected in Mindoro, undoubtedly on the Baco River.

The party on this trip left the coast at Subaan, about 10 miles north of Calapan, and followed a general southerly direction. Crossing a broad, interrupted, forested ridge of an altitude of about 300 meters, the Binabay River, a tributary of the Alag, was reached, and after traversing a narrower and slightly higher ridge, the Alag itself, a tributary of the Baco, was crossed at an altitude of about 70 m. This river was ascended to an altitude of about 300 meters, and again crossed above at an altitude of 400 meters after which, by a succession of more or less

interrupted climbs, camps were made at altitudes of 1,000, 1,100, 1,600, 1,800 and 2,400 meters.

From Mr. *Merrill's* reports, from the size of the rivers flowing from the mountain, and from the character of the vegetation of the lower and middle altitudes, it is safe to state that Halcon is a very wet mountain, more so than any other of approximately equal altitude in the Philippines. The party was out forty days, in this time there being more or less rain nearly every day, and for one period of thirteen days while they were at the higher altitudes they encountered constant heavy rain, day and night, and in this time never saw the sun. The moisture conditions of the rain-forest—that is, a high humidity essentially uninterrupted by any dry season—are thus brought down in sheltered or level places to approximately sea-level, and "high-forest" conditions are found nowhere except to a limited extent on the lower ridges, while the savannah-wood seems to be entirely wanting. By virtue of its lesser altitude, this lowland rain-forest undoubtedly enjoys an appreciably and constantly higher temperature than does the rain forest on Apo; for instance at an altitude of 1,200 meters, or at San Ramon at an altitude of 700 meters; still it has many species in common with both the above localities, and its bionomic character, so far as can be judged without a personal visit and a study of the fresh plants, is the same. As I have always found to be true elsewhere in the world, the temperature plays a most inconspicuous rôle as compared with the moisture in determining the bionomic character and the local distribution of plants, and furnishes no fit basis for the general classification of faunas or floras.

Halcon as compared with San Ramon, the locality where the fern bionony has been most studied, is notable for the absence of the savannah-wood and for the weak development of the high-forest as noted above; for the much greater development of the mossy forest, and for the presence above this of a montane brush. Here, as on Mount Apo (it is wanting on Mount Malindang), this brush degenerates in places to a mere heath the exceedingly limited vascular flora of which is a curious mixture of Australian and north-temperate pioneers. The Pteridophyte flora of these two heaths, 450 miles apart, and so far as known isolated by that distance, is almost identical.

Merrill's Halcon plants, as must be the case with so rich a collection, offer a very interesting contribution to our knowledge of the distribution of Philippine ferns. In view of the contiguity of Luzon and Mindoro it is but natural that a number of ferns, hitherto known only from the mountains of central Luzon, Maquiling, Mariveles, Banajao, etc., or at any rate not south of these mountains, should be found here. More considerable southward extensions of range are those of *Plagiogyria tuberculata* and *Lycopodium complanatum* var. *thyoides*. Halcon has also some Malayan plants hitherto known in the Philippines only from northern Luzon, such as *Saccoloma, Polypodium decrescens* and *Lycopodium casuarinoides;*

with these should be mentioned *Lomaria Fraseri* of New Zealand and central and northern Luzon. However, the range extensions are mostly in the direction that I have already shown to be the general one of migration[1]; that is, towards the north. Shorter steps of this kind are those of *Diplazium Palauanense* and *Gleichenia dicarpa,* known previously from Palawan, *Monogramme dareæcarpa* and *Polypodium tenuisectum* from Negros, and *P. mollicomum* and *P. Yoderi* from Panay. A considerable number of ferns, *Monachosorum, Oleandra Whitmeei, Diplaziopsis, Blechnun vestitum, Plagiogyria Christii, Acrosorus* (represented by a new species), *Polypodium Celebicum* and *Dicksonia chrisotricha* are here found for the first time north of Mindanao. With these notable internal extensions of range, it is rather striking, even when we remember that Halcon is near the center of the Archipelago, to find in this collection only eight additions to our flora, of plants already known elsewhere. Of these eight plants, two are cosmopolitan, and all are Malayan. Another factor in the present state of our knowledge is that here, as in Mindanao, the Celebes element seems to dominate over the Bornean; yet Mindoro is connected with Borneo through the Calamianes Islands, Palawan and Balabac, with only insignificant gaps, and Mr. *Merrill* believes that Halcon geologically is similar to Mount Kinabalu in north Borneo, it being a mass granite, marble, white quartz and with some schist-like rock. However this may be, the only Halcon fern in the collection which is apparently of Bornean origin is *Ophioglossum intermedium.*

While we are receiving the congratulations of our European friends and sometimes congratulating ourselves on the present progress in local pteridology, we must not forget that the picture we are painting is still almost wholly canvas. We know perhaps most of the ferns growing near one trail up Apo, one trail up Halcon, and we know something of the fern flora of Santo Tomas and Data in northern Luzon, and of the high plateau country between the two latter mountains. Of Malindang in Mindanao, Canlaon in Negros, Madiaas in Panay, and Banajao in central Luzon, all near 2,400 meters in altitude or higher, we know essentially nothing, having some half-dozen ferns from each. The highest mountain in northern Luzon has never been climbed by a botanist or a collector, nor has the second, Bulusan, neither have Malaya, nor Solis, nor any mountain of the high and extensive eastern *Cordillera.* The same is true of the high peaks of southern Luzon, of all mountains of Samar, of the high eastern range of Mindanao, of Roosevelt, Apo's great neighbor, and of the highest peak in Palawan. Yet it is on these great mountains, not on their summits, but on their large areas at intermediate elevations, that our fern vegetation reaches its most luxuriant development. Exploration has obviously not reached the point where statements as to inter-island range have more than a temporary interest.

[1] *This Journal* Botany (1907) 2: 3-10.

Dr. *H. Christ* is making a special study of our collections of *Dryopteris* (*Nephrodium*), and most of our *Selaginella* material is in the hands of Professor *Hieronymus*; therefore I have not undertaken to determine *Merrill's* material in either of the above genera. Of *Dryopteris* about 10 species are represented, and of *Selaginella*, about 6 species. Of the remaining Pteridophytes 20 species and varieties are described as new, and among these are some which emphasize the position of the Philippines as the richest region in the World in *Polypodium* and in several related genera such as *Prosaptia*, *Acrosorus* and *Plagiogyria*.

POLYPODIACEÆ.

ACROPHORUS Presl.

Acrophorus stipellatus (Wall.) Moore.

On ridges in the mossy forest at 1,300 m. alt. (No. 5962).

Already collected in the Philippines on Mount Apo, Mindanao, and on Mount Data, Luzon.

India, Malaya.

MONACHOSORUM Kunze.

Monachosorum subdigitatum (Bl.) Kuhn.

On ridges in the mossy forest at 1,300 m. alt. (No. 6164).

Previously known in the Philippines only from Mount Apo, Mindanao.

ASPIDIUM Swartz.

Aspidium leuzeanum (Gaudich.) Kunze.

In a damp ravine at 150 m. alt. (No. 5876).

Luzon and southward, variable, not common.

India to Polynesia.

Aspidium Barberi (Hook.) Copel.

Terrestrial in forests at 300 m. alt. (No. 5875).

Luzon, Majaijai (*Loher*).

Malaya.

A probable relative of *A. polymorphum*, with the form of *A. pachyphyllum* and unlike any other *Aspidium*, known to me, in the very long free excurrent included veinlets.

Aspidium polymorphum Wall.

On shaded cliffs at 150 m. alt. (No. 5874).

Throughout the Philippines and worthy of its name.

India to Malaya.

Aspidium macrodon (Reinw.) Keys.

On boulders in humid forests at 150 m. alt. (No. 5873).

Luzon, Mindoro.

Malaya, Fiji.

Aspidium Copelandii C. Chr. (*A. heterodon* Copel.).

Terrestrial in ridge forests at 700 m. alt. (No. 5872).

Very doubtfully distinct from *Aspidium decurrens* Presl, and the relative ranges of the two species uncertain, because many of the specimens in the herbarium are without stems.

Aspidium sp. near *A. decurrens* Presl.

In forests at 100 m. alt. (No. 5871).

A very tall fern with the lower pairs of pinnæ not quite connected by a wing and with but a very narrow wing decurrent a third of the way down the stipe.

Aspidium cicutarium (Linn.) Sw.

In the humid forests at 200 m. alt. (No. 5870), a very chaffy form which I have not seen before. On cliffs in dense forests at 1,800 m. alt. (No. 5869), a small flaccid form with rounded lobes, already known from Bontoc, Luzon. Other forms are not common in the Archipelago.

Pantropic.

POLYSTICHUM Roth.

Polystichum horizontale Presl (*P. aculeatum* var. *batjanense* Christ; *Aspidium batjanense* Christ.)

Epiphytic and terrestrial in the ridge forests at 800 m. alt. (No. 5879).

In the rain forest and submossy forest from Luzon south to Batjan.

Specimens of *Polystichum aculeatum* var. *batjanense* determined by *Christ*, are identical with the *P. horizontale* collected by *Cuming;* they do not agree well with *Blume's* description of *Aspidium moluccanum*, but I have no specimens of the latter species.

Polystichum amabile (Bl.) J. Sm.

On cliffs in forests at 2,000 m. alt. (No. 5878).

Throughout the Philippines at lesser altitudes.

Malaya to India and Japan.

Polystichum carvifolium (Kze.) C. Chr. (*P. coniifolium* Presl).

In ridge forests at 700 m. alt. (No. 5877).

Common throughout the Philippines in the upper high forest.

Natal to Hawaii.

CYCLOPELTIS J. Smith.

Cyclopeltis presliana (J. Sm.) Berk.

On cliffs in forests at 300 m. alt. (No. 5880), a form with small obtuse pinnæ.

Common in the Philippines; (perhaps *Aspidium Kingii* Hance).

Burmah to New Guinea.

LEPTOCHILUS Kaulfuss.

Leptochilus heteroclitus (Presl) C. Chr.

On boulders in humid forests at 100 m. alt. (No. 5883), pinnæ ovate and caudate, otherwise typical.

Common in the Philippines and exceedingly variable.

India to Melanesia.

Leptochilus Zollingeri (Kze.) Fée.

In a deep ravine at 300 m. alt. (No. 5884), specimens all simple.

Luzon, Mindanao.

Malaya.

DIPTERIS Reinw.

Dipteris conjugata (Kaulf.) Reinw.

In thickets at 2,000 m. alt. (No. 5882).

Common at higher altitudes throughout the Philippines.

India to Polynesia.

There is also in the collection an alpine form, collected on an open heath at 2,400 m. alt., dwarfed and distinct in appearance, but sterile (No. 5881), doubtless the var. *alpina* Christ in Bull. Herb. Boiss. II. 6 (1906) 991.

NEPHROLEPIS Schott.

Nephrolepis hirsutula Presl.

Abundant along the Alag River at 100 m. alt. (No. 5947).

Very common throughout the Philippines.

Pantropic.

Nephrolepis rufescens Presl.

In an old clearing at 700 m. alt. (No. 5948).

A species of doubtful status, not uncommon throughout the Philippines.

Nephrolepis barbata Copel.

Epiphytic in the ridge forests at 850 m. alt. (No. 5946).

Hitherto known only from a similar situation on Mount Apo, Mindanao, and from central Luzon, (*Loher*).

Nephrolepis sp. near *N. exaltata* (L.) Schott.

Epiphytic and on rocks at 1,000 m. alt. (No. 5945).

A plant unlike any other known to me, with dark shining stipe and rhachis; frond pale beneath but dark above, with many lime dots, the pinnæ glabrescent, serrulate and barbate; indusium attached by a deep sinus, submarginal; frond about 80 cm. long, one-fifth as wide.

OLEANDRA Cav.

Oleandra colubrina (Blanco) Copel.

Semiscandent, forming dense thickets in ridge forests at 1,100 m. alt. (No. 5937).

Intermediate between the typical form known from Luzon and Negros and the variety *nitida* of Mindanao.

Oleandra Whitmeei Bak.

On mossy cliffs in ridge forests at 2,100 m. alt. (No. 5936).

Hitherto unknown north of Mount Apo, Mindanao.

Celebes, Samoa.

LEUCOSTEGIA Presl.

Leucostegia immersa Presl, var. **amplissima** Christ.

On trees in the mossy forest at 1,800 m. alt., scandent, 3 m. high (No. 5939).

Luzon, Mindanao and Celebes.

The species in India and Malaya.

Leucostegia hymenophylloides (Bl.) Bedd.

Terrestrial in forests at 900 m., 1,000 m., and 1,500 m. (No. 5938).

All previous Philippine collections are from Luzon.

India to Polynesia.

HUMATA Cav.

Humata lepida (Presl) Moore.

On rocks at 150 m. alt., along the river (No. 5940), and epiphytic in the ridge forest at 1,400 m. alt. (No. 5941).

Previously known from Samar and Luzon, and also collected on Mount Halcon by *Merritt*, June, 1906, at 1,700 m. alt. (No. 4465).

DAVALLIA Smith.

Davallia divaricata Blume.

Epiphytic in humid forests at 100 m. alt. (No. 5943).

Previously collected in the Philippines in Luzon and in Mindanao.

India to Malaya.

Davallia embolostegia Copel.

Epiphytic in forests at 900 m. alt. (No. 5942).

Previously known only from Luzon.

TAPEINIDIUM (Presl) C. Chr.

Tapeinidium pinnatum (Cav.) C. Chr. (*Microlepia pinnata* J. Sm.).

Terrestrial in ridge forests at 700 m. alt. (No. 5944), both the typical pinnate form and the doubtless more primitive tripinnate form.

Common throughout the Philippines.

Malaya, Polynesia.

SACCOLOMA Kaulf.

Saccoloma moluccanum (Bl.) Mett.

Humid forests at 200 m. alt. (No. 5951) a very large form, the stipes nearly 1 m. tall.

Luzon, Benguet.

Malaya to Polynesia.

DENNSTAEDTIA Bernh.

Dennstaedtia scandens (Bl.) Moore.

Scandent in thickets and forests, sometimes 5 m. high, 570 to 1,000 m. alt. (No. 5953).

Previously collected in the Philippines in Luzon and Mindanao.

Malaya to Samoa.

The material is without any trace of an extrorse indusium and in my opinion would better be treated as both *Hooker* and *Smith* treated *Cuming's* Philippine plant[2] as *Hypolepis*, if *Hypolepis* were itself a natural genus. Very similar species in *Hypolepis* are already known from South Africa and Tropical America. In the latter region, as here, the separation is at present an arbitrary one. If the hypolepoid relatives of *Dennstaedtia* are not included in *Dennstaedtia*, they should receive a new generic name, since the original *Hypolepis*, *H. tenuifolia*, seems more nearly related to *Dryopteris*.

Dennstaedtia scabra (Wall.) Moore.

On ridges in the mossy forest at 2,100 m. alt. (No. 5950).

Already known from the Philippines from similar altitudes in northern Luzon, and on Mount Apo, Mindanao.

India to Japan and Celebes.

[2] Sp. Fil. 1: 66.

Dennstaedtia Merrilli Copel. n. sp.

Rhizomate repente, ca. 1.5 cm. crasso; stipite erecto, valido, deorsum 1 cm. crasso, 60 cm. alto, rhachique badiis, sub lente puberulis, oculo nudo glabrescentibus; fronde 100–120 cm. alta, ca. 50 cm. lata, quadripinnatifida, pinnis inferioribus diminutis, maximis 30 cm. longis, 11 cm. latis, acuminatis, rhachibus infra puberulis; pinnulis deltoideo-lanceolatis, valde acuminatis: pinnulis" deltoideo-lanceolatis, obtusis decurrentibus, substipitatis, infima acroscopica maxima usque 3 cm. longa, 1 cm. lata in segmenta subremota, utroque latere ca. 4, oblonga, obtusa vel truncata integra vel obscure serrata ad costam alatam, pinnatifida, glabris, herbaceis, infra pallidis; soris in sinubus marginalibus, immaturis interdum submarginalibus, fere 1 mm. latis, margine frondis saepe inflexa, cum indusio vero calicem imperfectam efficiente: receptaculo valido; annulo interrupto.

In silvas, 1,800 m. s. m. (No. 5949).

This plant might, with equal reason, be assigned to *Microlepia*, or *Dennstaedtia;* I should have preferred to have called it *Microlepia*, since that genus is at best undefinable, but its generic separation from *Dennstaedtia flaccida* would not be justifiable on any ground. Its probable relatives are *Dennstaedtias* of the *D. flaccida* group, *Microlepia hirta* and the so-called *M. dennstaedtioides* Copel.[3] with regard to which I am now prepared to accept *Christ's* judgment and call it *Dennstaedtia*, D. DENNSTAEDTIOIDES.

I have come to the conclusion that *Microlepia* will have to be abandoned, as a genus, and regarded in the future as the primitive section of *Dennstaedtia*.

ODONTOSORIA (Presl) Fée.

Odontosoria retusa (Cav.) J. Sm.

In old clearings at 700 m. alt. (No. 5935).

Known in the Philippines from Luzon and Mindanao.

Malaya to New Guinea.

Odontosoria chinensis (Linn.) J. Sm.

In forests at 900 m. alt. (No. 5934).

Common along small streams throughout the Philippines.

Madagascar to Japan and Polynesia.

LINDSAYA Dryand.

Lindsaya pectinata Bl.

Scandent, in humid forests at 200 m. alt. (No. 5933).

The most northern collections of this species are from Luzon.

India to Malaya.

Lindsaya hymenophylloides Bl.

Epiphytic in forests at 900 m. and at 1,800 m. alt. (Nos. 5931, 5932), remarkably luxuriant specimens.

Not uncommon from central Luzon southward.

Malaya and Polynesia.

[3] *This Journal* (1906) 1. *Suppl.* 146. *Tab.* 4.

Lindsaya gracilis Bl.

Terrestrial in humid forests at 200 m. alt. (No. 5928).

Throughout the Philippines, common.

Malaya.

Lindsaya pulchella (J. Sm.) Mett.

Epiphytic on *Alsophila* at 850 m. alt. (No. 5927).

Throughout the Philippines on tree-fern trunks.

Malaya to Samoa.

Lindsaya cultrata (Willd.) Sw., var. **minor** Hook. Sp. Fil. 1: 204.

On boulders along the Alag River at 375 m. alt. (No. 5930).

India to Madagascar and Australia.

This little plant, collected in Luzon by *Cuming*, and in Negros by *Whitford*, seems quite fixed in aspect, but the variety *varia*, common in Benguet, completely bridges the gap between it and the typical form.

Lindsaya davallioides Bl.

Terrestrial in ridge forests at 700 m. alt. (Nos. 5925, 5926), the specimens all bipinnate, also in ridge forests at 1,800 m. alt. (No. 5929) the specimens all simply pinnate; but *Merritt* collected bipinnate plants on Mount Halcon at 1,700 m. alt. in June, 1906 (No. 4463 *Merritt*).

Other Philippine collections are from Luzon and Palawan.

Malaya.

Lindsaya orbiculata (Lam.) Mett. ?

Terrestrial in forests at 200 and 700 m. alt. (Nos. 6010, 5924).

This plant is *Lindsaya javensis* Bl., a large form of which is *L. flabellata* var. *gigantea* Hook. I am doubtful of the propriety of regarding this as conspecific with the small fern already known from the Philippines as *L. orbiculata*.

ATHYRIUM Roth.

Athyrium toppingianum Copel. (*Asplenium*, Copel. in Perk. Frag. Fl. Philip. (1905) 184).

On damp ledges along streams at 375 m. alt. (No. 5921).

Common in similar situations throughout the Philippines. No. 1218c *Wichura*, distributed as a variety of *Asplenium Lasiopteris* Mett., is probably this species.

DIPLAZIUM Sw.

Diplazium japonicum (Thunb.) Bedd.

On damp ledges along streams at 375 m. alt. (No. 5920), a small plant, differing from *D. Oldhami* Christ in being less glabrous.

This plant is unquestionably congeneric with *Athyrium toppingianum* which is an unmistakable *Athyrium*. *Diplazium* in this part of the World has had a plural origin in *Athyrium*, and the naturalness of our classification would be promoted by ceasing to treat it as a distinct genus.

D. japonicum is supposed to occur from Japan to Mindanao and India.

Diplazium brachysoroides Copel. n. sp. (Pl. I, Fig. A.)

Stipite 20–40 cm. alto, 4 mm. crasso, basin versus nigro, supra sordide brunneo, rhachique paleis angustis horizontalibus 4–8 mm. longis sursum minoribus flexuosis sparsis vestitis; fronde 50–60 cm. alta, ca. 30 cm. lata, acuminata vix bipinnata apice pinnatifida; pinnis maximis medialibus, horizontalibus, acuminatis, nigro-pedicellatis, rhachin versus truncatis; segmentis oblongis, subacutis, maximis 3 cm. longis, 1 cm. latis, apices versus argute serratis, ala angusta connexis, rarius paucis ala carentibus (i. e., pinnulis), papyraceis, supra atro-viridibus, infra prasino-olivaceis, glabris; pinnis infimis paullo remotis diminutisque interdum deflexis, segmentis rhachis-apicis diminutis; costis venisque supra bilineato-praestantibus, venulis furcatis; soris infimis saepius diplazioideis, longioribus, aliis oblongis vix 2 mm. longis, venam tangentibus, vix ½ ad marginem attingentibus, indusiis persistentibus.

In silvis humidis, 200 m. s. m. (No. 5919).

In spite of the apparently total absence of athyrioid sori, I regard *Athyrium silvaticum* (Bl.) Milde (*Brachysorus*, Presl) as the nearest known relative of this fern. In *Diplazium* its closest affinity is probably to *D. sorsogonense* Presl.

Diplazium oligosorum Copel. n. sp.

Rhizomate erecto, 1 cm. crasso; stipitibus confertis, 40 cm. altis, 3 mm. crassis, basibus nigris incrassatis valde aculeatis paleis castaneis 1.5 mm. latis paucis vestitis, sursum rhachibusque levibus, glabris, et nisi ad pedes nigros pinnarum brunneo-stramineis; fronde 50–60 cm. alta, 40–50 cm. lata, acuminata, tripinnatifida; pinnis oppositis, inferioribus ca. 3-jugatis satis æqualibus, ca. 10 cm. inter se distantibus, stipitatis, horizontalibus, acuminatis; pinnulis subsessilibus, basibus truncatis, 2 cm. latis, 7 cm. longis, acuminato-caudatis, fere ad costam in segmentis oblongis 4–5 mm. latis obtusis subfalcatis interdum serrulatis pinnatisectis, glabris, tenue coriaceis, supra atro-viridibus, infra pallidis; venulis remotis, saepius simplicibus; soris margini quam costae propioribus, vix 1 mm. longis.

In silva muscosa ad terram, 1,800 m. s. m. (No. 5913).

A most distinct species, characterized by the swollen spiny bases, large and few opposite pinnæ, very acuminate pinnules and minute sori.

Diplazium Merrilli Copel. n. sp.

Rhizomate repente, 2 mm. crasso; stipitibus confertis, 2–5 cm. altis, paleis angustis horizontalibus 2–3 mm. longis nigro-brunneis dense vestitis; fronde 15–25 cm. alta, utrinque angustata, acuminata, profunde pinnatifida, sinubus angustis, vel segmentis saepius imbricatis; costa supra sparsissime infra densius paleacea; segmentis oblongis, fere horizontalibus, medialibus maximis, 1 cm. latis, apice rotunda interdum denticulata, herbaceis, venis infra minute paleaceis, aliter glabris, supra atrovirentibus, infra olivaceis; venulis furcatis; soris linearibus, longioribus interdum ad marginem nec usquam ad costam attingentibus.

Ad saxa in silva, 900 m. s. m. (No. 5914).

Near *D. zeylanicum* Moore, and *D. porphyrorachis* Diels, well characterized by the clustered fronds, short and very chaffy stipes, and imbricate segments, not separate even at the base of the frond.

Diplazium palauanense Copel.

In humid forests at 200 m. alt. (No. 5915).

Hitherto known from Palawan only.

One large frond has 16 pairs of pinnæ, the largest nearly 40 cm. long but only 3 cm. broad. A few "sori" are scolopendroid and some sori reach the margin; in spite of this modification of its character, however, the species is very distinct from *D. bantamense.*

Diplazium Woodii Copel. n. sp.

Rhizomate erecto, 1 cm. crasso; stipitibus confertis, 20–30 cm. altis, 2 mm. crassis, rectis, basibus brunneo-nigris, dense brunneo-falcato-paleaceis, sursum rhachibusque stramineo-brunneis, paleis paucis 2–3 mm. longis castaneis angustis distantibus vel deflexis sparsis vestitis; fronde 40–50 cm. alta, vix 20 cm. lata, acuminata, bipinnatifida; pinnis utroque latere 12–17, medialibus maximis horizontalibus brevi-pedicellatis 2 cm. latis, acuminatis, rhachin versus truncatis, profunde pinnatifidis, glabris, herbaceis, supra atro-viridibus infra olivaceis, costa castanea; segmentis oblongis, 5 mm. latis, subfalcatis, oblique truncatis serrulatis, ala 1.5 mm. lata connexis sinubus ca. 1 mm. latis interpositis, venulis simplicibus, 5–6-jugatis; soris 2–3 mm. longis, obliquis, rarissime diplazoideis costa remotis nec ad marginem attingentibus.

In silvis, 300 m. s. m. (No. 5917).

This would seem to be very near to *Blume's D. acuminatum,* so far as the original description of that species shows, but differs from *Christ's* undoubtedly correct interpretation of that species[4] in the paleaceous stipe, deeply cut pinnæ and in some minor details: I would not call it in the *Japonicum* group.

It is dedicated to Major-General *Leonard Wood.*

Diplazium asperum Bl.

In forests at 300 m. alt. (No. 5918). Fronds 2 to 3 m. high on stipes 1 m. high. The same fern has been found near the base of Mount Halcon by *McGregor* (No. 277).

India, Java.

ANISOGONIUM Presl.

I am by no means convinced that this genus will stand, but it is at least as distinct from *Diplazium* as the latter is from *Athyrium,* and as all our species already have names in *Anisogonium,* its use here does not encumber nomenclature.

Anisogonium cordifolium (Bl.) Bedd.

On ridges on the mossy forest at 1,400 m. alt. (No. 5909), fronds all simple; on forested ridges at 850 m. alt. (No. 5910), fronds all pinnate and the venation approaching that of *A. alismaefolium.*

Luzon and southward.

Malaya.

Anisogonium alismaefolium (J. Sm.) Christ.

In humid forests at 200 m. alt. (No. 5911), fronds all simple, upper surface with beautiful metallic luster.

Luzon, Mindanao.

Celebes.

[4] Ann. Jard. Bot. Buitenz. (1895) **12**: 220.

Anisogonium elegans Presl.

In humid forests at 200 m. alt. (No. 5912). We have the same fern from Mindanao, Surigao, leg. *Bolster*.

Unless my specimens are erroneously determined, *Christensen* has combined more than one species under *Diplazium fraxinifolium* Presl.

DIPLAZIOPSIS Christensen.

Diplaziopsis javanica (Bl.) C. Chr.

In damp ravines at 1,400 m. alt. (No. 5923).

Hitherto known in the Philippines only from Mount Apo, Mindanao.

India to Polynesia.

BLECHNUM Linn.

Blechnum orientale Linn.

In an old clearing at 700 m. alt. (No. 5908).

Throughout the Philippines.

India to Australia and Polynesia.

Blechnum egregium Copel.

In the ridge forests at 450 m. alt. (No. 5907), variable in the reduction of the fertile pinnæ.

Known also from Mount Maquiling and Mount Banajao, Luzon, and from Mindanao.

Blechnum Patersoni (Spreng.) Mett.

On boulders and cliffs in dense forests at 1,800 m. alt. (No. 5905).

At similar altitudes throughout the Archipelago.

Luzon to Tasmania.

Blechnum vestitum (Bl.) Kuhn.

On exposed ridge in forests at 1,300 m. alt., in similar situations at 1,800 and 2,100 m. alt., and in the mossy brush at 2,500 m. alt. (Nos. 5903, 5904). A more rigid, dwarfed form in the open heath at 2,400 m. alt. (No. 5902). A much more rigid form grows at an altitude of 2,400 m., on Mount Apo, Mindanao.

Malaya.

Blechnum Fraseri (Cunn.) Luerss., var. **philippinensis** Christ.

In the mossy ridge forest at 2,100 m., and at 2,550 m. alt. (No. 5906). The "trunk" at most 1 m. high.

Already known from similar altitudes in Luzon.

The species in New Zealand.

ASPLENIUM Linn.

Asplenium (Thamnopteris) cymbifolium Christ in Bull. Herb. Boiss. II. 6 (1906) 999.

Species nidiformis, frondibus sessilibus, 100–120 cm. longis, 15–20 cm. latis, abrupte brevissime caudatis, sub mediam frondem paullo angustata, baseos versus aequalibus vel paullo dilatatis 12–15 cm. latis, brunneis chartaceis humiferis, aliter subcoriaceis, glabris; costa supra deorsum concava sursum subtusque rotundata nec usquam acuta, nigra;

venis in lineam signatam intramarginalem anastomosantibus, inter se 1.4 mm. distantibus; soris a costa $\frac{9}{10}$ ad marginem attingentibus, inaequalibus, venulis 1–2 inter soros sterilibus, indusiis brunneis, 1 mm. latis.

In silva muscosa epiphyticum, 1,800 m. s. m. (No. 5901).

Type from Mount Maquiling, Luzon, leg. *Loher*.

This fern combines the fructification of *A. Phyllitidis* and the stature of *A. musaefolium*. My No. 1745 from Mindanao, doubtfully identified as *A. Phyllitidis*, is rather this species. More typical *A. Phyllitidis* occurs in Negros and Samar, but even this is by no means identical with the Indian plant.

Asplenium (Thamnopteris) colubrinum Christ in Bull. Herb. Boiss. II. 6 (1906) 999, var. **taeniophyllum** Copel. n. var.

Rhizomate 1 cm. crasso in substrato radicum foliorumque mortuorum immerso; stipitibus confertissimis 2–5 cm. longis; fronde ca. 80 cm. longa, 3 cm. lata, acuminata, deorsum per alam longam siccam ad stipitem sensim attenuata, integerrima, coriacea, glabra, costa supra biangulata deorsumque inter angulas concava, subtus valde carinata, fragile, pleurumque nigra; venulis inconspicuiis, inter se 1 mm. distantibus, intra marginem anastomosantibus; soris a costa vix mediam laminam transeuentibus, plerumque venulis 2 inter soros sterilibus, indusiis ca. 0.3 mm. latis.

In silvis epiphyticum, 700–900 m. s. m. (No. 5899).

The type collected by *Loher* in central Luzon.

Extast autem varietas hujus speciei frondibus interdum 4 cm. latis, soris $\frac{1}{2}$–$\frac{3}{5}$ ad marginem attingentibus, indusiis nigris, 0.6 mm. latis.

In silvis eisdem, 1,800 m. s. m. (No. 5900).

Conspicuous in the *Thamnopteris* group, and resembling a huge *Vittaria*, by the narrowness of the fronds, which are so crowded that with the narrowest possible bases they form dense "nests."

ASPLENIUM SCOLOPENDRIOIDES J. Sm., has been previously collected near the foot of Mount Halcon by *Merrill* (Nos. 1810, 4075) and by *McGregor* (No. 239).

Asplenium normale Don.

On ridges at 1,000 m. alt. (No. 5898), and in the mossy forest at 2,300 m. alt. (No. 5897).

Known from near the latter altitude on Mount Apo, Mindanao, and in northern Luzon.

India to Hawaii.

Asplenium heterocarpum Wall.

On cliffs in dense forests at 1,700 m. alt. (No. 5896).

Not previously reported from the Philippines.

India to Borneo and Formosa.

Asplenium unilaterale Lam.

On wet rocks at 1,400 m. alt. (No. 5895).

Already known in the Philippines from Luzon and Mindanao.

Africa to Hawaii.

Asplenium tenerum Forst.

In humid forests at 140 m. alt. (No. 5894).
Common in the Philippines from northern Luzon southward.
India to Polynesia.

Asplenium persicifolium J. Sm.

Epiphytic in humid forests at 140 m. alt. (No. 5893).
Previously accredited to Luzon and Bohol; Mount Banajao, (*Elmer* 7968).

Asplenium Lepturus J. Sm., Presl.

Epiphytic at 900 m. alt. (No. 5890), a small form; and in the ridge forests at 1,700 m. alt. (No. 5891), very large. Both numbers have more copious sori than are common in the *contiguum* group.
Already collected in Luzon and Palawan.

Asplenium laxivenum Copel. n. sp. (*A. contiguum* var. *bipinnatifidum* Christ in Bull. Herb. Boiss. 6 (1898) 152, non Baker).

Rhizomate breve, 6 mm. crasso, stipite griseo-brunneo, 30 cm. alto, sparse paleaceis; fronde 40–50 cm. alta, 12–15 cm. lata, acuminata; pinnis brevi-pedicellatis, ca. 7 cm. longis, 15 mm. latis, caudatis, plus minus profunde crenato-incisis, segmento primo acroscopico pleurumque libero, i. e., pinnulo, orbiculari-cuneato, supra inciso-crenato, subcoriaceis, glabris, venis furcatis, paucis, perobliquis, 1–2 mm. remotis, soris costae quam margini propioribus.

Insigniter variat., ad terram in silva muscosa, 1,350 m. s. m. (No. 5892).
The best constant character of this fern is the extremely lax venation. The stalked pinnæ distinguish it from *A. anisodontium*. Besides the above number, I would include here the following specimens:
LUZON, Mount Data, alt. 2,200 m. (*Copeland* 1857), with shorter subacute pinnæ; Baguio (*Elmer* 6012), with sometimes a free basiscopic pinnule. MINDANAO, Mount Apo (*Copeland* 1598*c*), deeply cut throughout.

Asplenium pellucidum Lam. (*A. hirtum* Kaulf.).

Terrestrial in forests at 700 m. alt. (No. 5888), and epiphytic in forests at 1,800 m. alt. (No. 5889).
More than one Philippine species has been included under this name. This form is rather widespread.

Asplenium cuneatum Lam.

Epiphytic in ridge forests at 900 m. alt. (No. 5887).
Mount Maquiling, Luzon (leg. *Loher*) and Mindanao.

Asplenium affine Sw.

Epiphytic in humid forests at 140 m. alt. (No. 5886).
Central Luzon and Mindanao.
East Africa to Polynesia.

Asplenium laserpitiifolium Lam.

Epiphytic at 900 m. alt. (No. 5885).
Throughout the Philippines.
Malaya, Polynesia and tropical Australia.

SYNGRAMMA J. Sm.

Syngramma alismaefolia (Presl) J. Sm.

On river banks at 200 m. alt. (No. 5922).

A rare fern from central Luzon southward.

Malaya.

PLAGIOGYRIA (Kze.) Mett.

Plagiogyria glauca (Bl.) Mett.

On ridges in the mossy forest at 1,800 m. alt., and above (No. 5963).

Already known from the Philippines from Mount Apo, Mindanao, and the highlands of northern Luzon where there are derived species.

India, Malaya, Formosa and Yunnan.

Plagiogyria tuberculata Copel. var. **latipinna** Copel. n. var.

Typo pinnis fertilibus latioribus, 3–4 mm. latis, diversa.

In dumetis, 2,400 m. s. m. (No. 5962).

In this collection are dwarfed, more coriaceous plants from more exposed ridges, and others, even more ample than the type, from more sheltered places.

The type of the species is from Bagnen, Lepanto, Luzon.

Plagiogyria Christii Copel.

On mossy ridge at 2,100 m. alt. (No. 5961).

The type is from Mount Apo, Mindanao. Possibly too near *P. euphlebia* Mett.

Plagiogyria falcata Copel. n. sp. (Pl. I, Fig. B.)

Caudice erecto, 5 mm. crasso, basibus stipitium persistentibus profunde obtecto; stipitibus confertis frondium sterilium 7–10 cm., frondium fertilium 16–24 cm. altis, glabris, deorsum triangulatis, pedibus incrassatis aerophoris carentibus; fronde sterile ca. 20 cm. alta, 6–7 cm. lata, brevi-acuminata, fere vel usque ad rhachin pinnata; pinnis proximis, adnatis nec basibus dilatatis, acutis, ca. 5 mm. latis, rectis vel subfalcatis, argute serratis vel rhachin versus integris, inferioribus valde deflexis, membranaceis, glaberrimis, nigro-viridibus; venulis conspicuis, fere semper simplicibus; fronde fertile 15–24 cm. alta, 2.5–4, cm. lata, acuminata, pinnata; pinnis falcato-arcuatis, 2–3 cm. longis, 2 mm. latis, inferioribus remotis.

In silva dumosa muscosa ad terram, 2,100 m. s. m. (No. 5960).

The most similar species previously known from the Philippines is *P. stenoptera* (Hance) Diels, of China, Formosa and Luzon.

PTERIS Linn.

Pteris quadriaurita Retz.

In forests at 250 m. alt. (No. 5959), a very ample form, *P. nemoralis* Willd., I believe, common in this part of the World. A similar form from a deep ravine at 700 m. alt. (No. 5957).

Pteris sp. near *P. quadriaurita* Retz., var. *setigera*.

On cliffs in forests at 600 m. alt. (No. 5958).

In the usually accepted sense, using *P. quadriaurita* as a catch-all, it will include this plant, which is distinguishable from other forms by its caudate pinnæ, narrow segments, basal basiscopic pinnules of several pinnæ usually enlarged, prominent veinlets and chaffy stipes.

Pteris tripartita Sw.

Along the Alag River at 200 m. alt. (No. 5956), stipe 2.5 m. high.
Common in the Philippines.
Africa to Polynesia.

HISTIOPTERIS (Agardh) J. Sm.

Histiopteris incisa (Thunb.) J. Sm.

In an old clearing at 700 m. alt. (No. 5955).
On many mountains in the Philippines.

PTERIDIUM Gleditsch.

Pteridium aquilinum (L.) Kuhn.

With the preceding species (No. 5954); a large form of this common cosmopolite.

MONOGRAMMA Schkuhr.

Monogramma trichoidea J. Sm.

On a rotten stump in ridge forests at 600 m. alt. (No. 5857).
Luzon and Mindanao, in the rain forest.

Monogramma dareæcarpa Hook.

On tree-trunks in humid forests at 250 m. alt. (No. 5856).
Also known from Negros in similar situations at 100 m. alt.

PLEUROGRAMME Presl.

Pleurogramme loheriana Christ in Bull. Herb. Boiss. II 6 (1906) 1006.

On mossy trees in the ridge forest at 1,400 m. alt. (No. 5855).
These specimens are rather more slender than those known from Luzon, and not forked. A stouter relative grows in Palawan and Mindanao.

VITTARIA Smith.

Vittaria amboinensis Fée.

Epiphytic in the ridge forest at 1,100 m. alt. (No. 5865).
Found also at the same altitude on Mount Mariveles.
Amboina, Burmah.

Vittaria elongata Sw.

Epiphytic at 1,800 m. alt. (No. 5864).
Throughout the Philippines.
India to Polynesia and New South Wales.

Vittaria sp. near *V. elongata* Sw.

Pendent from mossy trunks at 1,400 m. alt. (No. 5860).
Very similar to the preceding, but with the indusium thinner and not reaching the margin.

Vittaria sp. near *V. elongata* Sw.

Epiphytic in humid forests at 200 m. alt. (No. 5863).
This is possibly true *V. elongata;* it differs from the plant to which I have given that name, most conspicuously in the very lax venation of the latter.

Vittaria sp. near *V. elongata* Sw.

Epiphytic in dense thickets at 175 m. alt. (No. 5862). A very large form previously collected near the base of Mount Halcon by *Merrill* (No. 4044) and *McGregor* (No. 315).

Vittaria sp. near *V. elongata* Sw.

On mossy trunks at 1,100 m. alt. (No. 5859). Differing from *V. elongata* in its shorter, broader, thicker and falcate fronds. Also in Mindanao.

With the uncertainty that exists as to many species near *Vittaria elongata* Sw., and in view of the fact that we have now five species of this genus awaiting description by Dr. *Christ*, I prefer to let these specimens remain for the present unnamed. The Philippine flora is very much richer in species of *Vittaria* than has been imagined.

Vittaria scolopendrina (Bory) Thwaites.

On trunks in humid forests at 150 m. alt. (No. 5861).

Catanduanes (leg. *Baranda*), Mindanao.

Africa to Samoa.

ANTROPHYUM Kaulf.

Antropium reticulatum (Forst.) Kaulf.

On tree trunks at 100 m. alt. (No. 5867).

Very common in the Philippines.

Madagascar to Polynesia.

Antrophyum callaefolium Bl.

On boulders in damp forests at 1,900 m. alt. (No. 5866).

Luzon (leg. *Loher*), Mindanao.

Malaya and Polynesia.

TAENITIS Willd.

Taenitis blechnoides (Willd.) Sw.

Terrestrial in ridge forests at 700 m. alt. (No. 6003).

Common on forested ridges throughout the Philippines. The above specimen has the sori almost marginal.

India to Polynasia.

HYMENOLEPIS Kaulf.

Hymenolepis platyrhynchos (J. Sm.) Kze.

Epiphytic at 1,800 m. alt. (No. 6001).

Luzon to Celebes and Borneo.

The sterile margins of the fertile apex constitute a very perfect protection for the young bed of sporangia. The sterile margin reaches its whole width while very young, but the fertile portion remains a long time narrow, and completely covered by one of the sterile margins which is folded entirely across it. The fertile portion widens rapidly and becomes exposed just before maturity.

Hymenolepis spicata (Linn. f.) Presl.

Epiphytic at 150 m. alt. (No. 6000). A rather large form which is possibly *H. brachystachys* Hook., J. Sm., but not quite sessile.

The range of this "species" is from Madagascar to Polynesia.

Hymenolepis rigidissima Christ in Bull. Herb. Boiss. II. 6 (1906) 990.

Epiphytic at 1,800 m. alt. (No. 5999), with *H. platyrhynchos*, the specimens quite uniform, with the sterile portions not more than 1 cm. wide, 12 to 15 cm. long, and the fertile portions at most 2 mm. wide and 15 to 20 cm. long. The form occurs elsewhere in the Philippines at similar altitudes.

ACROSORUS Copel.

Acrosorus Merrilli Copel. n. sp.

Rhizomata ut in *A. exaltato;* frondibus 20–30 cm. longis, ca. 6 mm. latis vix ad costam glabram pinnatis, deorsum ad alam undulatam reductis, sessilibus, confertissimis, coriaceis, infra minute nigro-punctulatis; segmentis sterilibus obtuse inaequilateraliter triangularibus; fertilibus tortis, lamina reducta.

Mindoro, ad montem Halcon, ab arboribus permuscosis pendens, 1,300 m. s. m. (No. 6002).

This species has its nearest affinity in its geographically nearest relative, *A. exaltatus* of Mount Apo, from which it differs in being distinctly smaller and more slender, with narrower sterile segments and more reduced and twisted fertile segments, giving the fertile apical part of the frond an appearance of distinctness wanting in *A. exaltatus*. In this respect it approaches *A. Reineckei* (Christ), of Samoa.

PROSAPTIA Presl.

Prosaptia contigua (Forst.) Presl.

Epiphytic on mossy trees at 1,800 m. alt. (No. 5998). The typical form and the variety *monosora* are about equally abundant.

Malaya to Polynesia.

Prosaptia polymorpha Copel. n. sp.

Species *P. alatae* affinis, nec non illa segmentis latioribus minus truncatis, soribusque in dentes obtusis immersis nec ad apices segmentorum restrictis, diversa; frondibus usque ad 40 cm. longis, 3 cm. latis, acuminatis, subcoriaceis; segmentis oblongis, obtusis vel subacutis, deorsum sensim in undulas alae 5–8 cm. longae, demum integrae, in stipitem brevem attenuatae elatis; soris in segmento quoque ca. 5.

Apud fluminem Alag ad arbores saxaque epiphytica, 150 m. s. m. (No. 5997).

This material, although for the most part uniform in essentials, shows a few wide variations, one frond being *in its upper half* like *P. contigua monosora*, and a few so like *P. alata* that if they had been alone, they could have been referred to that species. *P. polymorpha* impresses me as being between *P. alata* and *P. Toppingi*, differing from the latter in the narrower, sharper sinuses, more rounded segments and less coriaceous texture.

POLYPODIUM Linn.

Polypodium Jagorianum Mett.

Epiphytic on exposed ridges at 1,400 m. alt. (No. 5966).

The known range of this species is from Mount Mariveles, Luzon, to Mount Silay, Negros, always in the same habitat.

Polypodium setosum Bl., var. **calvum** Copel. n. var.

Costa margineque superficieque pilis sparsissimis vestitis; aliter typicale.

Ad arbores moscosos, 2,400 m. s. m. (No. 5965).

This fern agrees with *Blume's* description and with specimens that I believe to be correctly determined, in texture, form, sori and pubescence of the stipe.

It agrees also with *Blume's* description [5] in the "Venis jam conspicuis, marginem fere attingentibus, semper furcatis, *ramulo uno plerumque item furcato.*" Since even a very considerable difference in the degree of pubescence would not be generally received as a good specific character, I treat this as a variety. However, it is unlike the plant already determined as *Polypodium setosum* from Mount Mariveles, and neither is it indentical with the typical Mount Apo material. A most interesting variation is the occurrence of anastomosing veins in certain plants with rather ample fronds.

Polypodium paucisorum Copel. n. sp. (Pl. III, Fig. B.)

Species *P. subcvenoso* Baker affinis illo frondibus gracilioribus et soris paucis, remotis, valde immersis distincta. Caudice erecto, paleaceo breve; frondibus confertis, sessilibus, 8–12 cm. altis 3–4 mm. latis, subacutis, deorsum valde attenuatis, integris, glabris, subcoriaceis; venulis occultis furcatis; soris supra valde prominentibus.

Ad arbores muscosos. 180–240 m. s. m. (No. 5964).

Merrill's material is very constant in the characteristic sparse distribution of the sori. In the insertion these are as figured by *Hooker*, not as described by *Christ*, in *P. sessilifolium*, being dorsal on the vein, which is very remarkable considering the deep immersion.

Polypodium cucullatum Nées et Bl.

On mossy trunks in the ridge forest, 1,150 to 1,400 m. alt. (No. 5968)..

Common in the Philippines.

Ceylon to New Calodonia.

Polypodium cucullatum var. **planum** Copel. n. var.

Typo segmentis planis amplis, proximis, oblongo-orbicularibus diversum; soro magno, subsuperficiale.

Ad arbores muscosos epiphyticum, 900 m. s. m. (No. 5967).

Except for a few intermediate forms, this would be regarded as a distinct new species, and might be considered as intermediate between *Eupolypodium* and *Calymmodon*, but it is probably derived from the latter and purely local.

Polypodium gracillimum Copel.

Epiphytic on mossy trunks at 1,400 m. alt. (No. 5858).

Throughout the Philippines.

Polypodium mollicomum Bl.

On mossy trees, exposed ridges at 1,400 m. alt. (No. 5973).

Previous Philippine collections of this species are *Yoder's* from Panay, and *Merritt's* from Mount Halcon, Mindoro.

Malaya.

While most of *Merrill's* copious material is typical *P. mollicomum*, it is variable in size and in the length of the stipe, and seems to me to intergrade completely with specimens, that by themselves, would be referred to *P. fuscatum* Bl. The latter thus becomes a synonym.

[5] Fl. Javae **2**: 116.

Polypodium subfalcatum Bl.

With the preceding (No. 5972).

India, Malaya.

These specimens are very small, but otherwise agree perfectly with *Blume's* description. The most similar fern hitherto known from the Philippines is *P. macrum*, which is more slender, more glabrous, repand instead of toothed, and has the sori decidedly farther from the costa.

Polypodium subfalcatum Bl., var. **semiintegrum** Copel. n. var.

Forma nana alpina a typo pinnis plus recurvis margine basicopicia pinnarum integra diversa.

Ad arbores muscosos, 1,900 m. s. m. (No. 5971).

This is very distinct in aspect, well marked by the margins of the pinnæ, the usually sterile basiscopic half being entire, while the acroscopic half is crenulate-toothed; but the material is too scant to warrant the description, as a distinct species, of a plant whose affinity is so clear and close. The longest fronds measure less than 10 cm.

Polypodium erythrotrichum Copel.

On mossy trees at 2,400 m. alt. (No. 5970).

Already known from Mount Apo and Mount Data. Very near *P. venulosum* Bl.

Polypodium obliquatum Bl.

On ledges in the ridge forest at 450 m. alt. (No. 5969).

In the Philippines this is the commonest fern in its group.

India, Malaya.

Polypodium celebicum Bl.

On mossy trees at 1,800 m. alt. (No. 5975).

Not hitherto known from north of Mindanao.

Malaya.

Polypodium decrescens Christ.

Terrestrial on ridges in the mossy forest at 2,200 m. alt. (No. 5974).

Previously known only from the type locality in Celebes and from Mount Data, Luzon. Related to *P. celebicum*.

Polypodium halconense Copel. n. sp. (Pl. II, Fig. B.)

Rhizomate ad terram repente, 15 mm. crasso, paleis griseo-fuligineis lanceolatis 2 mm. longis vestito; stipitibus inter se proximis, ca. 5 cm. altis, rhachibusque pilis rectis badiis 0.5–1 mm. longis vestitis; fronde 20 cm. alta, 45 mm. lata, utrinque abrupte acuminata, pinnata; pinnis rectis vel recurvis, decurrentibus nec nisi supremis confluentibus 25 mm. longis, 3.5 mm. latis, obtusis, $\frac{1}{2}$ vel $\frac{2}{3}$ ad costam in dentes divergentes 3 mm. inter se remotos incisis, infimis in auriculas abrupte diminutis coriaceo-papyraceis, infra pallidis, ad costas apicesque dentium setiferis, aliter glabris; venulis simplicibus, occultis; soris ad baseos venularum superficialibus pleurumque oppositis et trans costam confluentibus.

In silvis muscosis, 2,300 m. s. m. (No. 5976).

A relative of *P. solidum* as shown by the paleæ, color, texture and form, but much larger and in all respects more delicate.

Polypodium tenuisectum Bl.

On ridges in the mossy forest 2,200 to 2,500 m. alt. (No. 5978).

Already known from the Philippines, from Mount Apo, Mindanao, and Mount Canlaon, Negros.

Malay to Samoa.

Polypodium Yoderi Copel.

With the preceding (No. 5977). Hitherto known only from the type collection from Mount Madiaas, Panay.

Polypodium papillosum Bl.

On boulders in forests at 800 m. alt. (No. 5979); very large specimens.

Malaya.

Polypodium (Goniophlebium) integriore Copel. n. sp.

Rhizomate lignoso repente 3–4 mm. crasso, paleis castaneo-fulvis lanceolato-acuminatis 4 mm. longis vestito vel glabrescente et non calcareo; stipite glabro 15–25 cm. alto, stramineo-brunneo; fronde ca. 50 cm. alta; pinnis linearibus, 20 cm. longis, 15 mm. latis, sub apice valde caudata vix serrulatis aliter integris vel crenulatis, glabris, membraneceis, basibus anguste acuminatis, omnibus stipitatis; venis nigris, seriebus areolarum ca. 3; soris grandibus, leviter immersis.

Ad montem Halcon, epiphyticum 2,200 m. s. m. (No. 6005).

Nearly related to *P. persicifolium* Desv., which as at present construed seems to me to include two or more species. *P. integriore* can be recognized by its narrow, almost entire pinnæ, stalked, and with narrow bases. *P. persicifolium* occurs on Mount Apo[6] which is its most northern known habitat. The Apo form is much larger than the Javan, and membranaceous.

Polypodium (Goniophlebium) verrucosum Wall., has previously been collected on the Baco River, near the base of Mount Halcon by *McGregor* (No. 253). It differs from the preceding most conspicuously in its small and deeply immersed sori which resemble those of *P. papillosum* Bl.

Polypodium (Goniophlebium) subauriculatum Bl.

Epiphytic on mossy trees at 1,400 m. alt. (No. 6004).

These specimens are remarkable for the very coarsely serrate, strongly acuminate pinnæ, and for the very tardy and incompletely deciduous dense clothing of scales on the rachis. In the latter respect our Philippine plants usually differ from the Javan. This species is in urgent need of analysis.

Polypodium nummularium Mett.

Epiphytic on mossy trees at 1,800 m. alt. (Nos. 5994, 5995).

Hitherto known only from Luzon, on the mountains about Laguna de Bay.

Some specimens with fronds fertile towards the apex, but with broad sterile bases suggest *P. accedens*, and unmistakably approach *P. Whitfordi* Copel., of Mount Mariveles. This resemblance is strong enough thoroughly to establish the affinity of *P. nummularium* and *P. Whitfordi*. In Luzon both species are, as far as known, very stable and distinct in form. Regarding these species as, like our fern flora as a whole, migrants towards the north, Halcon may be regarded as the point of origin and separation; they being so distinct in Luzon that their close affinity escaped suspicion is to be ascribed to their (however recent) isolation there.

[6] Copeland: *This Journal* (1906) Suppl. 1: 162.

Polypodium lagunense Christ.

Epiphytic on exposed ridges 1,400 to 1,800 m. alt. (No. 5984, 5985).

This species is already known from Mount Maquiling, Luzon, and from Mount Silay, Negros (*Whitford* 1535). As a matter of affinity it unquestionably belongs in the *Crypsinus* group, the usual classification of "Phymatodes" according to the composition of the frond being in this case as unnatural as possible. The pinnæ of *Polypodium lagunense* resemble the whole simple frond of *P. nummularium*, the distinction of sterile and fertile parts being the same in both species. The simply pinnate frond form of *P. lagunense* is not yet quite stable, for *Merrill* found two plants (No. 5984) with large bipinnate fronds.

Polypodium accedens Bl.

Epiphytic in humid thickets at 150 m. alt. (No. 5092), typical form: and on rocks along the Alag River at 350 m. alt. (No. 5993).

Malaya and Polynesia.

The plants collected at an altitude of 350 m. are rather larger than is usual in this species. Some fronds are typical in form and fertile only at the constricted apex; others are narrower and fertile almost throughout, while entirely sterile ones are notably broad. This variation is in the direction of *P. nummularium* and *P. Whitfordi*, indicating the probability that both have descended from *P. accedens* in this locality.

Polypodium punctatum (L.) Sw.

On rocks along the Alag River at 350 m. alt. (No. 5991); dried fronds reddish in color.

Africa to Polynesia.

Polypodium monstrosum Copel. var. **integriore** Copel. in Elmer's Leaflets 1 (1906) 78.

On mossy trees at 1,950 m. alt. (No. 6104).

Previously known only from Luzon.

Polypodium validum Copel.

On prostrate logs in forests at 100 m. alt. (No. 5868).

Common in Mindanao.

Polypodium (Selliguea) macrophyllum (Bl.) Reinw.

On rocks in wet ravines at 150 m. alt. (No. 5990).

China to New Guinea.

Polypodium (Selliguea) calophlebium Copel. n. sp. (Pl. III, Fig. A.)

Rhizomate ad arbores muscosos late repente, 2–3 mm. crasso, paleis squarrosis fulvo-ferrugineis 3 mm. longis late lanceolatis basibus peltatis castaneis vestito; stipitibus erectis, glabris, frondium sterilium 3–6 cm. frondium fertilium 7–10 cm. altis; fronde sterile ovata, ca. 14 cm. longa, 5 cm. lata, valde acuminata, base rotunda, margine brunnea dura coriacea glaberrima, supra (sicca) umbrina, infra olivacea; venis primariis validis, rectis, marginem attingentibus, venulis infra occultis, supra reticulationem praestantem laxam pulchram efficientibus; fronde fertile lineare, ca. 16 cm. longa, 7–11 mm. lata, longe sursum attenuata, basi acuta, integra vel crenulata, soris saepius utroque latere costae nigrae uniseriatis, vel biseriatis at inter venas primarias saepe confluentibus.

Epiphytica 1,800 m. s. m. (No. 5989).

A species well distinguished by the conspicuous regular venation and by the fertile fronds so reduced in the majority of *Merrill's* specimens as to have a single round sorus between each two main veins. One frond, fertile towards the apex only, suggests a near affinity to *P caudiforme*. Another possible but not very intimate relative is *P. triquetrum*.

Polypodium commutatum Bl.

On boulders along the Alag River at 150 m. alt. (No. 5988).

This is identical with *Cuming* No. 97 from Luzon, and my No. 1582 from Mindanao. For the propriety of regarding it as *P. commutatum* Bl. (*P. affine* Bl.) see this JOURNAL (1906) 1 Suppl. 163–4.

Java.

Polypodium Phymatodes Linn.

Epiphytic along the Alag River at 100 m. alt. (No. 5986); the commonest Philippine form of this species, with acuminate segments.

Africa to Polynesia.

Polypodium palmatum Bl.

On boulders at 800 m. alt. (No. 5981); epiphytic at 2.100 m. alt. (No. 5982); epiphytic and terrestrial at 2.400 m. alt. (No. 5983).

Malaya.

Polypodium glauco-pruinatum C. Chr. (*P. glaucum* Kze.).

Epiphytic, mossy ridge forest at 1,100 m. alt. (No. 5987).

A most distinct species hitherto known only from Luzon.

Polypodium albido-squamatum Bl.

Epiphytic at 1.100 m. alt. (No. 5980).

Malaya to New Guinea.

This species and *P. glauco-pruinatum* have similar dark, long, harsh paleæ, and were growing together on the same tree.

Polypodium (Drynariopsis) heracleum Kze.

Epiphytic in river forest at 100 m. alt. (No. 6007), fronds 2.5 m. long.

Malaya to New Guinea.

DRYOSTACHYUM J. Sm.

Dryostachyum splendens J. Sm.

Epiphytic in ridge forests at 800 m. alt. (No. 6006).

This specimen has the expanded humus-collecting leaf-bases characteristic of the species and distinguishing it from *D. pilosum*.[7] The stipe has the vestige of an articulation, but even this is wanting in Luzon specimens terrestrial on cliffs. Its long confusion with *D. pilosum* makes the range of this plant questionable. I have it from Luzon and Mindanao.

CYCLOPHORUS Desv.

Cyclophorus angustatus (Sw.) Desv.

Epiphytic and on boulders along the Alag River at 150 m. alt. (No. 5996). Also collected by *McGregor* (No. 255) on the Baco River near the base of Mount Halcon, and in Luzon by *Cuming*, and by *Whitford* (No. 810).

India to Polynesia.

[7] Copeland: *This Journal* (1906) 1 Suppl. 165. *pl.* 26. 27.

LOMAGRAMMA J. Sm.

Lomagramma pteroides J. Sm.

Humid river forests at 150 m. alt. (No. 5854) "climbing 2 to 5 m. on tree-trunks, very abundant but rare in fruit." Also collected on the Baco River, near the base of Mount Halcon by *McGregor* (No. 235) and in Luzon by *Cuming*.

In view of the positive statement of *Presl*,[8] and of the evident distinctness of the plant he there described from the one we have, I can not believe that *Leptochilus lomarioides* Bl., is indentical with *Lomagramma pteroides* J. Sm. I am treating this plant as *L. pteroides* on an assumption based on geographical contiguity. I have another *Lomagramma* from Mindanao and Palawan, more coriaceous, more scaly and becoming red as it dries.

ELAPHOGLOSSUM Schott.

Elaphoglossum decurrens (Desv.) Moore.

On trunks in the ridge forest at 100 to 1,000 m. alt. (No. 5853).

Differs from the Javan plant as figured by Blume[9] only in being more acute. The margin is cartilaginous and glabrous.

Also known from Luzon.

Java.

PLATYCERIUM Desv.

Platycerium coronarium (Koenig.) Desv. (*P. biforme* Bl.)

Epiphytic in forests at 350 m. alt. (No. 6037).

Luzon, Negros, Masbate and Mindanao.

India, Malaya.

HYMENOPHYLLACEÆ.

TRICHOMANES Linn.

Trichomanes nitidum V. d. B.

On mossy trees at 900 m. alt. (No. 6067.)

Also known from Luzon, Baguio (*Elmer* 6023) det. *Christ*.

Java.

Trichomanes bimarginatum V. d. B.

On tree trunks in humid forests at 220 m. alt. (No. 6066).

India to Polynesia and Formosa.

Trichomanes pyxidiferum Linn.

On boulders along the Alag River at 150 m. alt. typical plants (No. 6064, 6065) and a very narrow form (No. 6063).

Common in the Philippines.

Pantropic.

Trichomanes sp. near *T. proliferum* Bl.

Narrower than *Blume's* species, the fertile and sterile fronds or parts of fronds distinct, the former very contracted and tall, each pinna usually consisting of one sorus and its stalk.

On damp ledges by streams at 350 m. alt. (No. 6062).

[8] Epim. Bot. (1848), 177.

[9] Fl. Javae 2: *pl. 10*.

Trichomanes Cumingii (Presl) C. Chr. (*T. Smithii* Hook.)

Epiphytic on *Cyathea* in ridge forest at 900 m. alt. (No. 6068).
"Philippines" *Cuming;* Mindanao.
Moluccas.

Trichomanes pallidum Bl.

(*a*) On mossy trees at 1,000 m. alt. (No. 6069); all sterile, the typical plant.
India to Polynesia.

(*b*) On mossy trees at 1,400 m. alt. (No. 6070). The slender greener form, *T. glauco-fuscum* Hook., which is perhaps a good species. This is the commoner form in the Philippines.

Trichomanes javanicum Bl. (*T. rhomboideum* J. Sm.).

On ledges along streams in forests at 250 m. alt. (No. 6061).
Very common in the Philippines.
India to Polynesia and Liu Kiu.

Trichomanes Pluma Hook.

Terrestrial, in forests at 1,400 m. alt. (No. 6072). Most of the specimens are very large and broad.

Mounts Maquiling, Banajao, Silay and Apo.
Perak to Samoa.

Trichomanes apiifolium Presl.

Terrestrial in ridge forest at 900 m. alt. (No. 6073); amply distinct from the heterogeneous jumble of plants in my herbarium determined by various botanists as *T. bauerianum*, and *T. meifolium*.

Throughout the Archipelago usually at greater altitudes and often epiphytic.
Java.

Trichomanes maximum Bl.

By streamlets in humid forests at 250 m. alt. (No. 6060).
Central Luzon and south.
Malaya, Polynesia and Queensland.

Trichomanes rigidum Sw.

On damp mossy boulders, terrestrial, and on cliffs in the ridge forests at 800 to 1,400 m. alt. (Nos. 6074, 6075, 6076, 6077). I can not distinguish *T. Cupressoides* Desv., to which on geographical grounds this plant would be ascribed.

Common in the Philippines.
Pantropic.

Trichomanes radicans Sw.

Terrestrial in damp ravines at 1,400 m. alt. (No. 6078), a remarkably large and stout form of this species.

Pantropic and in Ireland, Japan and Alabama.

HYMENOPHYLLUM Smith.

Hymenophyllum dilatatum Sw.

On banks in forests at 700 m. alt. (No. 6079).
Luzon, Mindanao.
Malaya to Polynesia and New Zealand.

Hymenophyllum fimbriatum J. Sm.

On trees at 1,800 m. alt. (No. 6081), different from *H. australe* Willd., in form, texture, and position of the sori; nearer to *H. dilatatum*.

Luzon (*Cuming* 218).

H. sp. near *H. polyanthos* Sw.

On trees in forests at 900 m. alt. (No. 6080), a remarkably lax plant with few pinnules.

H. sp. near *H. polyanthos* Sw.

On mossy trees in forests at 1,400 m. alt. (No. 6082), a large plant with long segments, very peculiar in the symmetry of the pinnæ; on mossy trunks at 900 m. alt. (No. 6083).

Hymenophyllum demissum Sw.

On mossy trees at 1,400 m. alt. (No. 6086).

Malaya to Polynesia and New Zealand.

Hymenophyllum halconense Copel. n. sp.

Rhizomate filiforme repente primo hirsuto, demum glabro; stipitibus filiformibus, 2–3 cm. altis, glabrescentibus, sursum laminis decurrentibus anguste alatis; lamina 4–9 cm. alta, 15–20 mm. lata, bipinnatifida et segmentis" interdum furcatis, margine integra, haud ciliata, costa ubique alata; costa venisque basibusque urceolarum infra rufo-pubescentibus, supra glabris; urceolis ad segmenta" infima, acroscopica apicalibus, sessilibus, vix ad mediam in lobos rotundatos fissis; receptaculis exsertis.

Ad arbores muscosos, 1,050 m. s. m. (No. 6084).

Related to *H. serrulatum* (Presl) C. Chr., rather than to *H. ciliatum;* differing from the former in being a smaller fern in all respects, and in its entire margins.

Very small specimens were also collected at 900 m. alt. (No. 6085), which I believe to be this species, especially as they are in large part sterile. Can this small form be *H. tenellum* Kuhn?

Hymenophyllum obtusum H. & A.

On mossy trees at 1,400 m. alt. (No. 6087).

This agrees exactly with my Hawaiian specimens (leg. *Hillebrand*) in pubescence and in the characteristic color, but usually has nothing like the truncate fronds of that plant, so that on the whole it is intermediate between *H. obtusum* and *H. ciliatum*.

We have identical specimens from Benguet, Luzon (*Elmer* 6021) det. *Christ*.

Hawaii, New Guinea.

Hymenophyllum tunbridgense (L.) Sm.

On mossy trees at 1,100 and 2,000 m. alt. (Nos. 6088, 6089); not typical, but hardly separable from this species.

England to New Zealand and Chile.

Hymenophyllum aculeatum (J. Sm.) Racib.

On trees in forests at 180 m. alt. (No. 6059). Some specimens can hardly be distinguished from *H. denticulatum* Sw.

Luzon.

Ceylon and Malaya.

Hymenophyllum denticulatum Sw.

On trees in forests at 900 m. alt. (No. 6071). This and the preceding species are too close together. *Merrill's* No. 6058 collected at 200 m. alt. is almost intermediate.

Luzon.

India, Malaya.

CYATHEACEÆ.

DICKSONIA L'Herit.

Dicksonia chrysotricha (Hassk.) Moore.

A small tree-fern on ridges at 1,100 m. alt. (No. 6009).

Also known from Mindanao (*Copeland* 1456), Mount Apo at 2,000 m. alt., and from San Ramon at 1,300 m. alt.

Java and (?) Celebes.

This species is notable for the usual restriction of the sori to the few lowest pinnules of the major pinnæ. It is the only arboreous *Dicksonia* occurring in this part of the World north of the equator.

CIBOTIUM Kaulf.

Cibotium Cumingii Kze.

On ridges in the mossy forest at 1,400 m. alt. (No. 6008); fronds 2 m. long; trunk wanting.

This is the usual Philippine form, already known on many Luzon mountains but previously from no other island. It has but one sorus on each side at the base of the segment, instead of two or more as is usual in the case of *C. Barometz.*

CYATHEA Smith.

Cyathea spinulosa Wall.

Abundant in forests between 800 and 1,100 m. alt. (No. 6056), the secondary rachises usually hirsute.

Luzon, Negros and Palawan.

Malaya to India and Japan.

Cyathea sp. near *C. Christii* of Negros and Mindanao and *C. spinulosa.* Stipe very short, rachis not very spiny nor very red, barely tripinnate with narrow pinnules II (or segments) and glabrous costæ. Stem 3 m. high.

Ridge forest at 1,800 m. alt. (No. 6055).

Cyathea tripinnata Copel. in Philip. Journ. Sci. 1 (1906) Suppl. 251.

In the ridge forest at 1,800 m. alt. (No. 6054). "Trunk 15 ft. high, leaves 8 ft. long. with about 18 pinnæ on each side."

Previously known only from the type locality, Mount Mariveles, Luzon.

ALSOPHILA R. Br.

Alsophila latebrosa Wall. (?).

"Tree-fern 20 ft. high, fronds 15 ft. long, pinnæ about 18 on each side, the lower ones about 6 inches long, very common in old clearings" (No. 6053).

The rachis and pinnæ are exactly like those of some of my Indian specimens of *A. latebrosa* Wall., but the bases of the stipes are very densely clothed (*a*) with long brown hairs; (*b*) with lanceolate, membranous, almost white scales 2 to 3 cm. long with narrow black bases, and (*c*) paleæ intermediate in form. It seems incredible that these should have escaped description if this is the real *A. latebrosa.*

Alsophila melanorhachis Copel. n. sp.

Arbor, caudice 2–3 m. alto, 5 cm. crasso; apice basibusque stipitum paleis lanceolatis 1 cm. longis rigidis castaneis pallido-marginatis vestitis; stipite 60 cm. longo, nigro, deorsum pulcherrime brunneo-punctato, ubique rhachique asperulis, supra paleis angustis crinitis fulvis vestitis; fronde 150 cm. longa, tripinnata pinnis utroque latere ca. 12, inframedialibus maximis, 50 cm. longis, 13 cm. latis, fere horizontalibus, acuminatis, pedicellis ca. 1 cm. longis; rhachibus supra paleaceis, infra potius nigro-glabrescentibus; pinnulis distinctis utroque latere fere 30, brevistipitatis 7 cm. longis, 1 cm. latis, serrato-acuminatis, deorsum ad rhachin infra perpaleaceam paleis biformibus, supra breve denseque villosum pinnatis; pinnulis'' anguste abovatis, obtusis, adnatis, minute nec lente crenato-serratis, tenue coriaceis, supra subnigris, glabris, infra atro-olivaceis, costis deorsum paleis bullatis ciliatis obtectis, aliter glabris; venulis sterilibus inferioribus plerumque furcatis, fertilibus simplicibus; soris costalibus; pinnulis prasertim fructuosis contractis, soris laminam totam complentibus.

In silvis muscosis, 1,800 m. s. m. (No. 6052).

A most striking species remarkable for the dark color of the whole frond and especially for the shiny black stipe and rachis.

SCHIZAEACEÆ.

SCHIZAEA Smith.

Schizaea dichotoma (L.) Smith.

On ridges in humid forests at 250 m. alt. (No. 6050).

Madagascar to Polynesia.

Schizaea digitata (Linn.) Smith.

In ridge forests at 450 m. alt. (No. 6049).

India to Polynesia.

LYGODIUM Sw.

Lygodium scandens (L.) Sw.

In old clearings at 700 m. alt. (No. 6051); a very round-leaved form found also in Formosa.

Throughout the Philippines, not common.

Africa to Polynesia.

Lygodium Merrilli Copel. n. sp. (Pl. IV.)

Species distinctissima pinnis sessilibus, pinnulis pinnatis, pinnulis'' haud articulatis, segmentis fertilibus sparsis, venulis ubique anastomosantibus.

In dumetis scandens, 300 m. s. m. (No. 6057).

In all probability a relative of *Lygodium flexuosum*, rather than of any other species with reticulate venation.

SCHIZAEACEÆ.

OSMUNDA Linn.

Osmunda bromeliæfolia (Presl) Copel. (*Nephrodium* (?) Presl, Rel. Haenk. 1 (1825) 33.)

On steep river banks at 100 m. alt. (Nos. 5852, 5851). *Cuming's* No. 173 from Luzon is this plant, and it seems to have been collected recently by *Loher*.

This species is distinguished from our common and widespread *Osmunda banksiæfolia*, exactly as was done by *Presl*[10] in that the latter "differt pinnis oppositis brevius petiolatis brevioribus inaequalateribus, basi cuneatis, undique inciso-serratis, serraturis convexis, venulis bifidis," only the position of the pinnæ is inconstant. *Bromeliæfolia* is the older of the two specific names.

GLEICHENIACEÆ.

GLEICHENIA Smith.

Gleichenia dicarpa R. Br.

In open heath-lands at 2,500 m. alt. (No. 6048), a small form, in some specimens approaching the var. *alpina*, found on Mount Apo and thence to New Zealand. We have this also from Mount Pulgar, Palawan.

New Caledonia, Tasmania.

There is also on Mount Apo a plant hardly as scaly as *Gleichenia vulcanica* Bl., but referable to it, which, according to my field notes, intergrades with *G. alpina:* if the notes are correct, these species are properly united in Species Filicum.

Gleichenia linearis (Burm.) Clarke.

Scandent on mossy ridge at 1,800 m. alt. (No. 6047), a very large handsome form.

Common in the Philippines.

Pantropic.

Gleichenia dolosa Copel.

On steep semi-forested slopes of ridge at 1,900 m. alt. (No. 6046). The costæ are decidedly scaly, but this is true of some specimens from Mount Mariveles as well.

Luzon, Mindanao.

Gleichenia glauca (Thunb.) Hook.

Scandent on exposed ridge at 2,000 m. alt. (No. 6045), the horizontal pinnæ about 6 ft. long.

Luzon.

India to Japan and Polynesia.

Gleichenia laevigata (Willd.) Hook.

In thickets at 700 m. alt. forming dense growths (No. 6044).

Luzon, Mindanao.

Malaya.

[10] Rel. Haenk. 1: 34.

MARATTIACEÆ.

ANGIOPTERIS Hoffm.

Angiopteris pruinosa Kge. (determined after *Christ*).

Common in forested bench-lands along the river at 100 to 200 m. alt. (No. 6043). Fronds 5 m. long; petiole 1.3 m. long, as thick as one's wrist; pinnæ about 13 on each side.

Negros (*Whitford* 1652) det. *Christ*. Panay, *Elmer*. Luzon, (*Mangubat* 1292).

Angiopteris angustifolia Presl.

In forests at 700 m. alt. (No. 6042); fronds about 8 ft. long, pinnæ about 8 on each side, costa broad, slightly scaly.

Java.

OPHIOGLOSSACEÆ.

OPHIOGLOSSUM Linn.

Ophioglossum pendulum Linn.

Epiphytic along the river at 100 m. alt. (No. 6041), a very large form, the lamina reaching a width of 6 cm. On mossy trunks at 1,800 m. alt. (No. 6040), a comparatively small form. The fertile segment is almost invariably forked in Philippine specimens.

Throughout the Archipelago.

Mauritius to Hawaii.

Ophioglossum intermedium Hook.

Terrestrial and on very rotten logs in forests at 900 m. alt. (No. 6039), variable in form, the sterile segment sometimes widened and bluntly lobed at the apex, always exceeded by the fertile segment.

Borneo, Java.

Ophioglossum sp. near *O. pedunculosum* Desv.

On very steep banks in shaded ravines at 900 m. alt. (No. 6038). This plant has a very fine costa and a narrower base and more lax venation than typical *O. pedunculosum*.

PSILOTACEÆ.

TMESIPTERIS Bernh.

Tmesipteris tannensis Bernh.

On tree-ferns at 1,800 m. alt. (No. 6024).

Found also on tree-fern trunks on Mount Data, Luzon, and on Mount Apo, Mindanao, at 1,800 to 2,100 m. alt.

New Caledonia, Samoa, New Zealand.

PSILOTUM Sw.

Psilotum complanatum Sw.

Epiphytic in humid forests at 150 m. alt. (No. 6021). Previously collected on the Baco River, near the foot of Mount Halcon by *Merrill* (No. 4081).

Java, New Guinea.

LYCOPODIACEÆ.

LYCOPODIUM Linn.

Lycopodium serratum Thunb.

On ridges in the mossy forest at 2,400 m. alt. (No. 6022), a very slender form, also in the mossy forest at 2,200 m. alt. (No. 6023).

At similar elevations in the mountains of northern Luzon.

India to Japan and Polynesia.

Lycopodium Hippuris Desv.

Terrestrial and erect on ridges in the mossy forest at 2,400 m. alt. (No. 6027).

Java to Samoa.

Lycopodium verticillatum Linn.

Pendent from mossy trees, 1,400 m. alt. (No. 6026).

Luzon (teste *Warburg*).

Pantropic.

Lycopodium squarrosum Forst.

Epiphytic near the Alag River at 100 m. alt. (No. 6028).

Throughout the Philippines, but nowhere abundant.

India to Polynesia.

Lycopodium Plegmaria Linn.

On mossy trees at 1,400 m. alt. (No. 6034) and at 1,900 m. alt. (No. 6035).

Central Luzon to Mindanao.

Africa to Polynesia.

Lycopodium filiforme Roxb. (non Swartz).

Epiphytic on mossy trees in humid forests at 200 to 300 m. alt. (No. 6036), a form with rather large leaves of the characteristic shape.

Central Luzon, Negros and Mindanao.

Monsoon region.

Lycopodium halconense Copel. n. sp.

Species *L. laterali* R. Br. affinis, differt statura minore, vix 10 cm. alta, foliis 3–4 mm. longis plerumque valde inflexis, ramis fertilibus ca. 7 mm. longis, deciduis.

Ad pseudericetum, 2,400 m. s. m. (No. 6031).

Altogether heath-like in aspect, being densely branched near the base, with close, erect branches.

Lycopodium cernuum Linn.

(*a*) On open heath at 2,400 m. alt., with the preceding (No. 6030).

The same form grows in similar habitats on Mount Apo, Mindanao, at 2,850 m. alt. An *L. vulcanicum* Bl. ?

(*b*) In forests on steep slopes, 2,200 m. alt. (No. 6029). The form with very fine leaves found throughout the Philippines and Tropics generally.

Lycopodium casuarinoides Spreng.

Scandent 10 to 15 ft. on exposed ridge at 2,100 m. alt. (No. 6032).

Also in Benguet, Luzon, at similar altitudes.

India, Malaya.

Lycopodium complanatum Linn., var. **thyoides** H. B. K.

In open heath-lands at 2,400 m. alt. (No. 6033).
Also in the mountains of northern Luzon.
The species throughout the North Temperate Zone.

EQUISETACEÆ.

EQUISETUM Linn.

Equisetum ramosissimum Desv.

Abundant along the Alag River at 100 m. alt. (No. 6025).
Northern Luzon.
Cosmopolitan.

ILLUSTRATIONS.

[Plate I by T. Espinosa, all others by Hugo Navarro.]

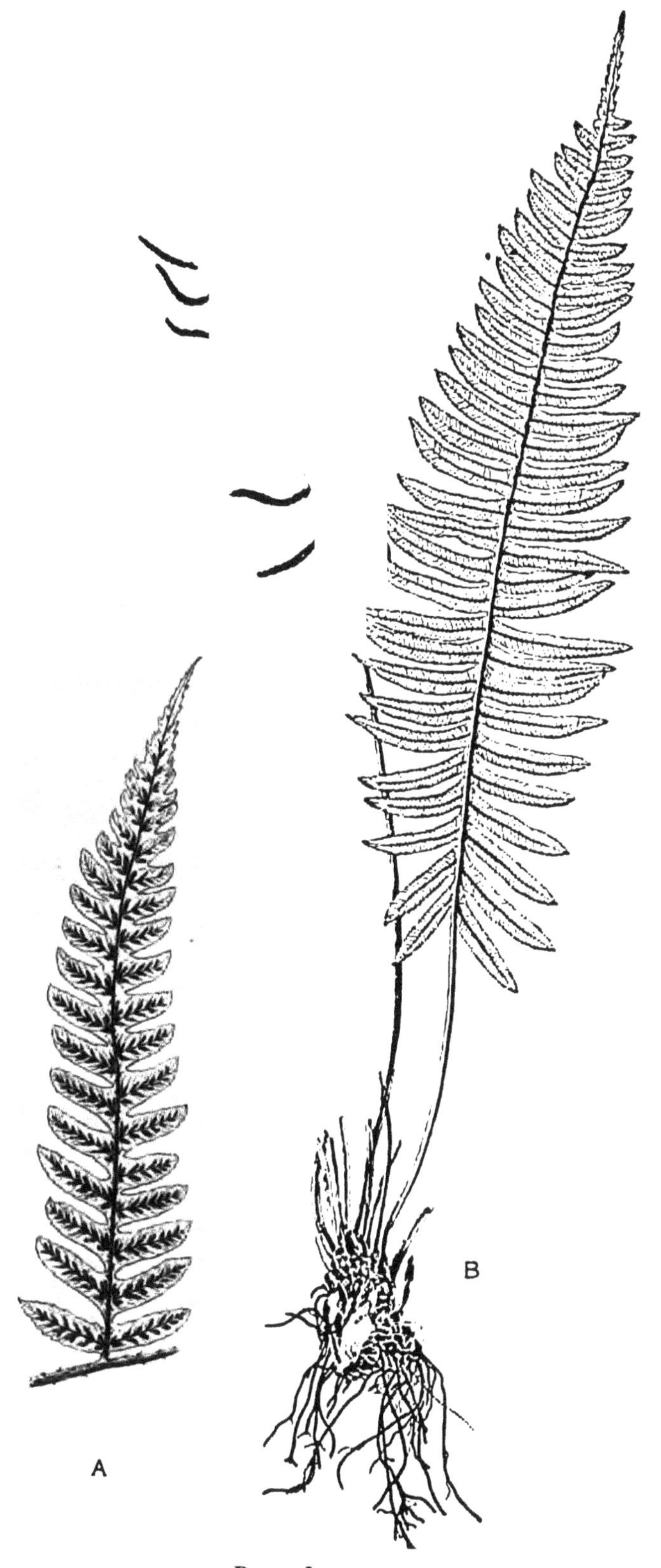

PLATE I.

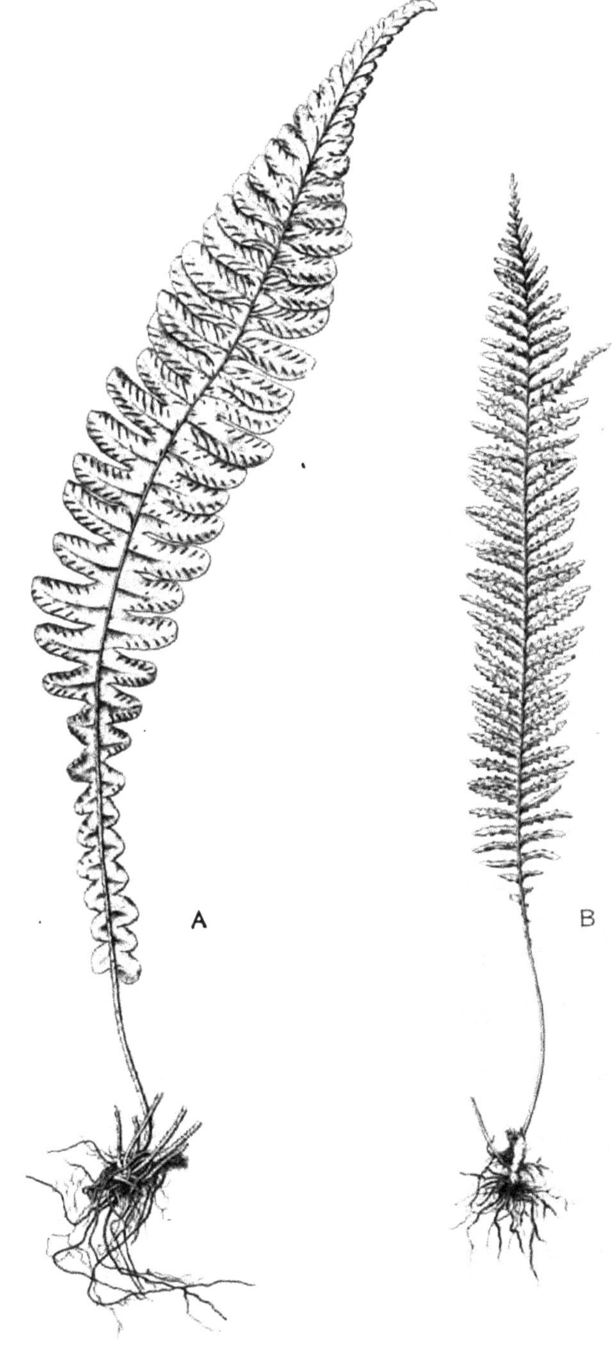

PLATE II.

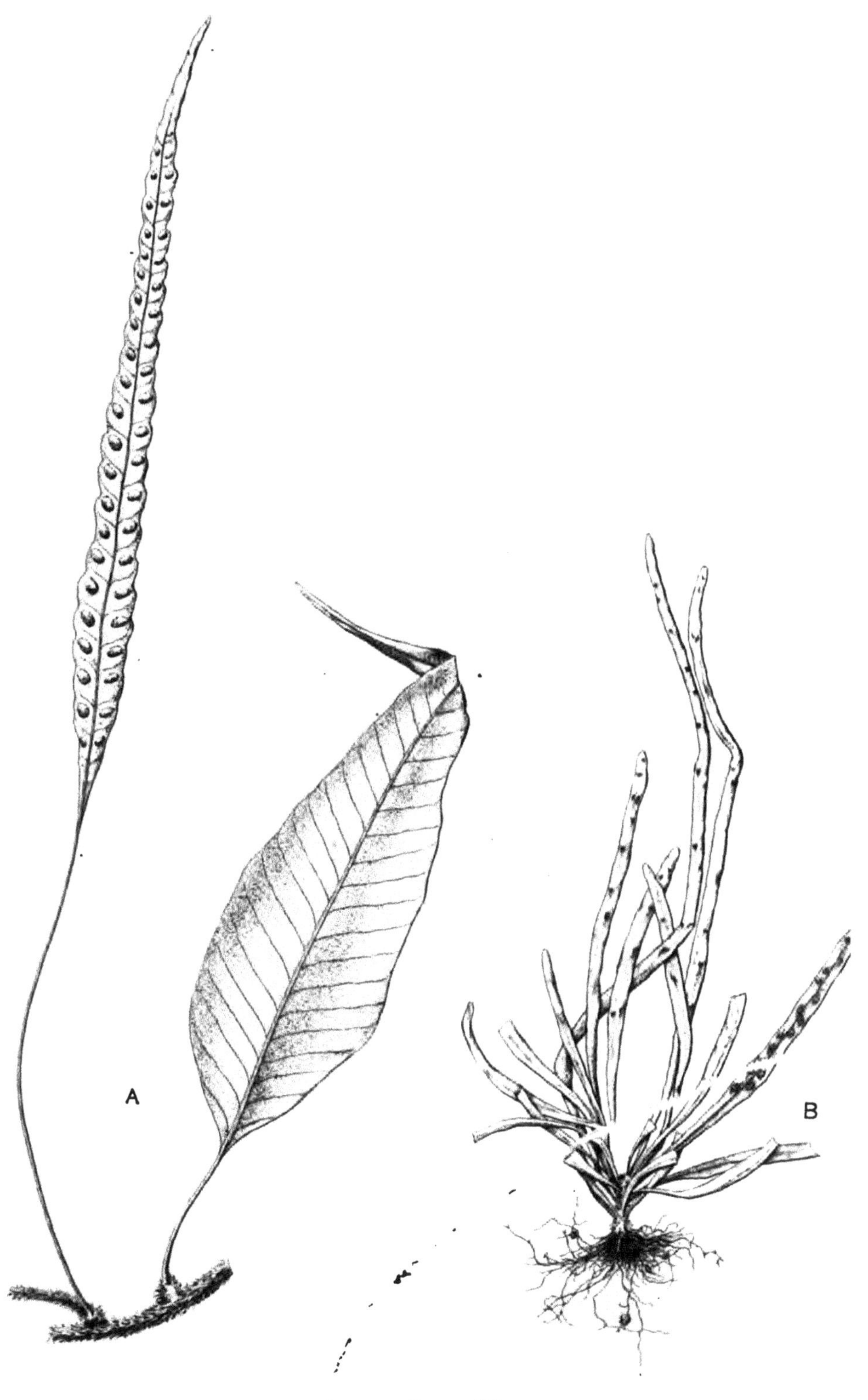

Plate III.

THE PHILIPPINE

JOURNAL OF SCIENCE

C. BOTANY

VOL. II | JUNE, 1907 | NO. 3

SPICILIGIUM FILICUM PHILIPPINENSIUM NOVARUM AUT IMPERFECTE COGNITARUM.

By H. CHRIST.
(*Basel, Switzerland.*)

In the collections of Philippine ferns sent me by *Elmer D. Merrill*, Botanist, of the Bureau of Science, and by Dr. *E. B. Copeland*, which have increased in an unexpected manner since the year 1903, I have distinguished very many forms not considered by Dr. *Copeland* in his papers on Philippine ferns,[1] and which are not included in my two works on the fern flora of the Archipelago.[2]

At the request of Mr. *Merrill*, I have prepared a list for publication in THE PHILIPPINE JOURNAL OF SCIENCE, containing the diagnoses of those species that appear to me to be undescribed, and observations on others which for one reason or another are of special interest. I wish to express my sincere thanks to Messrs. *Loher, Copeland* and *Merrill* for having with great liberality supplied me with the rich material that has enabled me to undertake this work.

Cuming's collection is the basis of our knowledge of the pteridophyte flora of the Philippines, and of which *J. Smith*[3] published a list of species, however, unfortunately, containing many almost *nomina nuda*.

[1] Ferns, in Perk. Frag. Fl. Philip. (1905) 175–194; Polypodiaceæ of the Philippine Islands, *Govt. Lab. Publ.* **28** (1905) 7–138; New Philippine Ferns. *This Journal* **1** (1906) Suppl. 143–166; New Philippine Ferns, II, *l. c.* 251–257; A New Polypodium and Two New Varieties, Elmer's *Leaflets Philip. Bot.* (1906) 78, 79.

[2] Filices Insularum Philippinarum, *Bull. Herb. Boiss.* **6** (1898) 127–154; 189–210; II. *l. c.* **6** (1906) 987–1011.

[3] Enumeratio Filicum Philippinarum in *Lond. Journ. Bot.* **3** (1841) 392–422.

Presl has described many of the species represented in *Cuming's* collection [4] in a masterly manner, but perhaps has carried the separation of species a little too far in some cases, although his observations are very exact and his descriptions very clear. In working on Philippine ferns, it is always necessary to consult this work in determining whether or not a species has been described.

HYMENOPHYLLUM Smith.

1. **Hymenophyllum Merrillii** n. sp.

Leptocionium, ex affinitate *H. holochili* (v. d. B.) C. Chr., Javanici, caespitosum, minus, laciniis brevioribus, colore atrofusco, textura crassiore.

Rhizomate filiformi repente caespitoso, cum stipite rhachique pilis rufis brevibus parce vestito, stipite filiformi 3 cm. longo, fronde ovata acuminata versus basin attenuata 6 cm. longa, 2 cm. lata, bipinnatifida, pinnis confertis ca. 8 utrinque, cuneato-ovatis antice acutis sessilibus nec adnatis infimis petiolulatis 6 mm. latis profunde pinnatifidis, segmentis cuneato-obtusis 3 utrinque, profunde laciniatis, laciniis lanceolatis 2 mm. latis serrulato-dentatis planis, rhachi haud alata, soris infimae laciniae anteriori pinnarum insidentibus, pro pinna solitariis, 3 aut 4 utroque rhacheos latere, ovatis, 2.5 mm. longis, apice bivalvatis serrulatis, receptaculo crasso valde exserto. Colore atrofusco. Textura rigidiuscula.

LUZON, Province of Pampanga, Mount Arayat (3927 *Merrill*) October, 1904; Province of Bataan, Mount Mariveles (*Loher*) March, 1897, alt. 1,400 m.

2. **Hymenophyllum serrulatum** (Presl) C. Chr. Ind. 367 (*H. Smithii* Hook. Sp. Fil. 1: 97. *Tab. 35 B*).

A species peculiar in the section *Leptocionium* by the valves of the sori being entire or very slightly dentate. A large species with ovate fronds 13 cm. long, 7 cm. broad, the stipes 8 cm. long. It appears to be one of the most widely distributed species of the genus in the Philippines.

LUZON, Province of Bataan, Mount Mariveles (3233 *Merrill*) October, 1903; (165, 443, *Whitford*) May, July, 1904; (208 *Copeland*) January, 1904; Province of Tayabas, Mount Banajao (918 *Whitford*) October, 1904; Province of Rizal, Angilog (*Loher*) March, 1906. NEGROS, Mount Silay (1509 *Whitford*) May, 1906.

The same species has been found in Perak (leg. *Hose*); the plant reported from Celebes under this name is doubtful.

3. **Hymenophyllum thuidium** Harringt. in Journ. Linn. Soc. Bot. 16 (1877) 26; Christ in Schum. und Lauterb. Nachtr. Fl. Deutsch. Schutsgeb. Südsee 1: 34.

A large very elegant species, all the foliaceous parts finely crisped and at the same time dentate-aristate. Very diaphanous, the sori small, globose, dark brown, terminal, the valves entire. Stipe 6 cm. long; frond 10 cm. long, 5 cm. wide.

MINDANAO, San Ramon at 800 m. alt. (1751 *Copeland*) April, 1905; Mount Apo, at 1,800 m. alt. (1441 *Copeland*) October, 1905.

Very nearly the same species is found in the Bismark Mountains, German New Guinea, leg *Schlechter* 14030, which I have called *H. Bismarkianum*.

[4] Epimeliae botanicae (1849).

4. **Hymenophyllum aculeatum** (v. d. Bosch Hym. Jav. *Tab. 31, Leptocionium*) Racib. Pter. Buitenz. 21.

LUZON, Province of Bataan, Mount Mariveles (209 *Copeland*) January, 1904; (3231 *Merrill*) October, 1903; Province of Tayabas, Mount Banajao (921 *Whitford*) October, 1904. MINDANAO, Province of Surigao (268 *Bolster*) March, 1906.

Identical with specimens from Mount Salak, Java, leg. *Raciborski*. It is the species enumerated in Bull. Herb. Boiss. **6**: 141, as *H. Neesii*.

5. **Hymenophyllum campanulatum** n. sp.

Leptocionium. Habitu omino *H. Tunbridgensis*, valvis integris, rhachi hispida.

Dense et late caespitosum, rhizomate tenui sed rigido ramosissimo. stipite rhachique nigris, pilis rigidis hispidis, stipite 1.5 cm. tenui, fronde 4 cm. longa 1 cm. lata bipinnatifida oblonga, basi et apice attenuata, pinnis ca. 6 utrinque arcuato-reflexis, alternis, 1 cm. longis, flabellato-partitis, laciniis 4 aut 5, linearibus, vix 1 mm. latis, parce aristato-serratis. Soris raris, prope basin costae positis, pedunculatis, campanulatis, valvis erecto-patentibus ovatis, 2 mm. longis. Textura rigidiuscula. Colore fusco.

NEGROS, Mount Silay (1549 *Whitford*) May, 1906, alt. 1,100 m.

6. **Hymenophyllum pycnocarpum** v. d. Bosch Hym. Jav. *Tab. 37*.

Laeve. Fronde ovata, tripinnatifida, rhachi alata, soris terminalibus paniculatis, valvis trigono-acutis.

LUZON, Province of Laguna, Mount Maquiling (*Loher*) January, 1890, January, 1906; Province of Tayabas, Mount Banajao (922 *Whitford*) October, 1904.

Hymenophyllum subdemissum Christ, Bull. Herb. Boiss. **6** (1898) 140, should be united with this species.

7. **Hymenophyllum Blumeanum** Sprengel, Syst. Veg. **4**: 131.

LUZON, Province of Bataan, Mount Mariveles (3232 *Merrill*) October, 1903, the lanceolate form figured by Van den Bosch Hymen. Jav. *Tab. 36, 2*. *H. polyanthos* Christ (non Sw.) Bull. Herb. Boiss. **6** (1898) 139, should be referred here.

8. **Hymenophyllum paniculiflorum** Presl, Hymen. (1843) 32, 55.

A small plant recognizable by its large sori which are ovoid or globose, terminal and occupying all the segments at the summit of the frond.

LUZON, Province of Benguet, Mount Tonglon, 2,250 m. alt. (*Loher*) April, 1904. Identical with specimens from Java, leg. *Giesenhagen* and *Raciborski*, and with specimens from Japan, leg. *Faurie*.

I am now of the opinion that *H. discosum* Christ, Bull. Herb. Boiss. **6** (1898) 140, should be united with this species, although the sori are much broader and more round than those in the Javan plant, leg. *Giesenhagen*.

9. **Hymenophyllum demissum** (Forst.) Sw. in Schrad. Journn. **1800**[2] (1801) 100.

LUZON, Province of Tayabas, Mount Banajao (*Loher*) February, 1906.

This plant is identical with the form found in New Zealand and Celebes, low, with small sori, the rachis winged only near the summit. *H. productum* Kze. Bot. Zeit. (1848) 305, Van den Bosch Hymen. Jav. *Tab. 45*, of Java, is a larger plant with triangular elongated valves and winged stipes.

10. **Hymenophyllum formosum** Brackenr. U. S. Explor. Exped. **16** (1854) 258. *t. 32. f. 3.*

MINDANAO, Mount Apo (1442 *Copeland*) October, 1904, alt. 1,800 m.; (340 *DeVore & Hoover*) May, 1903. Our plants match the form represented by *Tab. 47* Van den Bosch, Hymenophyllaceae Javanicae.

11. **Hymenophyllum Junghuhnii** Van den Bosch Hymen. Jav. 60. *Tab. 49.*

LUZON, District of Lepanto, Bagnen (1921 *Copeland*) November, 1905, alt. 1,950 m. MINDANAO, Province of Misamis, Mount Malindang (4638 *Mearns & Hutchinson*) May, 1906.

A form with very large sori and with broad segments.

The Philippine plants referred by various authors to H. dilatatum Sw., *are referable to the two above species.*

12. **Hymenophyllum australe** Willd. Sp. Pl. **5:** 527.

LUZON, District of Lepanto, Mount Data (1873 *Copeland*) October, 1905; Province of Benguet, Mount Tonglon (*Loher*); (5053 *Curran*) August, 1906. MINDANAO, Mount Apo (324 *DeVore & Hoover*) May, 1903.

The Philippine form has very narrow segments and is very compound, quadri-*pinnatifid.* Under the lens the margins are very finely denticulate.

TRICHOMANES Linn.

13. **Trichomanes parvulum** Poir. in Lam. Encycl. **8:** 46.

LUZON, Province of Bataan, Mount Mariveles (177 *Whitford*) May, 1904.

14. **Trichomanes diffusum** Blume, Enum. 225.

LUZON, Province of Laguna, Mount Maquiling (5137 *Merrill*) March, 1906, 300 m. alt. MINDANAO, Province of Zamboanga, San Ramon (1652 *Copeland*) April, 1905, 700 m. alt.

15. **Trichomanes nitidulum** Van den Bosch Hymen. Jav. 21.

LUZON, Province of Benguet, Baguio (6023 *Elmer*) March, 1904.

16. **Trichomanes rhomboideum** J. Sm. in Hook. Journ. Bot. **3:** 417; Van den Bosch Hymen. Jav. *Tab. 24.*

LUZON, Province of Bataan, Mount Mariveles (3121 *Merrill*) October, 1903; (200 *Copeland*) January, 1904.

This form, of the *Cephalomanes* group which is richly developed in the Philippines, is distinguishable by its rounded pinnæ, its elongated aristate teeth and long sori.

17. **Trichomanes Javanicum** Blume, Enum. 224; Van den Bosch Hymen. Jav. *Tab. 22.*

LUZON, Province of Rizal, Mabacal (*Loher*) March, 1906; Province of Laguna, Mount Maquiling (*Loher*) January, 1906; Province of Bataan, Mount Mariveles (203, 513 *Whitford*) July, 1904; (2307 *Borden*) January, 1905; (2420 *Meyer*) January, 1905; (207 *Copeland*) February, 1904. NEGROS, Gimagaan River (1603 *Whitford*) March, 1906. PALAWAN (578 *Foxworthy*) April, 1906.

Common and widely distributed in the Philippines.

Var. **intercalatum** n. var.

Soris margini anteriori, apici saepeque parti superiori marginis posterioris impositis numerosis (12) longe exsertis campanulato-clavatis ore non dilatato, dentibus marginis posterioris elongatis aristatis conspicuis. Pinnis saepe dissectis.

LUZON, Province of Laguna, Mount Maquiling (*Loher*) January, 1906; Province of Rizal, Orind (*Loher*) February, 1906; Province of Benguet, Sablan (6214 *Elmer*) April, 1904.

POLYSTICHUM Roth.

Polystichum, of the essentially Chinese group of *P. auriculatum*, is well represented in the Philippines. The first form known was *Phegopteris nervosa* Fée, Mem. **6**. 13. *Tab. 2. Fig. 4*, which I do not hesitate to reduce as a variety of *Polystichum deltodon*. The rich Chinese material that I have at my disposition has convinced me that there are but slightly marked differences by which the Philippine plant can be distinguished.

18. **Polystichum deltodon** (Bak.) Diels in Nat. Pflanzenfam. **1** [4] (1899) 191. *Aspidium deltodon* Baker in Gard. Chron. **14**: 494.

Var. **nervosum** (Fée Gen. 244; Mem. **6**: 13. *Tab. 2. Fig. 4, Phegopteris*).

Differt a typo montis Ōmi, Chinae occid., leg. *Faber* 1045 statura majori, pinnis magis numerosis (usque ad 50 utrinque) minus acutis.

LUZON, Province of Bataan, Mount Mariveles (*Loher*) November, 1894, alt. 1,400 m.; Province of Benguet, Baguio (5916 *Elmer*) March, 1904.

19. **Polystichum Copelandi** n. sp.

Rhizomate caespitoso foliis fasciculatis, stipite tenui 2.5 cm. longo cum rhachi squamis subulatis nigris parce vestito, viridi, fronde lanceolato-lineari, basi non attenuata, acuminata, usque ad 20 cm. longa, 2 cm. lata, pinnata pinnis densissime imbricatis inferioribus deflexis ca. 50 utrinque rhombeo-acutiusculis marginibus parallelis 1 cm. longis basi fere 0.5 cm. latis, postice cuneatis, antice truncatis acute auriculatis, postice integris antice minute dentatis dentibus ca. 10 haud aristatis, nervis furcatis, soris postice deficientibus antice ca. 6, 1 mm. a margine remotis minutis uniseriatis rufis impressis, indusio minuto fugaci rotundo umbonato. Textura herbacea, faciebus calvis, colore flavoviridi, opaco.

LUZON, District of Bontoc, Sagada (1901 *Copeland*) November, 1905, alt. 1,600 m.

Very near the small group of *P. hecatopteron* Diels, and strongly resembling *P. Dielsii* Christ in Bull. Acad. Mans. (1906) 238, from which it differs in its smaller size, its short stipe, the pinnæ not obtuse but pointed, the teeth more pronounced and the sori less marginal.

20. **Polystichum obliquum** (Don) Moore Ind. (1858) 87. *Aspidium obliquum* Don Prodr. 3. *A. caespitosum* Wall. Cat. No. 367.

Var. **Luzonicum** n. var.

Stipite debili 17 cm. longo, fronde 25 cm. longa 5.5 cm. lata, pinnis imbricatis oblongo-rhomboideis 3 cm. longis basi 12 mm. latis acutis acute auriculatis, margine fere integro sed dentibus parvis dejicientibus aristatis ciliato, soris biseriatis sed antice pluribus (ca. 10) medialibus minutis. Nervis occultis, textura herbacea, opaca.

LUZON, Province of Benguet, Trinidad (1812 *Copeland*) October, 1905, alt. 1,200 m.

Very close to the plant from China, India, Annam (leg. *Cadière*) and of Japan, but larger, the teeth very faint, scarcely visible but nevertheless aristate.

ASPIDIUM Swartz.

21. **Aspidium (Pleocnemia) Angilogense** Christ. in Bull. Herb. Boiss. II. **6** (1906) 1003.

I do not hesitate to refer here *Pleocnemia Cumingiana* Presl, Epim. Bot. 410, and regret that this specific name which has priority, is inappropriate in *Aspidium* because of the later use of the same specific name under *Aspidium* by several authors, *Kunze, Sturm*. In addition to the localities in Luzon cited by me in the original description of the species, *Dr. Copeland* has found it in Mindanao, San Ramon, April, 1905, No. 1698, alt. 600 m. He adds "Fronds 2 to 3 m. high, deltoid, stipe 2 m. high, stout, rhizome 10 to 20 cm. thick, ascending. *Presl*, without doubt after *Cuming's* notes says "arbor viginti-pedalis." In other characters I find our species to be very near *Nephrodium chrysotrichum* Baker from Samoa; Upolu, Apia leg. *Betche* 1880; Falwao leg. *Reinecke* 90, 94; Savai, leg. *Reinecke* 90b; Upolu, leg. *Reinecke* 94. The two collectors give the height of the Samoan plant as from 20 to 40 feet, with black trunks. It is certainly the largest arborescent fern outside of the *Cyatheaceae*.

22. **Aspidium profereoides** n. sp.

Fronde sterili longissime stipitata, stipite 45 cm. longa pennae cygni crassitie cum rhachi castaneo sulcato opaca squamis subulatis 0.5 cm. longis setisque brunneis sparso vestito, fronde ultra 50 cm. longa 32 cm. lata deltoideo-oblonga pinnata parce bipinnatifida, pinnis infimis nec abbreviatis nec postice auctis caeteris aequalibus, pinnis remotis recte patentibus ca. 15 utrinque infra apicem acuminatum inciso-lobatum, sessilibus, 18 cm. longis 3.5 cm. latis lanceolatis acutis usque ad mediam laminam incisis, ala utrinque 0.5 cm. lata et ultra, lobis angulo angusto obtuso separatis 12 mm. longis 1 cm. latis ovato-obtusis, ca. 18 utrinque, infimis aliquantum auctis et rhachin tegentibus, subintegris, nervis non prominentibus, in lobis pinnatis, 6 ad 8 utrinque, infimis areolam angulosam angustam secus costam, superioribus 1, 2 aut 3 areolas laterales secus costulam formantibus, ab areola costali ad sinum 3 aut 4 areolis intercalatis. Nervulis inclusis nullis. Textura flaccide herbacea, colore obscure viridi, opaco, faciebus imprimis costis furfuraceo-puberulis. Folio fertili sterili longiore; stipite 70 cm. et ultra, fronde 50 cm. longa, egregie contracta, pinnis valde remotis, 10 cm. longis, 2.5 cm. latis, ad alam angustam incisis, lobis lineari-lanceolatis falcatis 4 mm. latis, soris 1.5 mm. latis rotundis mediis brunneis, indusio flaccido atrobrunneo mox corrugato sine dubio reniformi.

Aspidium excellens Blume Enum. 120 differt ex descriptione Presliana l. c. frondibus monomorphis, segmentis acutis.

MINDANAO, District of Davao, Todaya (1467A *Copeland*) October, 1904, alt. 1,205 m. "Rootstock short, erect" *Copeland*.

A marked species with nerves of *Proferea* Presl, Epim. 619, and pronounced dimorphism.

LEPTOCHILUS Kaulfuss.

23. **Leptochilus heteroclitus** (Presl) C. Chr. Ind. Fil. (1905) 11, 385. *Acrostichum flagelliferum* Wall.

This species, widely distributed from British India, throughout Malaya, and in the Philippines especially, offers in the latter region a variability that approaches that of *Dryopteris canescens*. The apices of the sterile leaves are elongated into linear lash-like appendages which are proliferous and usually take root, the young plants emitting a fascicle of leaves entirely different from those of the adult ones. Often they are not simple linear leaves, with the lateral lobes more or less aborted, but frequently are singularly compound, enlarged and deeply incised. Often the little plants are fertile, and the fertile fronds offer irregularities analagous to the sterile ones. In other characters the differences are frequently strongly marked, and the various forms might readily be considered to represent distinct species. I have enumerated below the forms, that in my conception of this species, should be considered as varieties or subspecies, and although several of them have acquired a certain stability, certain characters are constant in all of them, showing their close relationship to *L. heteroclitus* and probable derivation from that species.

Var. **eurybasis** n. var.

Rhizomate repente, pennae anserinae crassitie, cum stipite basi squamis minutis dilute brunneis crispis sparso radicoso. Foliis sparsis sed appropinquatis, foliorum sterilium stipite 8 cm. longo flexuoso tenui griseo, fronde 13 cm. longo deltoideo-ovata, pinna terminali decurrente et cum pinna laterali proxima parva plus minus concreta, basi cuneata, acuminata, nec caudata nec prolifera, 8 cm. longa, 2.5 cm. lata oblonga parviloba, lobis 10 vel pluribus utrinque subrotundis 3 mm. latis acute denticulatis, pinnis lateralibus 3 aut 4 utrinque, mediis 3 cm. longis, 1 cm. latis cuneatis obtusis decurrentibus serrato-lobulatis, infimis valde postice auctis deltoideo-ovatis petiolatis nec decurrentibus, basi profunde incisis, caeterum serrato-lobatis. Nervis lateralibus rectis a costa ad marginem protensis 4 mm. distantibus, series plures areolarum valde irregulariarum, minores areolas nec nervulos liberos includentium continentibus. Colore atrato, textura herbacea, opaca. Foliorum fertilium lamina aequilonga sed angustiore, pinna terminali 5 cm. longa lanceolata acuta profunde lobata decurrente, pinnis lateralibus 4 utrinque obtusis, 2.5 cm. longis, 1.5 cm. latis, ovatis sessilibus aut adnatis, sed pinnis infimis petiolatis auctis, parte terminali late cuneato-ovata lobata, pinnulis lateralibus similibus minoribus.

MINDANAO, Lake Lanao, Camp Keithley (552 Mrs. *Clemens*) May, 1906, alt. 660 m.

Differing from the type by its aberrent fronds being in part bipinnate, deltoid at the enlarged base, posteriorly deeply incised; even the fertile fronds are sometimes bipinnate at the base.

Var. **Foxworthyi** n. var.

Pinna terminali lineari vel oblonga acuminata grosse dentata longe decurrente interdum radicante, pinnis duabus lateralibus rudimentariis minutis, 0.5 cm. aut ultra diametro, ellipticis aut rotundatis, pinna fertili terminali lineari-lanceolata duobus lobulis lateralibus ovatis suffulta. Planta vix 20 cm. alta.

LUZON, Province of Rizal, Bosoboso (68 *Foxworthy*) January, 1906.

Small, very narrow and very simple in comparison with the type, but closely related by the shoots developed along the upper parts of the leaves.

Var. **inconstans** (Copel. in Govt. Lab. Publ. **28** (1905) 43 pro specie).

In view of the fact that two varieties show a progressive intergradation between the type and the species described by *Copeland,* I have not hesitated to consider the latter as a variety of *Leptochilus heteroclitus,* a form still more reduced than the preceding, characterized by its very small size, the apices of the leaves often linear, the shoots very irregular. Var. *Foxworthyi* is almost exactly intermediate between the type and the variety *inconstans.*

LUZON, Province of Bataan, Lamao River (3128 *Merrill*) April, 1904; (251 *Copeland*) January, 1904; (1124 *Whitford*) March, 1905; Province of Rizal, Mabacal (*Loher*) March, 1906.

This variety seems to be rather widely distributed in the Philippines; I have specimens from Christmas Island (Straits Settlements) leg. *Ridley* that closely approach it.

Var. **Linnaeanus** (Fée Acrost. 87 pro specie).

I believe that this form can be reduced, with a sufficient degree of surety, as a variety of *Leptochilus heteroclitus,* as a derived form of that species, in spite of its uniformly elongated and very narrow leaves. Its texture, its nerves, although strongly simplified, its proliferous leaves and the strong resemblance of its offshoots to those of the normal form support this contention. We have then a series almost complete from the type, that seems to be always triphyllous in the Philippines, to the form with absolutely linear and undivided leaves.

LUZON, Province of Rizal, Manap River, near Montalban (*Loher*) 1892.

24. **Leptochilus diversifolius** Blume, Enum. 103 et Fil. Jav. *Tab. 12.*

MINDANAO, San Ramon (1543 *Copeland*) November, 1904, alt. 100 m.

Discovered by *Blume* in Java and generally confounded with *L. heteroclitus,* from which it is sufficiently distinct. Conf. *Raciborski* Pterid. Buitenz. 48.

ATHYRIUM Roth.

25. **Athyrium anisopteron** Christ, Ann. Acad. Mans. (1907).

LUZON, Province of Benguet, Mount Tonglon (*Loher*) 1894, alt. 2,250 m.; Pauai (1967 *Copeland*) November, 1905, alt. 2,150 m.

This is the plant that I have previously considered as *Aspidium Fauriei* var. *elatum* Christ,[5] and which *Copeland,* MSS., referred to *Nephrodium. Makino*[6] treats it with reason under *Athyrium.* After an examination of abundant material I have divided this species into several, the representative in the Philippines being *A. anisopteron,* a species of a Chinese group already known from Yunnan, leg. *Henry* et *P. Ducloux.*

[5] *Bull. Herb. Boiss.* **6** (1898) 193.
[6] *Bot. Mag. Tokyo* **17**: 160.

26. **Athyrium nanum** n. sp.

Rhizomate obliquo brevi aut subrepente, foliis subfasciculatis paucis (3 ad 4) stipite tenui fusco squamulis tenuissimis setulosis rufobrunneis patentibus cum rhachi pubescente fere 0.5 ad 1 cm. longo, fronde lanceolata 6 cm. longa, 7 mm. lata in longam apicem pinnatifidum prolongata, fere usque ad basin alata et utrinque 5 aut 6 pinnis remotiusculis praedita, pinnis segmentisque erecto-patentibus, 2.5 ad 5 mm. altis superioribus basi aequalibus integris oblongis inferioribus inaequalibus antice auriculatis subcrenatis infimis inaequalibus crenato-lobatis, puberulis, nervis furcatis vix pinnatis, soris ca. 5 pro segmento, vix 1 mm. longis, ovatis, ochraceis, indusio lanceolato rarius aspidioideo reniformi. Colore laete virente, textura flaccide herbacea.

MINDANAO, Lake Lanao, Camp Keithley (656 Mrs. *Clemens*) July, 1906.

The smallest species of the genus known, pinnatifid and in part only simply pinnate, distinguished by its rather strongly winged rachis.

27. **Athyrium drepanopteron** (Kze.) Moore. *Asplenium oxyphyllum* (Wall.) Hook. *Polypodium drepanopteron* Kze. Linnæa **23**: 278, 318. *Lastrea eburnea* J. Sm. Bot. Mag. **72** (1846) Comp. 34.

LUZON, Province of Benguet, Baguio (6498 *Elmer*) June, 1904; found also by other American collectors.

A continental type extending from Japan to Yunnan and north India.

28. **Athyrium Benguetense** n. sp.

Rhizomate ut videtur repente aut obliquo tenui nigro, stipitibus approximatis paucis tenuibus stramineis 25 cm. et ultra longis, basi squamulis lanceolatis brunneis 2 mm. longis sparsis vestitis, rhachi et costa puberulis, planta aliter nuda, fronde caudato-acuminata lanceolata 30 cm. longa 9 cm. lata basi non attenuata bipinnatifida, pinnis patentibus sessilibus (costa straminea) remotis, infimis subdeflexis, ca. 25 infra apicem lobatum, 5 cm. longis 16 mm. latis lanceolato-acuminatis fere ad rhachin incisis, segmentis subpectinatis ca. 18 utrinque 2.5 mm. latis, lineari-oblongis obtusiusculis subcrenatis aut integris, nervis 6 ad 8 utrinque simplicibus obliquis, soris mediis 6 ad 8 utrinque, rotundis vix 1 mm. latis ochraceis, indusio tenuissimo rotundo reniformi hyalino umbone obscure, mox evanido. Textura flaccida, colore atroviridi.

LUZON, District of Lepanto, Mount Data (*Loher*) February, 1894, 2,250 m. alt.; Province of Benguet, Pauai (1948 *Copeland*) November, 1905, alt. 2,150 m.

By its aspidioid sori a *Dryopteris*, but in all other respects a true *Athyrium*, very delicate and with segments almost entire.

29. **Athyrium Copelandi** n. sp.

Rhizomate ramoso breviter repente crassiusculo atrato, stipitibus plumbeo-stramineis basi squamis lanceolato-subulatis brunneis parce vestitis, supra cum rhachi furfuraceo-squamulosis, 9 cm. longis, fronde 14 cm. longa, 6 cm. lata oblonga, basi non attenuata longe acuminata, pinnis ca. 15 utrinque (rhachi versus apicem alata) sessilibus superioribus adnatis et decurrentibus, falcatis, inferioribus remotis, grosse lobatis 3 cm. longis 1 cm. latis lobis ca. 6 utrinque infimus antice maximis, 1 cm. longis, 3 cm. latis, triangulari-ovatis, acutiusculis subcrenatis, nervis in lobis pinnatis 3 ad 4 utrinque, obliquis, soris 2 ad 3 mm. longis 1 mm. latis, turgidis ovato-lanceolatis, omnino tectis, indusio leviter curvato, griseo, membranaceo, persistente, soris rufo-ochraceis. Textura coriacea, colore atrovirente subtus pallido, faciebus glabris.

LUZON, District of Lepanto, Mount Data (1909 *Copeland*) November, 1906, alt. 1,700 m.

A small alpine species intermediate between *Athyrium acrostichoides* (Sw.) Diels and *Diplazium japonicum* (Thunb.).

DIPLAZIUM Swartz.

30. **Diplazium bulbiferum** Brack. U. S. Expl. Exp. 16 (1854) 141. *Tab. 18. f. 1.*

Rhizomate pollicis crassitie, oblique erecto, radicibus nudis crassis et longis semi-supraterraneis suffulto, foliis fasciculatis sed paucis; stipitibus basi atratis squamisque subulatis nigris 0.5 cm. longis vestitis, superne raris setis nigris sparsis, 15 ad 25 cm. longis, tenuibus; fronde 20 ad 30 cm. longa, 13 cm. lata, ovato-oblonga apice pinnatifida-acuminata pinnata, ad basin apicis rhachi gemmifera et interdum vivipara, pinnis infimis aliquantum abbreviatis et deflexis, pinnis fere omnibus versus apicem usque egregie petiolatis, petiolo 3 ad 5 mm. longo, horizontali, pinnis 8 ad 10 utrinque infra apicem pinnatifidum, oblongis acutis nec caudatis ca. 4 cm. rarius magis longis 18 mm. latis basi inaequalibus postice cuneatis antice auriculato-truncatis grosse crenato-serratis lobulis brevibus raro ultra 3 mm. longis 3 ad 5 mm. latis decumbentibus, acutiusculis, nervis utrinque in lobulis pinnatis plerumque 2 aut 3 utroque costulae latere obliquis, soro plerumque duplici (diplazioideo) protenso 7 mm. longo brunneo anguste lineari nervulo anteriore imposito, soris irregularibus brevibus aliis nervulis insidentibus, indusio tenuissimo diaphano decolori. Textura herbacea, planta laevi, colore supra atroviridi, subtus pallido, opaco.

MINDANAO, Davao (701 *Copeland*) March, 1904; San Ramon (1673 *Copeland*) March, 1905; Lake Lanao, Camp Keithley (167, 252 Mrs. *Clemens*) February, 1906. LUZON, Province of Bataan, Mount Mariveles (6010 *Leiberg*) July, 1904; (234 *Whitford*) May, 1904; (238 *Copeland*) January, 1904.

A species of the *D. silvaticum* group, but smaller, proliferous, its pinnæ distant, generally short, shortly pointed and long petioled, unequally auriculate, slightly lobed, one long sorus and some short ones on each lobe, very pale beneath. The specimen from San Ramon is very large with 16 pairs of pinnæ 12 by 2 cm., and more deeply lobed than the others.

I think that the plant that I took for *D. silvaticum* Presl in my first paper on *Loher's* Philippine ferns,[1] and of which I have not a specimen at hand, is

[1] Bull. Herb. Boiss. 6: 153.

perhaps referable to the above species which seems to be widely distributed in the Philippines. *D. petiolare* Presl, Epim. 446, a species that I have not seen, differs, according to the description, in its linear pinnæ and pubescent rachis. *Diplazium silvaticum* is to me more and more a "collective species" worthy of being segregated into several distinct forms.

31. **Diplazium atratum** n. sp.

Stipite 25 cm. et ultra longo basi incrassito digiti diametro sulcato, ebeneo aut atroviolaceo, opaco uti tota planta, basi latere ventrali more cyathearum squamis subulatis 1.5 cm. longis nigris dense vestito, planta aliter subnuda aut minute furfuracea, fronde late deltoidea acuta 60 cm. et ultra longa basi 45 cm. et ultra lata, bipinnata, pinnis remotis petiolo 2 ad 3 cm. longo praeditis superioribus subsessilibus, recte patentibus, utrinque ca. 20, infimis haud reductis, usque ad 30 cm. longis, 10 cm. latis, longe acuminatis, basi attenuatis pinnulis ca. 20 infra apicem incisum ultra 1 cm. distantibus, recte patentibus inferioribus petiolulatis e basi lata lanceolato-acuminatis, ultra mediam laminam basique fere ad costam incisis, lobis ca. 15 utrinque porrectis 3 mm. longis trigono-falcatis acutis margine serrulato saepe reflexo pinnis superioribus pinnulis similibus, nervis 4 ad 7 utrinque obliquis crassis simplicibus, soris nervosa costa fere ad marginem segmentibus valde obliquis convexis atrobrunneis, linearibus simplicibus, indusio lineari coriaceo brunneo persistente. Fronde glabra, textura rigide, fere lignoso-coriacea, colore atrofusco, opaco.

PALAWAN, Victoria Peak (714, 683, 663 *Foxworthy*) March, 1906, alt. 600 to 1,100 m.

The plant with its hard woody texture and its dark color must be in sharp contrast to the surrounding vegetation. It is a very stiff coriaceous species with ample, deltoid, bipinnate fronds, the pinnæ petioled, the pinnules narrow and deeply incised, the lobes angular, narrow, their margins reflexed, the stipe and axial parts black, the frond itself blackish. The species has the general appearance of a *Dicksonia* or a *Cyathea*.

32. **Diplazium crenatoserratum** (Blume Enum. 177, *Asplenium*).

MINDANAO, San Ramon (1667 *Copeland*) March, 1905, alt. 650 m.

This species has not previously been reported from the Philippines; the above specimen matches material in my herbarium from Singapore leg. *Hose*, Java leg. *Lefebre*, Borneo leg. *Grabowsky*, *Hose*, *Niewenhuis*, Celebes leg. *Sarasin* and Sumatra leg. *Schneider*.

33. **Diplazium Smithianum** (Baker) Diels Nat. Pflanzenfam. 1⁴ (1889) 228. *Asplenium Smithianum* Baker Syn. 245.

I am unable to distinguish from this species No. 2667 *Merrill* from Bosoboso, Rizal Province, Luzon, determined by *Copeland* as *D. dolichosorum* Copel. The type of *Copeland's* species is from MINDANAO and he does not mention the LUZON plant in his diagnosis. *Loher* has previously found *D. Smithianum* in LUZON, and I have specimens from Celebes, leg. *Sarasin*, *Koorders* and *Warburg* (No. 15314) and from New Guinea, Sattelberg, *Weinland*, 1890. It is also found in Ceylon.

ASPLENIUM Linn.

34. **Asplenium exiguum** Bedd. Ferns. S. Ind. *t. 146.*

LUZON. Province of Benguet, Adouay (1845 *Copeland*) October, 1905, alt. 900 m.; Baguio (4887, 4860 *Curran*) August, 1906.

An Asiatic type.

I have specimens from Simla leg. *Blanford*, Massuri leg. *Hope*, Bhotan leg. *Griffith* No. 2812, Yunnan leg. *Delavay* and from The Nilgiris, southern India, leg. *Gamble*. It has not been found in Japan.

35. **Asplenium Elmeri** n. sp.

Rhizomate brevi radicoso, squamis castaneis subulatis coronato stipitibus fasciculatis 7 ad 12 cm. longis cum rhachi atrorufis subintentibus, squamulis patentibus subulatis brunneis vestitis, rhachi in parte superiore sed infra apicem prolifera, fronde oblonga acuminata basi attenuata 20 cm. longa 6 ad 8 cm. lata, bipinnata sive tripinnatifida, pinnis petiolatis, 16 ad 20 utrinque recte patentibus remotiusculis breviter petiolatis, costa libera supra breviter alata, pinnis ovato-elongatis acuminatis apice grosse dentato lanceolato, pinnis 2 ad 3 cm. longis, 1.5 cm. latis remotis paucis 2 rarius 3 utrinque, petiolulatis ovato-cuneatis obtusis grosse dentatis 8 mm. longis 3 mm. latis, pinnula basali anteriore aucta et rhachi approximata, nervis in segmentis flabellato-furcatis, soris 2 aut 3 in segmento, lanceolatis 4 mm. latis atrobrunneis indusio griseo angusto persistente. Colore opaco atroviridi, textura herbacea.

LUZON, Province of Benguet, Mount Santo Tomas (6538 *Elmer*) June, 1904: District of Lepanto. Mount Data (1858 *Copeland*) October, 1905.

Loher has found the same species previously on Mount Mariveles, Province of Bataan, LUZON, September, 1893, but his specimens are larger, 55 cm. tall. Very nearly the same species is found in Celebes (1322 *Sarasin*) November, 1895, but in *Sarasin's* plant the segments are longer and the rachis is also proliferous. A small species resembling *Asplenium cuneatum* Lam., and *A. praemorsum* Sw., which are frequently found in herbaria under the name *A. laserpitiifolium*, and sometimes under the name *A. contiguum*. A good species characterized by its stipe and rachis being covered with scales.

I no longer maintain as a Philippine species *Asplenium nitidum*, that I previously credited to LUZON.[a] I have not seen from the Archipelago the large plant with long pinnules which are auricled and deeply incised such as is represented in Malacca, by specimens leg. *Ridley* and in Borneo, leg. *Nieuwenhuis*, etc.

36. **Asplenium praemorsum** Sw. Prodr. 130.

LUZON, Province of Benguet, Pauai (*Copeland*) November, 1905, alt. 2,200 m.

A small form with narrowly cuneiform segments previously found in Luzon by *Loher*.

37. **Asplenium truncatilobum** (Presl) *Tarachia trunctailoba* Presl Epim. (1849) 437. *Asplenium arayatense* Christ Mss.

LUZON, Province of Pampanga, Mount Arayat (3816, 3909 *Merrill*) May, October, 1904: Province of Laguna, Mount Maquiling (*Loher*) 1906: Province of Zambales, Mount Pinatubo (*Loher*) February, 1906: Province of Benguet. Adouay (1857c *Copeland*) October, 1905.

[a] *Bull. Herb. Boiss.* (1898), 6, 153.

The specimens have been referred by the collectors with doubt, sometimes to *Asplenium hirtum* Kaulf., sometimes to *A. caudatum* Forst. They have more the appearance of *A. horridum* Kaulf., but differ from the latter in being smaller, with shorter pinnæ, the lobes less numerous, truncate at the apices, and the stipe particularly villous. I believe that the specimens cited above are identical with the species clearly described by *Presl.*

38. **Asplenium horridum** Kaulf. Enum. 173.

MINDANAO, District of Davao, Mount Apo (319 *DeVore et Hoover*) 1903.

Very typical and agreeing with specimens from the Sunda Islands and Polynesia.

39. **Asplenium militare** Copel. in Philip. Journ. Sci. 1 (1906) Suppl. 254.

MINDANAO, District of Davao, Mount Apo (321 *DeVore et Hoover*) May, 1903; (1505 *Copeland*), 1,800 m. alt., a specimen with more deeply lobed pinnæ.

A species remarkable for its resemblance to *Asplenium serra* Langsd. et Fisch., of tropical America, very large with broad lanceolate pinnæ which are lobed and finely denticulate, the sori short and close to the costa.

40. **Asplenium cuneatum** Lam. Encycl. 2: 309.

This group, difficult everywhere, is particularly polymorphous in the Philippines. A form that can be admitted to the Philippine flora, without doubt, is the following:

Var. **tripinnatum** Fourn. Fil. Nov. Caledon. 307.

Tripinnatifidum, segmentis brevibus ovato-cuneatis.

MINDANAO, District of Davao, Mount Apo (318 *DeVore et Hoover*) May, 1903: Province of Zamboanga, San Ramon (1728 *Copeland*) April, 1904, alt. 600 m. LUZON, Province of Rizal, Montalban (*Loher*) March, 1906.

41. **Asplenium laserpitiifolium** Lam. Encycl. 2: 310.

This species in the young state is perhaps sometimes confused with the preceding, but *Fournier* l. c., indicates a good distinctive character. In *A. cuneatum* the sori are narrow, flabellate and reach to the border of the pinnæ, while in *A. laserpitiifolium* they are convex and are confined to the middle of the pinnæ.

LUZON, Province of Benguet, Baguio (6029 *Elmer*): Province of Bataan, Mount Mariveles (176 *Whitford*) May, 1904.

Var. **subvenustum** n. var.

This is a reduced form 30 to 45 cm. high with very much divided fronds and slender stipes, the pinnules 5 mm. long, sometimes longer, triangular, flabellate, the sori small, short, two or three on a pinnule, resembling *Adiantum venustum* Don. The specimens are not young plants, but appear to be full grown and constant in the above characters.

LUZON, Province of Rizal, Bosoboso (1097 *Ramos*) July, 1906; (80 *Foxworthy*) January, 1906: Province of Cavite, Mendez Nuñez (1037 *Mangubat*) August, 1906.

42. **Asplenium affine** Sw.; Schrad. Journ. **1800**[2] (1801) 56.

After comparison with specimens from Bourbon and Africa, I admit to the Philippine flora as this species, a specimen with bipinnate fronds, the pinnules elongate-rhomboidal, 2 cm. long, unequal, irregularly dentate, slightly incised, the sori numerous, straight, narrow, elongated, and parallel. In texture this species is firmer than the preceding one.

MINDANAO, District of Davao, Todaya (1502 *Copeland*) October, 1904, alt. 725 m.

43. **Asplenium insititium** Brack. U. S. Explor. Exped. 161. *pl. 22. f. 2.*

LUZON, Province of Benguet, Baguio (6012 *Elmer*) March. 1904.

I have identified this plant after comparison with specimens from the Sandwich Islands, leg. *Hillebrand* and *Baldwin*, and from New Caledonia leg. *Franc.* It is the form that *Copeland* in his Polypodiaceæ of the Philippines. 84. supposes to be the variety *bipinnatifidum* of *A. contiguum* Kaulf., but it belongs evidently in the section with *A. cuneatum*.

STENOCHLAENA J. Sm.

The species of this genus often can not be determined with certainty without utilizing the characters shown by the secondary leaves. Unfortunately these secondary leaves are as yet imperfectly known in many species. for frequently when adult and soriferous leaves are found. the secondary leaves are not to be found, and without the three forms and without the certainty of their having come from the same plant it is often difficult if not impossible to identify these forms of *Stenochlaena* with trimorphous leaves.

Underwood[9] separates the species of *Stenochlaena* of the Old World, which have the veins springing directly from the midrib, into two groups: 1, *Teratophyllum* with trimorphous leaves and with spiny naked rhizomes, and 2, *Lomariopsis* with rhizomes covered with scales but spineless. However, our knowledge of the last group is not sufficiently complete to determine whether or not the secondary leaves are present or lacking, but I am of the opinion that they are present, at least in some species.

From the Philippines I am able to record the following species:

44. **Stenochlaena aculeata** (Blume) Kunze Bot. Zeit. **6**: 142. *Lomaria aculeata* Blume Enum. Pl. Jav. 205.

LUZON, Province of Rizal (*Loher*) March, 1906; (2695 *Ahern's collector*) conf. *Verhandl. Schweiz. Nat. Forsch. Gesellsch.* (1906) *Tab. 8;* Province of Benguet (6264 *Elmer*); northern Luzon (*Warburg*). MINDANAO, Mount Batangan (*Warburg*).

The secondary leaves of this plant agree very well with those figured by *Hooker* Sp. Fil. **1**: 56. B, for *Davallia achilleaefolia* Wall., which is cited by *Underwood* as a synonym of *S. aculeata.* I do not hesitate to identify with this species the form described by *Copeland* as *Asplenium epiphyticum* (Perk. Frag. Fl. Philip. (1905) 184), and Dr. *Copeland* himself admits in his Comparative Ecology of San Ramon Polypodiaceæ,[10] that this plant is "apparently identical with occasional immature forms of *S. aculeata.*" This form constitutes simply, as I have demonstrated in *Verhand. Schweiz. Nat. Forsch. Ges.* (1906) *Tab. 5,* the metamorphosis of the secondary leaves to the adult ones, combined with asplenioid sori which appear on the metamorphosed leaves as a reminder of the origin of the genus, which is from the vicinity of *Asplenium.* I have a specimen from Dr. *Copeland* which has beside scolopendriform leaves, a portion of the rhizome with spines and with tripinnate leaves similar to those figured by Hooker for *Davallia achilleaefolia.*

Copeland found his *Asplenium epiphyticum* without the adult form of *Stenochlaena,* which shows that the species of *Stenochlaena* are not always normally developed, but remain sometimes in a stunted condition. An analagous case is found in the Philippines in *Leptochilus heteroclitus.*

[9] *Bull. Torr. Bot. Club.* **33** (1906) 35.

[10] *This Journal, Bot.* **2** (1907) 69.

MINDANAO, Surigao (260 *Bolster*); Davao (699 *Copeland*); San Ramon (1572 *Copeland*).

The same plant with simple leaves but with their bases cut into irregular pinnate segments has been called by Bory, *Scolopendrium Durvillei* (Kunze Schkuhr Suppl. *Tab. 5.*).

MINDANAO, Mount Batangan (14111 *Warburg*).

45. **Stenochlaena Williamsii** Underw. in Bull. Torr. Bot. Club. **33** (1906) 41.

LUZON, Province of Bataan, Lamao River (368 *Barnes*), det. *Copeland*.

The specimens agree well with the description of the species. The secondary leaves, which I presume belong with the specimens cited, but without being able to determine this point with certainty, differ from those of the preceding species in their linear, more elongated segments and with a tendency of the frond to become gradually larger and to present auricles at the anterior base of the pinnules, a point of union with *Asplenium* of the *cuneatum* group, conf. *Verhandl. Schweiz. Nat. Forsch. Gesell.* (1906) *Tab. 6.* The secondary leaves mentioned above are those of specimens from Mindanao (*Warburg*), Luzon (*Warburg*) and North Celebes, Bojong (15321 *Warburg*).

46. **Stenochlaena arthropteroides** Christ in Bull. Herb. Boiss. II. **6** (1906) 998.

LUZON, Province of Rizal, Montalban (*Loher*) January, 1906, a very similar form, but slightly larger from the Lamao River, Province of Bataan (85 *Barnes*).

This species is distinguished by its very unequal and crenulate pinnæ, one specimen with secondary leaves bearing also some adult leaves, which, although small, are strongly crenulate. The secondary leaves are rather large, 10 cm. long, 4 cm. wide, nearly sessile, tripinnatifid, the rachis reddish, flexuous, the pinnæ ovate, obtuse, 2 cm. long, 1 cm. wide; the pinnules serrate, cut into linear segments which are obtuse, often bi- or tri-furcate, 2 mm. long, 0.5 to 1 mm. wide, the color very dark green.

47. **Stenochlaena subtrifoliata** Copel. in Philip. Journ. Sci. **1** (1906) Suppl. 152.

MINDANAO, District of Zamboanga, San Ramon (1749 *Copeland*), alt. 750 m.

Judging from the above authentic specimen this is a very distinct species, characterized by the cartilaginous borders of the adult pinnæ.

48. **Stenochlaena palustris** (Burm.) Bedd. *Polypodium palustre* Burm. Fl. Ind. (1768) 234.

MINDANAO, Davao (532 *Copeland*).

The typical form, identical with specimens in my herbarium from Java, Ceylon. Himalaya and Samoa.

49. **Stenochlaena** sp.

This is the plant described by me in *Bull. Herb. Boiss.* II. **6** (1906) 997, and which approaches *S. Milnei* Underw. ex descr., but which it is impossible for me to identify specifically because fertile fronds are lacking.

DAVALLIA Sm.

50. **Davallia decurrens** Hook. Sp. Fil. **1**: 167. *t. 94 B.*

This species, found by *Cuming*, is cited by *Copeland* in his Polypodiaceæ of the Philippines 54, without exact locality. It appears to be rare. I have specimens from Montalban, Province of Rizal, LUZON, collected by *Loher* in March, 1906, that agree exactly with *Hooker's* figure, except that *Loher's* specimens are smaller, and have bi- to tri-pinnatifid fronds instead of tri- to quadri-pinnatifid ones.

51. **Davallia vestita** Blume Enum. (1828) 233.

LUZON, Province of Pampanga, Mount Arayat (3878 *Merrill*) May, 1904: Province of Tayabas, Mount Banajao (*Loher*) February, 1906. NEGROS, Mount Silay (1516 *Whitford*) May, 1906.

This species does not seem to be rare in the Philippines. The specimens are less scaly than those of Java and Celebes.

52. **Davallia pusilla** Mett. Ann. Sc. Nat. IV. **15** (1861) 79.

MINDANAO, Province of Zamboanga, San Ramon (1665 *Copeland*) 1905.

Copeland[11] has identified this number as *Humata parvula* J. Sm., but my specimen agrees exactly with specimens of *Davallia pusilla* from New Caledonia leg. *Franc.*

MICROLEPIA Presl.

53. **Microlepia Sablanensis** n. sp.

Tripinnata, ampla, rhachi opaca fulvostraminea, brevissime et molliter puberula, pinnis 35 cm. et ultra longis, 9 cm. latis, elongato-caudatis, breviter petiolatis basi vix abbreviatis, pinnulis pectinato-confertis numerosis (40 et ultra utrinque) recte patentibus, fere sessilibus, acuminatis, basi inaequali, segmento infimo anteriore aucto, libero, ad rhachim adpresso, ad rhachim incisis, segmentis ca. 20 utrinque confertis angulo angusto separatis oblongis, obtusis, 0.5 cm. longis, 2.5 mm. latis, inaequalibus, postice subintegris cuneatis, antice truncatis, crenatis, lobulis plerumque 3 minutis, nervis manifestis, in lobulis bi- aut tri-furcatis, soris minutis, 2 aut 3 plerumque antice in sinubus lobulorum positis, globosis, indusio inconspicuo semicupuliformi tenuissimo. Textura herbacea, costis nervisque subtus pilosis, facie superiore laevi, sed opaca, colore obscure viridi.

LUZON, Province of Benguet, Sablan (6231 *Elmer*) April, 1904.

Differing from *Microlepia speluncae* (L.) Moore in its larger fronds, finer pubescence, the pinnules more numerous, the segments smaller, less unequal and the denticulations finer.

DENNSTAEDTIA Bernh.

54. **Dennstaedtia Smithii** (Hook.) Moore Index. 308.

LUZON, Province of Bataan, Mount Mariveles (1133 *Whitford*) March, 1904: Province of Rizal (91 *Foxworthy*) January, 1906.

The specimens cited above agree exactly with the figure given by *Hooker*,[12] but a species very close to this, recently described by *Copeland*, seems to be more common and widely distributed. It is larger, more pubescent, and with the basal pinnule of the III order anteriorly very much augmented. It is:

55. **Dennstaedtia Williamsi** Copel. in Philip. Journ. Sci. **1** (1906) Suppl. 148, of which I had prepared the following diagnosis before learning that *Copeland* had already described the species.

Amplissima, quadripinnatifida, stipite 2 m. alto, 2.5 cm. diametro, tereti, rufo-testaceo, dense cum rhachi pilis strigosis patentibus 2.5 mm. longis fulvis tomentoso postea glabrata, frondibus 2 m. altis, fasciculatis

[11] *This Journal* **1** (1906) Suppl. 147.

[12] Sp. Fil. **1**: *t. 28 D.*

(*Copeland*) pinnis petiolatis, 90 cm. longis, 30 cm. latis oblongis acuminatis basi haud attenuatis. Pinnulis ca. 35 utrinque, infimis remotis, reliquis approximatis, petiolulatis, infimis 16 cm. longis, basi 6 cm. latis, acuminatis, oblongis, antice basi valde auctis, i. e., pinnula^{III} 5 cm. longa lanceolata, rhachi adpressa et pinnularum^{II} proximam attingente et superante, basi pinnulae^{II} posteriore cuneata, pinnulis^{III} valde abbreviatis praedita. Pinnulis^{III} imbricato-confertis ca. 25 utrinque, lanceolato-obtusiusculis, basi inaequali antice aucta, brevissime petiolatis, 2 cm. longis, 0.5 cm. latis, usque ad costulam alatam incisis, segmentis ultimis oblongis obtusis inaequalibus antice crenatis 0.5 cm. longis, 2.5 mm. latis, costis nervisque adpresse rufo-pubescentibus, nervis in segmentis pinnatis furcatis manifestis, soris in sinu dentium positis, uno rarius pluribus pro segmento, marginalibus, 1 mm. diametro, globosus, brunneis, indusio superiore (marginali) manifesto deflexo, inferiore sporangiis conferto. Textura coriacea rigida, colore griseo-virente, opaco.

MINDANAO, Province of Zamboanga, San Ramon (1632 *Copeland*) February, 1905: District of Davao, Mount Batangan (14134 *Warburg*): Lake Lanao, Camp Keithley (375 Mrs. *Clemens*) March, 1906. LUZON, Province of Rizal, Mount Batay (*Loher*) April, 1905; Arambibi River (*Loher*) March.

56. **Dennstaedtia Hooveri** n. sp.

Amplissima, rhizomate repente, fronde 2 m. alta, subdeltoidea, quadripinnatifida, stipite 1.5 ad 2 m. (*Copeland*) rhachi digiti crassitie, cum costis costulisque pubescentia brevi strigosa rufa tecta, pinnis 60 cm. longis, 18 cm. latis breviter petiolatis basi vix attenuatis, acuminatis, pinnulis alternis approximatis petiolulatis, ca. 35 utrinque, e basi lata et antice aucta oblongis acutis, 9 cm. longis basi 3 cm. latis, costa haud alata, pinnulis^{III} ca. 18 utrinque, fere imbricato-confertis, ovato-obtusis inaequalibus, basi posteriori cuneatis, antice truncatoauctis 1 cm. longis, 0.5 cm. latis usque ad rhachin incisis, segmentis^{IV} 3 aut 4 utrinque, cuneato-oblongis obtusis, 3 mm. longis 2.5 mm. latis crenato-lobatis, iis pinnarum sterilium elongatis dentibus acutiusculis, nervis in segmentis pinnatis et furcatis, faciebus utrinque pilosis, pilis albidis tortuosis, soris 1 mm. latis rotundatis, uno rarius pluribus pro segmento, in dente obtuso brevi basilari posito, indusio tenuissimo mox evanido infero semicupuliformi membranaceo-griseo. Textura tenuiter herbacea, colore obscuro-viridi.

MINDANAO, District of Davao, Mount Apo (*DeVore & Hoover*) May, 1903: Province of Zamboanga, San Ramon (*Copeland*) May, 1905, alt. 850 m.

A species related to *D. flaccida* (Forst.) Bernh., of Samoa, characterized by its long and rather frequent villous hairs, the pinnules of the third order broad, irregularly parted and serrate-imbricate, and a rather pronounced dimorphism between the fertile and sterile fronds. Texture thin. Rachis and costæ covered with a reddish pubescence. It is distinguished from *D. flaccida* by its lobes and pinnules being narrower and in its shorter pubescence, otherwise very similar to that species.

HEMIGRAMMA nov. gen.

Foliis rosulatis simplicibus aut irregulariter pinnatipartitis, dimorphis, soriferis contractis, nervatione sagenioidea, i. e., pinnata, inter nervos laterales multifarie areolata, nervulis liberis inclusis, soris lineatis ramosis nervos anastomosantes sequentibus, ipsisque irregulariter anastomosantibus et reticulatis.

57. **Hemigramma Zollingeri** (S. Kurz) *Hemionitis Zollingeri* S. Kurz in Journ. As. Soc. Beng. **39**:[2] 90. *t. 5.*

I believe that some of the specimens on which Dr. *Copeland* based his new species, *Hemionitis gymnopteroidea*[13] should be referred to *Hemigramma Zollingeri*. *Copeland* states that his species can be distinguished from *H. Zollingeri* by the copious free veinlets in the sterile frond. In my specimens of the species from Batavia, Java, ex Herb. Hort. Bot. Bogor., and from Celebes, leg. *Sarasin*, the veinlets are also very numerous, even as *Kurz* himself shows in the figure cited above; moreover the constantly contracted fertile fronds of *Hemionitis gymnopteroidea* are not a peculiarity of that species, as in my specimens from Montalban, leg. *Loher*, they are 1 cm. wide and present the oblique lozenge-shaped soriferous bodies as do the specimens from Java.

Var. **major** (Copel) *Hemionitis gymnopteroidea* forma *major* Copel. l. c.

Rhizomate crasso obliquo rudimentis stipitum tecto, foliis fasciculatis paucis (3 ad 4) stipitibus, basi incrassatis usque ad 16 cm. longis basi squamis subulato-setiformibus patentibus nigris usque ad 1 cm. longis dense barbato, caeterum parcius sparso, lamina sterili 20 ad 25 cm. longa saepe basi aequilata, interdum simplice ovato-acuminata medio 7 cm. lata repanda aut obtuse lobata, saepius profunde pinnatifida aut basi pinnata, valde irregulariter lobata, apice aucto et elongato usque ad 8 cm. lata, pinnis usque ad 3 utrinque, ovatis repandis acutis aut obtusis, pinnis basalibus saepe postice cordatis, omnibus ala plus minus lata junctis, fronde saepe decurrente, lamina fertili valde contracta, 8 cm. longa, 4 cm. lata, 2 pinnis utrinque 5 cm. longis 3 mm. latis et apice irregulariter lobato, nervis lateralibus 12 ad 15 utrinque pro pinna, patentibus, rectiusculis, manifestis, fere ad marginem protensis, ca. 5 areolas rectangulas includentibus, quae nervulorum reti nec non nervulis crebris liberis furcatis repletae sunt, pinnis soriferis irregularibus, sporangiorum nervulos sequentibus fulvis tectis. Textura herbacea membranacea, fere diaphana, colore laete virente, opaca.

LUZON, Province of Bataan, Lamao River (2124 *Borden*) December, 1904: Province of Rizal, Montalban (*Loher*) September, 1891. MINDANAO, Province of Zamboanga, San Ramon (1780 *Copeland*) May, 1905.

It does not form a nearly sessile rosette like the species, but the leaves are fascicled and stipitate and the sterile fronds, like the fertile ones, are pinnatifid and even pinnate. The size is much larger.

This form imitates strangely *Leptochilus latifolius* (Meyen) (*Gymnopteris taccaefolia* J. Sm.) in the protean variation of the fronds and in its habit. Normally to rate the large plant as a variety would not be comprehensible, from

[13] Perk. Frag. Fl. Philip. (1905) 183.

the study of dried plants alone, without the evidence given by *Copeland*, and the absolute identity of the venation, tissue and its general structure. It is a striking example of the strange and luxuriant forms found in the Philippines.

Leptochilus latifolius is distinguishable from our plant by its firm texture, not diaphanous, its color, black when dry, its proliferous fronds and its areolæ supplied with a network of very irregular nerves.

Diel's procedure in placing *Hemionitis Zollingeri* in *Syngramme* is to me an unnatural arrangment. The ancestry of the plant is rather in the *Aspidicae*, analagous to *Stenosemia*, and accordingly the above new generic name is proposed for it.

CONIOGRAMME Fée.

58. **Coniogramme fraxinea** (Don.) Diels in Nat. Pflanzenfam. 1[4] (1899) 262. *Diplazium fraxineum* Don Prodr. Fl. Nepal. (1825) 12.

This genus is in need of revision and contains a plurality of forms which doubtless can be studied with better results in the field than in the herbarium. In addition to the ordinary form with bi- to tri-pinate fronds which are membranous and serrate, *Copeland*[14] indicates, without name, another one that usually has entire and simply pinnate fronds. This latter form is very close to one found in China that I have described as the variety *spinulosa*[15] but the Philippine form is larger and with nearly entire margins, and I call it:

Var. **Copelandi** n. var.

MINDANAO, Province of Zamboanga, San Ramon (1746 *Copeland*) April, 1905. LUZON, Province of Rizal, Mabacal (*Loher*) March, 1906. The same plant, but denticulate, has been found in Benguet Province, Baguio (6032 *Elmer*) March, 1904.

LINDSAYA Dry.

59. **Lindsaya falcata** Dry. Trans. Linn. Soc. 3 (1797) 41. *t.* 7. *f.* 2.

Negros, Gimagaan River (66 *Copeland*) January, 1904; (1568 *Whitford*) May, 1906.

The above specimens agree exactly with material from tropical America. It is rather remarkable that this species, like *Lindsaya lancea* (Linn.) Bedd., should be found in tropical America and again in the orient.

PTERIS Linn.

60. **Pteris quadriaurita** Retz. Obs. 6: 38.

Stipite cum rhachi plerumque glabro, stramineo, segmentis oblongis, basi conjunctis, nervis liberis, subtus manifestis, textura herbacea, colore laete virente.

LUZON, Province of Rizal (111 *Foxworthy*) January, 1906; Mabacal (*Loher*) March, 1906: Province of Zambales, Mount Pinatubo (*Loher*) February, 1906: Province of Bataan, Lamao River (239, 240 *Copeland*) February, 1904: Province of Union, Bauang (5619 *Elmer*) February, 1904.

The above specimens represent the typical form of this polymorphous species, being membranous in texture, the nerves manifest on the lower surface and not united, the segments oblong united at the base, the stipe and rachis generally smooth. The numerous derived forms of this group which are found in the Philippines can be grouped as follows:

[14] *Govt. Lab. Publ.* **28** (1905) 67.

[15] *Bull. Soc. Bot. France* (1905) 52,55.

61. **Pteris biaurita** Linn. Sp. Pl. (1753) 1076.

LUZON, Province of Benguet, Mount Tonglon (*Loher*): Buguias (1905 *Copeland*) October, 1905; Manila (*Usteri*) December, 1902. PALAWAN, (594 *Foxworthy*) April, 1906.

A subspecies resembling the typical form, but the nerves forming a costal areola which is sometimes narrow and difficult to find, sometimes rather large and distinct.

62. **Pteris Blumeana** Agardh Pterid. 22.

Stipite rhachique, saepe etiam costis, rufostramineis, castaneis aut atratis verrucis minutis asperis, segmentis linearibus basi plus minus liberis, numerosis, pectinato-confertis, nervis liberis subtus tenuibus minus manifestis, textura coriacea rigida, colore obscure viridi saepe glaucino.

MINDANAO, Lake Lanao, Camp Keithley (255 Mrs. *Clemens*) February, 1906: District of Davao, Mount Dagatpan (*Warburg*); Santa Cruz (254 *DeVore & Hoover*) April, 1903: Province of Zamboanga, San Ramon (1596, 1599 *Copeland*) January, 1905. LUZON, Province of Benguet, Sablan (6213 *Elmer*) April, 1904.

A subspecies with coriaceous texture, linear segments slightly or not at all united at the base, free veins which are very slender and scarcely visible above, its color dark and often somewhat glaucous, the stipe and rachis dark, reddish or blackish and somewhat verrucose. It is a Malayan form that I have also from Java, Tjibodas, leg. *Raciborski*, 1897; Perak, leg. *Ridley* No. 9543; Khasia, leg. Austen, det. *Clarke* "Var. *Khasiana*," and Yabim, near Limbang, German New Guinea, leg. *H. Zahn*, 1905. I believe it to be identical with *Pteris Blumeana* Agardh, as described by *Hooker* [16] and by *Raciborski*.[17]

Var. **asperula** J. Sm. in Hook. Journ. Bot. 3 (1841) 405, et Hook. Sp. Fil. 3: 181, *t. 135 A*; var. *setigera* Hook. l. c.

Costis setulis albis 1.5 mm. longis superne regulariter ciliatis.

BASILAN (93 *DeVore & Hoover*) April, 1903. MINDANAO, Province of Zamboanga, San Ramon (1651 *Copeland*) February, 1905.

63. **Pteris Whitfordi** n. subsp. (*P. Whitfordi* Copel. in Philip. Journ. Sci. 1 (1906) Suppl. 255, pro specie).

A typo differt colore atroviridi, textura coriacea, segmentis angulo acuto erecto-patentibus usque ad costam separatis, sinu fere nullo, stipite valido, paleis brunneis 0.5 cm. longis lanceolatis basi vestito, planta magna 60 ad 70 cm. alta, nervis manifestis.

NEGROS, Gimagan River (1660 *Whitford*).

64. **Pteris parviloba** subsp.; Christ in Bull. Sci. Fr. et Belg. 28 (1898) 264. *t. 12*.

Statura gracili, stipite tenui, rufostramineo, scaberulo, fronde late deltoideo, pinnis infimis auctis et valde deflexis, segmentis confertis lineari-oblongis, brevibus, basi plerumque liberis, nervis tenuibus subtus

[16] Sp. Fil. 3: 180.
[17] Pterid. Buitenzorg 156.

manifestis, textura papyracea rigidiuscula, colore dilute ochraceo-viridi, soris et indusiis saepius angustissimis, facie superiore saepius setis rigidis scabra.

LUZON. Province of Bataan, Lamao River (241 *Copeland*) February, 1904; Province of Benguet, Baguio (5816 *Elmer*) March, 1904; Ambuklao (1827 *Copeland*) October, 1905: Province of Pampanga, Mount Arayat (3814 *Merrill*) May, 1904; Province of Rizal, Montalban (*Loher*) October, 1890, November, 1905; Angilog (*Loher*) March, 1906. PALAWAN (741, 662 *Foxworthy*) March, April, 1906.

A slender small form, the fronds rather short and broadly deltoid, the stipes rough, slender, tawny, the segments pectinate, free at the base, papyraceous, yellowish in color, the sori and indusia generally very slender. This form is often found strongly reduced in size, resembling *Pteris Grevilleana*. I have the same form from Sikkim "*Pteris aspericaulis* Wall." leg. Dr. *Jerdon*, and from Cao Bang, Tonkin, leg. Dr. *Billet*, 1906.

Var. **pluricaudata** (Copel. in Philip. Journ. Sci. 1 (1906) Suppl. 156, pro specie).

Fronde minuta 10 cm. longa et lata pedata pentagona, segmentis imbricatis vix 1 cm. longis.

LUZON, Province of Bataan, Mount Mariveles (3755 *Merrill*) January, 1904.

65. **Pteris ensiformis** Burm. Fl. Ind. 230, var. **permixta** n. var.

An insular form, characterized by the apex of the fertile frond being elongated into a linear caudiform appendage much exceeding the plant, furnished with irregular short obtuse horizontal lobes as in *P. heteromorpha*.

PANAY, Capiz (56 *Copeland*) January, 1904.

66. **Pteris cretica** Linn. var. **stenophylla** Hook. et Grev. Ic. Fil. (1829) *t. 130*. *P. digitata* Wall. Cat. 91.

LUZON, Province of Rizal (53 *Foxworthy*) January, 1906; District of Bontoc, Sagada (1903 *Copeland*) November, 1905, identical with specimens from India, Gharwal, Bhatta Massuri, leg. *Hope;* Sikkim, Tendong, leg. *Gamble*, No. 10340.

Pteris cretica is exceedingly variable in the Philippines, I have not seen from the Archipelago forms resembling those of Europe, but the above specimens match closely *P. stenophylla*, with 3 to several pinnæ at the summit of the frond, more or less fan-like, the tips elongated. Texture normal, firm, shining, light green.

67. **Pteris intromissa** n. subsp.

Differt a typo *P. creticae* pinnis et segmentis angustioribus magis decurrentibus, obscure aut plumbeo-viridibus fere glaucinis, opacis, textura tenuiter papyracea soro angustissimo, nervis tenuissimis suboccultis. Rhizomate brevi, radicoso, stipitibus fasciculatis, tenuibus rufo- aut fusco-stramineis opacis, fronde 30 cm. longa et lata, apice ternata, infra pinnata, pinnis oppositis, mediis simplicibus infimis profunde bi- aut tri-partitis, decurrentibus lineari-lanceolatis aut linearibus caudatis, 20 cm. longis, sterilibus 12 mm. fertilibus 6 mm. latis, sterilibus a basi ad apicem, fertilibus apice acute serratis, nervis tenuissimis parum manifestis, simplicibus aut furcatis, obliquis soro indusioque angustissimo

0.7 mm. lato. Textura tenuiter papyracea, colore opaco obscure viridi subglauco.

MINDANAO, District of Davao (700 *Copeland*) March, 1904. LUZON, Province of Benguet, Sablan (6160 *Elmer*) April, 1904: Province of Rizal, Montalban (*Loher*) 1896; Province of Bataan, Lamao River (3122 *Merrill*) October, 1903.

The same species is found in China, Swatow, leg. *Henry*.

VITTARIA Sm.

68. **Vittaria Merrillii** n. sp.

Rhizomate pennae corvinae crassitie, horizontaliter et longe repente vix ramoso non caespitoso, setis tenuissimis 0.6 cm. longis erectis atrofuscus tecto, stipitibus remotis basi articulatis et setulosis, tenuibus, 1 mm. diametro pendentibus, flexuosis, 20 cm. longis, et tandem in laminam decurrentem sensim transeuntibus, fuscostramineis, lamina 60 cm. et ultra longa, 1 cm. usque ad 1.8 cm. lata lineari, acuminata sed saepe dilatata et irregulariter bi- ad quadrifida, lobis falcatis acutis 1.5 cm. longis, costa tenui saepe inconspicua, nervis valde elongato-obliquis tenuibus, seriebus 4 ad 5 utrinque, soro angustissimo, stricte marginali, indusio 0.7 mm. lato pallido tecto. Textura flaccida, colore dilute virente, opaco.

MINDANAO, District of Davao, Mount Apo (1516, 1192 *Copeland*) October, April, 1904, 1,200 to 1,550 m. alt.; Lake Lanao, Camp Keithley (104 Mrs. *Clemens*) January, 1906, alt. 660 m. MINDORO, Baco River (4044 *Merrill*) March, 1905, near sea level.

A species related to *Vittaria elongata* Sw., but with very slender and elongated stipes, the fronds very broad with a tendency to become enlarged and cleft at the apex, the rhizome elongated and running like that of *Polypodium*. The forking of the fronds reminds one of *Hecistopteris*, and of its affinity with *Vittaria* which *Goebel*, based on anatomical characters, has asserted.

69. **Vittaria pachystemma** n. sp.

Rhizomate repente more Polypodii, pennae anserinae crassitie, setulis crispatis raris nigris coronato, aliter nudo, stipitibus non fasciculatis sed approximatis basi articulatis cum costa rufostramineis, 6 cm. longis, 2 mm. et ultra latis, sensim in laminam transeuntibus, lamina 25 ad 30 cm. longa falcata lineari-lanceolata acuta nec longe acuminata 12 mm. lata, costa plana sed manifesta 1 mm. lata rufostraminea, soris angustissimis a basi laminae ad apicem continuis, submarginalibus, indusio vix 0.5 mm. lato pallido. Textura succulenta sicce coriacea, opaca, colore lurido-glaucina, planta nuda.

MINDANAO, Province of Zamboanga, San Ramon (1589 *Copeland*) January, 1905, alt. 500 m.

A species with succulent coriaceous opaque pointed falciform leaves and thick stipes.

70. **Vittaria Philippinensis** Christ in Bull. Herb. Boiss. II. 6 (1906) 1007.

I have a specimen, quite identical with this species, from the Liu Kiu Islands, Okinawa, *Matsumura* 213.

71. **Vittaria subcoriacea** n. sp.

Rhizomate uti videtur breviter repente, squamis tenuissimis setaceis nigrofuscis 0.5 cm. longis cum foliorum basi dense vestito, foliis fasciculatis (ca. 8) breviter stipitatis, i. e., lamina in stipitem 3 ad 5 cm. longam sensim decrescente, costa stipitis valida, lucida, castanea, sed in lamina sensim applanata minusque manifesta, fronde 55 cm. longa lineari 8 mm. lata acutiuscula, margine late (2 mm.) reflexo adpresso, in angulo sorum 1 mm. latium omnio tegente, soro margineque infra medium folium incipiente et ad apicem continuo, nervis facie superiore prominentibus fere longis simplicibus valde obliquis. Textura suberosa, folio sicco 1 mm. crasso, fragillimo, colore ad basin plantae atrocastaneo, supra ochraceo-viridi, facie minute rugulosa puberula.

PALAWAN, Victoria Peak (669 *Foxworthy*) March, 1906, alt. 1,100 m.

Characterized by being brittle succulent, thick, the costa very strong, shining below, flattened and not prominent above, the naked margins broadly reflexed, under which the soriferous line is completely hidden, nerves simple, long, very oblique.

PLEUROGRAMME (Bl.) Presl.

72. **Pleurogramme Loheriana** Christ in Bull. Herb. Boiss. II. 6 (1906) 1006.

LUZON, Province of Bataan, Mount Mariveles (1388 *Copeland*) August, 1904; (127 *Whitford*) May, 1904; District of Lepanto, Mount Data (1883 *Copeland*) November, 1905, a form with bifurcate pinnæ. MINDANAO, Province of Zamboanga, San Ramon (1763 *Copeland*) May, 1905; Province of Misamis, Mount Malindang (4619 *Mearns & Hutchinson*) May, 1906.

It is related to *P. pusilla* (Blume), (*Vittaria falcata* Kze.) differing from that species in having the soriferous line sunk in an exactly marginal groove, that is to say, placed in the thick tissue of the leaf, so that it presents a border of sporangia emerging from the groove, while in *P. pusilla* the sori are arranged in an intramarginal groove that does not reach to the border of the frond.

ANTROPHYUM Kaulfuss.

73. **Antrophyum Clementis** n. sp.

Folio elongato-lanceolato caudato in apicem et in stipitem alatum sensim excurrente, cum stipite 30 cm. longa, medio latissima ibique 22 mm. lata, costa flava vix 8 cm. longa mox omnino desineute, areolis fere 2.5 cm. longis 2 mm. latis fere 12 in folii diametro, verticalibus, soris filiformibus atrobrunneis submarginalibus (seriebus fere tribus utrinque) folio medio soris destituto, longitudine valde irregulari, ab 0.5 cm. ad 7 cm. longis, verticalibus, haud conjunctis, 2 mm. distantibus non convexis sed sulco minimo faciei superioris respondentibus. Textura subdiaphama herbacea, colore smaragdino, folio margine tenuissimo hyalino circumdato.

MINDANAO, Lake Lanao, Camp Keithley (119 Mrs. *Clemens*) January, 1906, alt. 660 m.

Characterized by its frond when dry being light green, the stipe with decurrent margins, the costa ceasing at the base of the frond, the apex sharp and elongated, the tissue diphanous, the sori vertical and not joined.

ELAPHOGLOSSUM Schott.

74. **Elaphoglossum Copelandi** n. sp.

Rhizomate repente lignoso, fere digiti minoris crassitie, nigro, setulis subulatis atrobrunneis parce vestito, stipitibus subarticulatis, approximatis nec fasciculatis basi atrofuscis squamis lanceolatis parvis brunneis sparsis, 25 cm. longis, rufostramineis, sulcatis, pennae corvinae crassitie, squamulis atratis fimbriato-laceratis vestitis, fronde sterili lanceolato-elongata usque ad 43 cm. longa, 2.5 cm. lata versus basin apicemque acutum attenuata nec decurrente, costa manifesta prominente rufa, nervis occultis horizontalibus 1 mm. distantibus ad marginem sensim protensis nec clavatis simplicibus aut basi furcatis. Textura chartacea firma, colore obscure brunneo, subnitente, facie superiore squamis sparsis notata, inferiore densius squamis vestita, margine costaque squamis majoribus valde fimbriatis cinnamomeis passim ciliata, squamis peltatis longe ciliato-fimbriatis ciliis numerosis (20 ad 30) squamis partim minoribus hyalinis nec coloratis partim majoribus cinnamomeis, i. e. cellulis nucleis rufo-fulvis impletis. Frondis fertilis stipite usque ad 40 cm. longa, lamina 35 cm. longa, 1 cm. lata lineari, latere superiore densissime squamoso, inferiore omnino sporangiis brunneis impleto.

MINDANAO, District of Davao, Mount Apo (1014, 1541 *Copeland*) April, October, 1904, alt. 1,800 m.

A large species related to *Elaphoglossum petiolatum* (Sw.). Scales large, peltate, deeply ciliate, some of them colorless, other larger and with dark brown centers.

75. **Elaphoglossum petiolatum** (Sw.) Urban Symb. Ant. 4 (1903) 61. *Acrostichum petiolatum* Sw. Prodr. (1788) 128; *A. viscosum* Sw.

LUZON, Province of Benguet, Baguio (6509 *Elmer*) June, 1904; (5125 *Curran*) August, 1906; Daklan (1838 *Copeland*) October, 1905.

A form with lacerate and fringed, reddish, very abundant scales, the fertile fronds 30 cm. long, the sterile ones 1.5 cm. wide.

Previously collected in the Philippines by *Cuming*, but not included by *Copeland* in his Polypodiaceæ of the Philippines.[13]

76. **Elaphoglossum laurifolium** (Thouars) Moore Ind. 16. *Acrostichum laurifolium* Thouars Fl. Trist. d'Acunha 31.

LUZON, District of Lepanto, Mount Data (1866 *Copeland*) October, 1905.

Differs from *E. latifolium* of tropical America by its long creeping rhizome, its distant long stipitate and sharply pointed leaves and its smaller size.

Ceylon through Malaya.

CYCLOPHORUS Desv.

77. **Cyclophorus Lingua** (Thunb.) Desv. Prodr. (1827) 224. *Acrostichum Lingua* Thunb. Fl. Jap. 330.

LUZON, Province of Benguet, Baguio (1816 *Copeland*) October, 1905. MINDANAO, Lake Lanao, Camp Keithley (118 Mrs. *Clemens*) January, 1906; District of Davao, Todaya (1303 *Copeland*) April, 1904.

[13] *Govt. Lab. Publ.* **28** (1905).

This plant of temperate and southern China and of Japan, Liu Kiu Islands and Formosa, has been indicated by some authors as extending to British India, but the very rich material in my herbarium, following *Giesenhagen's* revision, does not show this range. The most southern specimens I have seen are from Annam, leg. *Cadière*, and Tonkin, leg. *Billet*. The species appears again in the Philippines with many other characteristic Chinese forms. That even the rare species of the Philippine flora are found to be widely distributed in the different islands of the Archipelago, as is the case with the present one, is some evidence as to the unity of the group.

POLYPODIUM Linn.

78. **Polypodium phyllomanes** Christ in Bull. Acad. Mans. (1902) 210, var. **ovatum** (Wall.) *P. ovatum* Wall; Hook. et Grev. Icon. Fil. (1827) *t. 41.*

LUZON, District of Lepanto, Mount Data (1908 *Copeland*) November, 1905, alt. 1,800, m.

This essentially Chinese type, widely distributed in China and extending into British India only in the high valleys of Bhootan, Khasia and Sikkim, appears again in the mountains of northern Luzon with many other characteristic Chinese plants. The specimens agree perfectly with the slightly elongated form of China.

79. **Polypodium hemionitideum** Wall. Cat. (1828) 284.

LUZON, Province of Benguet, Baguio (967 *Barnes*) May to June, 1904, found previously in Luzon by *Warburg*.

A continental type, China, Yunnan, leg. *Henry;* Formosa, leg. *Faurie, Matsumura;* India, Khasia, leg. *Austin, Clarke, Blanford;* Sikkim, leg. *Jerdon;* also reported from the Nilgherries, southern India.

80. **Polypodium suboppositum** Christ in Bull. Herb. Boiss. II. **6** (1906) 995.

LUZON, District of Lepanto, Bagnen (1964 *Copeland*) November, 1905, the type from Mount Pinatubo, Province of Zambales, Luzon (*Loher*).

81. **Polypodium Sablanianum** n. sp.

I have previously treated the polymorphism of the *Microsorium*[19] type of which the best known representative is *P. punctatum* (L.) Sw. The present new form, from its venation is almost exactly intermediate between *P. punctatum* Sw., and *P. myriocarpum* Mett. In the former, the lateral nerves are very regular, oblique, reaching the margin and inclosing 4 or 5 areolæ, which are elongated, rectangular, between the rachis and the margins; these areolæ are about equal, and the network of nerves is weak, inclosed in and more or less hidden by the fleshy tissue. In *P. myriocarpum* on the contrary, the lateral nerves do not reach the margin of the frond, and inclose only one large areola along the rachis and another narrower one between it and the margin, and the veins are strong and rather prominent. *P. Sablanianum* has an even more delicate texture than *P. myriocarpum*, and its nerves are slender, not prominent and form a single very large square areola, at the side of which is another very narrow obscure one. The plant is larger, 73 cm. long, 5.5 cm. wide, the apex of the frond long caudate, the base long decurrent, the stipe very short. Texture membranous, color bright green, the costa light yellow, shining, the fronds which appear to be solitary and not fascicled have undulate margins, the rhizome is climbing, brown, as thick as a goose quill, covered with stiff subulate dark brown 4 mm. long scales.

LUZON, Province of Benguet, Sablan (6142 *Elmer*) April, 1904.

[19] *Bull. Herb. Boiss.* II. **6** (1906) 993.

82. Polypodium flaccidum n. sp. (*Phymatodes.*)

Rhizomate repente pennae anserinae crassitie, griseo-tomentoso nec glauco, radicoso, stipitibus solitariis articulatis basi parce squamis nigris praeditis aut glabris, tenuibus, stramineis, 6 ad 8 cm. longis sed ob laminam longe decurrentem fere usque ad basim alatis, fronde 40 cm. longa 18 cm. lata ovali, longe sensimque decurrente et lobo simplici caudato 18 cm. longo terminata, 4 aut 5 lobis lateralibus ala costali utrinque 12 mm. lata, lobis remotis 10 cm. longis, 1.5 cm. latis acutissime caudatis sinubus vastis rotundatis 2 ad 3 cm. latis interjectis, marginibus integris rhachi costisque stramineis manifestis, nervis lateralibus curvatis tenuissimis versus marginem evanidis, areas et areolas numerosas oblongas nervulos furcatos includentibus, soris irregulariter bi- aut triseriatis numerosis 1.5 mm. latis rotundatis fere planis ochraceis, facie superiore macula obscura vix impressa notatis, fovea marginata sorali deficiente. Textura flaccida papyracea, diaphana, colore dilute virente.

LUZON, Province of Rizal, Bosoboso (1095 *Ramos*) July, 1906.

A species near *P. phymatodes*, but distinguished by its long decurrent frond, thin and diaphanous texture and small irregular sori which are not immersed in pits.

83. Polypodium palmatum Blume Fil. Jav. 150.

Copeland[20] credits *Polypodium trifidum* Don to the Philippines, but as he presumes, i. e., all the specimens from the Archipelago are referable to *P. palmatum* Blume, also those that I have cited previously for *P. trifidum*.[21] Typical *P. trifidum* is an essentially Chinese species which does not extend to Japan nor to the Malayan Archipelago.

84. Polypodium productum n. sp.

Rhizomate breviter repente radicoso squamis minutis 2 mm. longis setaceis sparso, foliis approximatis nudis 63 cm. longis in stipitem stramineum 10 cm. longum sensim decrescentibus longe et acute acuminatis, medio 1 ad 12. cm. versus apicem 4 mm. latis, costa prominente, straminea margine anguste revoluto, nervis occultis, soris 30 ad 50 utrinque marginalibus impressis versus apicem ultra marginem protrusis ovatis 4 mm. longis, 2 ad 3 mm. distantibus. Textura rigide coriacea, colore flavovirescente sublucido.

MINDANAO, Province of Zamboanga, San Ramon (1585 *Copeland*) January, 1906, alt. 175 m.

A species related to *P. longifolium* Mett., which is also found in the Archipelago, PALAWAN (631 *Foxworthy*) April, 1906, differing from the latter in its firm texture, pale color, shining, the fronds narrower and long acuminate at the apex, the margins inflexed and the sori much elongated. *P. longifolium* is broader, the apex not sharp, the texture less firm, opaque, dark brown when dry, the sori often more rounded.

[20] *Govt. Lab. Publ.* **28** (1905) 129.
[21] *Bull. Herb. Boiss.* **6** (1898) 200.

GLEICHENIA Smith.

(§ *Diplopterygium.*)

On examination of the Philippine forms of this group with the aid of survey given by *Presl* [22] it is possible to distinguish the following forms:

85. **Gleichenia excelsa** J. Sm. in Hook. Journ. Bot. **3**: 420, nomen nudum; Hook. Sp. Fil. **1**: 5. *Tab. 4. B;* Presl l. c. 385.

LUZON, Province of Benguet, Baguio (6006 *Elmer*); (*Loher*) March, 1897.

This species is distributed from northern India to central China and Java. Shillong (*Clarke* 37478); Munipore (*Watt* 6139). Penang Hills (*Ridley*). Yunnan, Mengtze (*Henry* 9167); Moilim (*Egbert* 1897); Hongkong (*Faber* 1091); Ningpo Mountains (*Faber* 219). Java (*Schiffner*).

86. **Gleichenia glauca** (Thunbg. Fl. Jap. 338, *Polypodium*) Hook. Sp. Fil. **1**: 4 (Swartz Syn. 164, 390, *Mertensia*).

LUZON, Province of Tayabas, Mount Banajao (*Loher*) February, 1906, alt. 2,250 m.

Identical with specimens from Japan except for a covering of reddish hairs along the costæ in young plants. The discovery of this species in the Philippines augments the number of species of the temperate regions of the East that extend to the highlands of the Philippines. It is on the whole rather a remarkable distribution, for in this group it is the only species found in Japan, while from China I know *Gleichenia excelsa*, *G. gigantea* Wall. (Yunnan, *Delavay*) and the following:

87. **Gleichenia laevissima** Christ in Bull. Acad. Mans. (1902) 268.

LUZON, Province of Benguet, Pauai (1954 *Copeland*), 2,000 m. alt.

Still another Chinese type, of which I have specimens from Yunnan (*Delavay*) and Lu Mount (*Faber*) August, 1897.

LYGODIUM Sw.

88. **Lygodium Basilanicum** n. sp.

Axibus ochreis tenuibus vix ultra 1 mm. crassis, pinnis petiolatis, petiolo 3 cm. longo angustissime alato, pinna ambitu semirotunda 12 cm. longa et lata, dichotoma, partibus petiolulatis (petiolo 0.5 cm. longo alato) pedatifidis, 3 aut 4-lobis, centro indiviso, 1 ad 2 cm. longo et lato, lobis patulis, 9 cm. longis, 6 mm. latis, linearibus acuminatis integris tenuissime marginatis, lobo basali deflexo, costis tenuissimis manifestis rufostramineis, nervis prominulis obliquis 2 aut 3-furcatis, confertis, lobis fertilibus medio aut infra medium subito usque ad alam 1 mm. latam contractis, sporophyllis pectinato-confertis, 2.5 mm. longis, 1 mm. latis, utrinque circa 8 sporangia gerentibus brunneis munitis. Textura coriacea nec papyracea, colore sicce brunneo, opaco.

BASILAN (28 *DeVore et Hoover*) April, 1903.

A small species of the *circinatum* group, differing from *L. circinatum* by its pinnæ being dichotomous and with petioled pedately arranged pinnules, the segments not dimorphous but narrowed and bearing the sporophylls on the upper half. Dimensions of *L. Japonicum*.

[22] Epim. Bot., 384.

CYATHEA Sm.

89. **Cyathea rufopannosa** n. sp.

Stipite erecto arboreo, 2 m. alto aut altiore, anguloso, 3 cm. diametro, stipite digiti crassitie, cylindrico, 55 cm. alto, castaneo, basi verrucis numerosis brevibus sed huic inde pungentibus scaberrima et squamis 1.5 cm. longis lanceolatis acuminato-falcatis dure scariosis nec diaphanis lucidissimis castaneis tecta, undique cum rhachi et costis indumento furfuraceo fibrilloso spisso tecto, fronde late ovata usque ad 1.5 m. longa (*Copeland*) fere 60 cm. lata versus basin attenuata, rhachi digiti crassitie verrucis cabra rufotomentosa et setulis subulatis fibrillosa, pinnis remotis uti videtur ca. 15 utrinque, infra apicem pinnatisectum, infimis deflexis, mediis 37 cm. longis (infimis 22 cm. longis) 11 cm. latis, sessilibus, ad basin paulum attenuatis, acuminatis, pinnulis ca. 30 utrinque, approximatis 5.5 cm. longis 1.5 cm. latis lanceolatis acutiusculis nec caudatis, usque ad rhachim incisis, segmentis pectinatis ca. 20 utrinque, ligulatis, obtusiusculis, crenulatis, infimis posterioribus dentatis, 5 mm. longis, 2 mm. latis parce setulosis, nervis furcatis ca. 8 utrinque, costis pinnularum subtus squamis lanceolatis rufis nec non squamulis bullatis umbonatis rufis dense obtectis, soris parvis paucis costae pinnulae approximatis ultra medium segmenti raro protensis, confertis rufis 1 mm. latis, indusio brunneo-rufo irregulariter confracto more *Amphicosmiae*. Textura herbacea, colore partium frondosarum laete virente, partium axialium rufo-cinnamomeo.

MINDANAO, Province of Zamboanga, San Ramon (1730, 1735 *Copeland*) April, 1905, alt. 1,200 m.

A striking species, the base of the stipe with short sharp spines, the scales at the base of the stipe large, firm, shining, dark brown, the axial parts and even the costæ covered with a dense fibrillous brick-red pubescence, the fronds tripinnatifid, the pinnules and segments serrate, the latter small, slightly dentate, the sori small, reddish, borne near the costa.

90. **Cyathea Loheri** Christ in Bull. Herb. Boiss. II. 6 (1906) 1007.

This species was discovered by *Loher* on Mount Banajao and on Mount Maquiling, LUZON. A form occurs on Mount Tonglon (Santo Tomas); that is sufficiently distinct to warrant being described as a variety.

Var. **Tonglonensis** n. var.

Stipite rugoso et cicatricoso, squamis subulato-setaceis basi verrucosis flexuosis flaccidis atrobrunneis ultra 1 cm. longis patentibus, nec rigidis adpressis scarioso-argenteis tecto, costis densissime squamis bullatis brunneis tectis, segmentis minoribus dense pectinatis.

LUZON, Province of Benguet, Mount Tonglon (4991 *Curran*) August, 1906.

It is possible that *Alsophila lepifera* J. Sm. apud Hook. Sp. Fil. 1: 54, is the same as *Cyathea Loheri*, but the description of the former is too incomplete to verify this.

91. **Cyathea Negrosiana** n. sp.

Stipite digiti et basi fere pollicis crassitie, basi atro-, supra cum rhachi costisque rufo-castaneo, verrucis minutis creberrimis, scaberrimo, lucido, basi squamis subulato-setaceis flexuosis lucentibus atrofuscis 2 cm. longis dense vestito, ob pinnas abbreviatis inferiores valde descendentes solummodo 15 aut 20 cm. longo, fronde tripinnata ultra 1 m. longa 25 cm. lata, ovato-acuminata versus basin sensim attenuata, pinnis remotis, 10 ad 15 utrinque, erecto-patentibus mediis longissimis, 35 cm. longis, 14 cm. latis, caudato-acuminatis, sessilibus, versus basin non attenuatis, pinnulis remotiusculis, 1.5 cm. distantibus, sessilibus, ca. 20 utrinque infra apicem pinnatifidum, 7 cm. longis, acuminatis, 14 mm. latis, fere usque ad costam incisis, segmentis approximatis falcato-lanceolatis, acutiusculis, 7 mm. longis, 2.5 mm. latis, obtuse crenulatis, costis costulisque dense squamulis ovatis bullatis rufo brunneis more *C. Loheri* vestitis, nervis basi furcatis, 8 aut 10 utrinque, soris costulis adpressis, 3 aut 5 utrinque minutis, indusio griseo-brunneo primum globulari nitidulo mox confracto, frustulis squamiformibus irregularibus receptaculum nigrum elevatum circumdantibus. Textura herbacea, colore atroviridi, opaco.

NEGROS, Mount Silay (1536 *Whitford*) May, 1906, alt. 1,000 m.

This species was determined by *Copeland* as *Cyathea Christii*, but it is distinguished from the latter by its axial parts being richly covered with furfuraceous scales, by the costæ being covered with rounded inflated scales, by its membranous texture and very reduced size.

92. **Cyathea ferruginea** n. sp.

Stipite pennae cygni crassitie cum rhachi subnitido, anguloso, castaneo, floccoso-paleaceo sublaevi, pinnis 26 cm. longis, 11.5 cm. latis stipitatis ovato-acuminatis basi aliquantum attenuatis, i. e., pinnula infima abbreviata, costis pinnarum indumento floccoso paleaceo squamis rufis crispis e basi ovata subulatis crispulis patentibus rufis constituto tectis, pinnulis ca. 15 utrinque infra apicem lobatum, remotis, i. e., spatio 2 cm. lato separatis, lanceolatis, acutis nec caudatis sessilibus 6 cm. longis, 11 cm. latis, ad alam angustam incisis, segmentis ca. 14 utrinque confertis, angulo fere nullo interjecto, falcatis acutiusculis aut obtusis, ovatis, 0.5 cm. longis, 3 mm. latis, minute crenulatis, costulis pinnularum supra puberulis subtus omnino squamis rufo-griseis ovatis adpressis acutis 2 mm. longis tectis, faceibus fere glabris, nervis 6 plerumque furcatis, soris mediis 4 aut 5 utrinque, cinnamomeis, vix 1 mm. latis, confluentibus, indusio fugaci vix conspicuo, frustulis cum squamulis mixtis. Textura herbacea, colore dilute viridi, opaco.

PALAWAN, Mount Pulgar (560 *Foxworthy*) March, 1906, alt. 1,150 m.

A small species 2 m. high, acaulescent according to *Foxworthy's* notes, growing just below the summit of the mountain on an exposed ridge in the mossy forest.

ALSOPHILA R. Br.

93. **Alsophila calocoma** n. sp.

Caudice arborescente, stipite ad basin coma densissima et pulcherrima squamarum scariosum argenteo-lucidarum pallidorum sed apice rufarum, e basi 3 mm. lata ovatorum, longe subulato caudatarum aristatarumque, usque ad 4 aut 5 cm. longarum ornato, aliter inermi, sed verrucis minutis rugoso, rhachi tuberculis verrucosis creberrimis rugosissimo, squamulisque furfuraceis sparsa rufo-ochracea, opaca, fronde tripinnatisecta, pinnis amplis 75 cm. longis, 25 cm. latis acuminato-caudatis petiolatis (petiolo 2 ad 3 cm. longo) pinnulis confertis 20 ad 30 utrinque infra apicem lobatum, fere sessilibus, 15 cm. longis, 2.5 cm. latis e basi lata lanceolato-acuminatis, infimis haud abbreviatis, supremis late adnatis et decurrentibus, costis rufo-brunneis tenuibus, cum costularum parte inferiore subtus serie squamularum candidarum lucidarum ovatarum adpressarum ciliatorum elegantissime vestitis, partibus foliaceis plantae laevibus, pinnulis usque ad costam incisis, segmentis falcato-ligulatis obtusiusculis aut acutiusculis, 11 mm. longis, 3 mm. latis, fere integris rarius crenulatis, pectinato-confertis, inferioribus liberis, i. e., basi spatio separatis et aliquantulum angustatis, ca. 35 utrinque, nervis ca. 12 tenuissimis furcatis saepe tri- sive pluries-furcatis, soris 8 ad 10 utrinque, mediis 1 mm. latis sese tangentibus nec confluentibus cinnamomeis globosis receptaculo minuto elevato nigro, textura flaccide herbacea, colore glauco- aut plumbeo-viridi, supra obscuro, subtus pallidiore.

LUZON, Province of Rizal, Mount Alabut (*Loher*) February, 1904, alt. 1,900 m.; Angilog (*Loher*) March, 1906; Province of Benguet, near Baguio (*Loher*) March, 1897.

A beautiful species characterized by the shining hairs on the stipe and the shining white or metallic scales on the under surface of the segments.

This is the species that I had previously identified[23] as *A. lepifera* J. Sm., but I am now convinced that it is a distinct species. The short description of *A. lepifera* J. Sm., given by *Hooker*[24] is quite insufficient from which to identify *Smith's* species, except that it appears to be near, if not identical with *A. tomentosa* (Blume), as *Christensen* supposes in his Index Filicum 44, or perhaps the same as *Cyathea Loheri* Christ. *A. calocoma* is distinguished from *A. latebrosa*, with which it shares the character of the costal and costular scales, by its whitish scales and by its basal segments being free and attenuate toward the base, as well as by the marked glaucescence of the frond. In *A. latebrosa* I have never observed the exceptionally long scales which are shining, silvery or somewhat golden in color and strongly pointed, 4 to 5 cm. in length, such as are found in *A. calocoma*. It is a delicate species with trifurcate nerves and a very rugose rachis.

Var. **congesta** (*Alsophila lepifera* var. *congesta* Christ in Bull. Herb. Boiss. 6 (1898) 137).

This variety is identical with the type in having the same very large basal, scarious, silvery, subulate, 5 cm. long scales which are 4 mm. wide below,

[23] *Bull. Herb. Boiss.* 6 (1898) 137.
[24] Sp. Fil. 1: 54.

yellowish rhachis which is very rough with small spines, small whitish costular scales, and the basal segments free, remote and narrowed below, but it is readily distinguishable from the type by its shorter narrower segments which are more falcate and more strongly serrate. The pinnules do not exceed 8 cm. in length. In general appearance quite different from the type, but having the same essential characters. It appears to be an alpine form of the species.

LUZON, Province of Rizal, Arambibi (*Loher*) March, 1903; Province of Benguet (6504 *Elmer*) June, 1904.

94. **Alsophila latebrosa** (Wall.) Hook. Sp. Fil. 1: 37.

The ordinary form of this species so common in India, the Malayan Peninsula, Java, Borneo, Celebes and Amboina (herb. *Christ*), is not known to me from the Philippines, where the species is represented by a form notably larger. It is the same as with *Alsophila contaminans,* which is represented in the Philippines by the large variety *Celebica.*

Var. **major** n. var.

Rhachi fulvo straminea, laevi aut minutissime furfuracea, pinnis ca. 28 utrinque infra apicem, 55 cm. longis, 18 cm. latis fere sessilibus oblongo-acuminatis, pinnulis confertis recte patentibus sessilibus 9 cm. longis, 2 cm. latis oblongo-caudatis, usque ad costam tenuam nigram incisis, segmentis imbricato-confertis rotundato-obtusis rectis oblongis ca. 20 utrinque 1 cm. longis 2.5 mm. latis, crenulatis, nervis ca. 9 utrinque, tenuissimis basi furcatis saepe trifurcatis, costula squamulis rotundis peltato-umbilicatis 0.3 mm. latis flavis elegantissime vestitis, fronde caeterum glabra, tenuiter herbacea, colore obscure viridi subtus pallido, opaco.

MINDANAO, Province of Surigao (325 *Bolster*) May, 1906: District of Davao, Mount Batangan (*Warburg*).

The costular scales in *A. lepifera* are white, oval and larger than in the above variety.

DICKSONIA L.' Herit.

95. **Dicksonia Copelandi** n. sp.

Ampla, basi stipitis coma densa pilorum tenuissimorum 7 mm. longorum rufobrunneorum coperta, stipite plantae junioribus iisdem pilis vestito, rufostramineo, subnitente, pinnis deltoideo-acuminatis, petiolatis, 45 cm. longis 24 cm. basi latis inaequalibus, antice auctis, pinnulis^{II} ca. 20 utrinque confertis, inferioribus mediisque deltoideis, infimis 15 cm. longis 12 cm. latis petiolatis, acuminatis, pinnulis^{III} deltoideo-oblongis acuminatis ca. 12 utrinque, subinaequalibus, infimis 6 cm. longis 3 cm. latis, petiolulatis, pinnulis^{IV} infimis 1.5 cm. longis basi 8 mm. latis liberis oblongis acutis subinaequalibus profunde serratis, dentibus trigono-acutis mucronatis, nervis suboccultis in pinnulis^{IV} pinnatis furcatis, soris in dentibus terminalibus sed mucrone superatis, uno pro dente, praecipue antice positis, numerosis, brunneis 1 mm. latis globosis coriaceis irregulariter bivalvis. Textura coriacea, costis nervisque pilosis, facie superiore glabra subnitente, colore ochraceo-viridi.

LUZON, Province of Benguet, Baguio (*Loher*). March, 1897, alt. 1,400 m.; (6025 *Elmer*) March, 1904: District of Lepanto, Bagnen (1912 *Copeland*) November, 1905, alt. 2,000 m.

A species closely related to *D. straminea* Labill., but strongly pilose, larger, quadripinnatifid, the pinnæ and pinnules broader and more strongly serrate, broadly deltoid, the pinnules of the third order broader the lobes shorter and broader. Resembling the South American *D. coniifolia* Sw.

MARATTIA Smith.

All the Philippine forms of this genus that I have examined have been identified after the classification in the monograph of *DeVriese* and *Harting*, in which work the diagnoses are by no means comparative and in which the differences between related species are not noted. The number of species is so large and their characters so uniform that the distinctive characters of each species should have been emphasized. At any rate the group merits more attention than the successors to the two Dutch botanists have given it, for certainly the forms are very numerous and can not all be reduced to a single species. The morphological differences between young and adult fronds are very great, and the latter, even the fertile ones, frequently present characters that are ordinarily found only in young fronds. In diagnoses the adult parts only have been considered.

I have been able to elucidate here, with a fair degree of certainty, the following forms:

96. **Marattia sambucina** Blume. Enum. (1828) 256; DeVriese et Harting Monog. 6.

Textura firmiter chartacea, rhachi laevissima flava, pinnulis sessilibus basi acute cuneatis, acutis, adultis 7.5 raro 11 cm. longis, 1 ad 1.4 cm. latis, margine omnino dentatis, dentibus obliquis raro patentibus, synangiis 1.5 mm. a margine remotis 1.5 mm. longis non contiguis late ovatis, 6 ad 8-loculatis. Colore pallide viridi.

LUZON, Province of Benguet, Mount Tonglon (*Loher*) April, 1906; Baguio (*Loher*) January, 1893: Province of Union, Castilla (*Loher*) March, 1906.

This is the most widely distributed form, identical with specimens from Java leg. *Raciborski* and from Celebes leg. *Sarasin*. It is the form previously considered by me as *M. fraxinea*.[a]

97. **Marattia silvatica** Blume Enum. (1828) 256; DeVriese et Harting Monog. 6. III. 25.

Firmiter chartacea, rhachi laevissima flava, pinnulis petiolulatis, basi acute cuneatis acutis, adultis 9 cm. longis 13 mm. latis, margine omnino dentatis, dentibus brevibus patentibus, synangiis fere marginalibus, fere contiguis, 2 mm. et ultra longis, oblongis, 12 ad 15-locularibus. Colore pallido.

LUZON, Province of Benguet, Baguio (5833 *Elmer*) March, 1904.

98. **Marattia pellucida** Presl Suppl. Tent. Pterid. 10; DeVriese et Harting Monog. 6.

Herbacea, rhachi flava, laevissima, pinnulis petiolatis, basi abrupte cuneatis, apice abrupte acuminatis, adultis 10 cm. longis, 14 mm. latis,

[a] *Bull. Herb. Boiss.* 6 (1898) 207.

margine omnino dentatis, dentibus brevissimis apertis, synangiis 1 mm. a margine remotis non contiguis brevibus 1 mm. longis oblongis 8-locularibus. Colore dilute viridi-plumbeo, nervis egregie pellucidis.

MINDANAO, District of Davao, Mount Apo (1455 *Copeland*) October, 1904, alt. 1,550 m.

99. **Marattia vestita** n. sp.

Ampla, caudice 2.5 cm. crasso, opaco, rufobrunneo dense pustulis atque squamis ovatis flaccidis 1 cm. longis et 0.5 cm. latis minoribus et angustioribus mixtis scabro, rhachibus rufofuscis supremis ochraceo-rufis opacis et abunde cum costis squamis lanceolatis brunneis squamulisque fibrillosis vestitis, pinnis oblongis 70 cm. et ultra longis 30 cm. latis, petiolo 7 cm. longo praeditis, basi attenuatis, acuminatis, pinnulis petiolatis ca. 10 cm. remotis 20 cm. et ultra longis, 15 cm. latis basi attenuatis, pinnula terminali praeditis, pinnulis^{III} alternis valde (2.5 cm.) remotis basi articulatis infimis brevissime subpetiolulatis, basi acute cuneatis, acuminatis, 9 cm. longis, 12 mm. latis, lanceolatis, margine omnino dentatis, dentibus patentibus apertis, nervis conspicuis fere 2 mm. remotis nigris simplicibus, soris minutis remotis oblongis 1 mm. longis ochraceis 5-locularibus subclausis. Textura firmiter sed tenuiter chartacea, colore supra obscure, infra pallide viridi, opaco.

MINDANAO, District of Davao, Mount Apo (1179 *Copeland*) April, 1904.

A species peculiar in its axial parts being not polished or shining but dull, dark colored, rough, and with numerous scales. The denticulation is very open and the synangia are smaller than in any other species known from the Archipelago.

100. **Marattia Ternatea** DeVriese et Harting Monog. 4. *t. 3. 16.*

Chartacea fere coriacea laevissima, pinnulis petiolulatis, petiolis squamulosis, planta caeterum glabra, rhachi ochraceo-plumbea, pinnulis basi cuneato-ovatis, acutis, lanceolatis, 12 ad 20 cm. longis, 2 ad 2.4 cm. latis, minute denticulatis et ob marginem inflexum fere integris, nervis valde remotis (ultra 2 mm.) ochraceis, synangiis remotis, oblongis, ultra 3 mm. longis, 15-locularibus, 1.5 mm. a margine remotis. Colore supra obscure, infra palidissime viridi.

LUZON, Province of Bataan, Mount Mariveles (2082 *Borden*) September to December, 1904; (1116 *Whitford*) May, 1905.

In this species the pinnæ are larger than in any other one known from the Archipelago, their borders in part nearly entire and in part dentate.

ANGIOPTERIS Hoffm.

Angiopteris offers in a still greater degree than *Marattia* the lack of palpable differential characters, and the differences between the various forms, quite distinct to the practiced eye, are difficult to diagnose properly. I believe it possible to distinguish the following species:

101. **Angiopteris cartilagidens** Christ in Bull. Herb. Boiss. 6 (1906) 207.

LUZON, Province of Benguet, Baguio (*Loher*), alt. 1,400 m.

One of the most sharply defined species, characterized by its dentation, texture and scales.

Endemic.

102. **Angiopteris similis** Presl in DeVriese et Harting Monog. Maratt. 17.

MINDANAO, Lake Lanao, Camp Keithley (115 Mrs. *Clemens*) January, 1906, alt. 660 m.

A species with thinly papyraceous texture, shining, pale green, the nerves recurrent, slender, slightly visible, the sori small, dark brown, close to the margin, the pinnules large, 18 cm. long, 21 mm. broad, the teeth prominent only at the sterile apices.

Java.

103. **Angiopteris angustifolia** Presl ex DeVriese et Harting Monog. Maratt. 18.

LUZON, Province of Bataan, Lamao River (3791 *Merrill*) January, 1904. NEGROS, Gimagaan River (1659 *Whitford*) May, 1906. MINDANAO, Lake Lanao (Mrs. *Clemens*) April, 1906: Province of Surigao (240 *Bolster*) April, 1906: District of Davao, Todaya (1459 *Copeland*) October, 1904.

This seems to be the most widely distributed species in the Archipelago, and was first collected by *Cuming*. It is distributed from Annam, leg. *Cadière*, to Formosa, leg. *Faurie*, south to the Sunda Islands.

104. **Angiopteris caudata** DeVriese et Harting Monog. Maratt. 20.

LUZON, Province of Benguet, Mount Tonglon (*Loher*) April, 1906; Baguio (5126 *Curran*) August, 1906; (5930 *Elmer*) March, 1904: Province of Laguna, Mount Maquiling (*Loher*) June, 1906.

Pinnules strongly narrowed, the upper ones 8 mm. broad, very gradually narrowed into the long pointed apex. The pinnæ resemble those of *Pteris longifolia* Linn.

105. **Angiopteris pruinosa** Kze. Schkuhr Suppl. 1: *t. 91.*

NEGROS, Gimagan River (1652 *Whitford*) May, 1906.

Fronds bluish white beneath quite similar in color to those of *Lomaria glauca* Blume.

CHRISTENSENIA Maxon.

106. **Christensenia Cumingiana** n. sp.

Omnium reliquarum formarum adhuc cognitarum minima, rhizomate crasso, brevi, carnoso, radicoso, foliis approximatis, junioribus subfasciculatis, stipite usque ad 14 cm. longo cum costis nervisque rufostramineo furfuraceo, fronde tam simplici ovata breviter acuta basi subcordata 8 ad 13 cm. longa 3 ad 5 cm. lata, repanda aut grosse dentata, quam tripartita, pinna centrali late ovata 11 cm. longa 5 cm. lata acuta longe et anguste cuneata, lobato-repanda, pinnis lateralibus adnatis valde inaequalibus postice cordato-auctis, antice anguste cuneatis 9 cm. longis 3 cm. latis, nervis manifestis, ca. 10 utrinque, rectis patentibus interstitio ca. 1 cm. lato, facie superiore laevi inferiore albida, stomatibus rotundis dense tecta, laevi, synangiis vix 1.5 mm. latis (deciduis aut immaturis) brunneis inter nervos biseriatis, ca. 4 pro serie. Textura modice succulenta.

MINDANAO, Province of Zamboanga, San Ramon (*Copeland* s. n.) March, 1905, alt. 200 to 650 m.

All the other species of *Christensenia* (*Kaulfussia*) known to me have the lateral pinnæ rather strongly petioled. *DeVriese* and *Harting* have indicated in their Monograph of the Marattiaceae 14, that the *Kaulfussia* found by *Cuming* in the Philippines, which was considerably smaller than the other known forms, might perhaps prove to be a distinct species. In comparison with specimens from Assam, leg. *King;* Selangor, leg. *Ridley;* Java, leg. *Raciborski*, and Sumatra, leg. *Schneider*, the specimens from MINDANAO are very reduced. The fertile frond is often simple and when it is tripartite the lobes are joined at the base, not petiolate, and very unequal, whitish beneath.

BOTRYCHIUM Sw.

107. **Botrychium lanuginosum** Wall., var. **nanum** n. var.

LUZON, Province of Benguet, Bugias (1848 *Copeland*) October, 1905, alt. 1,550 m., with the large form.

Like our European *Botrychium* this large species has also a dwarfed form, about 17 cm. high, the fertile frond with its stipe about 10 cm. high.

SUPPLEMENT.

There are two forms hitherto confounded as *Aspidium coadunatum* Wall. or even as *A. cicutarium* Sw., which is a very different West Indian species. After a careful examination and comparison with other specimens of the Malayan and wider Asiatic area, I can indicate the following diagnostic points:

Aspidium coadunatum Wall. Cat. 377 non Hook. et Grev. Ic. Fil. 202. *Sagenia* J. Sm. Hook. Journ. Bot. 4: 184. Presl Epimel. 60.

Stipite rufo glaberrimo lucente, basi squamis ovatis acutis 0.5 cm. longis flaccidis brunneis sparso. Rachi laevi lucida flavido-rufa. Pinnulis ovato-lanceolatis acutis, lobis late ovatis obtusis sive subacutis. Faciebus pilis albidis brevibus pubescentibus, marginibus ciliatis. Nervis ad marginem protensis, nervulis luteo-brunneis tenuibus abunde anastomosantibus, nervulis inclusis clavatis frequentibus. Colore laete virente, textura diaphana. Indusio orbiculari-reniformi margine pallidiore.

LUZON, Bontoc, Sagada (1899 *Copeland*): Province of Rizal, Bosoboso (1033 *Ramos*).

A common species in tropical Asia.

Wynaad, Malabar, leg. *Bicknell;* Anamalays, Province of Madras, leg. *Beddome;* Mercara Coorg. 1,100–1,200 m., leg. *Richter;* Koon Beeling, Burma, leg. *Brandis;* Ceylon, leg. *Wall.;* Yunnan, leg. *Henry* 10341, 10354; Sze tchuen, Mount Omi, leg. *Wilon* 5376. Viti, Plewa River, leg. *Moore;* Tahiti, leg. *Radeond.*

Aspidium Malayense n. sp.

Stipite opaco vix subnitente brunneo, basi squamis ad 1 cm. longis lanceolato-subulatis rigidis atrobrunneis, caeterum cum rachi squamis setaceis atratis plus minus dense vestita. Rachi fere opaca fusco-aut olivaceostraminea pinnulis lnceolato-angustatis caudatis, lobis acutis ovato-lanceolatis grosse dentatis. Faciebus fere laevibus nec ciliatis. Colore sicce griseo-aut atroviridi, textura opaca. Nervis lateralibus ad marginem protensis, nervulis nigris crassis, solummodo secus costas anastomosantibus, nervulis inclusis clavatis raris aut nullis, parce indusio brunneo peltato.

LUZON, Laguna Province, Majayjay (*Loher*) 1891; Bataan Province, Lamao River (1959 *Borden*, 217, 1396 *Copeland*); northern Luzon (*Warburg*). MINDANAO, Todaya, Davao (1468 *Copeland*).

I have the same species from Malacca, Johor, leg. *Ridley* 10976; Singapore, leg. *Hose* (1894).

Aspidium melanorachis Bak. Journ. Bot. 1888. *Nephrodium*. 315 Sarawak, Borneo, leg. *Hose* is very near.

Differt squamis atropurpureis linearibus flaccidis flexuosis, rachi costisque atropurpureis, fronde ampliore, faciebus dense pubescentibus, soris magis numerosis, irregulariter sparsis nec stricte biseriatis, minoribus, indusio tenui griseo mox evanido, nervi valde anastomosantibus, nervulis inclusis multis.

Diplazium vestitum Presl Epimel. 87, 1849. Hook. Ic. II. 46.

By comparison with Hooker's figure of Cuming's specimen from Leyte, I have identified as this species the plant found by Loher at Mabacal and Angilog, Rizal Province, March, 1906. It is very much larger than the one represented by Hooker; but all the details correspond perfectly, especially the axial parts covered with scales and furfuraceous down. The lateral nerves and the sori converge in the sinus between two lobes, but they do not touch before the sinus. The stipe, which attains the size of a finger, is rough with small warty projections.

The same plant from Celebes leg. *Koorders* 16986.

I distinguish this from the form which I have identified in Bull. Herb. Boiss. VI. 1906, 1001, with *D. Smithianum* (Bak.) Diels by the shorter sori, which touch somewhere before the sinus, by the rough scaly rachis armed like the costæ with sharp prickles, and by the more truncate lobes. I have this from Ceylon, leg. *Wall.* 38/275, from Celebes, Bojong, leg. *Warburg* 15314, *Sarasin* 108, and from New Guinea, Sattelberg, leg. *Weinland* 1890.

I must say that these two forms are exceedingly similar and appear almost like one specific type in the wider sense. *Diplazium dolichosorum* Copel.[1] is intermediate between *D. vestitum* Presl and *D. Smithianum*; the sori are those of the former, but the imperfect pubescent covering and the truncate lobes are as in the second.

[1] New Phil. Ferns, l. c., 151.

THE PHILIPPINE SPECIES OF DRYOPTERIS.

By H. CHRIST.
(*Basel, Switzerland.*)

Some time ago Mr. *Elmer D. Merrill,* Botanist of the Bureau of Science, Manila, sent me all the Philippine material of the genus *Dryopteris* from the herbarium of that institution, in order to give me an opportunity to prepare a classified list of the species found in the Archipelago. The collection contains many of *Cuming's* plants, and a large number of specimens collected by the American botanists since the occupation of the Philippines by the United States. In addition to the above material I have also received from Dr. *E. B. Copeland* a notable collection, and Mr. *Loher* has had the kindness to furnish me with an additional and very interesting collection, supplementary to the one he sent me in 1897 and which was the basis of my work "Filices Insularum Philippinarum."[1] Since the publication of the above paper some important works of Dr. *Copeland* have notably advanced our knowledge of the ferns of the Philippines. In his *Polypodiaceæ* of the Philippine Islands,[2] Dr. *Copeland* admits 60 species of *Nephrodium,* compiling the descriptions of all the species credited to the Philippines, even of those species of which he had not seen specimens. In my present paper I have not attempted to account for all the species of the genus that have been credited to the Archipelago by various authors, but have considered only those of which specimens are before me. In a group so difficult as *Dryopteris* and so subject to diverse interpretation, it appears to me that the latter treatment is the surest, even if completeness is sacrificed.

I have limited *Dryopteris* in the sense of *Christensen's* Index Filicum; that is, excluding *Pleocnemia* and *Sagenia* and treating only *Lastrea* (including *Phegopteris*) and *Nephrodium* proper (including *Goniopteris, Mesochlaena* and *Meniscium*). As the Philippines are particularly rich in species and forms of *Dryopteris,* the task of treating all the species was sufficiently arduous. In regard to nomenclature I have followed *Christensen's* Index Filicum and accepted the generic name

[1] *Bull. Herb. Boiss.* **6** (1898) 127-154; 189-210.
[2] *Govt. Lab. Publ.* **28** (1906) 18-32.

Dryopteris, in spite of the sacrifice of personal opinions and in spite of being obliged to discard names that have been in constant use for a century and which are known to all botánists.

Nowhere else is the type of *Nephrodium* with anastomosing veins so diversified as in the Philippines. There are in the Philippines forms with very narrow pinnæ, and some special characters are found in the species of this region more often than in those of other parts of the world. These characters are: Pinnæ attenuated toward their bases, the lower ones deflexed, the pinnæ degenerating into auricles at the base of the frond, sometimes abruptly, sometimes gradually. In other equatorial regions species with these characters are rather rare. In tropical America, *Dryopteris sagittata* (Sw.) is almost the only known species of the group where the frond is abruptly reduced at the base, the lower pinnæ being represented by auricles, and *D. refracta* (Fisch. & Mey.) is one of the rare examples of a species with deflexed pinnæ. The Malayan region offers the most frequent examples of species presenting the two last characters, for example *D. sagittifolia* (Blume) of Java, but even in the Malayan region such species do not approach in number those of the Philippines.

There is in the Philippines a tendency to "insular" reduced types which is rather interesting. These reduced types elsewhere are rare, and abnormal. The irregularity and reduction of the fronds and even the dimorphism of the fertile fronds is normal in *Dryopteris canescens* (Blume) as found in the Philippines, and *D. glandulosa* has analogous tendencies. These variations do not as yet appear to be constant, and they offer some subspecies and varieties of doubtful value, which are discussed later under the two above species. In the Archipelago moreover are analogous variations in other genera, for example the singularly stunted forms that are grouped under *Leptochilus heteroclitus* (Presl) (*Acrostichum flagelliferum* Wall.), and some species of *Pteris,* such as *P. ensiformis* Burm., and *P. heteromorpha* Fée. In the West Indies, Cuba, Jamaica, Porto Rico, Santo Domingo, etc., analogous insular forms are found in *Polystichum, Fadyenia, Sagenia* and especially in *Dryopteris reptans* (Gmel.) which there offer multiple reduced forms. I am sure that the very prolonged isolation of these archipelagoes plays some rôle in the occurrence of these variations, although it is not possible at present to specify just what this influence is.

The wonderful variations of *Dryopteris canescens,* which are found in other parts of Malaya (Celebes) only as rarities, but which are developed in the Philippines into a bewildering series of forms, appear to me to throw some new light on the "aberrant forms" of the old school of pteridologists. By the variations of *Dryopteris canescens,* which present an unbroken and insensible transition from a true *Nephrodium* to a plant entirely achrostichoid as to the sori, the affinity of

Leptochilus, Gymnopteris, Polybotrya, Egenolfia, Stenosemia, and *Caenopteris* with *Aspidium,* in a broad sense, appears to me to acquire a new support; and what is more, although perhaps in the cases where it has not yet been possible to find the aspidioid type of all acrostichoid plants, it is probable that the aspidioid type has not been preserved or that it has been so modified as to be unrecognizable. Be that as it may, for *Stenosemia* one must admit the immediate descent of *Pleocnemia membranifolia* (*Dictyopteris Chattagramica* Clarke) as *Beddome* has asserted.[3] Likewise I now connect my *Gymnopteris Bonii*[4] from Tonkin, directly with *Aspidium repandum* Willd. The contention that "Acrostichum" is only "Aspidium" with reduced fertile pinnæ, appears to me to be better established than ever. Is this a step in advance in the development or a degeneration? The example of *Dryopteris canescens,* where the incontestable deformation of the pinnæ both fertile and sterile, is accompanied by the acrostichoid formation as to the soriferous parts, appears to me to point strongly to the latter; that is to say, an aberration and weakening of the type, which one can scarcely call only teratological, because the influences that have caused the changes are unknown.

OBSERVATIONS.

1. In my Filices Insularum Philippinarum[5] I have noted *Aspidium Fauriei* var. *elatius* Christ and *A. grammitoides.* Both belong in *Athyrium,* with aspidioid sori, as is the case with *Athyrium oxyphyllum* which is found in the Philippines with absolutely aspidioid sori.

2. In his Polypodiaceæ of the Philippine Islands,[6] *Copeland* includes *Nephrodium asperulum* (J. Sm.) Copel. The species was based on No. 63 *Cuming, Polypodium asperulum* J. Sm., and the specimen in the Herbarium of the Bureau of Science is to me *Microlepia speluncae* (Linn.) Moore, with submarginal sori.

3. *Copeland*[7] admits *Nephrodium rugulosum* (Labill.) Copel., but to me the plant indicated is *Hypolepis.* Species of *Hypolepis* with the sori more or less intramarginal give rise to some doubt as to their proper disposition. There is a form in the Philippines which has a rhizome often, if not always, creeping, which is generally a good character of true *Hypolepis* and which indicates the relationship of that genus with *Pteridium.* This form was considered by me at first as *Dryopteris setigera* (Blume) O. Ktz., and later as *Aspidium vile* Kunze, of Java, with which it has a great resemblance. It has been collected on Mount Apo, Mindanao, by *Copeland* (No. 1462) October, 1904, and on Mount

[3] Suppl. Ferns Brit. Ind. 48, 40.

[4] *Bull. Herb. Boiss.* II 4: 610.

[5] *Bull. Herb. Boiss.* 6 (1898) 193.

[6] *Govt. Lab. Publ.* 28 (1905) 25.

[7] L. c. 26.

Arayat, Luzon, by the *Bolster* (Nos. 79, 98.) The rhizome seems to be very slightly creeping, the sori are submarginal at the anterior base of the lobes and the texture of the plant is rather thin and not coriaceous as in specimens of *Aspidium vile* from Java leg. *Raciborski*. After examining the material at present available, I do not consider that this doubtful species can be referred with certainty to *Dryopteris*.

4. *Aspidium varium* Sw., is to me a *Polystichum*, and for this reason this Chinese type, which is also found in northern Luzon, is not considered in the following list:

DRYOPTERIS Adanson.

I. NEPHRODIUM (including *Mesochlaena*, *Goniopteris* and *Meniscium*).

1. **Dryopteris megaphylla** (Mett.) C. Chr. Ind. Fil. (1905) 277. *Aspidium megaphyllum* Mett. Ann. Lugd. Bat. 1 (1864) 233. *Aspidium pennigerum* Blume Enum. (1828) 153. *Nephrodium pennigerum* Bedd. Handb. (1892) 73.

Haud male quadrans cum specimine Javanico a *Raciborski* lecto et determinato, conf. *Raciborski* Pterid. Buitenz. 190, sed planta Mindanaensis gaudet rhizomate erecto, radicibus multis simplicibus et stipitum fasciculatorum basi oriundis suffulto, quum *Raciborski* plantae Javanicae rhizoma repens stipites que remotos attribuat. *Beddome* recte monet "Caudex erect."

MINDANAO, District of Davao, Todaya (1236 *Copeland*) April, 1904; Province of Zamboanga, San Ramon (*Copeland*) March, 1905.

Malaya.

2. **Dryopteris truncata** (Poir.) O. Ktze. Rev. Gen. Pl. 2 (1891) 814. *Polypodium truncatum* Poir. in Lam. Encycl. 5 (1804) 534.

Nervis 6 utrinque quorum 2 ad 3 junctis, lobis rotundatis aut convexe truncatis.

LUZON, Province of Laguna, Pagsanjan (1995b *Copeland*) February, 1906; Los Baños (*Alberto*) May, 1905: Province of Bataan, Mount Mariveles (391 *Topping*) May, 1904: Province of Benguet, Baguio (4948 *Curran*) August, 1906. MINDANAO, Province of Zamboanga, San Ramon (1674 *Copeland*) March, 1905. BASILAN (88 *DeVore & Hoover*) April, 1903.

Malaya.

3. **Dryopteris abrupta** (Blume) O. Ktze. Rev. Gen. Pl. 2 (1891) 812. *Aspidium abruptum* Blume Enum. (1828) 154. *Nephrodium abruptum* Hook. Sp. Fil. 4 (1862) 77. *t. 241. B.*

Nervis utrinque quorum 4 aut 5 junctis, lobis horizontaliter aut concave truncatis, apice denticulatis.

MINDANAO, Province of Zamboanga, San Ramon (*Copeland*) February to March, 1905.

Malaya.

Scarcely more than a subspecies of the preceding.

4. **Dryopteris adenophora** C. Chr. Ind. Fil. (1905) 251. *Nephrodium hirsutum* J. Sm. in Hook. Journ. Bot. 3 (1841) 412; Hook. Sp. Fil. 4: 70. *t. 140*, non Don, nec Bory.

LUZON, Province of Bataan, Mount Mariveles (419 *Topping*) May, 1904; (1312 *Whitford*) January, 1905: Province of Tayabas, Mount Banajao (968 *Whitford*) October, 1904: Province of Zambales, Mount Pinatubo (*Loher*) February, 1906: MINDORO, Baco River (276 *McGregor*) May, 1905. NEGROS, Gimagaan River (1658 *Whitford*) May, 1906.

Philippines and Celebes.

5. **Dryopteris ferox** (Blume) O. Ktze. Rev. Gen. Pl. **2** (1891) 812. *Aspidium ferox* Blume Enum. (1828) 153.

LUZON, without locality (172 *Cuming*) "*Gonioptcris aspera* J. Sm. *Polypodium asperum* Roxb. in herb. Linn. Soc." *J. Smith* in Hook. Journ. Bot. **3** (1841) 396: Province of Benguet, Sablan (6232 *Elmer*) April, 1904; Baguio (320 *Topping*) January to February, 1903: Province of Cavite, Mendez Nuñez (1355 *Mangubat*) August, 1906. MINDORO, Baco River (237 *McGregor*) May, 1905. MINDANAO, Zamboanga (1578 *Copeland*) 1905; Lake Lanao, Camp Keithley (107a Mrs. *Clemens*) January, 1906.

Malaya.

Var. **calvescens** n. var.

Pustulis et setis axialibus fere evanidis.

MINDANAO, Province of Zamboanga, San Ramon (1721 *Copeland*) 1905, alt. 800 m.

6. **Dryopteris Todayensis** n. sp.

Rhizomate oblique erecto, supraterraneo, radicoso, paucos (ca. 3) stipites emittente, fere nudo, atrobrunneo, digiti crassitie. Stipite crasso fere digiti minoris, solido griseo-brunneo nudo aut paucis squamulis lanceolatis brunneis parce obsito, 40 cm. longo, fronde 70 cm. et ultra longa 20 cm. lata oblongo-acuminata pinnata, versus basin vix attenuata sed abrupte secus stipitem utroque latere in 8 ad 10 auriculas breves rudimentarias transeunte, rhachi brunneo-grisea puberula, pinnis confertis sessilibus 40 ad 55 utrinque, basi subcallosis tuncato-cuneatis subinaequalibus, caudato-acuminatis, 11 cm. longis basi 11 mm. latis usque ad mediam partem incisis, ala utrinque 3 mm. lata relicta, lobis confertis ca. 45 utrinque lanceolato-acutis valde falcatis 3 mm. longis 2 mm. latis, nervis tenuibus non prominulis 7 ad 8 utrinque, areolam unam costalem formantibus, secundis in sinum acutum excurrentibus, caeteris liberis, facie inferiore puberula, superiore glabra, exceptis costulis costulisque adpresse pilosis, soris minutis, brunneis, mediis, indusio parvo reniformi brunneo integro puberulo. Textura flaccide herbacea, colore atroviridi. Differt a *D. truncata* lobis falcatis profundioribus, nervisque pluribus.

MINDANAO, District of Davao, Todaya, on the slopes of Mount Apo at 1,200 m. alt. (1463 *Copeland*) October, 1904. NEGROS, Gimagaan River (1658 *Whitford*) May, 1906.

A species of large size, the frond abruptly narrowed at the base, the stipe with numerous small auriculate pinnæ, the lobes narrow, the lower surface slightly pubescent, the nervules forming one costal areola, the sori very small, the color a very dark green. A similar plant, but the stipe without auricles, is represented by No. 607 *Copeland*, from Davao.

boanga, (1575, 1575a, 1575b *Copeland*) 1905; District of Davao (390 *Copeland*) March, 1904; Santa Cruz (218 *DeVore & Hoover*) April, 1903; Mount Apo (378 *DeVore & Hoover*) May, 1903.

Malayan region to the Seychelles.

It is impossible for me to follow *Christensen's* Index in treating this species as *D. unita* and renewing the confusion that has exsisted for a long time regarding *Nephrodium unitum* R. Br. (=*Dryopteris gongylodes*). *Blume's* name is here accepted for the species. There is a limit even to the virtues of priority!

Var. **mucronata** (J. Sm.) *Nephrodium mucronatum* J. Sm. in Hook. Journ. Bot. **3** (1841) 412.

Inter typum et *N. callosum* (Bl.) intermedia. *N. cucullato* typico major, pinnis 2 cm. et ultra latis, nervis utrinque plus minus 12, quorum 6 junctis, soris minutis submarginalibus, pinnis infimis versus basin angustatis, insertione pinnarum callosa, omnio puberula. *N. callosum* magnitudine et glabritie differt.

LUZON, without locality (182 *Cuming* nec 268). MINDANAO, Lake Lanao, Camp Keithley (107 Mrs. *Clemens*) January, 1906.

10. **Dryopteris arida** (Don) O. Ktze. Rev. Gen. Pl. **2** (1891) 812. *Aspidium aridum* Don Prodr. Fl. Nepal. (1825) 4.

LUZON, Province of Nueva Ecija, Carranglang (283 *Merrill*) May, 1902: Province of Benguet, Trinidad (212 *Topping*) January, 1903. MINDORO, Baco (879 *Merrill*) April, 1903. MINDANAO, District of Davao, Davao (447,326 *Copeland*) March, 1904: Province of Zamboanga (1576 *Copeland*) 1905: Lake Lanao, Camp Keithley (Mrs. *Clemens*) January, 1906.

No. 279 *Cuming* "*Nephrodium mucronatum* J. Sm." appears to differ from this species in its pinnæ being more strongly hastate at the base and in being more strongly villous. *Dryopteris arida* of the Philippines is usually more strongly villous and the pinnæ are shorter and more distant than in the form found in India (Dehra Dun, leg. *Blanford*).

Malaya.

11. **Dryopteris gongyloides** (Schkuhr) O. Ktze. Rev. Gen. Pl. **2** (1891) 812. *Aspidium* Schkuhr Krypt. 289. *Nephrodium unitum* R. Br.

LUZON, without locality (259 *Cuming*) "*Nephrodium unitum* R. Br." J. Sm. in Hook. Journ. Bot. **3** (1841) 411: Province of Cagayan (133bis *Bolster*) July, 1905: Province of Nueva Ecija, Carranglang (283 *Merrill*) May, 1902: Province of Bataan, Mount Mariveles (1239 *Borden*) June, 1904. MINDANAO, Province of Zamboanga, San Ramon (*Copeland*) March, 1905.

Tropics of both hemispheres, as far north as Algeria.

12. **Dryopteris hispidula** (Dcne.) O. Ktze. Rev. Gen. Pl. **2** (1891) 813. *Aspidium hispidulum* Dcne. Nouv Ann. Mus. **3**: 346.

LUZON, without locality (268 *Cuming* nec 182), "*Nephrodium mucronatum* J. Sm." in Hook. Journ. Bot. **3** (1841) 412.

Cum planta Borneensi a cl. *Niewenhuis* lecta exacte convenit, sed minus cum speciminibus aliter collectis.

A specimen from Baguio, Province of Benguet, LUZON (5108 *Curran*) August, 1906, appears to me to be intermediate between *Dryopteris basilaris* and *D. hispidula*.

Borneo and the Philippines; its other distribution in Malaya uncertain.

13. **Dryopteris basilaris** (Presl) C. Chr. Ind. Fil. (1905) 254. *Nephrodium basilare* Presl Epim. Bot. (1849) 258, nomen. *Nephrodium philippinense* Bak. Ann. Bot. **5** (1891) 327.

This species is one of the most distinct, most important and most widely distributed of the genus in the Philippines, and is characterized by *Baker* as follows:

"Rootstock and complete stipe not seen. Frond oblong-lanceolate, bipinnatifid, 2–3 ft. long, 1–1½ ft. broad, moderately firm, glabrous, rachis naked. Pinnæ lanceolate, acuminate, 8–9 in. long, ½ in. broad, cut down less than half way to the rachis into oblong erecto-patent lobes ⅛ in. broad, lower pinnæ not dwarfed, veins simple, 8–9 jugate. Sori medial, indusium firm, glabrous, persistent. Near *arbuscula*, but lower pinnæ not gradually dwarfed."

It was on specimens Nos. 10, 84 and 338 *Cuming* that *Baker* based his imperfect description, and of these I have before me a specimen of the second number. Based on this number, and the abundant material collected by the American botanists in the Philippines, the following detailed description is given:

Rhizomate obliquo subercto valde radicoso crasso, foliis subfasciculatis (4 aut 5) stipite basi sulcato-dilatato, squamis subulatis usque ad 2 cm. longis brunneis e basi lanceolata filiformi-elongatis vestito, rufostramineo, glabro lucente, tereti, basi pennae cygni crassitie, 45 cm. longo, fronde usque ad 65 cm. longa, 20 cm. lata late ovata acuminata, pinnata, basi abrupte terminata, pinnis infimis haud abbreviatis, sed stipite utrinque 8 aut 10 auriculis obtusis rudimentariis remotis instructo, pinnis alternis erecto-patentibus numerosis sessilibus aut brevissime petiolatis, pinna terminali saepius valde elongata aut basi bifida, pinnis lateralibus approximatis, ca. 35 utrinque, 14 cm. longis, 1 cm. latis caudato-acuminatis lanceolato-linearibus, basi antiore recte truncata, posteriore semicordata, marginibus lobatis usque ad tertiam partem, lobis decumbentibus 3 mm. longis oblongis subobtusis, costis pallidis manifestis, nervis 5 ad 8 utrinque, infimis aream unam formantibus, soris minutis mediis 5 utrinque, indusio griseo persistente. Colore brunneo-viridi, textura subcoriacea rigidiuscula, planta glabra.

LUZON, without locality (84 *Cuming*) "*Nephrodium caudiculatum* Presl" J. Sm. in Hook. Journ. Bot. **3** (1841) 411: Province of Rizal, Antipolo (*Guerrero*) June, 1903: Province of Cagayan (163, 175 *Bolster*) August, 1905: Province of Bataan, Mount Mariveles (407 *Topping*) May, 1904; (6666 *Elmer*) November, 1904; (224, 225 *Copeland*) February, 1904; (371, 108 *Whitford*) June, 1904; (2554 *Merrill*) June, 1903: Province of Tayabas, Sampaloc (12759 *Warburg*): Province of Benguet, Baguio (4915, 4946 *Curran*) August, 1906: Province of Cavite, Mendez Nuñez (1304 *Mangubat*) August, 1906: Province of Isabela, Malunu (*Warburg*). MINDANAO, Davao (637 *Copeland*) March, 1904: Province of Zamboanga (1685 *Copeland*).

Widely distributed in the Philippines; endemic.

14. **Dryopteris Luzonica** n. sp.

A species, resembling the preceding, and like it widely distributed in Luzon, but well characterized by its very thin texture, bright green color, its pinnæ horizontal, in rather remote pairs and dilated at the base, the lobes obtuse, often truncate, the pinnæ strongly elongated into a filiform apex, the apex of the frond rather pinnatifid and terminated by one pinna, the stipe having generally one or two pairs of auricles. I shall content myself with indicating here the characters by which it differs from *Dryopteris basilaris:*

Basi stipitis squamis destitutis aut minutis brevibus, stipite gracili sed pinnis infimis abbreviatis, et stipite auricula una, rarius pluribus instructa, flavostramineo, fronde acuminato apice pinnatifida minus abrupte terminata pinnis remotioribus inferioribus mediisque horizontalibus, oppositis, basi antice et postice dilatatis quasi utrinque stipulatis, lobis brevioribus apice truncato-obtusis sive abruptis et denticulatis, textura diaphano-tenui, nervis plerumque 5, soris ochraceis, indusio tenuissimo mox evanido, colore dilute smaragdino.

LUZON, Province of Rizal, Bosoboso (1083 *Ramos*) July, 1906; (89 *Foxworthy*) January, 1906; Antipolo (*Guerrero*) June, 1903; (*Loher*) March, 1906, March, 1893: Province of Bataan, Mount Mariveles (1239 *Borden*) June, 1904: Province of Cavite, Mendez Nuñez (1289, 1302 *Mangubat*) August, 1906: Province of Laguna, Los Baños (*Loher*) January, 1906; Mount Maquiling (*Loher*) January, 1906; Pagsanjan (1995a *Copeland*) February, 1906; (514 *Topping*) 1904: Province of Batangas, Mount Malarayat (2002 *Copeland*) February, 1906: Province of Cagayan (120 *Bolster*) July, 1905: Manila (*Rothdauscher*) 1897 in Herb. Monac: Province of Isabela, Malunu (11577 *Warburg*).

Var. **puberula** n. var.

Rhachi costis et nervis puberulis.

LUZON, Province of Cagayan, Tabug (175 *Bolster*) August, 1905.

Var. **polyotis** n. var.

Pinnis latioribus, basi 16 mm., brevius acuminatis, et stipite usque ad basin auriculis numerosis (ca. 20 utrinque) vestitis.

LUZON, Province of Rizal, Montalban (5064 *Merrill*) March, 1906. MINDANAO, Province of Zamboanga, San Ramon (1571 *Copeland*) December, 1904.

15. **Dryopteris parasitica** (Linn.) O. Ktze. Rev. Gen. Pl. 2 (1891) 811. *Polypodium parasiticum* Linn. Sp. Pl. (1753) 1090.

LUZON, without locality (83 *Cuming*) "*Nephrodium molle* R. Br." J. Sm. in Hook. Journ. Bot. 3 (1841) 412. Province of Rizal, Bosoboso (1084 *Ramos*) July, 1906: Province of Tayabas, Malicboi (*Ritchie*) May, 1903; Atimonan (8 *Gregory*) August, 1904: Province of Laguna, Pagsanjan, (1995 *Copeland*) February, 1906. MINDANAO, Province of Zamboanga (1605 *Copeland*) 1905: District of Davao (607 *Copeland*) March, 1904. PALAWAN (Paragua) Ewiig River (720 *Merrill*) February, 1903.

Tropics of both hemispheres.

Var. **falcatula** n. var.

Differt a typo pinnis inferioribus oppositis, refractis, basi antice stipulatis, segmentisque profundius incisis falcatis acutioribus, aliter typo conformis. An *Nephrodium molliusculum* Wall. Cat.?

MINDANAO, Province of Zamboanga (1677 *Copeland*) 1905.

16. **Dryopteris procurrens** (Mett.) O. Ktze. Rev. Gen. Pl. **2** (1891) 813. *Aspidium procurrens* Mett. Ann. Lugd. Bat. **1**: 231.

LUZON, Province of Laguna, Pagsanjan (1992 *Copeland*) February, 1903: Province of Rizal, Bosoboso (1094 *Ramos*) July, 1906: Province of Bataan, Mount Mariveles (226, 1389 *Copeland*) February, August, 1904. CULION (589 *Merrill*) December, 1902. MINDANAO, Province of Zamboanga (1693a *Copeland*).

Malaya.

From repeated examinations of material from the Philippines I have not been able to determine with certainty the form described by *Hooker*, Synopsis 292, as *Nephrodium latipinna*, as that species is represented by specimens from Hongkong and Tonkin (leg. *Cadière*).

17. **Dryopteris heterocarpa** (Blume) O. Kuntze Rev. Gen. Pl. **2** (1891) 813. *Polypodium heterocarpum* Blume Enum. (1828) 155.

LUZON, Province of Laguna, Mount Maquiling (2027 *Copeland*) March, 1906.

Sunda Islands.

18. **Dryopteris canescens** (Blume) C. Chr. Ind. (1905) 256. *Polypodium canescens* Blume Enum. (1828) 158. *Gymnogramme canescens* Blume Fil. Jav. 93. *t. 40*. *Aspidium canescens* Christ. Ann. Jard. Bot. Buitenz. **15**[1]: 130.

The Philippines share with Celebes a plurality of forms of this species, interesting because of the numerous more or less "insular" forms into which it is divided. I refer the reader to what I have said regarding it in *Ann. Jard. Bot. Buitenzorg* l. c., where I have shown its affinity to the group containing *P. parasitica* of which it appears to be a weakly derived species, weakly derived because of its generally reduced dimensions, the indusium frequently lacking, and its sori irregular, but above all in the variation in the form of the fronds which present all forms of pinnæ from those linear and elongated to those variously cut, lobed and dilated in a most bizarre manner, and finally in the dimorphism and narrowness of the fertile fronds which have much elongated stipes and the pinnæ so narrowed that the sori lose their distinctness and form a mass which entirely covers the narrow fertile pinnæ, in this latter respect resembling those of *Egenolfia appendiculata*.

In Celebes I have distinguished three forms—*nephrodiformis*, which is scarcely dimorphous; *gymnogrammoides*, with the fertile fronds somewhat reduced; and *acrostichoides* with the fertile pinnæ narrowly linear. For the species as it occurs in the Philippines, this distinction does not suffice, and it is necessary to distinguish a large number of forms, some of which have acquired the value of subspecies, or perhaps in some cases, of species. These forms I characterize as follows:

Var. **lobatum** n. var.

Statura minore, stipite longiore (20 cm., frondis 20 cm.) pinnis minus numerosis, latioribus, lobis latioribus, paucioribus, profundioribus nervis flexuosis, interdum irregularibus, aream unam costalem formantibus, pubescentia sensiore grisea imprimis costas nervosque tegente, et soris indusiis carentibus, saepe irregulariter elongatis.

LUZON, Province of Rizal, Mabacal (*Loher*) March, 1906: Province of Benguet, Baguio (1866 *Copeland*) November, 1905.

Java, Celebes.

This variety more or less resembles the large form figured by *Blume* and approaches a small *D. parasitica*, but the stipe is relatively longer, 20 cm., the frond 20 cm., the pinnæ less numerous, longer, the lobes longer and more numerous and more deeply divided, the nerves very undulating, forming one costal areola, pubescent.

Var. **degener** n. var.

Rhizomate elongato subrepente, pinnis ovatis saepe obovatis basi attenuatis, obtusissimis, obtuse crenato-lobatis, apice frondis elongato lato obtuse lobato, pinnis sterilibus brevius (6 cm.) fertilibus saepe longius (usque ad 25 cm.) stipitatis, pinnis fertilibus remotis, soris irregulariter sparsis plus minus rotundis. Tota planta a 20 cm. usque ad 42 cm. alta, textura crassiuscula, colore obscure fere atroviridi, pubescentia imprimis rhacheos densa, strigosa, brunnea.

LUZON, Province of Rizal, Angilog (*Loher*) February, 1906, the larger form; Montalban (*Loher*) March, 1906, the smaller form.

This is an accentuated variation of the normal form, the length of the merely lobed apex and the pinnæ, scarcely coarsely crenate, giving the plant a singular aspect.

Var. **subsimplicifolia** n. var.

Smaller, distinguished from the preceding by its one distinct terminal elongated pinna, nearly entire, and in the lateral pinnæ being very slightly developed and auricle-like.

LUZON, Province of Tayabas (Infanta) (784 *Whitford*) September, 1904.

19. **Dryopteris diversiloba** (Presl) n. subsp. *Nephrodium diversilobum* Presl Epim. (1849) 47; Mett. Aspid. 100. *Goniopteris asymmetrica* Fée Gen. 253.

Rhizomate debili, elongato, plus minus repente, stipitibus plus minus fasciculatis fere caespitosis aut subsolitariis, debilibus flexousis 8 cm. longis, fronde oblonga 10 cm. longa, 7 cm. lata, pinnata, pinnis subpetiolatis rhombeis aut lata ovatis obtusis aut in apicem lanceolatum prolongatis 5 usque ad 7 utrinque 4 cm. longis, 2.5 cm. latis basi saepe attenuatis sive hastulatis aut subcordiformibus, crenatis, apice lobatis lobis valde irregularibus, brevibus et usque ad 2 aut 3 cm. longis, lanceolato-obtusis 2.5 mm. latis mixtis. Apice frondis saepe lato, valde elongato, lobato; pinnis fertilibus vix contractis, saepe apice solummodo sorifero, soris minutis exindusiatis numerosis irregulariter sparsis rariter seriatis saepe elongatis. Tota planta griseo pubescente, textura herbacea, colore dilute viridi-griseo.

LUZON, Province of Nueva Viscaya, Quiangan (162 *Merrill*) June, 1902: Province of Rizal, Mabacal (*Loher*) March, 1906; northern Luzon (11611 *Warburg*) 1888. NEGROS, Gimagaan River (83 *Copeland*) 1904. MINDANAO, Province of Zamboanga, San Ramon (1547, 1774, 1754 *Copeland*) November, 1904, April, May, 1905: Province of Misamis, Mount Malindang (4613, 4710 *Mearns & Hutchinson*) May, 1906: District of Davao (698 *Copeland*) March, 1904: Province of Surigao (252 *Bolster*) April, 1906.

Apparently common and widely distributed in the Philippines; endemic.

This is a form of the *D. canescens* group, but so accentuated, and at the same time so widely distributed (it should be one of the most common ferns in the Archipelago), that it should be recognized as a subspecies. A small plant, almost turf forming with elongated, weak and often running rhizomes, the fronds not, or but little dimorphous, the pinnæ few, short, broad, nearly square and very irregularly lobed, the lobes sometimes short and obtuse, sometimes greatly elongated. The specimens with the elongated pointed pinnæ have the appearance of a sufficiently distinct species, but often the long and short pinnæ are found on the same plant. The frond is often terminated by a single simple pinna, but sometimes it is pinnatifid. The pinnæ are slightly petioled, somewhat hastate and slightly cordate at the base, slightly lobed toward the base, but nearly always with some strongly elongated and unequal lobes toward the apex which is abruptly truncate. The stipe is always slender and flexuous, about 8 cm. long, the frond about 10 cm. long, the lateral pinnæ 5 to 7 on each side, 4 cm. long, 2.5 cm. wide, the terminal one 5 to 10 cm. long. The sori are small, very irregular, sometimes few, sometimes very numerous often occupying only the terminal part of the frond.

Var. **acrostichoides** (J. Sm.) *Nephrodium acrostichoides* J. Sm. in Hook. Journ. Bot. 3 (1841) 411; Christ Ann. Jard. Bot. Buitenz. 15[1] (1898) 130.

The sterile frond is more or less that of *D. diversiloba*, but the fertile frond is very long stipitate and the pinnæ are strongly reduced in width, approaching those of *Gymnopteris* and *Egenolfia*. Two subvarieties are distinguishable:

Subvar. **rhombea**, n. subvar.

Frondis fertilis stipite debili flexuoso valde elongato, lamina 7 cm. longa, pinnis 5 utrinque, apice frondis elongato lobato acuminato, pinnis rhomboideo-lanceolatis usque ad linearibus, subpetiolatis, basi truncatis, obtusis, crenulatis, 8–4 mm. latis, soris confertis aut seriatis aut omnio confluentibus, areola una.

Sterile pinnæ with the form of those of *D. diversiloba*, the frond with the stipe 12 cm. long, the fertile frond, including the stipe 25 cm. long, the stipe being about 18 cm. long.

LUZON, without locality (149 *Cuming*); Province of Rizal, Bosoboso (1084 *Ramos*) July, 1906; Province of Bataan, Mount Mariveles (3130 *Merrill*) October, 1903; (*Copeland*) August, 1904; (6153 *Leiberg*) July, 1904; Province of Tayabas (Infanta) (784 *Whitford*) September, 1904. MINDANAO, District of Davao (503 *Copeland*) March, 1904.

Philippines and Celebes.

Subvar. **lanceola** n. subvar.

Differt a praecedente pinnis sterilibus lanceolatis margine fere integris apice obtusis aut acutiusculis, apice frondis valde elongato fere caudato, areola una huic inde duabus.

LUZON, Province of Bataan, Mount Mariveles (250 *Copeland*) January, 1904; Province of Rizal (140 *Foxworthy*) January, 1906. NEGROS, Gimagaan River (1600 *Whitford*) May, 1906. MINDANAO, Province of Zamboanga (*Copeland*) 1905.

A specimen from San Ramon (*Copeland*) April, 1905, offers the maximum reduction, the fertile pinnæ being reduced to a width of 2 mm., and the sori accordingly having the appearance of a string of beads as is the case in specimens from Celebes leg. *Sarasin*.

20. **Dryopteris acromanes** n. sp.

Rhizomate brevi crasso radicoso, foliis fasciculatis, stipite rufo-aut plumbeo-stramineo, tenui, 15 ad 30 cm. longo, fere nudo, fronde oblongo-acuminata, 17 cm. longa, 9 cm. lata, pinnata, pinnis ca. 7 utrinque infra apiceme longatum lobatum, petiolulatis, basi lata truncata sed pinnis infimis basi attenuatis et deflexis, haud abbreviatis, pinnis obovatis sive rhombeo-elongatis 4.5 cm. longis, 3 cm. latis versus apicem latissimis ad mediam laminam sive ultra incisis lobis ovatis rotundato-obtusis 3 cm. latis ca. 8 utrinque, versus apicem pinnae repente elongatis, fronde fertile conformi, nervis ca. 8 utrinque aream unam costalem formantibus, soris magnis brunneis rotundis ultra 1 mm. latis, in lobis submarginalibus (lamina media soris destituta) brunneis exindusiatis. Faciebus tenuissime puberulis, textura herbacea, colore obscure viridi.

LUZON, Province of Laguna, Mount Maquiling (*Loher*) April, 1906: Province of Rizal, Ampalit (*Loher*) April, 1906.

No. 51 *Cuming* "*Polypodium adfine* Reinw." in Herb. Bureau of Science, approaches *Loher's* specimens cited above, except that the sori are not confined entirely to the lobes and are less marginal.

In many respects similar to *D. canescens* var. *lobatum*, but the accrescence of the pinnæ toward the apex and the increasing length of the lobes toward the tips of the pinnæ is more accentuated. The sori are large, marginal, bordering the lobes in a single series and the plant has not the harshness and grayish color that distinguishes *D. canescens*, so that the present form can hardly be referred to the preceding as a subspecies.

21. **Dryopteris xiphioides** n. sp.

Rhizomate obliquo repente, stipitibus debilibus valde approximatis numerosis aequilongis 30 cm. longis rufostramineis, fronde pinnata 20 cm. longa 11 cm. lata, late ovato-elongata, ad basin vix attenuata, pinna terminali 10 cm. longa basi aut libera aut pinnis lateralibus valde abbreviatis vicina sive connata, 13 mm. lata acuminata lanceolata crenata, pinnis lateralibus 4 aut 5 similibus sed haud ultra 7 cm. longis, nervis 3, rarius 4 omnibus junctis, soris paucis minutis exindusiatis, tota planta parce griseo pubescente, textura herbacea, colore griseo-viridi.

MINDANAO, Province of Zamboanga, San Ramon (*Copeland*, s. n.) April, 1905. alt. 800 m.

A species of the *D. canescens* group, remarkable by its very elongated falcate pinnæ which are not reduced toward the base of the frond and but slightly lobed.

22. **Dryopteris Merrillii** n. sp.

Rhizomate erecto radicoso crasso, foliis fasciculatis numerosis, stipite rufostramineo flexuosa vix pennae corvinae crassitie basi squamulis minimis ruguloso aliter nudo (rachi facieque frondis parce puberulis) 18 ad 20 cm. longo, fronde ovato-oblonga acuminata, 20 ad 29 cm. longa, 12 cm. lata, apice elongato lobato, pinnata, pinnis confertis patentibus infimis interdum reductis et deflexis, egregie petiolulatis, falcato-lanceolatis, acutis, 6 cm. longis, 12 mm. latis, basi verticaliter truncata egregie hastata, antice plus minus aucta, pinnis dentato-serratis dentibus decumbentibus, nervis goniopteridis, pinnatis inter costam marginemque areas 4 ad 5 formantibus quaque area nervulum liberum porrectum includente. Fronde fertili longius stipitata, pinnis angustioribus magis remotis. Adsunt pinnae fertiles 6 mm. 5 mm. et 2 mm. latae! Soris confertis

4 ad 5 seriatis aut confluentibus minimis brunneis rotundis exindusiatis. Textura herbacea, colore laete virente.

PALAWAN (742, 862 *Foxworthy*) March, April, 1906. MINDANAO, Province of Surigao, Surigao (26 *Bolster*) March, 1906, the latter very small, about 23 cm. high. resembling a specimen from Borneo leg. *Ridley*, 1901.

This presents the appearance of a well-established species. It is large, fasciculate, with a definitely established dimorphism, and is readily recognizable by its numerous pectinate lanceolate pinnæ, stipitate and manifestly hastate at the base, the nerves forming several areolæ. In some respects it resembles *Egenolfia appendiculata* and might be mistaken for that species except that the bases of the pinnæ are equal.

23. **Dryopteris Philippina** (Presl) C. Chr. Ind. Fil. (1905) 284. *Physematium philippinum* Presl Epim. (1849) 34. *Lastrea exigua* J. Sm. in Hook. Journ. Bot. **3** (1841) 412.

LUZON, without locality (251, 272 *Cuming*): Province of Rizal. Montalban (*Loher*) March, 1906. MINDANAO, Province of Zamboanga. San Ramon (1705 *Copeland*) 1905: Province of Surigao (307. 327 *Bolster*) May, June. 1906.

This is a reduced form of *D. Merrillii* with obtuse pinnæ, more simple venation and the pinnæ auriculate only anteriorly.

24. **Dryopteris microloncha** n. sp.

Nana, rhizomate crasso obliquo atrobrunneo radicoso, foliis numerosis fasciculatis, stipite raris squamulis brunneis sparso rufostramineo 2 ad 3 cm. longo tenui, rhachi straminea parce furfuracea, planta aliter nuda, fronde oblonga 16 ad 24 cm. longa, 4 ad 6 cm. lata acuminata et in longam cuspidem lobatam excurrentem versus basin sensim auriculis obtusis rotundatis, demum 5 aut 3 mm. longis et latis decrescente, pinnis ca. 12 utrinque, mediis 3 cm. longis, 0.5 cm. latis sessilibus e basi hastulata sive utrinque sed antice magis auriculata sensim acuminatis, vix ad tertiam laminae partem incisis, lobulis truncatis, nervis in lobulis parce pinnatis, infimis irregulariter junctis, soris numerosis minutis brunneis irregulariter triseriatis, undusio atrobrunneo orbiculari, persistente.

LEYTE (317 *Cuming*) "*Nephrodium caudiculatum* Presl" J. Sm. in Hook. Journ. Bot. **3** (1841) 411. LUZON, Province of Rizal (54 *Foxworthy*) January, 1906; Morong (1381 *Ramos*) August, 1906: Province of Cavite (1304 *Mangubat*) August, 1906.

A small plant resembling *D. Amboinensis* (Willd. Sp. Pl. **5**: 228, *Aspidium*), but even smaller than that species, with numerous obtuse auricles on the stipe, narrow pinnæ which are scarcely incised, and more numerous sori.

25. **Dryopteris polycarpa** (Blume) *Aspidium polycarpum* Blume Enum. (1828) 156. *Mesochlaena polycarpa* Bedd. Ferns Brit. Ind. Suppl. 13.

SAMAR (327 ? *Cuming*) 1836–40. The interrogation point concerns only the number in *Cuming's* series, and not the plant itself, the identity of which is incontestable.

I do not consider that the elongated sori and their arrangement in horizontal lines merits the generic separation of this form, as in all other respects it is a true *Nephrodium*.

Malaya.

26. **Dryopteris chamaeotaria** n. subsp.

Rhizomate subrepente, stipitibus approximatis, tenuibus, flexuosis, 10 ad 13 cm. longis, parce puberulis et squamis pallide fuscis subulatis sparsis, rufostramineis, fronde ovata 13 cm. longa 7 cm. lata, magna pinna libera ovato-acuminata basi grosse lobata 6 cm. longa 2 cm. lata terminata, 2 vel 3 pinnis lateralibus utrinque, alternis, petiolulatis, supremis adnatis, similibus sive valde reductis ovato- aut rhombeo-obtusis, nervis pinnatis, 4 areolas inter costam marginemque formantibus, soris fere 1 mm. latis rotundis, usque ad 5 pro lobo utroque costulae latere, uti videter exindusiatis, rhachibus faciebusque minute puberulis, textura herbacea, colore laete virente.

LUZON, Province of Bataan, Lamao River, Mount Mariveles (1369 *Whitford*) September, 1905; (387 *Topping*) May, 1904; (6970 *Elmer*) November, 1904.

A small deformed plant connected with the type of *D. Otaria*, analogous to the relationship between *D. diversiloba* and *D. canescens*, an insular reduced form with feeble characters.

27. **Dryopteris Otaria** (Kunze) O. Ktze. Rev. Gen. Pl. 2 (1891) 813. *Aspidium Otaria* Kunze; Mett Aspid. 34. n. 73.

PALAWAN (764 *Merrill*) February, 1903.

Rare, but distributed across the Malayan region.

28. **Dryopteris Ramosii** n. sp.

Habitu cum *Meniscio triphyllo* v. elato valde conveniens, pinnis valde remotis, paucis, fere integris, caudatis, gemmaque minuta axillari peculiaris. Planta debilis, textura tenui.

Rhizomate breviter repente crasso radicoso brunneo, foliis paucis approximatis, stipite flexuoso basi incrassato squamulis paucis brevibus sparso rufo-stramineo, ad basin pennae anserinae, porro vix corvinae crassitie, 35 usque ad 60 cm. longa, frondem multum superante; tota planta nuda; fronde 20 ad 35 cm. longa, oblonga, pinnata, pinnis valde remotis, 5 cm. distantibus, alternis, paucis, 1 ad 4 utrinque cum pinna terminali longe petiolata, pinnis erecto patentibus basi cuneatis, fere sessilibus, ovato-oblongis 12 ad 15 cm. longis, 3 cm. latis, longe et abrupte caudato acuminatis, margine subintegris aut repando-cuneatis, in axilla rhachiali saepe gemma rotunda minima praeditis, costulis manifestis sed tenuibus a costa ad marginem protensis 6 mm. separatis, nervis ca. 8 ad 10 utrinque, omnibus more *Meniscii* junctis et nervulos intermedios longitudinaliter junctos emittentibus, soris minutis, brunneis, rotundis, 7 aut 8 utroque costulae latere, costulae approximatis, indusio nullo. Textura herbacea aut papyracea, colore obscure viridi, subtus pallidiore. *D. otaria* longe recedit pinnis lobato-serratis, indusio etc:

LUZON, Province of Rizal (1792 *Ramos*) January, 1907. MINDORO, Mount Halcon (6093 *Merrill*) November, 1906.

29. **Dryopteris pteroides** (Retz.) O. Ktze. Rev. Gen. Pl. **2** (1891) 813. *Polypodium pteroides* Retz. Obs. **6**: 39.

MINDANAO (293 *Cuming*) "*Nephrodium Cumingii* J. Sm." in Hook. Journ. Bot. 3 (1841) 411: Lake Lanao (254 Mrs. *Clemens*) February, 1906; Province of Zamboanga (1604 *Copeland*) 1905; District of Davao (636 *Copeland*) March, 1904. BALABAC (420 *Mangubat*): MINDORO, Calapan (984 *Merrill*) April, 1903. CULION (487, 594 *Merrill*) December, 1902. LUZON, Province of Rizal, Montalban (*Loher*) March, 1906: Province of Pampanga, Mount Arayat (54 *Bolster*) March, 1905. PALAWAN (271 *Bermejos*) December, 1905.

Throughout the Malayan region.

30. **Dryopteris extensa** (Blume) O. Ktze. Rev. Gen. Pl. **2** (1891) 812. *Aspidium extensum* Blume Enum. (1828) 156.

BALABAC (415 *Mangubat*) April, 1905. No. 391 *Cuming*, distributed in Cuming's Philippine series, was from MALACCA, fide J. Sm., Hook. Journ. Bot. 3 (1841) 411, sub *Nephrodium cumingii* J. Sm. It is referable to *Dryopteris extensa*.

Throughout the Malayan region.

31. **Dryopteris Bordenii** n. sp.

Rhizomate elongato obliquo crasso, radicoso, squamulis minutis lanceolatis crispis dilute brunneis sparso, foliis paucis subfasciculatis, stipite firmo usque ad 30 cm. longo, saepe breviore, plumbeo- aut castaneo-stramineo, puberulo auriculis parvis triangularibus subacutis saepe ad meros lobulos minimos aut ad callos reductis, circ. 10 utrinque, instructo, fronde ovata basi haud attenuata sed pinnis inferioribus valde deflexis, 25 usque ad 35 cm. longa, 15 ad 20 cm. lata, acuminata, pinnata, pinnis infra remotiusculis, supra confertis alternis, inferioribus ad basin valde attenuatis, acuminatis, lanceolatis, sessilibus, supremis, adnatis, ca. 15 utrinque infra apicem pinnatifidum recte patentibus, 10 cm. longis 16 mm. latis, basi truncatis, inferioribus attenuato-cuneatis, usque ad mediam laminae partem incisis, ala 0.5 cm. lata relicta, lobis obliquis subfalcatis pectinato-confertis, sinu fere nullo interjecto, obtusis, integris, ca. 25 utrinque, 3 mm. longis, 2.5 mm. latis, nervis parum conspicuis, 7 utrinque, una area costali et secunda sinu applicata, rhachi faciebus costis costulisque breviter pubescentibus, soris mediis parvis atrobrunneis, sporangiis laevibus, indusio minuto obscure griseo reformi mox evanido. Colore obscure viridi, opaco, textura herbacea.

LUZON, Province of Bataan, Lamao River, Mount Mariveles (1237 *Borden*) June, 1904; (6823 *Elmer*) November, 1904: Province of Rizal (66, 78 *Foxworthy*) January, 1906: Province of Pampanga, Mount Arayat (493 *Topping*) February, 1904: Province of Tayabas, Mount Banajao (*Loher*) February, 1906. PALAWAN (571 *Foxworthy*) April, 1906.

The frond has the configuration of that of *P. sagittaefolia*, the base of the frond being abrupt and the pinnæ being replaced by reflexed and pointed auricles which occupy the stipe to the base. *D. Bordenii* however does not belong, like *D. sagittaefolia*, in the group with *D. parasitica*, not having hairy sporangia. The rhizome is oblique, elongated, the roots strong, the scales very small lanceolate and twisted.

32. **Dryopteris moulmeinensis** (Bedd.) C. Chr. Ind. Fil. (1905) 278. *Nephrodium moulmeinense* Bedd. Ferns Brit. Ind. Correct. (1870); Hooker Synopsis 503.

MINDORO, Baco River (997 *Merrill*) April, 1903. MINDANAO, Province of Zamboanga (1613 *Copeland*) 1905; San Ramon (*Copeland*) May, 1904: District of Davao, Todaya (1240 *Copeland*) April, 1904: Lake Lanao, Camp Keithley (117 Mrs. *Clemens*) January, 1906.

Throughout the Malayan Region.

33. **Dryopteris urophylla** (Wall.) C. Chr. Ind. Fil. (1905) 299. *Polypodium urophyllum* Wall. Cat. (1828) 229; Hook. Sp. Fil. **5**: 9.

LUZON, Province of Bataan, Mount Mariveles (6090 *Leiberg*) July, 1904: Province of Laguna, Mount Maquiling (2025 *Copeland*) March, 1906. MINDANAO, Province of Surigao (223 *Bolster*) January, 1906: Lake Lanao, Camp Keithley (Mrs. *Clemens*) March, 1906: District of Davao (952 *Copeland*) April, 1904.

Widely distributed in Malaya.

Var. **pustulosa** Copel. MSS. pro specie.

"Nearest *N. moulmeinense*, from which it differs in the subhispid, rough-pustulous surface" *Copeland*.

LUZON, Province of Bataan, Lamao River (218 *Copeland*) February, 1904.

34. **Dryopteris cuspidata** (Blume) *Meniscium cuspidatum* Blume Fil. Jav. 102. *t. 45*.

MINDORO, Baco River (168 *McGregor*) April–May, 1905.

This is the typical form of *Blume's* species, of which I have identical material from Java, Tjipoes, leg. *Raciborski*, and from Perak, leg. *Hose*, and differs from the plant of northern India (*Meniscium longifrons* Wall.) in its fleshy-papyraceous texture, opaque, the areolæ less numerous (8 to 12, rarely more) more or less concealed under the membranous epidermis and not costellate and prominent. The proliferous shoots in the axils of the upper pinnæ are also present in the Philippine plant, as indicated by *Blume* in the Javan form.

Christensen in his Index Filicum unites this species, although with doubt, with *Dryopteris urophylla;* however the proliferation in *Blume's* species, and the membranous epidermis covering and in part concealing the areolæ sharply distinguishes *D. cuspidata* from *D. urophylla*. *D. longifrons* differs in having a very smooth shining surface, the areolæ in strong relief, and in the form of its pinnæ which are elongated and with their margins nearly parallel. In regard to the elongated sori, I have from Java a specimen with them nearly round. It goes without saying that diagnoses alone are of little value in indicating the differences in forms and the slight characters that distinguish these undivided species of *Dryopteris;* characters that strike the eye on examination of specimens are often difficult to express in words in this and parallel cases.

35. **Dryopteris glandulosa** (Blume) O. Ktze. Rev. Gen. Pl. **2** (1891) 812. *Aspidium glandulosum* Blume Enum. (1828) 144.

LUZON, Province of Rizal, Bosoboso (964 *Ramos*) July, 1906; Mabacal (*Loher*) March, 1906: Province of Bataan, Mount Mariveles (427 *Topping*) May, 1904. LEYTE (298 *Cuming*). MINDANAO, Province of Zamboanga (1718 *Copeland*).

Differing from the two preceding by its short erect rhizome which is not creeping. The villosity is slightly glandular.

Sunda Islands and eastern Malaya.

36. **Dryopteris lineata** (Blume) C. Chr. Ind. Fil. (1905) 275. *Aspidium lineatum* Blume Enum. (1828) 144.

MINDANAO, Province of Zamboanga, San Ramon (1218 *Copeland*) April, 1905, 700 m. alt.

Glabrous, the frond not reduced at the base. It has the appearance of *Cyclopeltis semicordata*, but the pinnæ are less numerous, broader and not articulate.

Malayan region.

37. **Dryopteris Spenceri** (Copeland MSS, *Nephrodium*) n. sp.

Rhizomate elongato, radicoso, foliis paucis (3) stipite 20 ad 30 cm. longo rufostramineo, pennae anserinae crassitie, fronde 50 cm. et ultra longa 12 cm. lata, oblonga, pinnata, pinna magna basi saepe petiolata 12 cm. longa 3.5 cm. lata acuminata grosse lobata terminata, pinnis lateralibus sessilibus, oblongis, falcatis, breviter acuminatis, plus minus crenato-lobatis (lobis 5 mm. latis) 8 cm. longis, 2 cm. latis basi inaequalibus antice truncatis postice semicordato-rotundatis rhachimque tegentibus, et versus stipitem in auriculas breves trigonas numerosas (usque ad 10 utrinque) abeuntibus. Tota planta pube brevi griseo parce obsita, nervis manifestis prominulis, pinnarum lateralium ca. 6 utrinque 3 aut 4 areolas inter costam et marginem formantibus, nervis pinnae terminalis usque ad 12, saepe furcatis, et 10 areolas cum areolis aliquot lateralibus (more *Pleocnemiae*) formantibus, soris numerosis, 4 usque ad 10 utrinque, costulis approximatis saepe ovatis exindusiatis. Textura herbacea, colore laete virente.

MINDANAO, District of Davao, Tòdaya (1464 *Copeland*) October, 1904, alt. 800 m.; Sibulan River (981 *Copeland*) April, 1904. A plant from San Ramon, MINDANAO (*Copeland* s. n.) April, 1905, from about the same altitude as the above is distinguishable by its shorter and more numerous pinnæ.

A very large species of the group of *D. Stegnogramme* (*Gymnogramme aspidioides* Blume Fl. Jav. *pl.* 98.) but very different from that species in having a large terminal pinna instead of a pinnatifid apex, and with auricles descending along the stipe.

38. **Dryopteris simplicifolia** (J. Sm.). *Nephrodium simplicifolium* J. Sm. in Hook. Journ. Bot. 3 (1841) 411.

A reduced insular form of the *D. glandulosa* type. The plant small, the terminal pinna only developed, the lateral ones reduced to auricles.

LEYTE (315 *Cuming*). This is the plant figured by *Hooker* Sp. Fil. 1: 19. A sterile frond with the above specimen shows that it is a larger plant than figured and demonstrates clearly that the species is more especially a reduction of *D. Spenceri*. Specimens from San Ramon, MINDANAO (*Copeland*) February, April, 1905, have the nerves less pronounced, the terminal pinna narrower and the pubescence more grayish.

39. **Dryopteris prolifera** (Retz.) C. Chr. Ind. Fil. (1905) 286. *Hemionitis prolifera* Retz. Obs. 6: 38.

LUZON, Province of Cagayan (133 *Bolster*) July, 1905: Province of Tayabas, Lucena (616 *Whitford*) August, 1904: Without locality (168 *Cuming*).

Throughout the Malayan region to tropical Africa.

40. **Dryopteris rubida** (J. Sm.) O. Kuntze Rev. Gen. Pl. **2** (1891) 813. *Goniopteris rubida* J. Sm. in Hook. Journ. Bot. **3** (1841) 395; *Polypodium rubidum* Hook. Sp. Fil. **5**: 12.

LUZON (415 *Cuming*): Province of Bataan, Mount Mariveles (272 *Whitford*) May, 1904. PALAWAN (675, 684 *Foxworthy*) March, April, 1906.

The base of the stipe, which is not described by *Hooker*, is as thick as one's finger, covered with subulate stiff dark brown scales 2 cm. long, and also pustular.

Endemic to the Philippines.

41. **Dryopteris triphylla** (Sw.) C. Chr. Ind. Fil. (1905) 298. *Meniscium triphyllum* Sw. in Schrad. Journ. **1800**[2]: 16.

LUZON (11609 *Warburg*). LEYTE (299 *Cuming*). NEGROS (76 *Copeland*); Gimagaan River (1606 *Whitford*) May, 1906.

Malayan region.

Var. **elata** n. var.

Majus, 60 cm. et ultra alta, pinnis saepius 5, remotis fere lanceolatis valde elongato-candatis, soris brevibus non junctis.

NEGROS, Gimagaan River (1608 *Whitford*) May, 1906. LUZON, Province of Rizal, Mabacal (*Loher*) March, 1906; Angilog (*Loher*) March, 1906.

II. LASTREA (including *Phegopteris*).

42. **Dryopteris Loheriana** (Christ) C. Chr. Ind. Fil. (1905) 275. *Aspidium Loherianum* Christ. in Bull. Herb. Boiss. **6** (1898) 191.

LUZON, Province of Rizal, Montalban (*Loher*) October, 1890; Mount Batay (*Loher*) April, 1905: Province of Laguna, Mount Maquiling (*Loher*) April, 1906: Province of Pampanga, Mount Arayat (3908 *Merrill*) October, 1904: District of Lepanto, Balili (1910b *Copeland*) November, 1905.

This species, which has all the appearances of a *Lastrea*, sometimes has the basal nerves joined.

Endemic to the Philippines.

43. **Dryopteris stenobasis** C. Chr. Ind. Fil. (1905) 294. *Lastrea attenuata* J. Sm. in Hook. Journ. Bot. **3** (1841) 412.

SAMAR (327 *Cuming*).

This species presents in the greatest degree the reduction of the lower pinnæ, a character common to so many of the Philippine species. The position of the species, with its numerous nerves in the very narrow lobes, is uncertain and seems to approach *Nephrodium*.

Endemic.

44. **Dryopteris orientalis** (Gmel.) C. Chr. Ind. Fil. (1905) 281. *Polypodium orientale* Gmel. Syst. **2**: 1312.

Var. **Webbiana** (Hook.) *Nephrodium Webbianum* Hook. Sp. Fil. **4**: 85.

Differs from *D. orientalis* (*Polypodium pectinatum* Forsk. and *Aspidium albopunctatum* Bory) in lacking the calcareous coating on the upper surface and the lobes more horizontal and more angular, but it is not more than a variety of the African species. It is found also in Amboina and the Viti Islands, and appears unexpectedly in the Philippines. The Island of Réunion is the intermediate place in its distribution.

MINDANAO, Province of Zamboanga, San Ramon (1712 *Copeland*) April, 1905, alt. 850 m.

Eastern Malayan region, rare and widely distributed.

45. **Dryopteris Beddomei** (Baker) O. Ktze. Rev. Gen. Pl. **2** (1891) 812. *Nephrodium Beddomei* Baker Synopsis 267.

LUZON, Province of Benguet, Baguio (6491 *Elmer*) June, 1904; (331 *Topping*) January, 1903; (4941, 5089 *Curran*) August, 1906; (1818 *Copeland*) October, 1905; (Dr. *Pond*) March, 1904; Tilad (*Loher*) February, 1904: Mount Tonglon (5010 *Curran*) August, 1906.

British India, China and Malaya.

46. **Dryopteris immersa** (Blume) O. Ktze. Rev. Gen. Pl. **2** (1891) 813. *Aspidium immersum* Blume Enum. (1828) 156.

LUZON (72 *Cuming*) "*Lastrea verrucosa* J. Sm." in Hook. Journ. Bot. **3** (1841) 412. MINDANAO, District of Davao (695 *Copeland*) March, 1904: Province of Zamboanga, San Ramon (1574 *Copeland*) December, 1904.

Malaya.

Var. **ligulata** (J. Sm.) *Lastrea ligulata* J. Sm. in Hook. Journ. Bot. **3** (1841) 412. *Aspidium ligulatum* Mett. Aspid. no. 213.

CEBU (343 *Cuming*) "*Lastrea ligulata* J. Sm." l. c.

In examining the above authentic specimen I find but slight differences between it and *D. immersa*. The rachis is atroviolaceous rather than of a pale straw color such as is usually the case with the latter species.

47. **Dryopteris Motleyana** (Hook.) C. Chr. Ind. Fil. (1905) 278. *Nephrodium Motleyanum* Hook. Syn. 266.

NEGROS, Gimagaan River (93 *Copeland*) January, 1904; (1485 *Whitford*) May, 1906. MINDANAO, Province of Zamboanga, San Ramon (1713 *Copeland*) April, 1905. PALAWAN (541 *Foxworthy*) April, 1906.

Sunda Islands, and probably other islands in the Malayan region.

48. **Dryopteris Luerssenii** (Harringt.) C. Chr. Ind. (1905) 276. *Nephrodium Luerssenii* Harringt. in Journ. Linn. Soc. Bot. **16** (1877) 29.

Ab *Aspidium xylode* Kunze differt textura magis coriacea, segmentis acutioribus, ala costali latiore, soris costalibus mox faciem inferiorem segmenti impleatibus, basi frondis vix aut abrupte attenuata.

LUZON, Province of Benguet, Baguio (Dr. *Pond*) March, 1904; (181, 236, 214 *Topping*) January, February, 1903; (6514, 6515 *Elmer*) June, 1904: District of Lepanto (1910 *Copeland*) November, 1905.

Endemic to the Philippines.

49. **Dryopteris Foxii** (Copeland MSS. *Nephrodium*) n. sp.

I transcribe here the manuscript diagnosis of the author:

"Rhizomate breve repente vel adscendente, stipitibus confertis 5 ad 10 cm. altis, stramineis, glabris, facie superiore canaliculatis, fronde lanceolata 20 ad 30 cm. alta, 6 ad 9 cm. lata utrinque angustata, bipinnata, pinnis lanceolatis, acuminatis, adscendentibus, infimis remotis minutis, pinnulis inferioribus lineari-lanceolatis, 8 mm. longis, 1 ad 1.5 mm. latis, acutis, obscure dentatis, adnatis, remotis, sequentibus confluentibus demum in caudam subserrantam coadunatis, membranaceis, supra glabris, infra sparse pubescentibus, pilis albis, brevibus, venulis in pinnulis maximis plerumque utrinque 5, soris medialibus indusiis reniformibus glabris, subpersistentibus."

"A representative of the chiefly American group of *Nephrodium oppositum* (Sw.) Diels (*N. conterminum* Desv.) from which it differs mostly in the medial instead of submarginal sori. It is very common on rocky banks submerged during floods."

MINDANAO, District of Davao, Catalonan (940 *Copeland*) April, 1904; Davao (*Copeland*) April, 1904: Province of Zamboanga, San Ramon (1555 *Copeland*) December, 1904. LUZON, Province of Rizal, Bosoboso (1084 *Ramos*) July, 1906; Arambibi River (*Loher*) March, 1893; Montalban (*Loher*) 1906: Province of Batangas, Santo Tomas (2000 *Copeland*) February, 1906: Province of Benguet, Baguio (167, 258 *Topping*) January, 1903; Sablan (6178 *Elmer*) April, 1904; Baguio (5010 *Curran*) August, 1906; (6577 *Elmer*) June, 1904: Province of Cagayan (*Warburg*); (119 *Bolster*) July, 1905: Manila (*Rothdauscher*) 1879 in Herb. Monac: Province of Zambales, Pinatubo (*Loher*) February, 1906.

This species has been known to me for a long time, but was considered as *Lastrea ligulata* J. Sm. It is described here from the most common form—that is, rather small specimens—although sometimes it attains a size three times as large as is indicated in the diagnosis, and even larger. The plant can always be readily distinguished from *D. immersa* and *D. ligulata* (which to me are not specifically distinct) by its very sharp segments, which are cuneate and decurrent, and by its light green color and more firm texture. Its affinity is with *D. Koordersii* Christ[8] of Celebes, but that species is distinguishable by its very peculiar indusium which I have described as follows: "Indusio subgloboso lateraliter inhaerente duro crustato valde convexo brunneo nitido adiaphano sorum margine deflexo (more *Matoniæ*) tegente."

50. **Dryopteris quadriaurita** n. sp.

Rhizomate uti videtur obliquo coma squamarum subulatarum fere 1 cm. longarum rigidarum opacarum castanearum coronato, frondibus subsolitariis aut paucis, stipite 33 cm. longo nudo griseo-stramineo tereti pennae corvinae crassitie, fronde 35 cm. longa 24 cm. lata deltoideo-oblonga versus basia vix attenuata, pinnis ca. 15 utrinque infra apicem pinnatifidum sessilibus inferioribus oppositis, infimis declinatis, acuminatis 12 cm. longis, 2.5 cm. latis ad rhachim incisis horizontalibus remotiusculis, pinnulis linearibus, sinu acuto interjecto, acutis, integris, 14 mm. longis, 3 mm. latis, inferioribus liberis, falcatis, infimis rhachi incumbentibus auctis incisis stipulaceis, costis brevissime puberulis, cum costulis stramineis, nervis liberis simplicibus obliquis 8 ad 10 utrinque, soris impressis mediis minutis exindusiatis. Textura coriacea, rigida, colore laete virente.

MINDANAO, Province of Zamboanga, San Ramon (1714, 1713 *Copeland*) April, 1905, alt. 850 m.

A species with the appearance of *Pteris quadriaurita* Retz., the fronds not fasciculate, bipinnate, deltoid-oblong, the pinnæ cut to the costa and furnished at the base with incised stipules, the pinnules coriaceous, linear, their margins entire, the nerves simple, the sori small. It differs from *D. patens* in its narrow linear segments.

[8] *Ann. Jard. Bot. Buitenz.* 15[1]: 128.

51. **Dryopteris flaccida** (Blume) O. Ktze. Rev. Gen. Pl. **2** (1891) 812. *Aspidium flaccidum* Blume Enum. (1828) 161.

LUZON, Province of Benguet, Baguio (157, 171 *Topping*) January, 1903.

Malaya.

52. **Dryopteris erubescens** (Wall.) C. Chr. Ind. Fil. (1905) 263. *Polypodium erubescens* Wall.

MINDANAO, Province of Zamboanga, San Ramon (1612 *Copeland*) January, 1905, at 75 m. alt.

Malaya.

53. **Dryopteris Metteniana** Hieronym. MSS. n. sp. sub *Nephrodium*. *Lastrea spectabilis* J. Sm. in Hook. Journ. Bot. **3** (1841) 412, sed *Aspidium spectabile* Blume Enum. 158 *D. syrmaticam* amplectitur.

Differt a *D. Syrmatico,* cui similis dente in sium loborum posito, amplitudine, pinnis fere sessilibus et pinnis basalibus postice egregie auctis bipinnatifidis, texture membranacea, colore atroviridi, soribus pluriseriatis.

Ampla, nuda, stipite plumbeo-stramineo valido, fronde 70 cm. longa 30 cm. lata ovata, acuminata, bi- et subtripinnatifida, pinnis patentibus remotis ca. 15 utrinque infra apicem pinnatifidum, inferioribus breviter petiolatis, reliquis sessilibus, versus basin postice attenuatis, ovato-oblongis supremis lanceolatis 18 cm. longis 4 cm. et ultra latis superioribus angustioribus caudato acuminatis, basi cuneato-truncatis subinaequalibus, usque ad mediam laminam incisis, lobis grossis sinu aperto rotundato separatis subfalcatis ovatis usque ad 3 cm. longis et 1 cm. latis serrato-crenatis acutiusculis, ca. 15 utrinque, dente in sinu posito, ca. 8 utrinque, pinnis infimis deorsum valde auctis, pinnula basali deflexa 7 cm. longa profunde lobata, costis costulisque prominentibus stramineis, nervis tenuibus bi-aut trifurcatis, liberis, soris mediis minutis brunneis, ramo anteriore basali nervulorum impositis saepe biseriatis, 6 aut 7 utrinque, indusio minimo rudimentario griseo. Textura tenuiter membranacea, colore atroviridi.

LUZON (13 *Cuming*) "*Lastrea spectabilis* J. Sm." in Hook. Journ. Bot. **3** (1841) 412. MINDANAO, Mount Batangan (*Warburg*) 1888. The same species is found in CELEBES, Maros Bantimurung, South Celebes (16586 *Warburg*); Takale Kadjo, 500 m. alt. (*Sarasin*) February, 1895.

54. **Dryopteris Syrmatica** (Willd.) O. Ktze. Rev. Gen. Pl. **2** (1891) 814. *Aspidium Syrmaticum* Willd. Sp. Pl. **5**: 237.

Var. **petiolosa** n. var.

Pinnis longe petiolatis (petiolo 1 cm. et ultra) supremis solummodo subsessilibus, 16 cm. longis, 3.5 cm. latis, soris mediis minutis indusio griseo tectis. Rhizomate monente *Copeland* erecto terrestri.

LUZON (14 *Cuming*) "*Lastrea spectabilis* J. Sm." in Hook. Journ. Bot. **3** (1841) 412: Province of Laguna, Los Baños (*Alberto*) May, 1905. MINDANAO, Province of Zamboanga, San Ramon (736, 1581 *Copeland*) May, December, 1904: District of Davao (953, 928, 669 *Copeland*); Mount Batangan (14122 *Warburg*).

The Philippine plant is distinguishable from those of southern China (leg.

Henry) and India by its very long petioled pinnæ and smaller size. The same variety has been found on Christmas Island, Straits Settlements (leg. *Ridley*).

The type is widely distributed in Malaya.

55. **Dryopteris Sagenioides** (Mett.) O. Ktze. Rev. Gen. Pl. **2** (1891) 813. *Aspidium Sagenioides* Mett. Aspid. 113, No. 269.

LEYTE (302 *Cuming*). MINDANAO, Lake Lanao, Camp Keithley (386 Mrs. *Clemens*) March, 1906: District of Davao, Todaya (1238 *Copeland*) April, 1904.

Eastern Malaya.

56. **Dryopteris Boryana** (Willd.) C. Chr. Ind. Fil. (1905) 255. *Aspidium Boryanum* Willd. Sp. Pl. **5**: 285.

Forma pinnulis ovatis ad tertiam aut quartam partem solummodo incisis, aliter typica.

LUZON, District of Lepanto (1731 *Copeland*) November, 1905, alt. 2,000 m.

Widely distributed in the Malayan region, reaching to Japan.

57. **Dryopteris viscosa** (J. Sm.) O. Ktze. Rev. Gen. Pl. **2** (1891) 814. *Lastrea viscosa* J. Sm. in Hook. Journ. Bot. **3** (1841) 412.

Rhizomate crasso erecto semisupraterraneo stipitibus vetustis abunde obtecto nigricante, foliis valde numerosis dense fasciculatis, stipitibus rigidis pennae corvinae crassitie 12 aut 20 cm. longis, cum rhachi pilis ochreo-griseis dense tomentosis et insuper squamis ovatis atrobrunneis 0.3 cm. longis vestitis, fronde 25 ad 35 cm. longa, 8 ad 12 cm. lata oblonga, acuminata, basi ob aliquot pinnas breviores attenuata, binipinnatifida, pinnis 6 cm. longis, 1.5 cm. latis breviter acuminatis patentibus, infimis deflexis, remotiusculis, 20 ad 25 utroque latere infra apicem lobatum, sessilibus, fere usque ad costam incisis, segmentis oblongis obtusis angulo acuto separatis ca. 15 utrinque, subcrenatis, nervis 6 utrinque simplicibus, soris 1 mm. diametro, mediis rufobrunneis, indusio persistente convexo coriaceo brunneo, costis costulis et facie imprimis inferiore pilis rigidis albidis pubescentibus. Textura carnosula, colore sicce atrobrunneo, opaca.

MINDANAO, District of Davao, Mount Apo (1022, 1044 *Copeland*); (327 *DeVore & Hoover*) May, 1903. LUZON, Province of Bataan, Mount Mariveles (1105 *Whitford*) February, 1905.

Malacca (401 *Cuming*) in herb. Bureau of Science.

The affinity of this species is with *D. polylepis* (Fr. et Sav.) of China and Japan. It is characterized by its thick erect rhizome, its double villosity consisting of large blackish scales and grayish-yellow pubescence, its somewhat fleshy texture, and its very large brown persistent coriaceous indusia. It has the general appearance of a small member of the *filix mas* group, but its nerves are simple. An alpine form.

At isolated points from Perak, Malacca and Borneo.

58. **Dryopteris erythrosora** (Eaton) O. Ktze. Rev. Gen. Pl. **2** (1891) 812. *Aspidium erythrosorum* Eaton in Parry, Narr. Exp. to China **2** (1856) 330.

LUZON, District of Lepanto, Bagnen (1929 *Copeland*) November, 1905, alt. 2,000 m.

This peculiarly Chinese and Japanese species was found previously in Luzon by *Loher* in 1894, Mount Tonglon, Province of Benguet, alt. 2,250 m. It is also found is Assam, leg. *Mann*.

59. **Dryopteris marginata** (Wall.) Clarke in Trans. Linn. Soc. 2: 521. *t. 71.* *Aspidium marginatum* Wall. Cat. (1828) 366.

LUZON, District of Lepanto, Mount Data (1906 *Copeland*) November, 1905, alt. 1,800 m.

A member of the group of *D. filix mas* sensu latiori.

The discovery of this essentially Chinese plant, also found in the Himalayan region in the Philippines, is significant of the continental influence in the flora of northern Luzon, indicated also by other ferns such as *Dryopteris varia, D. erythrosora* etc.

60. **Dryopteris hirtipes** (Blume) O. Ktze. Rev. Gen. Pl. 2 (1891) 813. *Aspidium hirtipes* Blume Enum. (1828) 148.

LUZON, District of Lepanto, Mount Data (1887 *Copeland*) October, 1905: Province of Benguet, Baguio (6529 *Elmer*) June, 1904; (302, 303 *Topping*) January, 1903.

China and Malaya.

61. **Dryopteris filix mas** (Linn.) Schott Gen. Fil. *t. 9.* *Polypodium filix mas* Linn. Sp. Pl. (1753) 1090.

Var. **parallelogramma** (Kunze) *Aspidium parallelogrammum* Kunze Linnæa 13 (1839) 146.

LUZON, District of Lepanto, Mount Data (1875 *Copeland*) October, 1905, alt. 2,250 m.

The tropical variety of the European species, closest to the variety *paleacea* Moore; also in Celebes (leg. *Sarasin*).

Widely distributed in the Tropics of both hemispheres.

62. **Dryopteris heleopteroides** n. sp.

Rhizomate brevi radicoso crasso, foliis fasciculatis stipite basi incrassito brunneo squamis pallide brunneis diaphanis subulatis 0.5 cm. longis vestito, stipite rufostramineo, folii sterilis 6 cm. longo parce fibrilloso, planta aliter nuda, fronde deltoidea 16 cm. longa et fere aequilata, bipinnatifida, pinnis approximatis, ca. 10 infra apicem lobatum infimis maximis petiolulatis profunde ad alam angustam pinnatisectis, 8 cm. longis basi 3.5 cm. latis oblongis obtusis, segmentis ovato-rhombeis obtusis ca. 8 aut 10, infimis maximis 2 cm. longis 1 cm. latis obtusis grosse et irregulariter crenato-serrulatis, pinnis superioribus sessilibus et adnatis, grosse lobatis, lobis obtusis trigono-arcuatis, nervis inconspicuis in lobis pinnatis et furcatis, folia fertili longius (17 cm.) stipitata, fronde deltoidea 10 cm. longa, 7 cm. lata, pinnis ca. 8 utrinque, remotis, segmentis rhombo-obtusis, aequalibus, subintegris, 1 cm. longis, 0.5 cm. latis soris fere marginalibus, confertis, ca. 4 utrinque, ochraceis, 1 ad 2 mm. latis, indusio pallido reniformi bullato subpersistente. Textura subcoriacea, omnino *D. cochleatae* aut *D. chryocomae,* colore pallide viridi, opaco.

LUZON, Province of Benguet, Bued River (1837a *Copeland*) October, 1905, alt. 1,100 m.

The affinity of this species is with *D. filix mas,* and more particularly with *Nephrodium cochleatum* Don Prodr. Fl. Nepal. 6, by its dimorphism. The pinnæ of the fertile fronds are much more reduced than those of the sterile ones. The plant is small (always ?), with deltoid fronds, the sterile ones irregularly lobed. It has the appearance of the forma *Heleopteris* of *D. filix mas*. The presence of this member of the *filix mas* group augments the continental and temperate element in the mountains of northern Luzon, already known to be of considerable magnitude.

63. **Dryopteris Balabacensis** n. sp.

Ampla, stipite pennae anserinae crassitie, 55 cm. longo, angulosa, nuda uti tota planta, cum rhachibus rufocastanea sive rufostraminea, fronde tripinnata deltoidea 50 cm. longa et aequaliter aut latiore, basi tripartita, pinnis 8 ad 10 infra apicem pinnatifidem valde remotis (primo interstitio 14 cm. longo) petiolatis, petiolo infimarum pinnarum 5 cm. longo, pinnis infimis 30 cm. et ultra longis basi 25 cm. latis, deltoideis, postice acutis, pinnula infima posteriore 13.5 cm. longa et 7 cm. lata, pinnulis III incisolobatis, ovato-oblongis 2 cm. latis 4 ad 5 cm. longis obtuse lobatis, pinnis superioribus sessilibus oblongis acuminatis, ad basin usque ad costam incisis versus apicem lobatis, lobis extremis postice decurrentibus, oblongis acutiusculis, 8 mm. latis sinubus acutis dentatis, dentibus decumbentibus, nervis in lobis pinnatis et biaut trifurcatis, manifestis. Textura coriacea, colore ochreo-viridi, nitidulo, costis rufostramineis faciebus glabris, soris minutis numerosis submarginalibus mediisve, indusio nigro çoriaceo-carnoso reniformi mox convoluto persistente praeditis.

BALABAC (392 *Mangubat*) March to April, 1906. PALAWAN (698, 712, *Foxworthy*) March to April, 1906. SIBUYAN (25 *McGregor*) July, 1904.

This species belongs to the *D. sparsa* group but is larger than that species, its pinnæ long stipitate, the base of the frond strongly tripartite, the basal pinnæ usually large and compound like the rest of the frond, stipe glabrous, indusium fleshy, convolute, black. It has the appearance of *Sagenia*, but the nerves are not united.

64. **Dryopteris sparsa** (Don) O. Ktze. Rev. Gen. Pl. **2** (1891) 813. *Nephrodium sparsum* Don. Prodr. Fl. Nepal. (1825) 6.

LUZON, Province of Benguet, Baguio (282 *Topping*) January, 1903. PALAWAN (672 *Foxworthy*) March to April, 1906. MINDANAO, Province of Zamboanga, San Ramon (1727 *Copeland*) April, 1905.

Widely distributed in tropical Asia.

65. **Dryopteris purpurascens** (Blume) *Nephrodium purpurascens* Blume Enum. 169; Mett. in Ann. Lugd. Bat. **1**: 227; Raciborski Pter. Buitenz. 174, non Hook. Sp. Fil. 4. *t. 262.*

Differt a *D. sparso* squamis basalibus subulato-angustatis, fronde quadripinnatifida, pinnis infimis decompositis, magnitudine quadrupla. Icon. Hook. cit. est *D. sparsa.*

MINDORO, Mount Halcon (6101 *Merrill*) November, 1906.

The species is also known from Java.

66. **Dryopteris subarborea** (Bak.) C. Chr. Ind. Fil. (1905) 295. *Nephrodium subarboreum* Bak. in Journ. Linn. Soc. Bot. **24** (1887) 259. *N. megaphyllum* Bak. l. c. **22**: 227. *N. incisum* Copel. Polypod. Philip. 26, non Hook. Sp. Fil. **4**: 133 quod est *D. Boryana.*

MINDANAO, District of Davao, Mount Apo (1136 *Copeland*) April, 1904; (1614a *Copeland*) October, 1904, alt. 1,800 m. LUZON, Province of Benguet, Baguio (*Loher*) 1897, alt. 1,400 m.

The same species has been found in Borneo, Sarawak, leg. *Hose,* 1894; Batjan, leg. *Warburg;* Celebes, Lokon, leg. *Sarasin* No. 719, 1894.

It is related to *D. filix mas,* sensu latissimo, in spite of its extremely decompound frond and large size. In authentic specimens from Sarawak the segments are almost entire, while in those from other localities they are strongly dentate.

67. **Dryopteris dissecta** (Forst.) O. Ktze. Rev. Gen. Pl. **2** (1891) 812. *Polypodium dissectum* Forst. Prodr. 31.

LUZON, Province of Cavite, Mendez Nuñez (1297, 1311 *Mangubat*) August, 1906: Province of Laguna, Los Baños (*Loher*) January, 1906; Pagsanjan (513 *Topping*) 1904: Province of Bataan, Mount Mariveles (369 *Topping*) May, 1904; (198 *Whitford*) May, 1904: without locality (36, 244 *Cuming*) "*Lastrea membranifolia* J. Sm." in Hook Journ. Bot. **2**. (1841) 412. MINDANAO, District of Davao, Mount Apo (1465a *Copeland*) October, 1904, alt. 1,200 m.

A plant with blackish hairs issuing from pustules.

Widely distributed in tropical Asia.

68. **Dryopteris obscura** (Fée) O. Ktze. Rev. Gen. Pl. **2** (1891) 812. *Phegopteris obscura* Fée Gen. Fil. 243; Christ, Bull. Herb. Boiss. **6** (1898) 196. *t. 5.*

LUZON, Province of Laguna, Mount Maquiling (*Loher*) January, 1906: Province of Zambales, Mount Pinatubo (*Loher*) February, 1906: Province of Rizal, Montalban (*Loher*) March, 1903: Manila (*Usteri*) February, 1903.

The same species is found in Annam, Quang Binh leg. *Cadière* 1894, Herb. Mus. Paris 91, 126.

69. **Dryopteris Preslii** (Bak.) O. Ktze. Rev. Gen. Pl. **2** (1891) 813. *Nephrodium Preslii* Baker Syn. Fil. 272.

Baker's diagnosis is sufficiently clear to satisfactorily identify this plant. It is an insular dwarfed form that seems to be related to *D. obscura* from its general appearance, although smaller in all its parts.

BOHOL (354 *Cuming*) "*Lastrea spectabilis* J. Sm." in Hook. Journ. Bot. **3** (1841) 412. There has been an error, apparently, in copying the label, as *Lastrea spectabilis* J. Sm. = *Dryopteris Syrmatica,* our specimen being entirely different from the latter species. (*Baker* indicates No. 255 *Cuming* as the type of *Nephrodium Preslii.*)

Endemic.

70. **Dryopteris brunnea** (Wall.) C. Chr. Ind. Fil. (1905) 255. *Polypodium brunneum* Wall. Cat. (1828) 333. *P. distans* Don Prodr. Fl. Nepal. 2.

LUZON, Province of Benguet, Baguio (959 *Barnes*) May, June, 1904: District of Lepanto, Bagnen (1931 *Copeland*) November, 1905, alt. 2,000 m.

Widely distributed in tropical Asia.

71. **Dryopteris crenata** (Forsk.) O. Ktze. Rev. Gen. Pl. **2** (1891) 811. *Polypodium crenatum* Forsk. Fl. Aeg. Arab. 185.

LUZON, Province of Benguet, Twin Peaks (6480 *Elmer*) June, 1904; Baguio (6595 *Elmer*) June, 1904; Bugias (1851 *Copeland*) October, 1905: Province of Rizal, Montalban (*Loher*) October, 1903.

Widely distributed from China across tropical Asia; Cape Verde Islands.

72. **Dryopteris setigera** (Blume) O. Ktze. Rev. Gen. Pl. **2** (1891) 813. *Cheilanthes setigera* Blume Enum. (1828) 138.

LUZON, without locality (1, 75, 412 *Cuming*) "*Polypodium trichodes* Reinw." J. Sm. in Hook. Journ. Bot. **3** (1841) 394: Province of Rizal (1084 *Ramos*) July, 1906; Antipolo (*Guerrero*) June, 1903; Tanay (2266 *Merrill*) May, 1903; (90 *Foxworthy*) January, 1906: Province of Cavite, Mendez Nuñez (1355 *Mangubat*) August, 1906: Province of Laguna, Cavinti (*Loher*) February, 1906: Province of Benguet, Daklan (1837 *Copeland*) October, 1905; Baguio (178 *Topping*) January, 1903. PALAWAN (282 *Bermejos*) January, 1906. MINDANAO, Province of Zamboanga (1614, 1691 *Copeland*) March, 1904: District of Davao (611 *Copeland*) March, 1904.

I have previously indicated[9] some Philippine specimens as *Phegopteris ornata* (Wall.) Bedd., which *Loher* found at Montalban and on Mount Maquiling. They now appear to me to be strongly developed forms of *Dryopteris setigera*, and I can not identify the Philippine form with certainty with *D. ornata* as represented by specimens from Darjeeling (7465 *Gamble*) 1897.

Widely distributed in China and Malaya.

73. **Dryopteris setosa** (Presl) C. Chr. Ind. Fil. (1905) 292. *Lastrea setosa* Presl Epim. (1849) 40. *Polypodium setosum* Presl Rel. Haenk. **1**: 27, non Sw. *Phegopteris hirta* Christ Bull. Herb. Boiss. **6**: 195.

LUZON, Province of Rizal, near Montalban (*Loher*) 1897, in herb. Kew.

Endemic.

74. **Dryopteris intermedia** (Blume) O. Ktze. Rev. Gen. Pl. **2** (1891) 813. *Aspidium intermedium* Blume Enum. 161. *Dryopteris rhodolepis* C. Chr. Ind. Fil. (1905) 288, ex parte, nec Clarke Trans. Linn. Soc. **2**: 526, *Nephrodium*.

LUZON, without locality (80, 151 *Cuming*) "*Lastrea propinqua* J. Sm." in Hook. Journ. Bot. **3** (1841) 412: Province of Bataan, Lamao River (1240, 1241 *Borden*) June, 1904; (199 *Whitford*) May, 1904; (363, 370 *Topping*) May, 1904. NEGROS, Gimagaan River (1605 *Whitford*) May, 1906. MINDANAO, Province of Zamboanga, San Ramon (1465c, 1765d, *Copeland*) April, 1905.

Widely distributed in tropical Asia.

Var. **Mannii** (Hope) *Lastrea Mannii* Hope in Journ. Bot. **28** (1890) 145.

Fronde facie fere *D. filicis maris Europeae,* valde elongata oblonga bipinnatifida pinnis regulariter lobatis, lobis simplicibus, pinnis infimis solummodo bipinnitifidis et postice auctis.

MINDANAO, Province of Zamboanga, San Ramon (1588 *Copeland*) January, 1905; (1649 *Copeland*) February, 1905, alt. 500 m.

Assam, leg *Mann.*

Var. **microloba** n. var.

Stipite rhachique purpureis, pilis atrorubris patentibus densissime tectis, pinnulis minoribus 1.5 ad 2 cm. longis et 0.5 cm. latis, lobis confertis 6 utrinque, angustis 2 ad 4 mm. latis.

MINDANAO, Province of Zamboanga (1702 *Copeland*) 1905, alt. 850 m.

In general appearance quite different from the type.

[9] *Bull. Herb. Boiss.* **6** (1898) 196.

75. **Dryopteris rhodolepis** (Clarke) *Nephrodium rhodolepis* Clarke in Trans. Linn. Soc. 2: 526. *t.* 72.

Major magisque composita quam *D. intermedia*, pilis atropurpureis basi pustulatis fere nullis sed squamis rufobrunneis subulatis, lanceolatis et ovatis mixtis, stipite rhachi costisque abunde vestitis.

LUZON, District of Lepanto, Bagnen (1920 *Copeland*) November, 1905. MINDANAO, Province of Zamboanga (1773 *Copeland*) 1905, alt. 1,000 m.

C. Christensen has erroneously identified *Dryopteris intermedia* (Bl.) with *D. rhodolepis* (Clarke) in his index Filicum, 288, as *Clarke* has expressed very clearly l. c. 527, distinctive characters of the latter, "primary, secondary and tertiary rachises with ovate acute subadpressed hyaline rose-mauve scales."

China and British India, Himalayan region.

76. **Dryopteris Copelandi** n. sp.

Differt a *D. intermedia* absentia pilorum atropurpureorum basi pustulatorum, indumento squamato, fronde postice non aucta, pinnis angustis minoribus, segmentis minoribus denticulatis. Potius *D. spinulosae* quam *D. intermediae* appropinquanda. Rhizomate uti videtur obliquo pauca folia emittente, stipite pennae corvinae crassitie rufostramineo, 30 cm. longo, cum rhachi costisque abunde squamulis minimis setiformibus strigillosis rufobrunneis obtecto nec non squamis ovatis obtusis 0.5 cm. et ultra longis et latis diaphanis dilute brunneis vestito, fronde deltoideo-oblonga 40 cm. longa 30 cm. lata tripinnata, pinnis ovato-oblongis inferioribus breviter petiolulatis remotis (infimo intervallo 7 cm.) acuminatis basi vix attenuatis, utrinque ca. 15 infra apicem, pinnis infimis postice vix auctis, pinnula basali posteriore sequente breviore, pinnulis ca. 15 utrinque, approximatis, 4 cm. longis, 1.5 cm. latis, valde regulariter fere usque ad costam pinnatis, segmentis III subinaequalibus, basi subdecurrentibus, rhombeo-oblongis 6 mm. longis 3 mm. latis obtusissimis, 10 utroque latere, regulariter dentatis, dentibus ca. 5 utrinque acutiusculis, nervis pinnatis, furcatis, soris rufobrunneis numerosis ca. 3 utrinque, exindusiatis, faciebus pilis albidis numerosis pubescentibus, textura flaccide herbacea, colore laete virente.

LUZON, District of Lepanto, Mount Data (1887 *Copeland*) October, 1905, alt. 2,250 m.

77. **Dryopteris Rizalensis** n. sp.

Rhizomate brevi crasso, squamarum rigidarum 1 cm. longarum setiformium brunnearum coma dense vestito. Stipitibus subfasciculatis pennae corvinae crassitie, sulcatis, 25 cm. longis atrocastaneis squamis setiformibus atropurpureis patentibus flexuosis 6 mm. longis dense vestitis, fronde 32 cm. longa, basi 20 cm. lata elongato-deltoidea, basi bipinnatifida caeterum pinnata, pinnis falcatis acutis 7 ad 8 utrinque infra apicem incisum, infimis petiolulatis remotis, basi postice auctis 12 cm. longis, 5 cm. latis deltoideo-elongatis, caeteris lanceolato-oblongis 10 cm. longis, 3.5 cm. latis, superioribus decurrenti-adnatis, ad basin profunde, supra ad mediam et tertiam partem lobatis, costis omnibus late alatis, lobis obtusis subintegris, ca. 10 utrinque 1.5 ad 2 cm. longis 1 cm. latis, rhachi costis nervisque squamulis brevibus setiformibus aut lanceolatis brunneis pubescentibus, nervis in lobis pinnatis furcatisque flexuosis, soris in lobis pluribus usque ad 6 utrinque medialibus minutis, 0.5 mm. latis, pallide fuscis, indusio reniformi flaccido griseo mox evanido. Textura herbacea, colore pallide virente.

MINDANAO, Province of Zamboanga, San Ramon (1649 *Copeland*) February, 1905: District of Davao, Mount Apo (1465b *Copeland*) October, 1904. A smaller form from Mabacal, Province of Rizal, LUZON (*Loher*) March, 1906.

The affinity of this species is with *Dryopteris intermedia* and *D. obscura*, but is less compound, the basal pinnæ being only bipinnatifid, the pinnæ and lobes broad.

SUPPLEMENT.

27a. **Dryopteris granulosa** (Presl) C. Chr. Ind. Fil. 269. *Polypodium granulosum* Presl Reliq. Haenk. 1. 24 t. 4 f. 2. 1825.

Differt a *D. otaria* (Kze. Mett.) pinnis crenato-dentatis nec profunde lobatis, dentibus integris nec spinuloso-serrulatis, nervis conspicuis fere omnibus junctis et nurvulum rectum sursum emittentibus, soris minutis exindusiatis. Facie rugosa. *D. otaria* differt pinnis lobatis, lobis aristato-ciliatis aut serratis, nervis magis abliquis inconspicuis, inferiorbus solummodo junctis, soris majoribus manifeste indusiatis. Facie glabra.

PALAWAN (863 *Foxworthy*) May, 1906. BALABAC (413 *Mangubat*) March, 1906.

The same plant but larger from Indo-China leg. *P. Eberhardt*.

NOTES ON PHILIPPINE PALMS, I.

By Dr. ODOARDO BECCARI.
(*Florence, Italy.*)

An enumeration of the palms growing in the Philippine Islands was recently published by me,[1] but as the extensive botanical explorations now in progress in the Archipelago are continually bringing to light numerous new forms of this fine group of plants, it is my purpose to describe them in these "notes" as material, courteously transmitted to me at Florence by *Elmer D. Merrill,* Botanist of the Bureau of Science at Manila, becomes available.

ARECA Linn.

Areca Whitfordii Becc. n. sp.

Major, caudice circ. 10 m. alto, 20 cm. crasso. Folia amplissima, limbo 2.3–2.5 m. longo; petiolo brevi, 15 cm. longo, 2.5–3 cm. spisso, superne profunde sulcato, marginibus acutis, vagina 1 m. longa; segmentis numerosis inaequidistantibus et faciculatis, costulis 2–3 validissimis et superne valde prominentibus percursis, basi argute 2–3-plicatis; segmentis intermediis circiter metralibus, 4.5–5 cm. latis, falciformibus, longe acuminatis; superioribus sensim brevioribus et apice obtuse dentatis; duobus terminalibus basi unitis, pluricostulatis. Spadices 45 cm. longi, 3-plicato-ramosi. Flores ♂ . . . . . Fructus perianthio cyathiformi-obconico 15 mm. alto suffulti, elongato-elliptici, 40–42 mm. longi, 18 mm. crassi, fere aequaliter utrinque sensim attenuati, apice truncato, 3 mm. lato, mammillaeformi et in medio breviter mucronulato; mesocarpii fibris tenuissimis, numerosissimis, pluriseriatis.

A rather large palm, about 10 m. high. Stem 20 cm. in diameter. (*Whitford*). *Leaves* very large; the leaf-sheath about 1 m. in length, coriaceous, its inner surface silvery white and more strongly striate than the outer; the petiole comparatively very short (15 cm. long in one specimen), 2.5–3 cm. thick, deeply channeled above, its margins sharp; the pinniferous part 2.3–2.5 m. long, its rachis very robust, round near the base beneath and somewhat depressed and flattish in the intermediate portion. The leaflets numerous, rather closely set, inequidistant and more or

[1] Le Palme delle Isole Filippine, in *Webbia* (1905) 281–359. See also: *Palmae* in Perkins, Frag. Fl. Philip. (1904) 45–48.

less fascicled, falciform, very long, furnished with 2–3 very robust and in the upper surface very prominent and sharp ribs, very slightly or not at all narrowing below, where the blade, on each side of the rib, is very strongly plicate downwards; intermediate leaflets about 1 m. long, 4.5–5 cm. broad and very long-acuminate; the upper ones are shorter and obtusely toothed at the summit, the two apical leaflets have many very approximate ribs, but they are not broader than the others. *Spadix* 45 cm. long (in one specimen) 3 times branched, with a broad lunate embracing base and a very broad and short peduncular part (2 cm. long, 3.5 cm. broad); the primary branches divided into several alternate and sinuous secondary ones; these bear the flowering branchlets which are also sinuous and their lower part bear 1–2 female flowers; *Male flowers* . . . . . *Fruiting perianth* broadly obconic-cyathiform, 15 mm. long; its sepals ovate, obscurely keeled on the back; its petals longer by one-third than the calyx and terminating in a broad triangular point. *Fruit* elongate-elliptical, 4–4.2 cm. long, 18 mm. broad, almost equally tapering to both ends, but with a truncate disciform, mammillate, 3 mm. broad apex and besides shortly mucronulate because of the permanent remains of the stigmas; the epicarp and endocarp are very thin; the mesocarp is composed of many layers of innumerable very thin fibers. *Seed* seen only in a young state.

In the semi-swampy forests called "guipa" by the natives, near sea level Bongabon River, MINDORO, No. 1372. *H. N. Whitford*, January 16, 1906. Vernacular name (Tagalog) "*Bungan Gubat.*"

A very fine species, probably the largest, or at least the most robust, of the genus, the trunk being considerably thicker than that of *Areca Catechu*, if not so high. *A. Whitfordii* is related to *A. borneensis* Becc., but is a much larger plant and with smaller fruit. On account of this obvious affinity, I presume that the male flowers of *A. Whitfordii* are 3-staminate.

Areca mammillata Becc. n. sp.

Gracilis 2 m. alta, caudice 3 cm. diam. Folia metralia, petiolo longiusculo superne anguste profundeque sulcato, segmentis nonnihil numerosis, subaequalibus, anguste falcato-sigmoideis, unicostatis, acuminato-caudatis, subaequidistantibus 3–4 costulatis, basi connatis, apice truncatis et obtuse dentatis. Spadices breves, dense lateque paniculati, duplicato-ramosi. Flores ♂ unilaterales gemini, inter se valde approximati et concinni 2.2–2.5 mm. longi, lanceolato-sigmoidei, acuti; staminibus 6; ovarii rudimento profundissime 3-partito, staminibus subaequilongo vel etiam longiori. Fructus ellipsoideo-oblongi, basi acutati, circiter ad tertiam superiorem partem aliquantum coarctati, vertice 3 mm. lato, discoideo-mammillato et in medio obtuse mucronato, 17–19 mm. longi, 6.5 mm. lati, extus sublente minute granulati; pericarpio tenui; mesocarpii fibris rigidis, parallelibus, 1–2 seriatis; endocarpio crustaceo, fragili; semine anguste ovato-conico basi planiusculo, 9–9.5 mm. longo, 5 mm. spisso, raphidis ramulis circiter 10, laxe anastomosantibus.

A slender palm about 2 m. high (non soboliferous?). *Stem* 3 cm. in diameter. *Leaves,* judging from the fragment seen by me, about 1 m. in length; the petiole rather long, 8–10 mm. thick, finely striate, subterete but deeply and rather narrowly furrowed on its upper surface (the margins of the furrow very sharp) obsoletely longitudinally keeled underneath, finely striate (when dry); the rachis convex beneath with a salient angle above and two side facets where the leaflets are attached; leaflets rather numerous, thin in texture, membranous-subherbaceous, all of about the same breadth, narrowly falcate-sigmoid and unicostate, slightly narrowed at the base, and slightly decurrent along the rachis, very gradually narrowing upwards into a long linear caudate tip; the mid-costa strong, raised and sharp above; the secondary nerves numerous, causing both surfaces to be finely striate; usually two of the nerves on each side of the mid-costa are sligthly stronger than the others, especially near the base; the margins sharp slightly strengthened by a secondary nerve; the lower leaflets 45 cm. long and 18–25 mm. broad; the two apical ones united by their bases into a deeply furcate flabellum; each of them 25 cm. long and about 5 cm. broad, 4–5-costulate, truncate and almost præmorse at the summit and with as many pairs of very short and obtuse teeth as there are costæ. *Spadix* erect, densely panicled, twice branched, about 20 cm. long, and 12–14 cm. broad, with 10–12 ascendent-fastigiate main branches which form a dense and broad panicle, almost without a peduncular part, as the lowest branches spring from just a little above the insertion of the general spathe. The spathe is lanceolate, 24 cm. in length and 4.5 cm. in width, somewhat narrowing upwards into an obtuse tip which is ciliate-fringed on the margins; main branches 14–15 cm. long divided into 14–15 flexuose fastigiate branchlets, which carry, in notches of their rather stout angular base, from 2 to 12 female flowers, suddenly becoming very slender, very closely and minutely notched in the remaining and longer portion which is covered with male flowers. *Male flowers* very regularly and closely packed together, very distinctly secund and in pairs at each notch of the branchlets, inconspicuously bracteolate, 22–25 mm. long, sigmoid-lanceolate, somewhat flattened or 3-gonous, acute-apiculate; the calyx very small, its sepals triangular, acute, sharply keeled; the corolla many times longer than the calyx, its petals lanceolate-acuminate, strongly striate externally; stamens 6, somewhat shorter than the petals, the filaments very short, the anthers narrowly sagittate, acuminate-apiculate, the cells almost parallel but deeply separated at the base; rudimentary ovary as long as or a little longer than the stamens, represented by 3 subulate bodies united only in their basal part. *Female flowers* broadly ovoid; the sepals very broad, twice broader than long, slightly apiculate, ciliolate on the margins; the petals a little more than one-third longer than the calyx, imbricate and broadened at both sides in their basal part into a rounded ciliolate small auricle, the central part gradually passing into a thick triangular, valvate point. *Fruiting perianth* slightly accrescent 8 mm. high. *Fruit* 17–19 mm. long, 6.5 mm. thick, ellipsoid-oblong with an acute base, somewhat contracted or rather suddenly narrowed below the summit, which is truncate, circular, 3 mm. in diameter, and with the remains of the stigmas forming a small blunt mucro in its center; the outer surface of the fruit is dull, of a rather light color and finely grained under the lens; the pericarp on the whole is rather thin, the epicarp being thinly crustaceous, the mesocarp formed by only two layers of rigid parallel fibers, and the endorcarp also crustaceous and very thin, brittle and polished inside. *Seed* narrowly ovoid with a conical and obtuse top and a flat base, 9–9.5 mm. long, 5 mm. thick; the branches of the raphis are about 10 erect from the base, of which those of the middle pass undivided over the top of the seed and the others bend laterally and slightly anastomose on the opposite side. The rumination is formed by brown lamellæ and does not reach the center. The embryo is basal.

In swampy places along the Saribun River, PALAWAN; No. 3816, *H. M. Curran* February, 1906.

This species seems to be most closely related to *A. oxycarpa* Scheff., from North Celebes on account of its 6-staminate flowers and the size of the fruit, but the fruit of *A. oxycarpa* is equally narrowed and acute at both ends, while in *A. mammillata* the upper end is narrower than the lower one, and terminates in a broad discoid mammilate summit.

Areca Vidaliana Becc. n. sp.

Gracilis. Folia metralia vel paullo ultra, petiolo subtus rotundato, supra canaliculato, marginibus obtusis; segmentis membranaceis utrinque 8–10, subaequidistantibus, 2.5–4 cm. inter se remotis e basi lata falcato-acinaciformibus, 2–3-costulatis et crebre acuteque nervoso-striatis, apice acuminato-caudatis, circiter 40 cm. longis, 3–5 cm. latis, duobus terminalibus 3–4-costulatis et apice profunde inciso-dentatis. Flores asymmetrice lanceo-lati, acuminati, 3-goni, 3–3.5 mm. longi, staminibus 6, ovarii rudimento profundissime 3-partito, staminibus longiori. Fructus oblongo-elliptici, 18 mm. longi, 7 mm. lati, utrinque aequaliter attenuati, vertice 3 mm. lato, discoido-mammillato et in medio obtuso mucronato.

Apparently a slender palm about the size of *A. mammillata*. *Leaves* about 1 m. in length; the petiole round beneath and somewhat channeled above, its margins very obtuse; the rachis indistinctly ribbed beneath along the middle in its lower portion and roundish higher up, not very prominently acute above; the segments 9–10 only on each side of the rachis, membranous, subequidistant, 2.5–4 mm. apart, broadly falcate-acinaciform, about 40 cm. in length and 3–5 cm. in width, very slightly or not at all narrowed at the base, very slightly decurrent along the rachis, gradually acuminate to a falcate point and besides produced into a linear caudate appendage (this 3–4 cm. in length) usually with

3, and occasionally, 2–4 not very strong costæ in the upper surface, the secondary nerves very numerous and rather sharp on both surfaces, which therefore appear strongly striate; the terminal leaflets are of unequal width in the only leaf seen by me and not broader than the others, having 3–4 costæ, are united only at their bases, and terminate in as many pairs of narrow teeth as there are costæ; each pair of teeth is separated by a sinus 15–20 mm. deep. *Male flowers* lanceolate, acuminate, 3-gonous, more or less asymmetric, 3–3.5 mm. long; the calyx very small about 0.5 mm. long, its sepals triangular, acute, sharply keeled; the corolla many times longer than the calyx, its petals lanceolate, acuminate, strongly striate externally; stamens 6, with short filaments and very narrowly sagittate acuminate-apiculate anthers; the rudimentary ovary is a little longer than the stamens and is formed by 3 slender subulate bodies united only at their basal part. *Fruiting perianth* 8 mm. high. *Fruit* oblong-elliptical, almost equally narrowed at both ends, 18 mm. long, 7 mm. thick, terminating in a sharply defined disciform, orbicular, 3 mm. in diameter, mammillate surface, the remains of the 3 stigmas forming a small obtuse mucro in its center, otherwise the fruit is very similar to that of *A. mammillata*.

PALAWAN, No. 3955 *Vidal*, in Herb. *Beccari*.

I have described this species from very fragmentary material consisting of an entire leaf, detached male flowers and only one fruit, nevertheless *A. Vidaliana* seems to me to be a very well characterized species on account of its relatively large hexandrous male flowers and small elliptical fruit; it is related to *A. mammillata* from which it differs in the broader 3-costulate leaflets, larger male flowers and in the fruit equally tapering to both ends.

PINANGA Blume.

Pinanga insignis Becc. n. sp.

Major, caudice subelato, 12–15 cm. diam. Folia amplissima; vagina circiter metrali, petiolo subnullo, rachi crassissima, basi 5 cm. spissa et in facie superiori profunde sulcata, in parte media supra bifaciali et subtus rotundata; segmenta numerosa, aequidistantia, alterna in parte media 6–7 cm. inter se remota, late ensiformia, rectissima, rigidule papyracea, subtus glabra et secus nervis secundariis minutissime scabridula, 1–2-costulata, basi aliquantum attenuata, ibique argute 1–2-plicata, circiter 1.2 m. longa, in acumen rectum symmetricum sensim attenuata, cuando unicostata apice integra et 4–5 cm. lata, si 2-costata apice profunde bifido et 8–9 cm. lata; segmenta terminalia 3–4-costulata, secus costulas inferiores apice profunde fissa, partitionibus rectis acute bidentatis. Spadix metralis, fructifer deflexus, parte pedicellari crassa brevissima, ramulis numerosis spiraliter insertis, 40–50 cm. longis et 5–6 mm. latis, basi subtrigonis, caetero compressis. Fructus concinne biseriato-pectinati, valde approximati, late ovoido-elliptici, utrinque attenuati, 24–25 mm. longi, 13–14 mm. lati, in summo vertice acutiuscule mammillati; semine ovato, vertice conico obtusiusculo, basi obliqua non caudiculata. Perianthium fructiferum depresse cupulare, truncatum, ad faucem non constrictum, 3 mm. altum, 7 mm. latum.

Stem about 15 cm. in diameter. *Leaves* very large; leaf-sheath nearly 1 m. in length, thickly coriaceous, covered externally with appressed radiate scales of a chestnut brown color; petiole very short or almost obsolete, the lowest leaflets springing from only a few centimeters above the mouth of the leaf-sheath; rachis very strong, 5 cm. thick at its base, rounded beneath in the lower portion and with a deep and broad furrow above, the margins of this rather sharp, in the intermediate portion the rachis is convex beneath and above bears the leaflets attached on board flattish facets on each side of a sharp salient angle. The leaflets are very numerous, equidistant and more or less alternate; in the intermediate portion of the leaf they are 6–7 cm. apart, broadly ensiform, very straight, papyraceous and very rigid, 1- or 2-costate, somewhat narrowed at the base, and when 1-costate 4–5 cm. in width and gradually acuminate into a straight symmetrical entire point, when 2-costate 8–9 cm. in width and deeply bifid; the costæ are very robust, close together at the base, where the blade is strongly bent downwards once or twice according to the number of the costæ; the two surfaces are glabrous and finely striate with numerous secondary nerves; the upper surface is almost shining and polished, the lower very slightly paler and furnished with very small asperities (visible only under a lens) along the secondary nerves; the upper leaflets are gradually shorter but of the same description; the two terminal ones have 3–4 ribs, are deeply parted in as many divisions as are the costæ; each of these divisions is straight, acuminate and again more or less deeply and acutely cleft at its summit. *Spadix* about 1 m. long, composed of a very short, robust and deflected peducular part and several floriferous branchlets which are spirally inserted at different heights along the central axis; the branchlets are 40–50 cm. long, 5–6 mm. in width, almost trigonous in their basal part and flattened in the remainder, and bear exactly 2-seriate glomeruli of flowers; the bract subtending each glomerule is very short, scale-like, lunate and with the margin entire. *Male flowers* . . . . *Female flowers* . . . . *Fruiting perianth* depressed-cupular with a flat base, truncate and not contracted at its mouth, 3 mm. long, 7 mm. broad; its petals and sepals similar, smooth externally, their margins round more or less cleft. Fruits horizontally biseriate and very regularly set, approximate, when quite ripe, 24–25 mm. long, 13–14 mm. thick, broadly ovoid-elliptical, tapering towards both ends but a little more towards the summit which ends in a conical mammilate rather more acute point. *Seed* ovoid, with a conic bluntish point, 16 mm. long, 11 mm. broad, the base broad, oblique and not caudiculate, the branches of the raphis numerous, anastomosing and forming a network of long and narrow loops.

Not common along the Bongabong River, MINDORO, No. 1388 *H. N. Whitford*, January 19, 1906. Vernacular Visayan name "*Sarauag;*" Tagalog, "*Prunbai.*"

It is one of the largest and most beautiful species of the genus. *Whitford* in his field notes gives 5 m. as the height of the entire plant, but from the dimensions of the leaves we should estimate a greater height. From the size of the leaf-sheaths the trunk may be judged to be 15 cm. in diameter.

Pinanga modesta Becc. n. sp.

Gracilis, 1.5–3 m. alta, caudice 10–15 mm. diam. internodis 3–6 cm. longis. Folia 0.6–1 m. longa, vagina circiter 18 cm. longa, petiolo 12–20 cm. longo, basi 6–8 mm. crasso, supra profunde canaliculato, marginibus obtusis, segmentis utrinque 3–4 et duobus terminalibus, chartaceis, rigidulis, subtus pallidis, subglaucescentibus et sub lente punctulato-puberulis, superne striatis et minutissime (saepe inconspicue) puntulato-scabridis, lateralibus acinaciformibus inaequalibus per paria oppositis, sensim longeque falcato-acuminatis, 3–6-costatis, 30–40 cm. longis, 4–7 cm. latis, duobus terminalibus basi connatis et apice acute duplicato-dentatis. *Spadix* fructifer brevis, 10–12 cm. longus, abrupte deflexus, parte pedicellari brevissima, in ramis perpaucis (3–4) valde compressis digitatim partitus; fructubus horizontalibus exacte et concinne pectinato-biseriatis, 8–10 mm. inter se remotis, angustissime ovatis, basi coarctatis et acutis, in parte superiori sensim attenuatis, summo apice papillaeformi, 15–16 mm. longis, 6 mm. crassis; semine e basi ovata conico-acuto, 11 mm. longo, 5 mm. crasso, basi caudiculato. Perianthium fructiferum 2 mm. altum, 4 mm. latum, e basi plana cupulari, in ore nonnihil coarctatum.

A slender species 1.5–3 m. high. *Stem* 10–15 mm. in diameter, its joints 3–6 cm. long, finely dotted with small brown radiate scales in the apical young parts. *Leaves* 0.6–1 m. long, not including the sheath; this about 18 cm. long, finely and sharply striate, more or less scaly furfuraceous; the petiole 12–20 cm. long, 6–8 mm. thick in its basal part, rounded beneath, deeply channeled above, the margins of the channel obtuse; the rachis very obscurely keeled beneath in its lower portion and flattish higher up; above bifacial from the middle upwards and appressed scaly-furfuraceous throughout as is the petiole. The leaflets few, usually only 3–4 on each side of the rachis and two at the summit, rather firmly papyraceous, dull, subglaucescent and under the lens finely punctulate-puberulous beneath, finely striate and punctulate above; the lateral leaflets are acinaciform, about 40 cm. long, 4–7 cm. broad, usually opposite, very slightly or not at all narrowed at the base, gradually falciform-acuminate, with 3–6 acute acostæ; the terminal leaflets are united by their bases, 5–10 cm. broad, with many (6–12) costæ and an equal number of main teeth at their summit, these separated by deep and acute reëntering angles, the resulting secondary teeth cleft again at their apices with their two divisions more or less acute or acuminate. *Fruit-spadix* deflexed, 10–12 cm. long, the peduncular part very short with only 3–4 digitate strongly flattened, slightly sinuous branches; bracts subtending the flowers inconspicuous. *Flowers* . . . . . *Fruiting perianth* shallowly cupular, 2 mm. high, 4 mm. across, truncate, narrower at the mouth than at the base; the sepals and petals similar, very closely imbricated, smooth outside, their margins rounded, entire, non-ciliate, the sepals somewhat thickened and gibbous at the base. *Fruits* exactly horizontally bifarious, 14–15 on each side of the rachis, very regularly but not very closely set, 15–16 mm. long (including the perianth), 6 mm. thick, conically acuminate from an ovoid base, suddenly broadening when just outside the mouth of the perianth, the base acute included; the apex of the fruit is nipple-like and this is surmounted by the remains of the stigmas in the shape of a small obtuse mucro. *Seed* ovoid, conic and acute above, 11 mm. long, 5 mm. thick, caudiculate at the base; the branches of the raphe are very few, curved on the sides and slightly anastomosing.

MINDANAO, San Ramon, District of Zamboanga, on ridges at an altitude of about 600 m. *E. B. Copeland*, March 5, 1905; also at Camp Keithley, Lake Lanao, No. 487, *Mary Strong Clemens*, April, 1906.

This is a small species with the habit and size of the Indian *Pinanga gracilis*, readily recognizable by its few acinaciform and underneath pubescent-scabrid leaflets, the spadix with very few digitate branches and the distichous rather remote conically ovoid fruits.

Pinanga Curranii Becc. n. sp.

Mediocris, subelata, caudice 7 cm. diam. Folia monnihil ampla, segmentis numerosis valde inaequalibus, inaequidistantibus attamen non congestis, subtus fugaciter puberulo-tomentellis, fere rectis, marginibus parallelibus, intermediis circiter 70 cm. longis, validissime 1–3-costulatis, 2–6 cm. latis, basi vix vel non attenuatis, in parte superiori leviter curvato-falcatis, apice profunde dentatis, dentibus triangularibus bifidis, partitionibus rectis acuminatis; segmentis superioribus brevioribus rectioribus; terminalibus latioribus pluricostulatis, apice horinzontaliter truncatis, breviter obtuseque dentatis. Spadicis ramuli fructiferi 18 cm. longi, valde compressi, sinuosi; fructibus horizontalibus exacte biseriatis, ovato-ellipticis, apice obtusissimo, basi parum acutatis, 15 mm. longis, 9 mm. latis; semine ovato in vertice rotundato, basi nonnihil acutato, 10 mm. longo, 7 mm. lato. Perianthium fructiferum depresso-cupulare, 2 mm. altum, 6 mm. latum, in ore non constrictum, immo paullo expansum.

A species of moderate size, 7 m. high. *Stem* 6 cm. in diameter. *Leaves* rather large with numerous leaflets; the rachis in the intermediate and apical portions triangular in cross section, flat and striate beneath, more or less bifacial above, puberulous-furfuraceous; the segments very unequal, inequidistant but not in groups, almost straight, with parallel margins, glabrous and finely striate by numerous secondary nerves above, paler and fugaceously puberulous-furfuraceous beneath; the intermediate leaflets about 70 cm. long with 1–3 very robust and in the upper surface very sharp costæ, 2–6 cm. broad, very slightly or not at all narrowed at the base and not at all sigmoid, slightly curved-falcate at the summit, where they terminate in as many pairs of teeth (or main teeth) as there are costæ, each pair being separated by an acute reëntering angle, 19–30 mm. deep; the single teeth are triangular and also cleft at their summit into two acuminate straight points; the upper leaflets are shorter and straighter than the lower ones, also with their margins exactly parallel but with the terminal teeth less acuminate; the two apical leaflets are the broadest, have 7–8 costæ, are horizontally truncate at the summit, where the main teeth are very short and their divisions very obtuse or rounded and convergent. Of the *spadix* I have seen only a few detached fruit-bearing branches; these are 18 cm. long, strongly flattened, carrying on each side 18–20 exactly bifarious fruits; the bract subtending the fruit is very narrowly semiannular, deflexed apiculate in the middle. *Fruiting perianth* depressed-cupular, 2 mm. high, 6 mm. across, with a flat base and a slightly broadened mouth; the sepals and petals similar, their margins round, ciliolate under a lens; the sepals very faintly striate outside, slightly thickened but not gibbous at the base. *Fruit* very regularly ovoid-elliptic, and very obtuse at the summit, 15 mm. long, 9 mm. broad. *Seed* ovoid, with a round top and a more acute base, 10 mm. long, 7 mm. thick; the branches of the raphe are few and slightly anastomosing.

PALAWAN, in the level forest at about 10 m. above sea level. No. 3515, *H. M. Curran*, January 11, 1906.

This is well characterized amongst the species with biseriate fruits by its large size, the straight, very broadly linear leaflets which are puberulous beneath and by the small ovoid-elliptical obtuse fruit. In *P. speciosa* the leaflets are also broadly linear and not falcate, but are quite glabrous on both surfaces.

Pinanga Barnesii var. **macrocarpa** Becc. n. var.

Fructibus 27–32 mm. longis, 17 mm. latis, semine ovato-oblongo, 2 cm. longo.

The *leaves* have 7–9 leaflets on each side of the rachis, instead of about 15, are broader, with more costæ and therefore with more falcate divisions at their summit than in the typical specimens. The *spadix* and *fruiting perianth* are as in the LUZON plant, but the *fruit* is somewhat larger than in this (27–32 mm. long, 17 mm. thick), regularly ovoid-elliptical, equally tapering at both ends and rather acute and mammillate at the apex.

Balete, Baco River, No. 275, MINDORO, *R. C. McGregor*, April 22, 1905.

Very similar to the typical form from LUZON, but apparently a larger plant with the trunk about 2 m. high and 6 cm. in diameter.

ONCOSPERMA Blume.

Oncosperma gracilipes Becc. n. sp.

Gracilis, caudice subelato, 4 cm. diam. in parte apicali rubiginoso-furfuraceo, spiculis 1–4 cm. longis undique horrido. Folia ampla, vagina 50 cm. longa spiculis nigris armata, petiolo brevi, parte pennifera circiter 2 m. longa, rachi subinermi, segmentis numerosis aequidistantibus, 4–5 cm. inter se remotis, rectissimis, ensiformibus, apice acuminato-subulatis, basi attenuatis, subtus secus costam mediam paleolis nonnullis praeditis; intermediis 60–70 cm. longis, 28–30 cm. latis. Spadix pro rata inter species affines brevis, 50 cm. longus, simpliciter-ramosus, ramulis ad 10 in apice partis peduncularis circ. 12 cm. longae, congestis, rigidis, crassis, 25–30 cm. longis, 8 mm. spissis, profunde crebreque scrobiculatis; fructibus mediocribus, perianthio circ. 12 mm. lato suffultis.

A slender palm, 7–10 m. high. *Stem* 4 cm. in diameter with internodes 5 cm. long, rusty furfuraceous at its summit, covered throughout with very slender needle-like unequal, brittle, black, shining spiculæ, which vary in length from 1 to 4 cm. and point in different directions. *Leaves* about 2 m. long not including the sheath; the sheath is rather thickly coriaceous, 50 cm. long, rusty-furfuraceous and strongly striated externally and more or less covered with the kind of spiculæ described above, but smaller; the petiole is very short (about 15 cm. in one specimen) round beneath, flattish above where it is very densely covered with spiculæ, which here however are appressed; its margins acute, the rachis more or less fugaceously rusty-furfuraceous with a few scattered spiculæ in its lower part, smooth from the middle upwards, with a flat facet above which gradually narrows into a salient angle separating the two side facets where the leaflets are inserted, underneath the rachis is convex at first and flattish from the middle upwards. Leaflets papyraceous, very numerous, equidistant, 4–5 cm. apart, quite smooth (not spinulous) on both surfaces, almost shining above, dull and slightly paler beneath, straight, ensiform, somewhat narrowed to a rather acute base (where the margins are strongly deflexed), gradually acuminate to a subulate point from the middle upwards, longitudinally plicate along the 2–3 secondary nerves on each side of the mid-costa; the costa is rather strong and prominent in the upper surface where the secondary nerves are slender; underneath the secondary nerves are prominent and as strong as the mid-costa, which is furnished with a few linear scales; the intermediate leaflets are 60–70 cm. long, 28–30 mm. broad; those near the summit are gradually shorter and narrower; the ultimate ones narrowly linear and though acuminate have the extreme apex obtuse. *Fruiting perianth* about 12 mm. in diameter; the divisions of the calyx and of the corolla broadly triangular and almost equal. *Fruit-spadix* about 50 cm. long, divided into about 10 simple branches almost grouped at the summit of a rather short, undivided, rather robust, rusty-furfuraceous peduncular portion, this is smooth above the insertion of the first spathe and densely spiculiferous below; the branches are all of about the same length, 25–30 cm. long and 8 mm. thick throughout, very closely and regularly hollowed by 4–5 vertical lines of orbicular, shallowly cupular, scrobiculi, which have below them a rather conspicuous leaf-like, broadly triangular, acute, deflexed bract. *Fruit,* when not quite mature, globose, 12 mm. in diameter, with a dull and finely granulate surface and with the scar left by the stigmas almost apical, orbicular, 2 mm. in diameter.

In a ravine in forests about 30 m. above the sea between Pagbilao and Atimonan, Province of Tayabas, LUZON, No. 4010, *Elmer D. Merrill,* March, 1905.

This differs from the few other species of the genus in its very slender trunk, in its spadix with few tufted, comparatively short and thick branches, which are all of about the same length and separate from the main axis at almost the same level. The fruits I have seen are still very young, but certainly they are larger than those of *O. filamentosa* and considerably smaller than those of *O. horrida.*

ARENGA Labill.

Arenga Ambong Becc. n. sp. *Wallichia oblongifolia* (non Griff.) Becc. in Webbia (1905) p. 48.

Caudex crassus et brevis, 30 cm. diam. (*Copeland.*) Folia amplissima, segmentis numerosis, intermediis circiter 70 cm. longis ambitu elongato-lanceolatis, in utroque margine sinibus 5–6, saepius oppositis, profunde excavatis, in sinibus 6–7 cm. latis, in partibus latioribus usque ad 8–9 cm., basi longe cuneatis, margine inferiore at basin auricula brevi rotundata praedito. Spadices simpliciter ramosi, ramis (dum flores ♂ adstant) 6–7 mm. spissis, fructiferis valde auctis, teretibus, glabris, digiti minoris crassitie. Flores ♂ in alabastro bene evoluto ovati, distincte abrupteque apiculati, 15 mm. longi; staminibus circiter 100. Fructus leviter depresso-globosus 3 cm. diam.

Stem short and thick (non soboliferous?). *Leaves* very large; the pinniferous part, according to *Copeland,* is 3 m. and the petiole alone 1–2 m. in length; the rachis in the terminal part is very slender, and in the intermediate portion about 2 cm. thick, not very regularly trigonous in cross section and with the upper salient angle not very sharp; the leaflets are numerous, very firmly papyraceous, those attached to a rachis of 2 cm. in thickness (probably the intermediate ones) are 70 cm. long, with 5–6 very large superposed indentations on both margins; the indentations of one side, generally, almost exactly opposite to those of the other side; an entire leaflet, therefore, has 5–6 places where they are narrower (6–7 cm. in width) than elsewhere (8–9 cm.) and appear as formed by several superposed reverted truncated cones; on the whole the leaflets are elongate-lanceolate in outline, slightly and rather gradually narrowing upwards into a rather broad and not very acute point; the basal part is rather narrowly, long and somewhat asymmetrical cuneate, which at its summit is 7–9 cm. broad: the margins of those portions between two indentations are straight and almost smooth in their lower part, but otherwise they are sharply dentato-serrate and almost spinulous; just at the base, the leaflets have their lower margin lengthened into a short and rounded ear-like lobe, while this is wanting in the upper margin; the mid-costa is very robust, much more prominent on the lower than on the upper surface; the secondary nerves are very numerous and very slender and do not run parallel to the mid-costa but slightly diverge from it, giving to both surfaces, but especially to the lower one, a finely striate appearance: both surfaces are glabrous, but the lower one is slightly paler than the upper, and in the young leaves looks as if it were covered with a very tenuous and adherent ashy indumentum which disappears with age; near the summit of the leaf the rachis is very slender and laterally flattened and the leaflets are shorter and more obtuse than the lower ones: the terminal leaflet is flabellate-cuneiform and unequally 3-lobed at its summit, being apparently formed by the union of 3 leaflets. The *spadix* (which I have not seen entire) is, from what I can judge, composed of an axial part carrying several simple spreading branches, which when the male flowers are full grown and the female one scarcely visible, are 6–7 mm. thick in their basal part and narrower higher up; but later when the same branches are covered with sessile, horizontal and spirally arranged fruits they attain the thickness of 17 mm. and the length of 70 cm., are terete, glabrous and wrinkled in the spaces between the fruits. The full grown *male flowers*, when not yet open, are ovoid, distinctly apiculate, 15 mm. long; the calyx is shortly cupular, its sepals are broader than high, with round glabrous and entire margins and are more or less gibbous-calcarate at the base; the petals are ovate-elliptical, concave or boat-shaped, distinctly apiculate; the stamens are about 100 (those of the center apparently abortive), their anthers are very narrowly linear subulate-aristate. *Female flowers* . . . . *Fruiting perianth* 2.5 cm. across, formed by coriaceous and also very dark colored segments, of which the sepals are narrowly lunate and strongly crenate on the margin and the petals broadly triangular, at least 3 times as long as the calyx, thickened and obscurely keeled on the back and with their margin entire or 2–3 dentate and sometimes more or less cleft at the apex. The *fruit* is globular, slightly depressed, 3 cm. in diameter slightly concave in its upper part, where there are 3 prominent lines radiating from 3 small central clefts, the remains of the stigmas. When the fruits have fallen the place of their insertion on the branches is marked by large and circular scars, 12–13 mm. in diameter, surrounded by 2 very narrowly lunate, blackish or dark chestnut brown coriaceous bracts, which together form a slightly concave or pateriform caliculum. The *seeds* are 3 in each fruit, but those seen by me were not quite mature.

MINDANAO, District of Zamboanga; San Ramon, collected by *E. B. Copeland* with full grown male flowers and unripe fruit March 16, 1906. Vernacular name in Chapocano "*Cabonegro*" and in Moro "*Ambung.*"

The collector's "field notes" state that the plant which supplied the specimen had a trunk 1 m. high and 30 cm. in diameter, adding, however, that it "becomes a tree" and that it is "a source of sago" as is the case with its near ally *A. undulatifolia* Becc.[2] from which it differs in the shape of the male flowers, these being oblong with a round top in *A. undulatifolia*, ovoid and suddenly apiculate in *A. Ambong;* furthermore the stamens in the latter are 100, while in the former they are about 150; the stem of *A. undulatifolia* is taller and thinner, and the leaflets at their base have two auricles one on each margin instead of one, otherwise the leaflets are extremely alike in the two species.

The leaves of young plants of *A. Ambong* in their terminal part have a striking resemblance to those of *Wallichia oblongifolia* Griff., to which I had[3] doubtfully referred a sterile and very incomplete specimen collected also at Zamboanga by *R. Garcia*, No. 653 *Ahern*. In *A. Ambong* the scars of the fallen male flowers are orbicular, about 6 mm. in diameter and show the punctiform marks of 10–12 fibro-vascular bundles; in *A. undulatifolia* there is at least double that number of these marks and this fact is apparently in correlation with the more numerous stamens in the latter species.

LIVISTONA R. Brown.

Livistona microcarpa Becc. n. sp.

Gracilis, elata, caudice 20 m. longo, 20 cm. crasso. Folia flabellato-orbicularia, multifida; petiolo gracili marginibus tantum prope basin spinis brevibus rectis dentiformibus horizontalibus vel raro reversis crebre armato, caetero subinermis; segmentis centralibus circiter usque ad tertiam superiorem partem connatis, circiter 80 cm. ab apice petioli metientibus, in parte libera 20–30 cm. longis et basi 4–4.5 cm. latis, ab hinc sensim sursum attenuatis prope apicem abrupte angustatis; summo apice rigido, breviter bifido, partitionibus secundariis breviter acuminatis et fere pungentibus; segmentis lateralibus angustioribus, in parte libera 40–50 cm. longis et apice quam centralibus profundius divisis; divisionibus longe acuminato-setiferis. Spadix, gracilis elongatus, parte axili subteretei digiti minoris crassitiae; inflorescentiis partialibus pluribus, duplicato ramosis, ramulis floriferis 3–5 cm. longis, 1 mm. crassis. Fructus sphaericus parvus, 11–12 mm. diametro.

A tall species. *Stem* slender, 20 m. high and 20 cm. in diameter (*Curran*). *Leaves* forming an almost complete multifid circle; the petiole long and slender, 3 cm. broad in its basal part, where it is flattish above and convex beneath and with the margins acute and very closely armed with small horizontal conical-dentiform spines of which the largest are 5–6 mm. long, usually straight, horizontal and occasionally retrorse; higher up the spines gradually become smaller and more scattered and

[2] Malesia 3: 93; see also Beccari "Nelle Foreste di Borneo." (Firenze 1902) pp. 329, 385, 593, 596.

[3] Webbia (1905), 48.

finally subtuberculiform; the central divisions of the leaf are united for about two-thirds their length and occasionally somewhat more, measuring 75–80 cm. from the apex of the petiole to their summit and only 25–30 cm. in their free portions which are 4–4.5 cm. broad at the base; from the base the segments gradually narrow towards the summit, where they are suddenly contracted into a short straight firm not drooping point; this is cleft to the extent of 4–6 cm.; each of the two resulting divisions is straight, acuminate and with an almost pungent apex; the lateral divisions are gradually narrower, 2–2.5 cm. broad reach nearer to the petiole, are gradually more acuminate and more deeply divided into two long straight setiform points, the filaments in the sinuses are obsolete or very rudimentary; the upper main costæ are rather prominent and sharp on both surfaces, which are almost equally green and almost shining; the secondary nerves and the transverse veinlets are very slender, and scarcely visible. *Spadix* elongated, slender, with several rather distant branches or partial inflorescences; its main axis rigid, straight, subterete, the size of a man's little finger. Primary spathes thinly coriaceous, dry, reddish brown, glabrous, tubular-cylindraceous, rather closely sheathing, obliquely truncate at the mouth, produced at one side into a blunt point, the margins at the apex more or less crenate and glabrous. The partial inflorescence is twice branched and forms a small lax, ovate, 20–25 cm. long panicle, with 2–3 primary basal branches and not many floriferous branchlets; the peduncular part of the panicle is concave on the inner side, and convex externally with very sharp edges, erect, and is almost entirely sheathed by its own spathe, the branchlets are 3–5 cm. long, subterete, about 1 mm. thick at the base. The flowers are spirally arranged all around the branchlets and rest on very small descoidal slightly prominent ebracteolate pulvinuli. Perfect *flowers* not seen. *Fruiting perianth* hard, very short, 2 mm. thick and 1.5 mm. high. *Fruit* exactly spherical, 11–12 mm. in diameter, black (at least when dry) with a finely wrinkled surface (when not quite mature) and without a visible mark of the remains of the stigma; the pericarp 1.5 mm. thick; the mesocarp grumose; the endocarp very thin. *Seed,* as far as can be judged from the immature fruit, exactly spherical with a broad orbicular hilum.

PALAWAN, in swamps along the Carañugan River. No. 3784 *H. M. Curran.*

Distinct from the only two other yet known Philippine species of this genus,[1] by the segments of the leaves which have their summit divided into two short straight not drooping points, by the petiole armed only at its base with small closely set spines and by the small round fruit.

The specimens examined by me consist of portions of the leaf and spadix with not quite mature fruit.

[1] *Livistona Merrillii* Becc. and *L. Whitfordii* Becc.; probably *L. Vidalii* Becc., must be referred to the genus *Pholidocarpus.*

CALAMUS Linn.

Calamus mollis Blanco var. **palawanicus** Becc. n. var.

Gracilis subinermis, ocrea spinulosa.

Stem, with the sheaths attached, 12–15 mm. in diameter. *Leaf-sheaths* remaining green when dry, more or less longitudinally striate, spineless or with some scattered small, short and broad-based, sometimes retrorse prickles. *Ocrea* dry, reddish-brown, spinulous, or at least tubercled-spinulous. *Leaflets* more or less inequidistant, often approximate in parsi on each side of the rachis or almost equidistant, otherwise as in the typical form but the margins not so strongly spinulous. *Spadix* with smooth primary spathes.

PALAWAN; No. 3613 *H. M. Curran*, January, 1906, in old clearings, locally quite common. (Male plant) and No. 191, 196 *J. Bermejos* (female plant) December, 1905.

It differs from typical *C. mollis* Blanco of LUZON in its leaf-sheaths being smooth or almost so, and in being less spinous in every part. It is scarcely distinguishable from *C. Meyenianus* Schauer in the ocrea being more or less spinulous and the leaflets having more strongly spinulous margins. *C. Meyenianus* itself is scarcely a distinct species, and had better be considered as a variety of *C. mollis* Blanco.

Calamus Merrittianus Becc. n. sp.

Scandens, amplissimus, caudice vaginato 6–7 cm. diametro. Folia amplissima, 3 m. et ultra longa; vagina spisse lignosa, spinis gracilibus valde inaequalibus, 2–4 cm. longis, laminaribus, saepe laciniatis, solitariis vel confluentibus, inter perplurimas spinulas parvas setiformes immixtis densiuscule armata. Segmenta numerosissima, aequidistantis, auguste lanceolata vel ensiformia, supra in costa media et in costulis duabus, subtus tantum in costa media, setosa. Spadices latissime paniculati, spathis primariis et secundariis spinis brevibus vel 10–12 mm. longis, nigrescentibus, rectis, horizontalibus vel deflexis, armati; spicis foemineis 7–8 cm. longis, crebre bifarie floriferis, parte pedicellari gracili, 2–2.5 cm. longa in fundo spathae suae insertis. Fructus parvi, squamis per orthostichas 21 ordinatis in dorso convexis et non sulcatis, margine minute fimbriatis.[5]

Scandent and very large. *Sheathed stem* 6–7 cm. in diameter. *Leaf-sheaths* woody, 3–4 mm. thick, gibbous above, rather densely armed with

[5] This species comes near *C. Merrillii* Becc., and belongs to Group XIV, which in my Monograph of the Genus *Calamus* in vol. 11 of the *Annals of the Royal Botanic Garden, Calcutta*, in press, is characterized as follows: "*Leaves* prolonged into a long and clawed cirrus. *Leaf-sheaths* not flagelliferous. *Spadix* not flagelliferous at its apex, usually shorter than the leaves. *Male spadix* ultra-decompound. *Female spadix* simply decompound, quite different from the male one. *Male* and *female* spikelets stalked or inserted inside at the base of their respective spathes on distinct pedicels which are as long as their respective spathes. *Fruiting perianth* explanate. *Seed* with ruminate or equable albumen."

feeble but very unequal spines, some of them having a swollen and broad base and a slender 2–4 cm. long, flat, laminar, elastic, dark brown, sometimes laciniate blade; of this kind some are solitary and others confluent or comb-like; intermingled with the large spines are many other very small, bristle-like, scattered ones unequal, and insident on a tubercled base. *Leaves* very large, in one specimen 3.2 m. long in the pinniferous part, which is regularly armed at intervals of about 10 cm., with half whorls of extraordinarily strong black claws; the petiole is very robust, 25 cm. long, 3 cm. broad at its base, where it is densely prickly on the back and on the margins, flattish and smooth above; rachis at first with a flat surface above, which is bordered on either side with a prickly ridge gradually becoming narrower and the ridges closer together until they unite to form a salient smooth angle; underneath the rachis is armed at first with solitary, and upwards with 2–5-nate rather remote, extraordinarily robust, black claws. *Leaflets* very numerous, equidistant, 3–3.5 cm. apart, narrowly lanceolate-ensiform, papyraceous, green on both surfaces, slightly paler beneath, slightly narrowing towards the base, very gradually and long-acuminate to a subulate and at the sides bristly tip, faintly 3-costulate, with the mid-costa very remotely and sparingly bristly only near the summit, the side-costæ are tenuous and bear very few, very long, spadiceous bristles; underneath the mid-costa alone is sparingly bristly; the margins are not very closely ciliate with short bristles; transverse veinlets not very conspicuous; the largest leaflets, the intermediate ones, 40 cm. long, 25–28 mm. broad; the upper and lower ones somewhat smaller. *Male spadix* . . . . . *Female spadix* 2.5 m. long, forming a large elongate pyramidate panicle, which terminates in a small and short tail-like prickly appendix and is composed of many partial inflorescences; primary spathes tubular, thinly coriaceous, rather closely sheathing, slightly enlarged above, strongly armed with straight, very short, or 10–12 mm. long, blackish, horizontal or slightly deflexed spines, prolonged at the summit into a triangular subulate point, usually split on the ventral side, glabrous at the mouth; partial inflorescences furnished with a long peduncular part and inserted far inside their respective spathes, arched downwards; the lower ones, the largest, 50 cm. long, with 14–15 spikelets on either side; the others gradually shorter and with fewer spikelets; secondary spathes tubular, closely sheathing, very slightly infundibular, usually longitudinally split, otherwise truncate and entire at the mouth, produced at the summit into a broadly triangular point, armed with many small slender scattered, solitary, horizontal spinules. Spikelets 7–8 cm. long, with a distinct slender pedicular part, which is inserted at the bottom of their respective spathes, arched downwards, with 20–22 flowers on each side; the upper spikelets shorter and with fewer flowers; spathes infundibuliform, exactly truncate and entire at the mouth, slightly apiculate on the outer side, glabrous and nonstriate;

involucrophorum cyathiform, narrowing towards the base and inserted at the bottom of its own spathel, flattened, two-keeled and bidentate on the side next to the axis; involucre cupular, almost entirely inclosed in the involucrophorum, truncate, bidentate on the side of the areola of the neuter flower, the areola deep, broadly lunate and sharply bordered. *Female flowers* 4 mm. long; the calyx with a polished base, and deeply parted into 3, ovate, externally striolate, acute lobes; corolla as long as the calyx, its segments acute, slightly narrower than the lobes of the calyx. *Fruit* seen only in its very young stage, and then ovate and stoutly conically beaked, with the scales in 21 longitudinal series, yellowish, with a rusty, minutely fringed margin, shining, convex and not channelled along the middle.

MINDORO: Bongabong River. No. 3912 *M. L. Merritt*, March, 1906.

Very closely related to *C. Merrillii*, from which it differs especially in its primary and secondary spathes, which are prickly in *C. Merrittianus* and smooth in *C. Merrillii.* It is also a near ally of *C. Zollingerii*, which, however, has quite differently armed leaf-sheaths.

Calamus Mindorensis Becc. n. sp.

Alte scandens, caudice vaginato 4 cm. diam. Folia in parte pinnifera circiter 1.5 m. longa; vagina lignosa spinis dimidiato-conicis, minutissimis, sparsis, 3–4 mm. longis armata; petiolo valido brevissimo vel subnullo; rhachi in angulo superiori inermi; segmentis aequidistantibus utrinque circiter 25, spatio nudo 6–7 cm. longo inter se dissitis, utrinque viridibus, auguste elliptico-lanceolatis, basi et apice acutis, plicato-pluricostulatis, superne in costa media et in costulis duabus spinulosis, subtus levibus. Spadices late laxeque paniculati; spathis secundariis auguste tubuloso-infundibuliformibus, levibus; spicis foemineis vermicularibus, gracilibus, 10–12 cm. longis, utrinque bifarie 20–22-floris, patentibus vel horizontalibus, nonnihil arcuatis, exacte ad faucemearum spatharum insertis. Perianthium fructiferum distincte et si breviter pedicelliforme. Fructus globosus, parvus, pisiformis, conspicue rostratus; squamis per orthostichas 18–20 ordinatis, nitidis, convexis nec in medio sulcatis, luride stramineis, apiculo rubello, margine scarioso minute erosulo-denticulato. Semen globosum.[6]

Rather robust and high climbing. *Sheathed stem* 4 cm. in diameter. Leaf-sheaths woody, 3 mm. thick, greenish with a smooth surface, very thinly covered with a fugaceous ashy indumentum when young, strongly

[6] This species is closely related to *C. Moseleyanus* Becc., and belongs to Group XV, of my Monograph, characterized as follows: "*Leaves* prolonged into a long and clawed cirrus. *Leaf-sheaths* not flagelliferous. *Spadices* usually shorter than the leaves, not or slightly flagelliferous at the apex. *Spikelets* not stalked, inserted near the mouth of their respective spathes. *Fruiting perianth* pedicelliform or almost explanate. *Seed* with more or less superficial intrusions of the integument or distinctly ruminate, embryo basilar or slightly shifted to one side."

gibbous above, feebly armed with very small, scattered, 3–4 mm. long, horizontal, semiconical, straight spines, which have the tip slightly darker than the surface of the sheath and the base lighter and tumescent. *Ocrea* very short, axillary, liguliform. *Leaves* cirriferous, large; petiole very short and robust or almost obsolete, flattish and smooth above, 3 cm. broad, armed along the margins with rather stout, 8–10 mm. long, straight, horizontal spines; rachis in the intermediate portion obtusely trigonous, fugaceously ashy-furfuraceous, with an obtuse smooth salient angle above, armed beneath with rather remote solitary and binate (upwards probably 3-nate) black-tipped claws. *Leaflets* rather numerous, (about 50 on the whole), equidistant, not very approximate (6–7 cm. apart) rigidly papyraceous, green on both surfaces, slightly paler beneath; narrowly elliptical-lanceolate, narrowing almost equally towards both ends, plicate-pluricostulate, base and apex acute, the latter spinulous; the mid-costa alone rather prominent and sharp above and spinulous as are two other lateral costulæ which are near the upper margin; the other costulæ are very tenuous and smooth; underneath the nerves are numerous, but devoid of bristles or spinules; the intermediate leaflets are 45–47 cm. long and 4.5–5 cm. broad; the lower ones are smaller, 20–25 cm. long and proportionally narrower. *Male spadix* . . . . . *Female spadix* decompound forming a large diffuse panicle; primary *spathes* . . . . . ; partial inflorescences 40–50 cm. long (the few I have seen) with 10–12 spikelets on each side; secondary spathes (the spathes of the partial inflorescences) narrowly tubular-infundibuliform, unarmed, striolate, very thinly and fugaceously furfuraceous, produced at the summit into a broadly triangular acute point; the mouth ciliate with small paleolæ; spikelets (when bearing the fruit) spreading or horizontal, slightly arched, with a distinct axillary callus, inserted just at the mouths of their respective spathes, 10–12 cm. long (the upper ones somewhat shorter) with 20–22 distichous flowers on each side; spathels shortly, very broadly and asymmetrically infundibuliform, obsoletely striately-veined, slightly produced and apiculate at one side, truncate and deciduously ciliolate at the mouth; involucrophorum very shallowly cupular, immersed in its own spathel, bidentate and laterally adnate to the base of the spathel above its own; involucre shallowly and irregularly cupular; areola of the neuter flower very depressedly lunate. *Fruiting perianth* shortly but distinctly pedicelliform; the calyx parted down to almost the middle into 3 triangular, slightly striately-veined, acute lobes, and with a smooth base; the segments of the corolla triangular, barely shorter than the teeth of the calyx. *Fruit* small, spherical, abruptly and comparatively stoutly beaked, 6.5 mm. in diameter (when not quite ripe) with a small basal acute caudiculum, which penetrates into the perianth; scales in 18–20 longitudinal series, shining, convex and not channeled along the middle, of a dirty straw-yellowish color and with a reddish slightly produced tip, and scarious erosely toothed margins. *Seed* small, globose (not quite mature).

Balete, Baco River, MINDORO, No. 309 *R. C. McGregor*, April, 1905.

A very near ally of *C. Moseleyanus* Becc., from which it differs in its larger dimensions, in the larger and more diffuse spadix with much longer spikelets and especially in the smaller fruit with more numerous scales, these being arranged in 18–20 longitudinal series.

Calamus Reyesianus Becc. n. sp.

Scandens mediocris. Spadix foemineus diffuse paniculatus (non cirrifer), spathis primariis elongato-infundibuliformibus, superne spinis rectis horizontalibus parvis dense armatis; spathis secundariis spinulis paucis quoque praeditis; spicis foemineis erecto-patentibus, 3–5.5 cm. longis, utrinque bifarie 7–12-floris, ad faucem earum spatharum insertis. Perianthium fructiferum latissime obconicum et subexplanatum. Fructus sphaericus, 15 mm. diam. brevissime abrupteque rostratus, squamis per orthostichas 18 ordinatis, stramineo-rubellis, linea intramarginali augustissima saturatiore notatis, in dorso convexiusculis et longitudinaliter profunde sulcatis, margine scarioso minute erosulo-denticulato, apice triangulari obtusiusculo. Semen globosum, 10–11 mm. diam., ruminatum.[7]

Apparently scandent and of moderate size. *Female spadix* (non cirriferous) diffusely paniculate, terminating in a small tail-like flattened appendage, this a few centimeters in length and spinous at its apex; primary spathes very closely sheathing, thinly coriaceous, elongate-infundibuliform, densely armed in their upper part with small straight horizontal spines, flat on the axial side, obliquely truncate and entire at the mouth, where they are produced at one side into an elongate triangular point, this keeled on its back. Partial inflorescences (those of the upper part of the spadix) spreading, 18–20 cm. long, with 5–6 spikelets on each side; secondary spathes narrowly infundibular with a few horizontal straight spines on their back at the summit, obliquely truncate, entire and fringed with deciduous paleolæ at their mouths and produced at one side into a triangular acute point; spikelets erecto-patent, inserted just above the mouth of their respective spathes, 3–5.5 cm. in length with 7–12 distichously arranged flowers on each side; spathes shallowly obliquely infundibular, shortly apiculate at one side, their margins entire and fringed with deciduous paleolæ; involucrophorum concave, very shallow, immersed in its spathel, produced externally into a triangular point, which subtends the neuter flower; involucre shallowly and asymmetrically cupular, bidentate and lunately excavate on the side of the neuter flower, of which the areola is comparatively large, lunate and sharply bordered. *Fruiting perianth* not

[7] This species comes near *C. palustris* Griff; and belongs, with the preceding, to the Group XV of my monograph.

forming a pedicel, very broadly obconic or almost explanate. *Fruit* spherical surmounted by a very short beak, 15 mm. in diameter; scales in 18 longitudinal series, of a reddish straw-yellow color, with a narrow, darker intro-marginal line, and scarious finely erosely toothed margins, rather convex, broadly and rather deeply channeled and with a triangular rather obtuse point. *Seed* globular, 10–11 mm. in diameter, with a not very closely pitted surface; the chalazal fovea indistinct and very superficial; albumen ruminate; embryo basal.

Unisan, *Province of Tayabas*, LUZON. *C. Reyes. October, 1904.*

The type specimen is the terminal part of a fruiting spadix only. The leaves which, I suppose, belong to this spadix, are much like those of *Calamus palustris*, to which this species is related, but from which it is distinguished by its perfectly spherical fruit. The leaves mentioned above belong, very probably, to the lower part of the stem, as some of them are peripinnate and one terminates with a short rudimentary cirrus, the leaflets are not numerous, are very inequidistant, approximate in groups of 2 or 3 on each side of the rachis, with long internodes between each group; elliptic or elliptic-lanceolate, 15–28 cm. long, concavo-convex, suddenly contracted into a bristly tip, shining above, paler and dull beneath, with 5 tenuous but acute costæ above and with numerous sharp, subparallel transverse veinlets.

DAEMONOROPS Blume.

Daemonorops Curranii Becc. n. sp. (Sect. *Piptospatha*).

Non alte scandens (?). Vaginae . . . . . Folia in cirrum gracilem solito modo semiverticillatim acileatum abeuntia; rachide superne spinulosa. Segmenta numerosa aequidistantis, circiter 4 cm. inter se remota, auguste lanceolata vel lanceolato-ensiformia; intermedia 33–40, cm. longa, 17–20 mm. lata, aerum pars latior paullo supra basin, in costa media prope apicem et secus nervos duo superne spinulosa; subtus in costa media tantum minute spinuloso-setosa. Spadix circiter 60 cm. longus, inapertus auguste cylindraceus, parte pedicellari brevi spinosa suffultus. Spathae coriaceae, elongato-spatulatae; inferior extus undique spinis brevibus deflexis plus minusve digitato-seriatis armato, superiores in dorso prope apicem tantum spinosae. Panicula fructifera elongato-oblonga, spicis majoribus 3.5–4.5 cm. longis, utrinque distice 6–9-floris; earum axis acute zig-zag sinuosa. Involucrophorum breviter pedicelliforme, 1–2 mm. longum, in axilla callosum apice truncatum. Involucrum apice planum orbiculare, limbo angustissimo anulari cintum, involucrophorum breviter superans. Florisa neutri areola parva concava aedicolaris non tumescens, perianthium fructiferum late obconicum. Fructus sphericus breviter conice rostratus 12 mm. diametro, squamis per orthostichas 12–14 ordinatis, stramineis. Semen globosum, leviter depressum, tuberculato-alveolatum 8 mm. diametro.

Apparently scandent and of moderate size. *Leaf-sheaths* . . . . . *Leaves* terminated with a not very long and slender cirrus, which is very regularly armed with approximate half-whorls of very sharp confluent claws; petiole . . . . .; the leaf-rachis (in the intermediate portion) is slightly convex beneath, where it is strongly and regularly armed with half-whorls of 5-nate claws; and with a very sharp and spinulous salient angle above and flat sided facets; leaflets numerous, equidistant, about 4 cm. apart, green and subshining on both surfaces, papyraceous, very narrowly lanceolate or lanceolate-ensiform, broadest not very far above the base and thence shortly narrowing downwards, gradually acuminate to a subulate and at the sides spinulous tip; in the upper surface the mid-costa is tenuous and sharp, spinulous only near its summit, and accompanied on each side by a tenuous secondary nerve; this stronger than some other nerves of the same kind and spinulous; underneath the mid-costa alone is minutely bristly spinulous; transverse veinlets very tenuous and sharp especially in the upper surface; the intermediate leaflets 33–40 cm. long, 17–20 mm. broad. *Female spadix* before flowering very narrowly cylindric and elongate, slightly arched; primary spathes at first tubular very obliquely truncate at the mouth and produced at the summit into a triangular point, later longitudinally split; the outermost spathe, after flowering, elongate-spathulate, gradually narrowing towards the base into a rather short, flattened, prickly, pedicellar part, totally and very densely armed externally with solitary or more or less seriate and confluent, deflexed, short, unequal spines, which have a reddish brown tip and a lighter swollen base; inner spathes prickly only on the back, especially near their summit, smooth on the margins at the mouth; when in flower or fruit, the female spadix is tenuously rusty furfuraceous in every part, about 60 cm. long, slender, rigid, with 6–7 partial inflorescences; the peduncular part of the spadix is 7–8 cm. long, 7–8 mm. broad, slightly flattened, very slightly enlarged upwards, armed with deflexed, solitary or confluent and sugiditate, straight, rather short, deflexed spines; the main axis (of the spadix) is straight with its lowest (2–3) internodes slightly flattened, the others obsoletely angular; secondary and tertiary spathes inconspicuous; partial inflorescences triangular in outline, the lower ones, the largest, 11–12 cm. long; the upper ones shorter, with 5–7 bifarious regularly alternate spreading spikelets on either side; the axis of the partial inflorescences straight, very acutely 3–4-gonous; the lower spikelets, the largest, 3.5–4.5 cm. long, with 6 to 9 bifarious flowers on each side, their axis very acutely angular and zigzag sinuous; upper spikelets shorter and with fewer flowers; spathels scarious, very shortly annular and embracing, produced at one side into a triangular spreading acute point; involucrophorum shortly but distinctly pedicelliform, 1–2 mm. long, angular, very spreading or horizontal when bearing the fruit, distinctly callous in its axilla, truncate and with a very short triangular point at one side at its summit; involucre slightly raised above the involucrophorum, its limb represented by a very narrow annular rim round the flat orbicular scar left by the flower. Areola of the neuter flower rather small, concave, niche-like, not callous. *Female flowers* 5 mm. long, when in bud, with an ovate base and a trigonus apex; the calyx very shallowly cupular with 3 broad acute teeth; the corolla 4 times as long as the calyx, parted down almost to the base into 3 elongately triangular, sharply striately veined segments. *Fruiting perianth* very broadly obconic and therefore not distinctly pedicelliform. *Fruit* small, spherical, very shortly and broadly conically beaked, 12 mm. in diameter when quite ripe; scales in 12 to 14 longitudinal series, polished, narrowly and sharply channeled along the middle exactly rhomboid, with an obtuse tip, straw yellow with very narrow almost entire margins. *Seed* globular, slightly depressed, 10 mm. broad, 8 mm. high, its surface pitted and tubercled.

PALAWAN, No. 3791 *H. M. Curran*, February, 1906.

D. Curranii is a near ally of *D. elongatus Bl.*, from which it differs in the leaves with equidistant leaflets, in the rachis spinulous on the salient angle above, and in the slightly larger spherical fruit.

INDEX TO PHILIPPINE BOTANICAL LITERATURE.

By Elmer D. Merrill.
(*From the botanical section of the Biological Laboratory, Bureau of Science.*)

The literature bearing directly or indirectly on Philippine botany is so extensive and so widely scattered that it has been considered advisable to prepare and publish from time to time lists of useful or essential works containing references to Philippine plants. Such lists will include short reviews of monographs of various genera and families that are represented in the Philippines, short articles, individual diagnoses, etc., and an attempt will be made to review obscure and rare papers referring to Philippine botany in the widest sense of the word. Special attention will be given to recent publications, but the older ones will not be ignored, if there is any special object in reviewing them. The ultimate object of this work is the preparation of a complete bibliography of Philippine botany. The list will be continued from time to time in this Journal.

Ames, Oakes. Descriptions of New Species of Acoridium from the Philippines. (*Proc. Biol. Soc. Wash.* **19** (1906) pp. 143–154.)

Eighteen new species of *Acoridium* are described, all, with the exception of *A. williamsii*, based on material collected by employees of the Bureau of Science. The species are as follows: *Acoridium williamsii*, *A. graminifolium*, *A. tenuifolium*, *A. tenue*, *A. parvulum*, *A. venustulum*, *A. strictiforme*, *A. anfractum*, *A. recurvum*, *A. philippinense*, *A. turpe*, *A. oliganthum*, *A. ocellatum*, *A. merrilli*, *A. longilabre*, *A. graciliscapum*, *A. cucullatum*, and *A. copelandii*.

Beccari, O. Le Palme delle Isole Filippine (in Martelli's *Webbia* (1905) pp. 315–359.)

An enumeration of all the palms definitely known from the Philippines, in which the following species and varieties are described for the first time: *Pinanga speciosa*, *P. copelandi*, *P. barnesii*, *P. elmerii*, and *P. chinensis* (from China); *Caryota merrillii; Orania paraguanensis; Livistona whitfordii*, *L. vidalii; Calamus mollis* var. *major*, *C. merrillii*, *C. siphonospathus* vars. *sublevis*, *oligolepis* (*major*), *oligolepis* (*minor*) and *polylepis*.

Beccari, O. Systematic Enumeration of the Species of Calamus and Daemonorops, with Diagnoses of the New Ones. (*Records Bot. Surv. India* **2** (1902) pp. 197–230.)

In this paper 164 species of *Calamus* are recognized and 77 species of *Daemonorops*, of which the following are credited to the Philippines: *Calamus spinifolius* n. sp., *C. mollis* Blanco, *C. Blancoi* Kunth, *C. cumingianus* n. sp., *C. mosleyanus* n. sp., *C. vidalianus* n. sp., *C. siphonospathus* Mart., *C. microcarpus* n. sp., *C. manillensis* H. Wendl., and *C. dimorphacanthus* n. sp. *Daemonorops fuscus* Mart., and *D. gaudichaudii* Mart. *Calamus discolor* Mart.,

C. curag Blanco and *C. meyenianus* Schauer are considered as doubtful, imperfectly known or unrecognizable species. All the species enumerated from the Philippines are endemic to the Archipelago.

Brand, A. Symplocaceae. (*Das Pflanzenreich*, **6** (1901) pp. 1–100.)

A single genus, *Symplocos*, is recognized, containing 281 species, of which the following are credited to the Philippines: *Symplocos patens* Presl, *S. patens* var. *ciliata* (Presl) Brand, *S. floridissima* Brand, *S. polyandra* (Blanco ?), Brand *S. ferruginea* Roxb., *S. oblongifolia* (Presl) Vidal, *S. cumingiana* Brand, and *S. luzoniensis* Rolfe, all endemic with the exception of *S. ferruginea*. (See also Brand in Perkins Frag. Fl. Philip. pp. 36–37.)

Brotherus, V. F. Contributions to the Bryological Flora of the Philippines, **I.** (*Öfversigt af Finska Vetenskaps-Societetens Förhandlingar* (1904–05) **47**, No. 14, pp. 1–12.)

Forty species of Philippine mosses are enumerated, mostly from the collections of *Merrill*, *Copeland* and *Elmer*. The following species are described as new: *Dicranoloma perarmatum*, *Macromitrium merrillii*, *Orthomnium loheri*, *Entodon longidens*, *Sematophyllum piliferum* and *Hypnodendron copelandii*.

Buchenau, Fr. Alismataceae. (*Das Pflanzenreich* **16** (1903) pp. 1–66.)

No species of the family is credited to the Philippines, but since the publication of the monograph the widely distributed *Sagittaria sagittifolia* L., has been found in Mindanao. The other two families considered by *Buchenau* in the same work, *Scheuchzeriaceae* and *Butomaceae*, are not represented in the Philippines.

Buchenau, Fr. Juncaceae. (*Das Pflanzenreich* **25** (1906) pp. 1–284.)

No species of the family is credited to the Philippines, but the widely distributed *Juncus effusus* L., is found on the mountains of Luzon and Mindanao, and at least one other species of the genus is found in northern Luzon.

Christ, H. Zur Farnflora von Celebes. (*Ann. Jard. Bot. Buitenz.* II. **4** (1904) pp. 33–44.)

Forty-nine species are enumerated, many of which extend to the Philippines.

Christ, H. Filices Borneenses. Fougeres receuillies par les expéditions des Messieurs Nieuwenhuis et Hallier dans la partie équatoriale de Bornéo. (*Ann. Jard. Bot. Buitenz.* II. **5** (1905) pp. 92–140, plates 1.)

An enumeration of 155 species, many of which are described as new, with numerous references to species extending to the Philippines.

Christ, H. Filices Insularum Philippinarum. (*Bull. Herb. Boiss.* **6** (1898) pp. 127–154; 189–210, plates 3.)

Two hundred and seventy-one species of ferns and fern allies are enumerated, the list being based on the collections made by *A. Loher*. A number of species are credited to the Philippines for the first time and the following described as new: *Alsophila lepifera* J. Sm., var. *congesta*, *A. fuliginosa*; *Hymenophyllum subdemissum* and *H. discosum*; *Lindsaya loheriana* and *L. capillacea*; *Lomaria fraseri* Cunn., var. *philippinensis*; *Plagiogyria glauca* Kunze var. *philippinensis*; *Asplenium loherianum*, *A. contiguum* Kaulf., var. *bipinnatifidum*; *Athyrium sarasinorum* Christ, var. *philippinense*; *Aspidium loherianum*, *A. grammitoides*, *A. fauriei* Christ var. *elatius*; *Polypodium loherianum*, *P. subobliquatum*, *P. sagitta*, *P. anomalum*, *P. lagunense*; and *Angiopteris cartilagidens*.

Christ, H. Filices Insularum Philippinarum, II. (*Bull. Herb. Boiss.* II. **6** (1906) pp. 987–1011.)

Like the preceding paper based also on material collected by *A. Loher*, with some references to specimens secured by other collectors, 102 species being enumerated, some reported from the Philippines for the first time and the following described as new: *Christopteris copelandi, Hymenolepis rigidissima, Cyclophorus argyrolepis, Selliguea flexiloba* Christ, var. *loheri, Polypodium elmeri* Copel., var. *separatum, P. mindanense, P. subirideum, P. subdrynariaceum, P. subopposítum, Aspidium batjanense, Stenochlaena arthropteroides, Asplenium cymbifolium, A. colubrinum, Diplazium acrotis, D. inconspicuum, Athyrium loheri, Dryopteris rizalensis, Aspidium biseriatum, A. angilogense, Leptochilus stolonifer, L. rizalianus, Saccoloma moluccanum* Mett., var. *stenolobum, Pleurogramme loheriana, Vittaria philippinensis, V. crispomarginata, Cyathea loheri, C. callosa, C. adenochlamys, Gleichenia loheri,* and *G. linearis* Burm., var. *stipulosa.*

Christ, H. Die Farnflora von Celebes. (*Ann. Jard. Bot. Buitenz.* **15** (1897) pp. 73–186, plates 5.)

An enumeration of the ferns known from Celebes, 308 species being listed, with numerous references to species growing in the Philippines.

Copeland, Edwin Bingham. Outline of a Year's Course in Botany. (*Bureau of Education* (*Manila*) *Bull.* **24** (1906) pp. 1–18.)

An outline of the work in botany given at the Philippine Normal School, Manila, and in the secondary schools of the Archipelago.

Copeland, Edwin Bingham. Key to the Families of Vascular Plants in the Philippine Islands. (*Bureau of Education* (*Manila*) *Bull.* **24** (1906) pp. 19–32.)

An analytical key to the families of vascular plants known to be represented in the Philippines, following the system of *Engler and Prantl*, followed by a systematic enumeration of the families, 199 families being listed.

Copeland, Edwin Bingham. Fungi esculentes Philippinenses. (*Ann. Mycol.* (1905) **3**: pp. 25–29.)

Twenty-one species described in the following genera: *Lycoperdon, Coprinus, Panaeolus, Agaricus* and *Lepiota.* In *Govt. Lab. Publ.* **28** (1905) pp. 141–146, the above paper is reprinted in English, with the addition of three half-tone plates under the title "New Species of Edible Philippine Fungi."

Diels, L. Droseraceae. (*Das Pflanzenreich* **26** (1906 pp. 1–136.)

A single genus, *Drosera*, is represented in the Philippines by the following species: *D. indica* L., British India to Australia, *D. spathulata* Labill., southern Japan and China to East Australia and New Zealand, and *D. peltata* Smith, British India to Australia.

Engler, A. Araceae-Pothoideae. (*Das Pflanzenreich* **21** (1905) pp. 1–330.)

The following species are credited to the Philippines: *Pothos longifolius* Presl, Philippines to Java, Sumatra and the Moluccas; *P. scandens* Linn., British India to Malaya; *P. inaequilaterus* (Presl) Engl., Philippines and Sumatra; *P. ovatifolius* Engl., endemic; *P. philippinensis* Engl., endemic; *P. luzonensis* (Presl) Schott, endemic; *Pothoidium lobbianum* Schott, a monotypic genus, Philippines, Celebes and the Moluccas. No form of *Acorus* is credited to the Archipelago, but the genus is represented in Luzon by forms referred to *A. calamus* L., but of which I have seen only sterile specimens. The Philippine form is possibly referable to *A. gramineus* Soland.

Graebner, P. Typhaceae and Sparganiaceae. (*Das Pflanzenreich*, **2** (1900) *Typhaceae* pp. 1–18; *Sparganiaceae* pp. 1–26.)

Of the *Typhaceae*, *Typha angustifolia* L. subsp. *javanica* Schnizl. is the only form credited to the Philippines. The *Sparganiaceae* are not represented in the Philippine flora.

Harms, H. Einige neue Arten der Gattungen Cynometra und Maniltoa (*Notizblatt Kgl. Bot. Gart. und Mus. Berlin* **3** (1902) pp. 186–191.)

Several species are described in both genera, including two from the Philippines, *Cynometra simplicifolia* and *C. warburgii*.

Hayata, B. Compositae Formosanae. (*Journ. Coll. Sci. Tokyo* **18** (1904) No. 8, pp. 1–45, plates 2.)

An enumeration of all the *Compositae* known from Formosa with analytical keys to genera and species. The same species are again enumerated by *Matsumura* and *Hayata* in their "Enumeratio Plantarum in Insula Formosa sponte crescentium," etc.

Hayata, B. Revisio Euphorbiacearum et Buxacearum Japonicarum. (*Journ. Coll. Sci. Tokyo* **20** (1904) No. 3, pp. 1–92, plates 6.)

The article contains analytical keys to the genera and species, with descriptions of both. Formosan species are included, 24 genera and 65 species of *Euphorbiaceae* and 2 genera and 3 species of *Buxaceae* being recognized, many of the former extending to the Philippines.

Hayek, August von. Verbenaceae novæ herbarii Vindobonensis. (*Fedde's Repertorium* **2** (1906) pp. 86–88.)

Several species of *Verbenaceae* are described as new including one, *Callicarpa elegans* Hayek n. sp. l. c. 88, from the Philippines, the type being No. 1460 *Cuming*.

Koehne, E. Lythraceae (*Das Pflanzenreich* **17** (1903) pp. 1–326.)

The following species are credited to the Philippines: *Rotalia mexicana* Cham. et Schlecht., var *spruceana* (Griseb.) Koehne, *R. ramosior* (L.) Koehne, *R. leptopetala* Koehne, *R. indica* (Willd.) Koehne, all widely distributed; *Ammannia coccinea* Rottb., subsp. *longifolia* Koehne, *A. baccifera* Linn., forma *typica*, Koehne, subf. *contracta* Koehne et subsp. *viridis* (Hornem.) Koehne, all widely distributed; *Pemphis acidula* Forst., a strand-plant extending from Africa, tropical Asia to Malaya, Polynesia and Australia; *Lagerstroemia indica* L. (introduced and cultivated only!) *L. speciosa* (L.) Pers., *L. batitinan* Vid., *L. piriformis* Koehne, *L. paniculata* (Turcz.) Vidal, the last three endemic; *Lawsonia inermis* Linn., introduced and cultivated.

Laguna, Maximo. Cien Helechos de Filipinas dispuestos con arreglo á la última edición (1874) de la "Synopsis Filicum" de Hooker y Baker. (*Ann. Soc. Esp. de Hist. Nat.* **7** (1878) pp. 1–19.)

An enumeration of 102 species of Philippine ferns, collected by *Baranda*, containing no descriptions and apparently no changes in nomenclature, but calling attention to the validity of some of *Cavanilles'* species, notably *Lygodium semihastatum*.

Maiden, J. H. On the Identification of a Species of Eucalyptus from the Philippines. (*Proc. U. S. Nat. Museum* **26** (1903) 691–692.)

One of the few species of this characteristic Australian genus found north of Australia, was collected by the botanists of the Wilke's U. S. Exploring Expedition near Zamboanga, Mindanao, in January, 1842, and described by *Asa Gray* under the name given it by the collector, *Eucalyptus multiflora* Rich. After examining the type Mr. *Maiden* reduces it to *Eucalyptus naudiniana* F. v. Müller, a species of the Bismark Archipelago.

Martelli, U. Le Composite raccolte dal Dottor O. Beccari nell' arcipelago Malese e nella Papuasia. (*Nuovo Giorn. Bot. Ital.* **15** (1883) pp. 281–305.)

An enumeration of the *Compositae* collected by Dr. *Beccari*, many of which extend to the Philippines.

Matsumura, J., and **Hayata, B.** Enumeratio Plantarum in Insula Formosa sponte crescentium hucusque rite cognitarum adjectis descriptionibus et figuris specierum pro regione novarum. (*Journ. Coll. Sci. Tokyo* **22** (1906) pp. 1–702, plates 18.)

An enumeration of all the plants known to the authors from Formosa with the descriptions of some new species, about 1,912 species being enumerated of which about 775 are known to extend to the Philippines. Undoubtedly the former number will be considerably increased as more extensive explorations are made in Formosa, and the latter will be increased when we shall have obtained a more thorough knowledge of the flora of northern Luzon and of the Batane Islands. The work on the Formosa flora is being prosecuted by Dr. *Hayata* and lists of additions are being published by him from time to time in the *Botanical Magazine, Tokyo.*

Merrill, Elmer D. Botanical Work in the Philippines. (*Bureau of Agriculture* (*Manila*) *Bull.* **4** (1903) pp. 1–53.)

An historical account of the work accomplished on the Philippine flora by various authors, with an account of the Manila Botanical Garden, herbaria and botanical libraries in Manila, Philippine botanical material in Europe and America and a partial bibliography relating to Philippine botany.

Merrill, Elmer D. Report on Investigations Made in Java in the Year 1902. (*Forestry Bureau* (*Manila*) *Bull.* **1** (1903) pp. 1–84.)

In the enumeration of the Philippine plants identified at Buitenzorg (Plantæ Ahernianæ, pp. 15–55) 66 families, 225 genera and about 400 species are listed, several genera, *Wallaceodendron, Erythroxylon, Walsura, Actephila, Gynotroches, Lepiniopsis,* and *Couthovia,* are reported from the Philippines for the first time and 5 species are described as new, *Evodia mindanaensis* (=*E. latifolia* DC.!), *Semecarpus macrophylla, Palaquium ahernianum, Vitex philippinensis* and *Timonius philippinensis.* Various errors in identifications have been corrected in later publications, but others remain to be considered. Among the apparent errors in identifications *Pinus khasia* is a form of *P. insularis,* as is the species following enumerated without name. *Quercus philippinensis,* is not *DeCandolle's* species but is *Q. celebica, Artocarpus blumei* is probably incorrectly identified and the specimen may be referable to *A. communis* Forst. *Ailanthus moluccana* is not that species but *A. philippinensis* Merr., *Canarium commune* is not the Linnean species but is *C. ovatum. Toona ciliata* is doubtful as to the species, the material being sterile it is impossible to be sure of the identification. *Walsura robusta* is not *Roxburgh's* species but distinct, *W. aherniana* Perk. *Pterospermum blumeanum,* whether or not *Korthal's* species, the specimens are referable to the earlier *P. obliquum* Blanco. *Saurauia reinwardtiana* Bl., specific identity very doubtful., *Arthrophyllum diversifolium* Bl., should be excluded, as the specimen cited is *Oroxylum indicum* Vent.!, leaf specimens only. *Trachelospermum,* the generic identification is doubtful, fruits only. *Ixora amboinica* can be excluded as the specimen cited is apparently referable to *Phaleria.* Undoubtedly other errors in identifications will be found later as the material is more thoroughly worked over and carefully compared.

Mez, Carl. Myrsinaceae. (*Das Pflanzenreich*, 9 (1902) pp. 1–437.)

Of this large and widely distributed family the following species are credited to the Philippines: *Maesa laxa* Mez, *M. haenkeana* Mez, *M. manillensis* Mez, *M. denticulata* Mez, *M. cumingii* Mez, *M. gaudichaudii* Mez, all endemic; *Ardisia corniculatum* (L.) Blanco, *A. floridum* R. & S., both widely distributed in the Indo-Malayan region; *Ardisia tomentosa* Presl, *A. philippinensis* A. DC., *A. disticha* A. DC., *A. mindanaensis* Mez, *A. marginata* Bl., *A. sulcata* Mez, *A. scabrida* Mez, *A. humilis* Vahl, *A. boissieri* A. DC., *A. pirifolia* Mez, *A. verrucosa* Presl, *A. grandidens* Mez, *A. perrottetiana* A. DC., *A. serrata* (Cav.) Pers., *A. castaneifolia* Mez, *A. candolleana* (O. Ktz.) Mez, *A. scalaris* Mez, *A. cumingiana* A. DC., *A. proteifolia* Mez, *A. warburgiana* Mez, *A. saligna* Mez, *A. crispa* (Thunb.) A. DC., *A. pardelina* Mez, *A. sinuato-crenata* Mez, *A. jagorii* Mez, all endemic except three species; *Discocalyx philippinensis* (A. DC.) Mez, *D. vidalii* Mez, *D. effusa* Mez, *D. minor* Mez, *D. cybianthoides* (A. DC.) Mez, *D. angustifolia* Mez, all endemic; *Embelia porteana* Mez, *E. philippinensis* A. DC. both endemic; *Rapanea philippinensis* (A. DC.) Mez, endemic. Since the publication of the monograph representatives of 2 other genera, *Amblyanthopsis* and *Labisia* have been discovered in the Philippines, and some species of genera listed above have been described. (See Mez, *This Journal* 1 (1906) Suppl. pp. 271–275.)

Palla, E. Scleria luzonensis Palla sp. nov. (*Allgemeine Bot. Zeitschr.* (1907).

The above new species described, to be issued in *Kneucker's* "Cyperaceae exsiccatæ," the type from Mount Arayat, Luzon.

Pax, F. Aceraceae. (*Das Pflanzenreich*, 8 (1902) pp. 1–89.)

Of the single genus in the family, *Acer* Linn., 114 species and many varieties are recognized, but none are credited to the Philippines. Since the publication of the above monograph 2 species have been discovered in the Archipelago, both undescribed by *Pax*, thus adding an additional family to the list of those previously known from the Philippines.

Pax, F., and Knuth, R. Primulaceae. (*Das Pflanzenreich* 22 (1905) pp. 1–386.)

Of this family *Androsace saxifragifolia* Bunge, northern India to China, Japan, Formosa, and Luzon, and *Lysimachia japonica* Thunb., with about the same distribution but extending to Java, and *L. ramosa* Wall., var. *typica* R. Knuth, Himalaya, Burma, Java, and Luzon, are the only forms credited to the Philippines. A few more species are, however, found in northern Luzon.

Perkins, J. Fragmenta Floræ Philippinæ. (Contributions to the Flora of the Philippine Islands, Leipzig, Gebrüder Borntraeger (1904–05) pp. 1–212, plates 4.)

This work was issued in three fascicles, I, pp. 1–66, March 12, 1904; II, pp. 67–152, June 30, 1904, and III, pp. 153–212, February 20, 1905. It was prepared by Dr. *Perkins* with the assistance of various specialists and was based largely on the Philippine collections of *Warburg*, *Ahern*, and *Merrill*. The chief groups treated are *Leguminosae*, *Dipterocarpaceae*, *Anacardiaceae*, *Meliaceae*, *Pinaceae*, *Taxaceae*, *Marantaceae*, *Gonystylaceae*, *Burseraceae*, *Elaeocarpaceae*, *Tiliaceae*, *Malvaceae*, *Bombacaceae*, *Sterculiaceae*, *Rosaceae*, and *Rutaceae*, by Perkins; *Symplocaceae* by A. Brand; *Acanthaceae* by G. Lindau; *Fagaceae* by O. von Seeman; *Typhaceae* by P. Graebner; *Orchidaceae* by R. Schlechter; *Palmae* by O. Beccari; *Sapindaceae* by L. Radlkofer; *Asclepiadaceae* by R. Schlechter and O. Warburg; *Myristicaceae*, *Pandanaceae*,

Begoniaceae, Ulmaceae, Moraceae, Urticaceae, Balanophoraceae, Aristolochiaceae, Magnoliaceae, Thymeliaceae, and *Ericaceae,* by O. Warburg; *Eriocaulonaceae* by W. Ruhland; *Gramineae* by C. Mez and R. Pilger; *Piperaceae* by C. de Candolle and ferns by E. B. Copeland. In this work no less than 2 genera and 219 species and varieties are described as new and 1 family, several genera and many species credited to the Philippines for the first time.

Perkins, J. Zwei neue Meliaceen. (*Notizblatt Kgl. Bot. Gart. und Mus. Berlin* (1903) pp. 78–79.)

Aglaia harmsiana and *Cipadessa warburgii* are described, the descriptions being translated into English in Perk. Frag. Fl. Philip. (1904) 30, 32.

Perkins, J., and Gilg, E. Monemiaceae. (*Das Pflanzenreich* 4 (1901) pp. 1–122.)

Thirty-one genera are recognized, but no representative of the entire family is cited from the Philippines. Since the publication of the work *Kibara ellipsoidea, K. depauperata, K. grandifolia,* and *Matthaea chartacea* have been described by *Merrill. Kibara coriacea* was previously credited to the Philippines by *Rolfe,* and *Matthaea sancta* by *Ceron.*

Pfitzer, E. Orchidaceae-Pleonandrae. (*Das Pflanzenreich* **12** (1903) pp. 1–132.)

Neuwiedia veratrifolia Blume and *N. zollingeri* Reichb. f., are credited to Luzon with doubt, and *Apostasia wallichii* R. Br., as perhaps growing in Luzon; *Paphiopedilum rothschildianum* (Reichb. f.) Pfitz., var. *elliotianum* (O'Brien) Pfitz., *P. philippinense* (Reichb. f.) Pfitz., and the variety *platytaenium* Desb., *P. roebbelinii* (Reichb. f.) Pfitz., *P. haynaldianum* (Reichb. f.) Pfitz., *P. argus* (Reichb. f.) Pfitz., and *P. ciliolare* (Reichb. f.) Pfitz., all endemic except the last one, which is also found in the Malayan Peninsula.

Pilger, R. Taxaceae. (*Das Pflanzenreich,* **18** (1903) pp. 1–124.)

Of this family the following species are credited to the Philippines: *Dacrydium falciforme* (Parl.) Presl, Borneo and Mindoro; *D. elatum* (Roxb.) Wall., Malaya, Mindoro; *Podocarpus imbricatus* Blume, var. *cumingii* (Parl.) Pilger, the variety endemic, the species from Burma to Malaya; *D. costalis* Presl, endemic; *Phyllocladus protractus* (Warb.) Pilger, Philippines, Moluccas and New Guinea; *Taxus baccata* subsp. *wallichiana* (Zucc.) Pilger, British India to Malaya and Celebes, other forms widely distributed in tropical and temperate regions of the World. Since the publication of the monograph several additional species of *Podocarpus* have been found in the Philippines, some identical with previously described species of the Malayan region, others undescribed.

Prain, D. Novicae Indicae XVIII.—The Asiatic Species of Dalbergia. (*Journ. As. Soc. Beng.* **70** (1901) part 2, pp. 39–65.)

Seventy-four species of the genus are recognized, of which the following five are credited to the Philippines: *Dalbergia polyphylla* Benth, endemic; *D. tamarindifolia* Roxb., Himalayan region to Malaya and Luzon; *D. candenatensis* (Dennst.) Prain (*D. torta* Grah.) western India to Malaya, Polynesia and Australia; *D. ferruginea* Roxb., Malaya, Philippines, New Guinea, and *D. cumingiana* Benth., endemic.

Prain, D. A List of the Asiatic Species of Ormosia. (*Journ. As. Soc. Beng.* **69** (1900) part 2, pp. 175–186.)

Twenty-two species of the genus are recognized, of which only one is found in the Philippines, the endemic *Ormosia calavensis* Blanco. A second Philippine species, *Ormosia paniculata* Merr., has since been described, *Philip. Journ. Sci.* (1906) **1.** Suppl. **64.**

Prain, D. Report on the Indian Species of Pterocarpus. (*Stray Leaves from Indian Forests;* issued with *Indian Forester* **26** (1900) No. 10, pp. 1–16.)

Five species are considered, especially with a view to the identity of the species yielding the padouk timber of commerce. But one species considered extends to the Philippines, *P. indicus* Willd., but in a footnote on page 10, the Philippine *P. vidalianus* Rolfe is reduced to *P. echinatus* Pers., a species previously known only from south Celebes.

Rendle, A. B. Najadaceae. (*Das Pflanzenreich* 7 (1901) pp. 1–21.)

A single genus, *Najas* Linn., is recognized, containing 32 species and many varieties, of which the following are found in the Philippines: *Najas foveolata* A. Br., *N. falciculata* A. Br., and *N. graminea* Del.

Robinson, C. B. The History of Botany in the Philippine Islands. (*Journ. N. Y. Bot. Gard.* 7 (1906) pp. 104–112.)

A sketch of the history of Philippine botany from the year 1587 to the year 1906, including some data not included by *Merrill* in his "Botanical Work in the Philippines."

Robinson, C. B. Some Affinities of the Philippine Flora. (*Torreya.* 7 (1907) pp. 1–4.)

A review of the introduction to *Merrill's* "New or Noteworthy Philippine Plants, V" *Philip. Journ. Sci.* 1 (1906) Suppl. pp. 169–246.

Ruhland, W. Eriocaulonaceae. (*Das Pflanzenreich*, **13** (1903) pp. 1–294.)

Nine genera are recognized of which but one, *Eriocaulon* Linn., with 193 species, is represented in the Philippines. The species credited to the Philippines are *E. truncatum* Ham., which should be excluded as the specimen cited from the Philippines, No. 2326 *Cuming*, was not collected in the Archipelago but in Malacca; *E. sexangulare* Linn., British India to China and the Philippines, and *E. sieboldianum* Sieb. et. Zucc., British India to China, Japan, the Philippines and Java. One species has been described from Philippine material since the publication of the above monograph, *E. merrillii* Ruhl.

Schindler, Anton K. Halorrhagaceae. (*Das Pflanzenreich* **23** (1905) pp. 1–133.)

Of this family but a single species, *Gunnera macrophylla* Blume, Java, Sumatra, Luzon and New Guinea, is credited to the Philippines. Since the publication of the monograph however the following species have been added to the Philippine flora: *Myriophyllum spicatum* L., widely distributed; *Halorrhagis micrantha* (Thunb.) R. Br., *H. philippinensis* Merr., *H. scabra* var. *elongata* Schindl., and *H. halconensis* Merr.

Schumann, K. Musaceae. (*Das Pflanzenreich*, **1** (1900) pp. 1–45.)

Five genera are recognized, two being represented in the Philippines, one *Musa* by many forms, the other *Ravenala* by occasional cultivated specimens. The only species of *Musa* credited to the Philippines by Schumann are *M. textilis* Née, the source of abacá, Manila hemp, and forms of *M. paradisiaca* L., the common banana and plantain. No attempt is made to reduce the numerous forms described by *Blanco* in his "Flora de Filipinas." *Ravenala madagascarensis* Sonn., the "traveller's palm" has been introduced into the Archipelago and is occasionally cultivated for ornamental purposes.

Schumann, K. Zingiberaceae. (*Das Pflanzenreich*, **20** (1904) pp. 1–458.)

The following species are credited to the Philippines: *Hedychium philippinense* K. Schum., endemic; (*Brachychilus* a genus of two species apparently erroneously credited to the Philippines in note on generic distribution and in conspectus of species, but under the species the Philippines are not mentioned); *Globba brevifolia* K. Schum., *G. gracilis* K. Schum., *G. campsophylla* K. Schum., *G. parviflora* Presl, *G. pyramidata* Gagnepain, *G.*

ectobolus K. Schum., *G. ustulata* Gagnepain, *G. barthei* Gagnepain, *G. heterobractea* K. Schum., all endemic; *Ammomum loheri* K. Schum., *Alpinia leptosolenia* K. Schum., endemic; *A. pubiflora* (Benth.) K. Schum., New Guinea, Caroline Islands and Mindanao; *A. brevilabris* Presl, *A. pulchella* K. Schum., New Guinea and Mindanao; *A. cumingii* K. Schum., *A. galanga* (L.) Sw., Malaya; *A. trachyascus* K. Schum., *A. macroscaphis* K. Schum., *A. haenkei* Presl, *A. elegans* (Presl) K. Schum., *A. rufa* (Presl) K. Schum., *A. parviflora* (Presl) Rolfe, *A. rolfei* K. Sch., *A. mollis* Presl, all endemic; *Costus speciosus* (Koenig) Smith, var. *leiocalyx* K. Schum., widely distributed. In addition to the above species definitely credited to the Philippines others are more or less common in the Archipelago such as *Hedychium coronarium* Koenig, species of *Kaemphera*, *Curcuma longa* Linn., *C. zeodaria* (Berg.) Rosc., *Zingiber officinale* Rosc., *Z. zerumbet* (L.) Sm., etc. (See also Ridley in *Govt. Lab. Publ.* **35** (1905) pp. 83–87.)

Schumann, K. Marantaceae. (*Das Pflanzenreich* **11** (1902) pp. 1–184.)

Of the 26 genera recognized, 4 are found in the Philippines, represented by the following species: *Donax arundastrum* Lour., British India to the Malayan Peninsula, Tonkin and the Philippines; *Monophrynium fasciculatum* (Presl) K. Schum., a monotypic endemic genus; *Phacelophrynium interruptum* (Warb.) K. Schum., *P. bracteosum* (Warb.) K. Schum., both endemic. *Maranta arundinacea* Linn., introduced from tropical America, the source of arrowroot, is commonly cultivated and subspontaneous in the Philippines. All the above species are considered and figured by *Perkins* in her Fragmenta Florae Philippinae (1904) pp. 67–73, plates 3.

Scribner, F. Lamson. Notes on the Grasses in the Bernhardi Herbarium, collected by Thaddeus Haenke, and described by J. S. Presl. (*Rept. Mo. Bot. Gard.* **10** (1899) 35–59, plates 54.)

Critical notes on the types of some of *Presl's* species, with illustrations, including many based on Philippine material, deposited in the herbarium of the Missouri Botanical Garden.

Solms-Laubach, H. Graf zu. Rafflesiaceae and Hydnoraceae. (*Das Pflanzenreich* **5** (1901) *Rafflesiaceae* pp. 1–19; *Hydnoraceae* pp. 1–9.)

The *Rafflesiaceae* are represented in the Philippines by *Rafflesia schadenbergiana* Goeppert, from Mindanao, and *R. manillana* Teschem., from Leyte, Samar and Luzon. To the latter species are reduced *R. cumingii* R. Br., *R. lagascae* Blanco and *R. philippinensis* Blanco. The *Hydnoraceae* are not represented in the Philippines.

Underwood, Lucien Marcus. A Summary of our Present Knowledge of the Ferns of the Philippines. (*Bull. Torr. Bot. Club.* **30** (1903) pp. 665–684.)

A consideration of the most important works treating the ferns of the Philippines and an account of the most important collections made in the Archipelago, with analytical keys to the families and genera of vascular cryptogams known to be represented in the Philippines, with some proposed changes in nomenclature, the final summary of vascular cryptogams being families 15, genera 105 and species 633.

Underwood, L. M. The Genus Stenochlaena. (*Bull. Torr. Bot. Club.* **33** (1906) pp. 35–50.)

The entire genus is considered, 23 species being recognized, of which the following are credited to the Philippines: *Stenochlaena laurifolia* Presl, endemic; *S. palustris* (Burm.) Bedd., widely distributed; *S. williamsii* n. sp., *S. aculeata* (Blume) Kunze, Tenasserim to Java and Borneo; *S. leptocarpa* (Fée) Underw., Java and the Philippines and *S. smithii* (Fée) Underw., endemic.

Vidal y Soler, Sebastian. Catálogo metódico de las plantas leñosas silvestres y cultivadas observadas en la provincia de Manila. (1880) pp. 1–48. (Reprint from *Revista de Montes* 4 (1880).)

In this work 531 species are enumerated, in which the following new names appear: *Pittosporum fernandezii*, *Aegle decandra*, *Dysoxylum blancoi*, *Parinarium racemosum*, *Medinilla lagunae*, *Homalium barandae*, and *Clerodendron navesianum*, but most of them are scarcely more than nomina nuda. Many of the errors in identifications were corrected later by Vidal in his Rev. Pl. Vasc. Filip. (1886). The introduction contains much of interest regarding the types of Philippine forests.

Usteri, Alfred. Beiträge zur Kenntnis der Philippinen und ihrer Vegetation. mit Ausblicken auf Nachbargebiete. (Inaugural-Dissertation zur Erlangung der philosophischen Doktorwürde der Universität Zurich. 1905. pp. 1–166. with 29 text-figures and 2 plates.)

This work contains a sketch of the geology and plant geography of the Philippines, meterology and climatology, notes on the islands of Guimaras, Cebu and Negros, a consideration of the different plant formations, notes on various cultivated plants, sugar culture, etc., and a catalogue of the plants collected, 1,431 species being enumerated from the Philippines, 1,303 from Java, 3 from Penang, 65 from Labuan, and about 100 from Singapore. In the enumeration of Philippine plants *Piper usterii* C. DC., and the variety *plurifistulosum* C. DC. are described, and in the *Orchidaceae* the name *Dendrobium usterii* Schltr. appears as a nomen nudum, later described in *Bull. Herb. Boiss.* II. 6 (1906) 458. Another new species is *Selaginella usterii* Hieron., with a very imperfect description. Under the Algae the following new species are described: *Phormidium usterii* Schmidle, and *Myxobactron usterianum* Schmidle, the latter the type of a new genus. Many species are credited to the Philippines for the first time.

Warburg, O. Pandanaceae. (*Das Pflanzenreich*, 3 (1900) pp. 1–97.).

Three genera, *Sararanga*, *Freycinetia* and *Pandanus* are recognized, the first with but a single species confined to the Solomon Islands and New Guinea, the other two genera widely distributed, *Freycinetia* with 62 species and *Pandanus* with 156. The Philippine forms recognized are *Freycinetia ferox* Warb., *F. luzonensis* Presl, *F. sphaerocephala* Gaudich., *F. vidalii* Hemsl., *F. jagorii* Warb., *F. philippinensis* Hemsl., and *F. scabripes* Warb., all endemic, and *Pandanus tectorius* Sol. (*P. odoratissimus* L. f., *P. spiralis* Blanco, *P. blancoi* Kth.). The species described by *Blanco*, *P. exaltatus*, *P. sabotan*, *P. gracilis*, *P. malatensis* and *P. radicans* are all considered doubtful species. Since the publication of *Warburg's* monograph a second species of *Sararanga* (*S. philippinensis*) has been found in the Philippines, and several species of *Freycinetia* and *Pandanus* have been described as new, while most of the species described by *Blanco* have been satisfactorially disposed of. (See, *Govt. Lab. Publ.* 17, 27, 29; *Philip. Journ. Sci.* 1 (1906) Suppl; Elmer, *Leaflets Philip. Bot.* 1 (1906).)

Williams, R. S. Notes on Luzon Mosses. (*The Bryologist*, 8 (1905) pp. 78–80.)

A popular account of some of the species observed while on a collecting trip in Luzon, including notes on forms observed about Manila, in the Province of Bataan, and in the Province of Benguet.

THE PHILIPPINE

JOURNAL OF SCIENCE

C. BOTANY

VOL. II | JULY 15, 1907 | No. 4

THE FLORA OF MOUNT HALCON, MINDORO.

By ELMER D. MERRILL.

(From the botanical section of the Biological Laboratory, Bureau of Science.)

Mount Halcon is perhaps the third highest mountain in the Philippines, it is situated in the north central part of Mindoro and near the geographical center of the entire Archipelago. Although it is within 100 miles of Manila and within 15 of Calapan, the capital of Mindoro, it has, so far as we have been able to determine, remained unascended up to the year 1906. In the latter part of that year a biological and zoölogical expedition was organized under the direction and with the support of Maj. Gen. *Leonard Wood*, the object being to explore Mount Halcon, to determine a feasible route to the mountain, to ascend the highest peak and to secure as much information as possible regarding it, as well as to make botanical and zoölogical collections. The expidition was successful in all respects although undertaken at the worst season of the year—that is, in the midst of the rainy season—and the highest point on Halcon was reached on November 22, 1906, twenty-one days after leaving the coast. The reader is referred to my account of the ascent of Halcon [1] for a narrative and geographical account of the trip, a description of Halcon and a summary of previous attempts made to ascend the mountain.

Before this time Halcon was but little known botanically, although the English ornithologist *John Whitehead* had made a small collection in the year 1895 of plants on Dulangan, a spur of the mountain. This collection yielded several species of special interest, some undescribed and

[1] *This Journal, Sec. A. Gen. Sci.* (1907), **2**, 179.

others previously known only from Borneo. Bornean types, considering the proximity of the Philippines to that large island, are rather rare in the Archipelago. *Whitehead's* plants have been considered by Rendle.[2]

Most of the species collected by *Whitehead* on Halcon were also brought in by me on the expedition under discussion. *Hugh Cuming* collected in Mindoro, between the years 1836 and 1840. He undoubtedly worked in the vicinity of Calapan and on the Baco River, at the north base of Halcon, although he probably did not penetrate far into the interior of the island. *Cuming,* in most cases, did not give localities for his plants; he was never more definite than to give the province or island, so that we are not absolutely certain as to just what parts of Mindoro he visited. In 1903 and 1905 I made short collecting trips up the Baco River, and in 1905 Mr. *R. C. McGregor,* of this Bureau, collected a considerable number of plants in the same region. In June, 1906, Mr. *M. L. Merritt,* of the Philippine Forestry Bureau, accompanied Lieut. *T. H. Jennings* on his attempt to ascend Halcon. The party reached an altitude of 7,250 feet and brought in 165 numbers of plants representing about 150 species. The plants collected by myself in November, 1906, are represented by 742 numbers and comprise about 600 distinct species. The greater part of this material was gathered within a period of thirty days, from November 2 to December 2, 1906, under very unfavorable conditions. The weather, nearly every day, was more or less rainy, and for thirteen days in succession, while the party was at and above an altitude of 4,500 feet, the rain did not cease day or night. As a result of these conditions much of the collecting was accomplished in the wet, and all specimens were of necessity dried by means of fire. Material once dried could only with difficulty be preserved, and constant alertness was needed to protect our collections against moisture when we were in camp, while packing and moving in the pouring rain, and as we were fording streams. The material secured by Mr. *Merritt* was prepared under scarcely more favorable circumstances.

The present paper is based on the plants collected by Mr. *Merritt* and on those secured by myself, 271 species and varieties being considered; these are distributed into 83 families and 168 genera. Two genera are proposed as new and two families are added to the number previously known from the Philippines, one of these, *Centrolepidaceæ* being quite new to the Archipelago, and one, *Iridaceæ,* was previously known to be represented in the Philippines only by introduced and cultivated species. Seven genera and fifteen species are reported from the Philippines for the first time, while thirty-nine species are described as new. The above summary is based only on the material considered in the present paper. The vascular cryptogams collected on Halcon by me have already been

[2] *Journ. Bot.* (1896), **34,** 355–358.

considered by Dr. *E. B. Copeland* under the title *Pteridophyta Halconenses*,[3] 206 species and varieties being represented in the collection, of which twenty were described as new, and eight reported for the first time from the Archipelago. Of the *Orchidaceæ* of the Halcon area, 101 species are known of which about 42 are new. These were all sent to Mr. *Oakes Ames* and are considered by him in a following paper. The mosses collected on the expedition have been enumerated by *V. F. Brotherus*, Helsingfors, Finland, his paper also following this one. No attempt has been made to determine the rather extensive collections of scale-mosses, lichens and fungi collected on the expedition, but it is hoped that arrangements can be perfected with various specialists which will result in having these groups considered at a later date. Nearly all the other material collected by Mr. *Merritt* and myself has been discussed in the present paper, but in one or two families, scantily represented on Halcon, specific identifications have not been made for one reason or another, and at least three species are not mentioned in this paper, the material representing them being insufficient to refer them with certainty to their respective genera.

Halcon is perhaps the most humid mountain in the Philippines, the rainy season continuing practically for nine months of the year, from May to January, without interruption, while the remaining three months are by no means free from precipitation, as we know from Lieutenant *Lee's* experience in the vicinity of Halcon in April, 1904. The enormous amount of rain in the Halcon area is shown by the relatively very large rivers flowing from the range. Although these in the maximum are not more than 25 or 30 miles in length, and perhaps they may be shorter, they have a breadth of from 100 to 300 yards in their lower parts, and a constant flow of a large volume of water. In the rainy season the volume of the latter is greatly augmented, all the rivers being subject to sudden and enormous floods, as is shown by the experience of *Whitehead, Lee* and our own party. *Whitehead* recorded that the Catuyran River, 200 yards wide at the place where his camp was located, rose over 20 feet within a period of less than twelve hours.

Epiphytic orchids, ferns and other plants, mosses, lichens, etc., which in other parts of the Philippines are usually found only at considerable altitudes above the sea, are in the vicinity of Halcon encountered at comparatively low elevations, along the Alag and Binabay Rivers, 65 to 200 meters above sea level, and along the Baco River at approximately sea level, thus showing that the relatively high humidity is not confined to Halcon itself but affects the surrounding low country to a considerable extent. As a result of this high humidity the open grass lands and savannah forests are entirely wanting on the north side of the Halcon Range,

[3] *This Journal, Sec. C, Bot.* (1907), 2, 119–151.

although open grass country is visible from the high ridges to the south of the mountain. The high forest, *Dipterocarpus* type is fairly well developed, extending up to an altitude of at least 500 meters, but even this type is quite changed by its environment; terrestrial ferns, orchids and herbaceous plants being comparatively abundant and epiphytic plants numerous. The prevailing species of *Dipterocarpus, Shorea,* etc., disappear above an altitude of 500 meters, and *Quercus llanosii,* various species of *Lauraceæ, Acer philippinum, Aralia, Casuarina sp., Englehardtia spicata, Artocarpus, Unona, Polyalthia, Weinmannia, Elæocarpus* and many other aborescent genera appear, and epiphytic plants become more abundant. No less than twenty-two species of the genera *Hymenophyllum* and *Trichomanes* are known from Halcon.

At and above an altitude of 1,000 meters, the mossy forest type is encountered on the exposed ridges, at first confined entirely to the crest line, but as altitude is gained, extending down the lateral slopes for a greater or less distance. These crest-line forests are characterized by arborescent genera such as *Agathis, Podocarpus, Dacrydium, Phyllocladus, Pinanga. Myrica, Drimys. Illicium, Neolitsea, Homalanthus, Ilex, Elæocarpus, Eurya, Ternstroemia, Adinandra, Mearnsia, Clethra, Vaccinium, Rhododendron, Symplocos,* and others, numerous species of epiphytic orchids, ferns and other plants, and some terrestrial species, notably *Burmannia longifolia,* the ground, tree trunks and branches being densely covered with thick masses of mosses, lichens, etc., forming an ideal habitat for the abundant species of epiphytic and pseudo-epiphytic plants. The trees are more or less stunted and as altitude is gained this character becomes emphasized. Scandent or semiscandent species of *Vaccinium, Diplycosia, Schefflera, Smilax, Calamus* and *Nepenthes* clamber everywhere through the dense ridge thickets and the fern *Oleandra colubrina* Copel., alone forms such dense masses that it is frequently difficult for the traveler to force his way through them. There is a constant change in the vegetation of these ridges as altitude is gained, some genera such as *Drimys, Podocarpus, Nepenthes, Phyllocladus, Agathis, Symplocos,* etc., persisting unaltered from an altitude of 1,000 meters to the summit of the highest peak, but terrestrial and epiphytic orchids, ferns and other plants entirely change, those at the higher altitudes being quite different from those at the lower ones. Mosses and lichens become more abundant and form much thicker and denser masses on the ground and trees, whereas *Sphagnum* appears in the ground cover. There is less diversity in constituent species on the highest ridges above 2,400 meters than at lower altitudes, but the trees and shrubs on them are greatly stunted, being reduced to montane brush which rarely exceeds a height of 3 meters. Epiphytic and terrestrial plants become reduced to comparatively few species and individuals, while mosses and scale mosses correspondingly increase in abundance and diversity of forms.

On the main ridge at an altitude of 2,400 meters the montane brush of the exposed ridges becomes reduced to a mere heath, characterized by open lands with a scant cover of grasses and sedges, with scattered dwarfed undershrubs and bushes and some very characteristic herbaceous plants, a mixture of northern or continental, Bornean and Australian types. These heath lands cover considerable areas on the south slopes of the main range of Halcon, but do not extend down to the north slope. The characteristic species are *Lycopodium halconense*, Copel, *L. cernuum* Linn., forma, *Gleichenia dicarpa* R. Br., *Dipteris conjugata* var. *alpina* Christ, *Miscanthus sinensis* And., *Isachne beneckei* Hack., *I. myosotis* Nees, *Schoenus melanostachys* R. Br., *Cladium latifolium* Merr., *Gahnia javanica* Mor., *Centrolepis philippinensis* Merr., *Eriocaulon brevipedunculatum* Merr., *Dianella ensifolia* DC., *Liriope brachyphylla* Merr., *Patersonia lowii* Stapf., *Drosera spathulata* Labill., *D. peltata* Sm., *Rubus rolfei* Vid., *Halorrhagis halconensis* Merr., *H. micrantha* R. Br., *Didiscus saniculæfolius* Merr., *Vaccinium banksii* Merr., *V. villarii* Vid., *V. whitfordii* Merr., *Rapanea retusa* Merr., *Utricularia orbiculata* Wall., *Hedyotis montana* Merr., *Leptospermum amboinense* Bl., *Rhododendron quadrasianum* Vid., and *Adinandra* sp.

The botanical exploration of Halcon has added representatives of two families new to the Archipelago to our knowledge to the Philippine flora, seven genera new to the Islands, fifteen species previously described from surrounding regions, and many new to science. Considering the proximity of Borneo to the Philippines, and the connecting chains of islands, the Sulu Archipelago at the south, Balabac, Palawan, the Calamianes, and the Mindoro chain extending to the north, Bornean types in the Philippine flora are comparatively rare, but it is not at all surprising to find a considerable number of characteristic Bornean plants on Halcon, although *Copeland*[4] in the 206 species and varieties of vascular cryptogams known from Halcon considered but one *Ophioglossum intermedium*, to be of probable Bornean origin. *Dacrydium falciforme Pilger*, *Patersonia lowii* Stapf, and *Didiscus saniculæfolius* Merr., are known only from Mindoro and Borneo, *Schoenus melanostachys* R. Br., from Mindoro, Borneo and Australia, *Burmannia longifolia* Becc. from Mindoro, Borneo and Malayan Peninsula, *Symplocos adenophylla* Wall., from Mindoro, Borneo, Banca, Singapore and Penang, while *Hedyotis eucapitata* Merr., is closely related to a species known only from North Borneo. On the other hand there is a rather remarkable assemblage of Australian types on Halcon, all at high altitudes. Among these may be mentioned *Schoenus melanostachys* R. Br., Australia, Borneo and Mindor, the genus being largely developed in Australia, with few species occurring in the Northern Hemisphere, *Centrolepis philippinensis* Merr.,

[4] *Loc. cit.*, 121.

perhaps the most remarkable find in the Philippines in the history of recent botanical exploration of the Archipelago, as this small family is almost entirely Australian, six genera being generally recognized, of which four, *Juncella*, *Brizula*, *Aphelia* and *Alepyrum*, are confined to Australia, New Zealand and Tasmania, and a fifth, *Gaimardia* of two species, confined to New Zealand, Cape Horn and the Falkland Islands. The sixth genus, *Centrolepis*, is represented by about twenty species, of which one is found in southern Asia (Cambodia), one on Mount Halcon and the remainder in south Australia and Tasmania. *Dianella caerulea*, the genus being a characteristic Australian one, extends from Australia through New Guinea to Luzon. *Patersonia lowii* Stepf, known from Borneo and Mindoro, is essentially an Australian type, two species of the genus being found on Mount Kinabalu, North Borneo, one extending to Mount Halcon, Mindoro, the remaining species, about nineteen, being confined to Australia. *Halorrhagis halconensis* Merr., the fourth species of the genus to be found in the Philippines, is also an Australian type, the genus being largely developed in Australia and for the greater part confined to that continent. *Didiscus saniculæfolius* Merr., of Mindoro and Borneo, is also an Australian type, twelve species of the genus being Australian, one New Caledonian, and one Mindoro and Bornean. *Cladium latifolium* Merr., is one of the comparatively few species of the genus found outside of Australia. The Australian element in the Philippines has previously been known to be rather large and characteristic, and it is considerably augmented by the species enumerated above.

I have advanced elsewhere[5] the belief that Mindoro is probably the one part of the group which has remained continuously above water for a longer period of time that any of the surrounding islands, and a part of it at least may have been so from the time that it was connected with the great land-mass of the ancient Malayan continent. The geological structure of the island, especially that part of it in the Halcon area, seems to be quite similar to that of Mount Kinabalu, North Borneo, so far as I can determine from available descriptions of the latter, and entirely different from that of the islands in closest proximity to it, namely Luzon and others. The presence only in Mindoro of the one large mammal in the Philippines (*Bubalus mindorensis*) is evidence in favor of the above hypothesis. Much is known of the avifauna of the island and in this character Mindoro is apparently related with Borneo through Palawan, rather than with its nearer neighbor, Luzon. The presence of a decided Bornean and Australian element in the flora at the higher altitudes on Halcon also indicates previous and close relationships with the great land-masses to the East and South, but I find that this botanical

[5] *This Journal, Sec. A. Gen. Sci.* (1907), **2**, 201.

evidence is not confined to the higher altitudes of Mindoro, for the characteristic Malayan or Indo-Malayan genera represented by *Chrysophyllum roxburghii* and *Ochtocharis javanica*, both species of low elevations, are at present known in the Philippines only from this island. However, as other Bornean and Australian types are known in the Philippines only in Luzon, the botanical evidence alone is not conclusive.

In the present paper and in those by *Copeland, Brotherus* and *Ames* on the Halcon flora, considerably over 600 species are considered, for the greater part collected on two short expeditions, both made under very unfavorable conditions for collecting and preserving botanical material because of the prevailing rains, Mr. *Merritt's* expedition having been made in June, and my own in November. Considering that in this region plants flower throughout the year, different species at different seasons, it seems very probable that we at present know considerably less than one-half the species actually growing on Halcon, and that future exploration will yield much material and many data of value. Halcon then, like Kinabalu, must still be considered to be very imperfectly known botanically, the former rather better than the latter, for in *Stapf's* paper on the flora of the latter only about 450 species are enumerated.

PINACEÆ.

AGATHIS Salisb.

Agathis philippinensis Warb. Monsunia 1 (1900) 185, *t. 8. f. E.*

In forests 700 to 2,500 m. alt., abundant, mature cone only collected.

Widely distributed in the Philippines, from northern Luzon to southern Mindanao. Endemic.

TAXACEÆ.[6]

DACRYDIUM Soland.

Dacrydium falciforme (Parl.) Pilger in Engl. Pflanzenreich **18** (1903) 45. *Podocarpus falciformis* Parl. in DC. Prodr. **16**² (1868) 685.

In forests at 1,800 m. alt. (No. 5744); also collected by *Merritt* in June, 1906, at 160 m. alt. (No. 4425).

Borneo.

This species was previously collected on Halcon (Dulangan) by *Whitehead*, and reported by *Rendle*.[7] The above specimens exactly match fragments of No. 1697 *Beccari*, from Mount Mattang, Sarawak, Borneo, kindly supplied me by Dr. *Beccari*, except that the leaves of the Halcon specimens are slightly smaller than in the Borneo plant. An interesting Bornean type in the Philippine flora.

Dacrydium elatum (Roxb.) Wall. ex Hook. Lond. Journ. Bot. **2** (1843) 144, *t. 2;* Pilger l. c. 51. *Juniperus elata* Roxb., Fl. Ind. **3** (1832) 838.

In forests on exposed ridges at 1,300 m. alt. (No. 5789); also collected by *Merritt* at 1,600 m. alt. (No. 4419).

[6] By F. W. Foxworthy, Bureau of Science, Manila.

[7] *Journ. Bot.* (1896), **34**, 355.

Tonkin to Singapore, Sumatra, Borneo and the Viti Islands.

This species is widely distributed, at least in the southern Philippines, on the higher mountains. NEGROS, Mount Silay (4543 P. del *Villar*) June, 1906; (4227 *Everett*) February, 1906. PANAY, Mount Madiaas (*Yoder*) April, 1905. MINDANAO, Mount Malindang (4547, 4548, 4731 *Mearns & Hutchinson*) May, 1906. It was first collected in the Philippines by *Whitehead*, on Mount Halcon in 1895. and reported by *Rendle*.[8]

Dacrydium sp. near *D. Beccarii* Parl.

In thickets, exposed ridges at 2,600 m. alt. (No. 5714), sterile specimens. Possibly a young form of the preceding.

PODOCARPUS L'Hérit.

Podocarpus imbricatus Blume var. **cumingii** (Parl.) Pilg. in Engl. Pflanzenreich. **18** (1903) 56.

In thickets on exposed ridges at 2,500 m. alt. (No. 5563); also collected by *Merritt* at 2,200 m. alt. (Nos. 4446, 4471).

The variety widely distributed on the higher mountains of the Philippines, endemic; the species from Burma to the Malayan Archipelago.

Podocarpus blumei Endl. Sn. (1847) 208; Pilger l. c. 60.

In forests at 1,800 m. alt. (No. 5728).

Previously known in the Philippines only from Mount Mariveles, Luzon.

Java to New Guinea.

Podocarpus amarus Blume Enum. Pl. Jav. (1827) 88; Pilger l. c. 68.

In forests at 1,800 m. alt. (No. 5703). Not previously reported from the Philippines.

Java and Sumatra to east Australia.

Podocarpus neriifolius D. Don. in Lamb. Pin. (1824) 21; Pilger l. c. 80.

In forests along the Alag River at and below 10 m. alt. (No. 5768). Sterile material.

Previously known in the Philippines only from Luzon.

British India to southern China through Malaya to New Guinea.

Podocarpus rumphii Blume Rumphia **3** (1847) 214; Pilger l. c. 81.

In forests at about 300 m. alt. (No. 5553). Only sterile material but probably this species which is new to the Philippines.

Celebes to the Moluccas and New Guinea.

Podocarpus glaucus Foxworthy n. sp. § *Stachycarpus?*

Arbor parva 5 ad 6 m. alta, ramulis congestis, foliis congestis, ascendente-patentibus, coriaceis, nitidis, glabris, subtus pallidis, oblongis, 9 ad 17 mm. longis, 3.5 ad 5.5 cm. latis, obtusis, basi sensim angusto-decurrentibus; flores masculi spicati, spicis cylindraceis, 1 ad 1.5 cm. longis, circa 3 mm. latis, dense multifloribus.

A small tree 5 to 6 m. tall, much branched, the branches terete, glabrous, gray or yellowish, the branchlets very numerous, short, crowded towards the ends of the branches. Leaves crowded towards the ends of the twigs, erect-spreading, often appearing subopposite on account of their contiguity, coriaceous, glabrous, smooth, shining, paler beneath and the younger ones very glaucous, oblong, elliptic-oblong or spatulate,

[8] *Loc. cit.*

9 to 17 mm. long, 3.5 to 5.5 mm. wide, the apex rounded or obtuse, the base gradually narrowed and somewhat decurrent, the midrib not prominent above, very prominent beneath, the margins thickened; petioles broad, 1 to 2 mm. long. Staminate spikes solitary in the upper leaf-axils, few, cylindrical, 1 to 1.5 cm. long, 3 mm. in diameter, densely many flowered, glaucous when young. Pistillate flowers and fruit not seen.

Borders of thickets on the margins of open heaths at 2,400 m. alt. (No. 5672).

Podocarpus pilgeri Foxworthy, nom. nov. *P. celebicus* Warb. Mons. **1** (1900) 192; Pilger l. c. 78, non *P. celebica* Hemsl. in Kew Bull. (1896) 39.

In forests at 2,100 m. alt. (No. 5754).

The specimens are sterile but match closely material with fruit, collected on Mount Malindang, MINDANAO, by *Mearns & Hutchinson*, May, 1906, No. 4673, which I have referred to the form described by *Warburg*, both agreeing closely with a fragment of the type kindly supplied me by Dr. *Engler*. Both *Warburg* and *Pilger* overlooked the fact that *Hemsley* had previously utilized the specific name *celebica*, and *Hemsley's* species is not included by the later author in his recent monograph of the family.

Podocarpus sp. § *Eupodocarpus*.

In forests at 900 m. alt. (No. 5615). Material too imperfect for accurate identification at this time.

PHYLLOCLADUS Rich.

Phyllocladus protractus (Warb.) Pilger in Engl. Pflanzenreich **18** (1903) 99. *P. hypophylla* var. *protracta* Warb. Monsunia. **1** (1900) 194.

In forests, exposed ridges at 1,300 m. alt. (No. 5788).

Widely distributed on the higher mountains of the Philippines from northern Luzon to southern Mindanao. The specimens reported from Mount Dulangan, a spur of Halcon, by *Rendle*,[9] collected by *Whitehead*, as *P. hypophylla* are probably referable to *Warburg's* species.

Moluccas and New Guinea.

PANDANCEÆ.

FREYCINETIA Gaudich.

Freycinetia multiflora Merrill n. sp. § *Oligostigma*.

Scandens, ramis ca. 7 mm. crassis; foliis lanceolatis, 20 ad 40 cm. longis, 1.5 ad 2 cm. latis, apice acutis vel acuminatis, supra basin et versus apicem denticulatis; inflorescentiae terminales; spadices ♀ quini vel sexi, oblongo-cylindrici, 8 ad 10 cm. longi, 1.5 cm. crassi; pedunculis 3 cm. longis, scabriusculis; stigmata 2 vel 3.

Scandent, the branches about 7 mm. thick. Leaves lanceolate, 20 to 40 cm. long, 1.5 to 2 cm. wide, the apex acute or acuminate, the base slightly narrowed, clasping, the margins below and towards the apex serrulate, in the median portion entire, the midrib glabrous on both surfaces or beneath with very few teeth in the upper portion. Inflorescence

[9] *Journ. Bot.* (1896), **34**, 355.

terminal; pistillate spadices 5 or 6, oblong-cylindrical, 8 to 10 cm. long, about 1.5 cm. thick, the peduncles 3 cm. long, ferruginous, strongly scabrous. Fruits very numerous, the free portions subpyramidal, strongly ridged. Stigmas 2, rarely 3.

Scandent in forests at 900 m. alt. (No. 5647).

A species possibly as closely related to *Freycinetia luzonensis* Presl, as to any other, differing from that species in its longer leaves and more numerous and much larger spadices. No. 2994 *Ahern's collector* from the Province of Rizal, Luzon, is apparently the same.

Freycinetia globosa Merrill, n. sp. § *Pleiostigma.*

Gracilis, scandens, ramulis 2 ad 3 mm. latis, foliis late lanceolatis vel oblongo-lanceolatis, 2.5 ad 7 cm. longis, 5 ad 15 mm. crassis, apice breviter acuminatis, basi abrupte angustatis, denticulatis; inflorescentiae terminales, spadices ♀ terni, fructiferi globosi, ca. 2 cm. diametro, pedunculis glabris, 1 cm. longis; fructus ca. 1 cm. longus; semina 2.5 mm. longa, anguste linearia; stigmata 4.

Slender, scandent, the branches reddish-brown, glabrous, 2 to 3 mm. thick. Leaves broadly lanceolate or oblong lanceolate, 2.5 to 7 cm. long, 5 to 15 mm. wide, shortly acuminate, the base rather abruptly, narrowed into a 5 mm. long clasping petiole, not auricled, the margins slightly denticulate throughout, the teeth small, often obscure; nerves about 15, parallel, nearly as prominent as the midrib. Inflorescence terminal, pistillate spadices globose, red, fleshy, about 2 cm. in diameter, the peduncles glabrous, about 1 cm. long; fruits nearly 1 cm. long, fleshy, ovoid; seeds numerous, white, narrowly linear, 2.5 mm. long.

Scandent in forests at 1,150 m. alt. (No. 5791).

A species apparently related to *Freycinetia ensifolia* Merr., from Mount Mariveles and to *F. sphaerocephala* Gaudich., differing from the former in its relatively broader, shorter leaves and from the latter in its longer, quite differently shaped and less strongly denticulate leaves which are not auriculate at the base.

GRAMINEÆ.

MISCANTHUS Anders.

Miscanthus sinensis Anders. Oefv. Vet. Akad. Forhandl. Stockh. (1855) 166; Merr. in Philip. Journ. Sci. 1 (1906) Suppl., 323.

In open heaths at 2,400 m. alt. (No. 5704).

On most Philippine mountains.

Japan and China to Cochin China, Borneo and Celebes.

POLLINIA Trin.

Pollinia sp. near *P. monantha* Nees; Merr. in Philip. Journ. Sci. 1 (1906) Suppl., 327.

In an old clearing at 700 m. alt. (No. 5627).

Apparently identical with specimens from Luzon that *Hackel* has indicated in lit. as an underscribed species.

PANICUM Linn.

Panicum sarmentosum Roxb. Fl. Ind. **1** (1820) 308; Merr. in Philip. Journ. Sci. **1** (1906) Suppl., 360.

In old clearings at 750 m. alt. (No. 5558).

Widely distributed in the Philippines.

India to southern China and Malaya.

Panicum palmæfolium Koenig in Naturforsch. **23** (1788) 208; Merr. l. c. 361.

In an old clearing at 900 m. alt. (No. 5585).

Widely distributed in the Philippines.

Tropical Africa to India, Japan and Malaya.

ISACHNE R. Br.

Isachne beneckei Hack. in Oesterr. Bot. Zeitschr. **51** (1901) 459; Merr. in Philip. Journ. Sci. **1** (1906) Suppl., 350.

In an open heath at 2,400 m. alt. (Nos. 6203, 6221).

In the Philippines previously known only from Luzon.

Java.

Isachne myosotis Nees in Hook. Kew Journ. **2** (1850) 98; Merr. l. c. 349.

In an open heath at 2,400 m. alt. (No. 6167); also collected by *Merritt*, No. 4405, at 1,500 m. alt. June, 1906.

Endemic to the Philippines.

ICHNANTHUS Beauv.

Ichnanthus pallens (Sw.) Munro in Benth. Fl. Hongk. (1861) 414; Merr. in Philip. Journ. Sci. **1** (1906) Suppl. 263.

In an old clearing and in forests at 900 m. alt. (No. 5538); on semishaded cliffs overhanging the Alag River at 380 m. alt. (No. 5498).

Previously known in the Philippines only from Luzon.

Tropics of both hemispheres.

LOPHATHERUM Brongn.

Lophatherum gracile Brongn. in Duperry Voy. Coqu. Bot. (1829) 50. *t. 8;* Merr. in Philip. Journ. Sci. **1** (1906) Suppl. 368.

In an old clearing at 700 m. alt. (No. 5543).

Previously known in the Philippines only from Luzon.

British India to southern China and Japan and Malaya.

BAMBUSA Schreb.

Bambusa sp. near *B. pygmaea* Miq.

In sphagnum in dense thickets at 2,600 m. alt. (No. 6222).

A remarkably small species, full-grown plants never exceeding 1.5 m. in height, and frequently less than 1 m. tall. I have the same form from similar habitat in northern Luzon (Pauai, *Merrill* (No. 4733), November, 1905). Unfortunately both numbers are without flowers or fruits.

CYPERACEÆ.

KYLLINGA Rottb.

Kyllinga intermedia R. Br. Prodr. (1810) 219; Clarke in Philip. Journ. Sci. Bot. 2 (1907) 78.

At 1,500 m. alt. (No. 4404 *Merritt*) June, 1906.

Previously known in the Philippines only from northern Luzon.

Formosa, the Philippines, north and east Australia and (?) the Fiji Islands.

CYPERUS Linn.

Cyperus diffusus Vahl. Enum. 2 (1806) 321; Clarke in Philip. Journ. Sci. Bot. 2 (1907) 83.

In an old clearing along the Alag River at 100 m. alt. (No. 5692).

Previously known in the Philippines from Luzon and Mindanao.

India to Malaya and New Guinea.

TORULINIUM Desv.

Torulinium confertum Desv. in Hamilt. Prodr. Ind. Occ. (1825) 15; Clarke in Philip. Journ. Sci. Bot. 2 (1907) 89.

With the preceding (No. 5693).

Widely distributed in the Philippines.

Cosmopolitan in the Tropics.

MAPANIA Aubl.

Mapania humilis (Hassk.) F.-Vill. Nov. App. (1883) 309; Clarke in Philip. Journ. Sci. Bot. 2 (1907) 109.

In forests along the Alag River at 200 m. alt. (No. 6130).

Previously known in the Philippines from Luzon, Mindoro and Mindanao.

Malayan peninsula and archipelago.

CLADIUM R. Br.

Cladium latifolium Merrill n. sp.

Planta robusta usque ad 1 m. alta, caespitosa, glabra, foliis rigidis, coriaceis, glabris, ecostatis, basi equitantibus, inferioribus sensim reductis, planis, lineari-lanceolatis, 2.5 ad 80 cm. longis, 1 ad 1.6 cm. latis, margine glabris, apice sensim acuminatis; paniculae oblongae, circiter 10 cm. longae, 4 cm. latae, densiflorae, spiculis numerosis, 3-floris, purpurascentibus, 6 ad 7 mm. longis.

Densely cæspitose, perennial, the culms about 1 m. high, glabrous. Leaves equitant at the base, the lower ones gradually reduced and the lowermost almost scale-like, all radical except one which is borne below the middle of the culm, 2.5 to 80 cm. long, 1 to 1.6 cm. wide, plane, glabrous throughout, rigid, coriaceous, ecostate, the margins smooth, the apex gradually acuminate or merely acute, the sheathing lower portions often purplish, otherwise green, the one culm-leaf about 4 cm. long. Culm erect, terete, striate, glabrous, 3 to 4 mm. in diameter. Panicle oblong, about 10 cm. long, 4 cm. in diameter, densely flowered, the branches in alternate fascicles, each fascicle subtended by a broad, inflated often purplish bract 1 to 2.5 cm. long, the apex contracted and produced as an oblong appendage about 1 cm. long. Spikelets purple, mostly sessile, rather crowded, usually 3-flowered, the flowers perfect, 6 to 7 mm. long; three lower glumes empty, ovate to oblong ovate, 2.5 to 4 mm. long, short acuminate, the three succeeding glumes oblong-ovate, 4 mm. long, short sharp acuminate, each with a perfect flower, the seventh glume usually empty. Nut immature, trigonous, with a trigonous glabrous beak, the style 3-cleft. Stamens 3; anthers lanceolate, 2.2 mm. long, mucronate-acuminate.

On an open heath at 2,400 m. alt. (No. 5562) (sterile). The type of the species is No. 2386 *Foxworthy* from Mount Banajao, Province of Tayabas, Luzon, March, 1907, alt. 2,250 m. The second species of the genus to be found in the Philippines.

Undoubtedly most closely related to the Hongkong *Cladium ensigerum* Hance, but differing from that species in its smooth leaf-margins and much smaller spikelets. An Australian type.

SCHOENUS Linn.

Schoenus melanostachys R. Br. Prodr. (1810) 231; Benth. Fl. Austr. **7** (1878) 370; Stapf in Trans. Linn. Soc. Bot. II. **4** (1894) 245.

In open heaths at 2.400 m. alt. (No. 6173).

North Borneo and Australia.

This species, new to the Philippines, is widely distributed in Australia, where the genus is largely developed, and has also been found on Mount Kinabalu in British North Borneo at an altitude of 1,700 m. It must therefore be considered both an Australian and a Bornean representative in the Philippine flora. *Schoenus apogon* R. & S., has been found in northern Luzon by *Loher* [10]

GAHNIA Forst.

Gahnia javanica Moritzi Verz. Zoll. Pfl. (1845–46) 98; Clarke in Philip. Journ. Sci. Bot. **2** (1907) 103.

In rather open thickets, exposed ridge at 2,250 m. alt. (No. 6162).

Previously known in the Philippines from Mount Banajao, Luzon, and Mount Apo, Mindanao.

Yunnan and Penang to New Guinea and the Viti Islands.

SCLERIA Berg.

Scleria chinensis Kunth Enum. **2** (1837) 357; Clarke in Philip. Journ. Sci. Bot. **2** (1907) 105.

At 1,700 m. alt. (No. 4439 *Merritt*) June, 1906.

Previously known in the Philippines from Luzon, Negros and Mindanao.

China to Singapore, Malaya and Queensland.

CAREX Linn.

Carex filicina Nees in Wight Contrib. (1834) 123; Clarke in Philip. Journ. Sci. Bot. **2** (1907) 107.

In exposed ridge-thickets at 2,250 m. alt. (No. 6,200); also collected by *Merritt* at 1,300 m. alt., in June, 1906 (No. 4384).

Previously known in the Philippines only from the mountains of northern and central Luzon.

India to China.

[10] Clarke, *This Journal*, Sec. C (1907), **2**, 102.

HYPOLYTRUM L. C. Rich.

Hypolytrum latifolium L. C. Rich. in Pers. Syn. **1** (1805) 70; Clarke in Philip. Journ. Sci. Bot. **2** (1907) 108.

In forests at 500 m. alt. (No. 4345 *Merritt*) June, 1906.

Widely distributed in the Philippines.

British India to Formosa, Malaya, Queensland and the Viti Islands.

PALMÆ.

PINANGA Blume.

Pinanga elmerii Becc. in Webbia (1905) 323.

In forests at 900 m. alt. (No. 5555).

Previously known only from the mountains of northern and central Luzon.

Pinanga sp.

In forests at 1,100 m. alt. (No. 5680); also collected by *Merritt* at an altitude of 1,500 m. in June, 1906 (No. 4468).

Pinanga maculata Porte, was observed in the forest at various places but all the specimens seen were without fruits or flowers.

Other than species of *Pinanga*, no other palms, except *Calamus*, were observed on the trip. *Calamus* is represented on Halcon by several species, but no specimens were found with fruit or flowers.

ARACEÆ.

ARISAEMA Linn.

Arisaema polyphylla (Blanco) Merr. in Govt. Lab. Publ. **27** (1905) 90.

In forests at 1,800 m. alt. (No. 6155).

Endemic in the Philippines.

SPATHIPHYLLUM Schott.

Spathiphyllum commutatum Schott. in Oest. Bot. Wochenbl. (1857) 158.

In forested ravines along a small stream at 700 m. alt. (No. 5486).

Luzon to Mindanao.

Celebes and Amboina.

CENTROLEPIDACEÆ.

CENTROLEPIS Labill.

Centrolepis philippinensis Merrill n. sp.

Perennis, dense caespitosa, multifoliata, foliis setaceis, ad 1 cm. longis, basi pilosis, pedunculo 1 ad 2 cm. longo, glumis 2, inaequalibus, lanceolatis, 4 ad 5 mm. longis, spiculis solitariis, 4-floris; stamen; ovarium 2-loculare.

Perennial, densely cæspitose forming close mats or tufts, the stems somewhat branched, the individual plants 3 to 4 cm. high, the usually hemispherical tufts frequently 10 cm. or more in diameter. Leaves very numerous, setaceous, glabrous above, about 1 cm. long, the lower portions pilose with weak white hairs. Peduncles 1 to 2 cm. long, glabrous, terminal, each bearing 1 spikelet about 5 mm. long. Spikelets 4-flowered, the glumes unequal, lanceolate, blunt, one 4 mm. long, one. 5 mm. long. Flowers 4, hermaphrodite, each with one hyaline oblong-obovate acute bract about 4 mm. long. Stamen 1; filament 3 mm. long; anther oblong, 1-celled, 1.5 mm. long. Ovary 2-celled, the ovules superposed; styles two, 2 to 2.5 mm. long.

In open heaths at 2,400 m. alt. (No. 6160).

Perhaps the most interesting discovery in recent botanical exploration of the Philippines, the family and genus being new to the Philippine flora, both being largely developed in Australia and New Zealand. Of the six genera at present recognized in the family, four, *Juncella*, *Brizula*, *Aphelia*, and *Alepyrum* are confined to Australia, New Zealand and Tasmania, a fifth genus of two species, *Gaimardia*, is represented in New Zealand, Cape Horn and the Falkland Islands, while the sixth genus, *Centrolepis*, is represented by about twenty species, of which one is found in Cambodia, and all the others, except the one described here, in south Australia and Tasmania. Following *Hieronymus'* treatment of the family in *Engler und Prantl's* Natürlichen Pflanzenfamilien, the present species would perhaps fall in the genus *Alepyrum*, but as *Alepyrum* Hiern., is invalidated by *Alepyrum* R. Br., it has been thought best to describe the present species under *Centrolepis*.

ERIOCAULONACEÆ.

ERIOCAULON Linn.

Eriocaulon brevipedunculatum Merrill, n. sp.

Planta densissime caespitosa, 5 ad 8 cm. alta, caulibus brevissimis, simplicibus, foliis congestis, anguste lanceolatis, acuminatis, 2 ad 2.5 cm. longis, 3 ad 4 mm. latis, glabris, pedunculis perbrevibus 0.5 ad 1 cm. longis; capitula semiglobosa, 5 ad 6 mm. lata, bracteis involucrantibus latiusculis, membranaceis, obtusis, ad 3 mm. longis; flores normaliter evoluti; perigonia 3-mera.

Densely cæspitose, 5 to 8 cm. tall, forming dense tufts. Stems short, simple; leaves very numerous, densely disposed throughout the entire length of the stem, the lower ones marcescent, narrowly lanceolate, acuminate, glabrous, 2 to 2.5 cm. long, 3 to 4 mm. wide, spreading, shining, flaccid. Peduncles solitary, the peduncles 0.5 to 1 cm. long. longis; capitula semiglobosa, 5 ad 6 mm. lata, bracteis involucrantibus mambranaceous, obtuse, about 3 mm. long, obovate. Staminate flowers 3-merous; sepals free, spatulate, slightly ciliate at the apex, about 2.5 mm. long; petals narrowly ovate, acuminate, 1 mm. long, the gland prominent; stamens 6, the filaments about 1 mm. long. Pistillate flowers 3-merous; sepals free, narrowly oblong or spatulate, obtuse, slightly ciliate above, about 2.5 mm. long; petals equaling the sepals, somewhat narrower; style about 1.5 mm. long, the three style-arms nearly 2 mm. long.

In an open heath at 2,400 m. alt. (No. 6214).

A species well characterized by its densely cæspitose habit and short peduncles, the heads being solitary and included in the densely disposed leaves, not exserted.

COMMELINACEÆ.

CYANOTIS Don.

Cyanotis moluccana (Roxb.) *Commelina moluccana* Roxb. Hort. Beng. (1814) 81; Fl. Ind. **1** (1820) 172, (ed. Wall. **1**: 176.) *Cyanotis uniflora* Hassk. Commel. Ind. (1870) 104; Clarke in DC. Monog Phan. **3** (1881) 242; Merr. in Philip. Journ. Sci. **1** (1906) Suppl. 34.

In open wet places near Subaan (No. 6227).

Widely distributed in the Philippines.

Malayan Archipelago.

FORRESTIA Lesson.

Forrestia philippinensis Merr. in Govt. Lab. Publ. **35** (1906) 5.

In forests along small streams at 300 m. alt. (No. 6152).

Endemic to the Philippines, the type from the Baco River near the base of Mount Halcon.

LILIACEÆ.

DIANELLA Lam.

Dianella ensifolia (L.) DC. in Red. Lil. (1802) *t. 1.*

In open heaths at 2,400 m. alt. (No. 5504).

On the higher mountains in the Philippines from northern Luzon to southern Mindanao.

British India to southern China, Malaya, Australia, Polynesia, and the Hawaiian Islands.

Dianella caerulea Sims. Bot. Mag. *t. 505.*

In forests (4469 *Merritt*) June, 1906.

Not previously reported from the Philippines, but apparently represented by the following specimens: LUZON, Province of Benguet (6030 *Elmer*); (4441, 4683 *Merrill*); Province of Bataan, Mount Mariveles (226 *Whitford*); Province of Pampanga, Mount Arayat (76 *Bolster*); Province of Laguna, Mount Maquiling (5125 *Merrill*).

New Guinea and Australia.

LIRIOPE Lour.

Liriope brachyphylla Merrill n. sp.

Glabra, usque ad 20 cm. alta, foliis confertis, membranaceis, glabris, anguste oblongis, 2.5 ad 3.5 cm. longis; scapi 1 ad 3, simplices; flores rosei, racemosi, ad bracteas solitarii; ovarium superum, 3-loculare, loculis pluri-ovulatis (ad 15).

A glabrous perennial herb. Leaves membranous, narrowly oblong, 2.5 to 3.5 cm. long, 3 to 6 mm. wide, somewhat narrowed and hyaline-sheathing below, the apex acute, all crowded at the base of the scape. Scapes 1 to 3 from each rosette of leaves, 20 cm. high or less, leafless, simple, glabrous, 10 to 15-flowered, the flower bearing portion 4 to 5 cm. long. Bracteoles linear-lanceolate, 3 to 3.5 mm. long, the flowers solitary in the axil of each bract, the pedicels about 3 mm. long. Perianth 5 to 6 mm. long, pale pink or salmon colored, the lobes 6, equal, lanceolate, acute or blunt, free or very slightly united at the very base, 3-nerved, about 1.5 mm. wide. Stamens 6; filaments 2 mm. long;

anthers 1 mm. long. Ovary free, superior, ovoid, 3-celled, each cell about 15-ovuled; style simple. Capsule 3-valved, ovoid, membranous, glàbrous, 4 mm. long. Seeds many, narrowly ovoid, 0.8 mm. long, reticulate.

On seepy slopes, open heath lands at 2,400 m. alt. (No. 5710).

The second species of the genus known, the other, *Liriope graminifolia* (Linn.) Baker, being known from Japan to northern Luzon, China and Cochin China. *L. brachyphylla* is distinguished from *L. graminifolia* by its comparatively short leaves, smaller size, solitary, not fascicled flowers and many ovules.

SMILAX Tourn.

Smilax china Linn. Sp. Pl. (1753) 1459.?

On exposed ridges in thickets 1,300 to 2,600 m. alt. (Nos. 6140, 6211, 6126).

The same form is represented by Nos. 4497, 4749 *Merrill* from northern Luzon. *Smilax china* has previously been credited to the Philippines by *F.-Villar*. Because of the lack of pistillate flowers I am not certain of the correctness of the above identification, but the speciments agree well with the description.

Japan to southern China and Formosa.

Smilax vicaria Kunth Enum. 5 (1850) 262.

In an old clearing at 900 m. alt. (No. 5579).

Endemic to the Philippines.

AMARYLLIDACEÆ.

CURCULIGO Gaertn.

Curculigo glabra Merrill n. sp.

Glabra; foliis usque ad 50 cm. longis, 13 cm. latis, breviter acuminatis, basi acutis; petiolo usque ad 40 cm. longo; pedunculo ad 20 cm. longo; baccis 8 mm. longis.

Glabrous throughout. Leaves up to 50 cm. long, 13 cm. wide, membranous, the apex short acuminate, the base acute, somewhat inequilateral, the nerves prominent; petioles up to 40 cm. in length, glabrous, somewhat inflated below. Peduncles about 20 cm. long, recurved above. Flowers not seen; bracts ovate, acuminate, 7 to 10 nerved; pedicels 1 cm. long. Fruits many, narrowly ovoid, not beaked, about 8 mm. long, fleshy and smooth when fresh, rugose when dry, somewhat crowded in a 6 to 8 cm. long head; seeds many, 1.5 mm. in diameter; the somewhat persistent perianth lobes 6 mm. long.

In forests at 275 m. alt. (No. 5750).

Well characterized by being entirely glabrous throughout.

DIOSCOREACEÆ.

DIOSCOREA Linn.

Dioscorea nummularia Lam. Encycl. 2 (1789) 331.

In an old clearing at 700 m. alt. (No. 5657).

Apparently widely distributed in the Philippines.

Malaya.

STENOMERIS Planch.

Stenomeris dioscoreæfolia Planch, in Ann. Sc. Nat. III, **18** (1852) 320.

In an old clearing at 700 m. alt. (No. 5775).

Endemic to the Philippines.

IRIDACEÆ.

PATERSONIA R. Br.

Patersonia lowii Stapf. in Trans. Linn. Soc. Bot. II. **4** (1894) 241. *pl. 20. f. 7–9.*

Open heaths at 2,400 m. alt. (No. 5507). Common, but very few specimens in flower in November.

Borneo.

The above is the first indigenous representative of this family to be found in the Philippines, several genera of the *Iridaceæ* being credited to the Philippines by *F.-Villar* and other authors, but all based on introduced and cultivated species. *Patersonia lowii* has previously been known only from Mount Kinabalu, North Borneo, its occurrence in the Philippines being an addition to the comparatively small Bornean element already known in the Archipelago.

ZINGIBERACEÆ.

ALPINIA Linn.

Alpinia brevilabris Presl Rel. Hænk. **1** (1830) 110. *t. 17.*

In forests at 2,000 m. alt. (No. 4458 *Merritt*) June, 1906.

Widely distributed in the Philippines. Endemic.

Alpinia sp. near *A. parviflora* Rolfe.

In forests at 900 m. alt. (No. 4361 *Merritt*) June, 1906.

Material too imperfect and scanty for accurate identification.

MARANTACEÆ.

PHACELOPHRYNUM K. Sch.

Phacelophrynum bracteosum (Warb.) K. Sch. in Engl. Pflanzenreich. **11** (1902) 123.

In forests along the Alag River at 100 m. alt. (No. 6132).

Widely distributed in the Philippines, endemic.

BURMANNIACEÆ.

BURMANNIA Linn.

Burmannia longifolia Becc. Malesia **1** (1878) 244; Rendle in Journ. Bot. **34** (1896) 355.

Terrestrial in the very mossy ridge forests 1,300 to 1,900 m. alt. (No. 5741).

Previously collected on "Dulangao," or more correctly "Dulangan," a spur of Mount Halcon, by *Whitehead*, and reported by *Rendle*, l. c.

Malayan Peninsula and Borneo.

Burmannia sp. near *B. clementis* Schltr.

In dense forests at 900 m. alt. (No. 5598). The third species of the genus to be found in the Philippines.

CASUARINACEÆ.

CASUARINA Linn.

Casuarina sp.

Gregarious over an area of about 100 acres at the head of a rivine, 700 m. alt. (No. 5779).

PIPERACEÆ.

PEPEROMIA Ruiz & Pav.

Peperomia recurvata Miq. Syst. Pip. (1843–44) 107.

On mossy trees and terrestrial, 1,200 to 1,800 m. alt. (Nos. 6107, 6147, 6184). Malaya.

PIPER Linn.

Piper rhombophyllum C. DC. Prodr. **16**[1] (1869) 352.

In forests 350 to 1,500 m. alt. (Nos. 5645, 5773); also collected by *Merritt* at 1,300 m. alt. (No. 4293).

Endemic to the Philippines.

Piper sp.

In forests at 1,800 m. alt. (No. 5592).

Piper sp.

In forests, altitude not given (*Merritt* No. 4474) June, 1906.

CHLORANTHACEÆ.

CHLORANTHUS Swartz.

Chloranthus brachystachys Blume Fl. Jav. Chloranth. (1828) 13. *t.* 2.

In forests at 700 m. alt. (No. 5644); also collected by *Merritt* at 1,200 and 1,650 m. alt. (Nos. 4380, 4416).

Widely distributed in the Philippines.

British India to southern China and Malaya.

Chloranthus officinalis Blume Enum. Pl. Jav. (1830) 79.

In forests at 900 m. alt. (No. 5574).

Widely distributed in the Philippines.

Distribution of the preceding species.

MYRICACEÆ.

MYRICA Linn.

Myrica esculenta Buch.-Ham. in G. Don. Fl. Nepal. (1825) 56; Chevalier Monog. Myric. (1901) 120. var. *farquahariana* (Wall.) Chev. l. c. *Myrica rubra* Merr. in Philip. Journ. Sci. **1** (1906) Suppl. 41, non S. & Z.

In forests at 1,700 m. alt. (No. 4433 *Merritt*) June, 1906.

An exceedingly variable form, identical with material previously reported from Mount Mariveles as *Myrica rubra*, the monograph by *Chevalier* not being available at the time the identification was made. For variation in the leaf-form of this species, see *Whitford*, Vegetation of the Lamao Forest Reserve II, Philip. Journ. Sci. **1** (1906) pl. *44*. *f. A. 1–3*.

British India to Malaya., the var. *farquahariana* in the Malayan Peninsula.

Myrica javanica Blume Bijdr. (1826) 517; Fl. Jav. Myric. 7. *t. 1;* Chev. Monog. Myric. (1901) 129.

In dense ridge-thickets at 2,600 m. alt. (No. 5708).

Previously known in the Philippines from Mount Apo, Mindanao.

Java.

JUGLANDACEÆ.

ENGLEHARDTIA Leschen.

Englehardtia spicata Blume Bijdr. (1826) 528.

In forests at 1,350 m. alt. (No. 5760).

Not common in the Philippines.

British India to Cochin China and Java.

FAGACEÆ.

QUERCUS Linn.

Quercus llanosii A. DC. Prodr. 16[2] (1864) 97.

In forests at 700 m. alt. (No. 5695).

Widely distributed in the Philippines.

Endemic.

ULMACEÆ.

GIRONNIERA Gaudich.

Gironniera celtidifolia Gaudich. Voy. Bonite Bot. (1844–66) *t. 85.*

In forests below 200 m. alt. (Nos. 4325, 4320 *Merritt*) June, 1906.

Not common in the Philippines.

Endemic.

MORACEÆ.

FICUS Linn.

Ficus hauili Blanco Fl. Filip. (1837) 684.?

In old clearings at 700 m. and in forests at 1,800 m. (Nos. 5748, 6127).

Common and widely distributed in the Philippines.

Endemic?

Ficus rubrovenia Merr. in Philip. Journ. Sci. 1 (1906) Suppl. 44.

In forests at 200 m. alt. (No. 4326 *Merritt*) June, 1906.

Previously known only from Luzon.

In addition to the above species, *Ficus mindoroensis* Merr., was abundant in forests below 200 m. alt., and *F. minahassae* Miq., was abundant along streams up to an altitude of 1,000 m.

ARTOCARPUS Forst.

Artocarpus sp.

In forests at 750 m. alt. (No. 5557).

Material fragmentary, from fallen branches of a large tree, quite different from any of the other species represented in our herbarium.

LORANTHACEÆ.

LORANTHUS Linn.

Loranthus halconensis Merrill n. sp. § *Dendropthoë.*

Foliis oblongo-lanceolatis coriaceis verticillatis usque ad 15 cm. longis; floribus ad 2.5 cm. longis, 6-meris, puberulis, dense fasciculatis; fasciculis axillaribus, pedunculis ad 1 mm. longis, 3-floris.

Scandent, glabrous except the puberulent inflorescence, the branches stout, terete, light gray. Leaves in whorls of 6, oblong lanceolate, coriaceous, glabrous, brown when dry, 10 to 15 cm. long, 3.5 to 5 cm. wide, acute or obtuse, the base acute, the midrib stout, the lateral nerves 5 to 6 on each side of the midrib, obscure, the reticulations obsolete; petioles 1 cm. long or less. Inflorescence on the branches below the leaves in the axils of fallen leaves, fasciculate, puberulent, 5 to 6 or more peduncles in a fascicle, each about 1 mm. long and bearing three flowers, the pedicels about 1 mm. long. Flowers yellow, about 2.5 cm. long, not inflated. Calyx densely puberulent, cylindrical, truncate, 3 mm. long, the basal bract broadly ovate, small. Corolla 6-merous, the lobes united for the lower 1.5 to 3 mm., puberulent outside, linear, 1.5 mm. wide, the reflexed portion above the insertion of the stamens lanceolate, acute, 6 mm. long. Filaments 1.5 mm. long; anthers 3 mm. long.

Parasitic on *Ficus minahassae* along the Alag River at 100 m. alt. (No. 5664). Well characterized by its whorled leaves, densely fascicled flowers from the larger branches and 6-merous corolla.

Loranthus mearnsii Merrill n. sp. § *Dendropthoë.*

Foliis oppositis vel subalternis, elliptico-ovatis, obtusis, basi acutis, 5 ad 8 cm. longis, glaberrimis; racemis solitariis, axillaribus, ferrugineo-puberulis, 3–5-floris; floribus usque ad 2.7 cm. longis; calyce 4-dentato; corolla irregulariter 4-lobata, tubo gibbo angulato.

Glabrous except the inflorescence. Branches brownish gray, terete, glabrous, lenticellate. Leaves coriaceous, glabrous, opposite or subalternate, elliptical-ovate, the apex obtuse, sometimes rounded, the base acute, 5 to 8 cm. long, 3 to 5 cm. wide; lateral nerves 3 to 4 on each side of the midrib, ascending; petioles 1 cm. long. or less. Racemes few, solitary, from the leaf-axils or from axils of fallen leaves, few-flowered, the rachis, pedicels and calyces ferruginous puberulent; pedicels 2 mm. long, the basal bract of the calyx ovate, acute or acuminate about as long as the calyx tube. Calyx tube cylindrical, 2 mm. long, the limb 1.5 mm. long, spreading, 4-toothed. Corolla 2.5 cm. long, green except the tips of the lobes which are red, the tube somewhat inflated, 4-angled, about 10 mm. long, the lobes 4, irregular, the reflexed portion above the insertion of the anthers linear, 11 mm. long. Filaments 5 mm. long; anthers basifixed, 4 mm. long.

Parasitic on various trees in forests at 1,800 m. alt. (No. 5733).

Loranthus sp.

On trees in forests at 1,700 m. alt. (4434 *Merritt*) June, 1906. Apparently an undescribed species, but with immature fruits only.

BALANOPHORACEÆ.

BALANOPHORA Forst.

Balanophora sp.

On roots of trees in forests at 1,800 m. alt. (No. 6156). Material too scanty and imperfect for specific identification.

POLYGONACEÆ.

POLYGONUM Linn.

Polygonum chinense Linn. Sp. Pl. (1753) 363.

Abundant on recent "slides" at 1,900 m. alt. (No. 5780).

Throughout the Philippines at higher altitudes.

India to Japan and Malaya.

MENISPERMACEÆ.

STEPHANIA Lour.

Stephania hernandifolia (Willd.) Walp. Repert. 1 (1842) 96.

In forests at 1,800 m. alt. (No. 5701).

Widely distributed in the Philippines.

Tropical Africa, Asia, through Malaya to Australia.

MAGNOLIACEÆ.

DRIMYS Forst.

Drimys piperita Hook. f. Icon. Pl. *t. 896.*

On forested ridges at 1,300 m. alt. (No. 6134), and in thickets on exposed ridges at 2,600 m. alt. (No. 6206); also collected by *Merritt* at 1,300 and 1,600 m. alt. (Nos. 4383, 4407).

On most of the higher mountains of the Philippines.

Borneo, New Guinea and New Caledonia.

ILLICIUM Linn.

Illicium sp.

In forests at 1,600 m. alt. (No. 4411 *Merritt*) June, 1906.

Material very scanty and with fruit only. Perhaps most closely related to *Illicium evenium* King from Perak, although clearly distinct from that species, *ex descriptione.* The genus is new to the Philippine flora and its occurrence in the Philippines must be considered as evidence of previous connection with the Asiatic continent, the genus being represented in North America, India, China and Japan, and with three species extending southward to the Malayan Peninsula.

ANONACEÆ.

UNONA Linn. f.

Unona mindorensis Merrill n. sp. § *Stenopetalum*.

Foliis oblongo-lanceolatis acuminatis 10 ad 17 cm. longis, basi acutis; pedunculis axillaribus solitariis ad 1.2 cm. longis; petalis 6, subaequalibus, ovato-lanceolatis, acutis, 1.5 ad 1.7 cm. longis, pubescentibus; ovulis 3 uniserialibus; carpellis maturis ovatis, acutis, ad 1.5 cm. longis.

A small tree about 6 m. high; branches slender brownish gray, slightly pubescent, terete. Leaves submembranous, oblong lanceolate, 10 to 17 cm. long, 2.5 to 4.5 cm. wide, the base acute, the apex gradually acuminate, dull above, somewhat paler and slightly shining beneath, glabrous or nearly so; nerves about 9 on each side of the midrib, ascending, distinct beneath, the reticulations few, indistinct; petioles stout, rugose, 5 mm. long or less. Peduncles axillary, solitary, about 1.2 cm. long, slightly pubescent and with one or two small basal bracts. Flowers greenish-white. Sepals 3, pubescent, triangular ovate, acute, about 3.5 mm. long. Petals 6, free, spreading in flower, ovate-lanceolate, acute, pubescent, the outer three about 1.7 cm. long, 7 mm. wide, the inner three 1.5 cm. long, 5 mm. wide. Stamens many, glabrous, 2 mm. long, the connectives truncate, overlapping. Ovaries about 12, densely hirsute, oblong, 3 mm. long; ovules 3, parietal in one row; styles 1.5 mm. long. Carpels ovoid, about 1.5 cm. long, acute, narrowed below into a short stout stipe, brown when dry, slightly pubescent.

In forests at 300 m. alt. (No. 5568). A closely related species is represented by No. 4060 *Merrill*, from the Baco River, near the base of Halcon, March, 1905.

OXYMITRA Blume.

Oxymitra sp. near *O. glauca* Hk. f. et Th.

In forests at 180 m. alt. (No. 5629). Specimens with fruit only, but undoubtedly referable to this genus.

PHAEANTHUS Hook. f. et Th.

Phaeanthus cumingii Miq. Fl. Ind. Bat. 1[2] (1859) 51.

In forests at 700 m. alt. (No. 5648).

Widely distributed in the Philippines. Endemic.

Phaeanthus acuminatus Merr. in Govt. Lab. Publ. **35** (1906) 11.

In forests at 150 m. alt. (No. 4321 *Merritt*) June, 1906.

Known only from Mindoro and Palawan.

GONIOTHALAMUS Blume.

Goniothalamus elmeri Merr. in Govt. Lab. Publ. **29** (1905) 13.

In forests at 900 m. alt. (No. 4354 *Merritt*) June, 1906.

Widely distributed in the Philippines. Endemic.

MYRISTICACEÆ.

HORSFIELDIA Willd.

Horsfieldia merrillii Warb. in Perk. Frag. Fl. Philip. (1904) 49.

In forests at 100 m. alt. (No. 5772).

Known only from Mindoro.

MONEMIACEÆ.

KIBARA Endl.

Kibara ellipsoidea Merr. in Philip. Journ. Sci. **1** (1906) Suppl. 56.

In forests at 450 m. alt. (No. 4313 *Merritt*) June, 1906.

Described from material collected on Mount Mariveles, Luzon. Nearly or quite the same species is represented by material collected near Lake Lanao, Mindanao, by Mrs. *Clemens*. Endemic to the Philippines. Mangyan, "*Barao-barao*."

LAURACEÆ.

NEOLITSEA (Benth.) Merr.

Neolitsea zeylanica (Nees) Merr. in Philip. Journ. Sci. **1** (1906) Suppl. 56.

In forests at 1,800 m. alt. (No. 5666); also collected by *Merritt* in June, 1906, in forests at 950 m. alt. (No. 4369).

CRYPTOCARYA R. Br.

Cryptocarya acuminata Merr. in Philip. Journ. Sci. **1** (1906) Suppl. 192.

In forests at 450 m. alt. (No. 4342 *Merritt*) June, 1906.

Known only from Mindoro.

NEPENTHACEÆ.

NEPENTHES Linn.[11]

Nepenthes sp.

On bowlders in the river bed (Alag River) at 350 m. alt. (No. 5790).

Nepenthes sp.

In thickets along the Binabay River at 200 m. alt. (No. 5785).

Nepenthes sp.

In thickets on exposed ridges 1,500 to 2,600 m. alt. (No. 5774).

DROSERACEÆ.

DROSERA Linn.

Drosera spathulata Labill. Nov. Holl. Pl. Spec. **1** (1804) 79. *t. 106. f. 1;* Diels in Engler's Pflanzenreich **26** (1906) 83. *f. 31. A, B.*

In open heaths at 2,400 m. alt. (No. 5784), locally abundant.

Previously collected in the Philippines by *Cuming*, locality not given, probably Luzon.

Southern Japan and China, Borneo, Australia and New Zealand.

[11] As most of our Philippine *Nepenthes* material, including one number from Halcon, collected by *Merritt*, is in the hands of *Dr. Macfarlane*, who is monographing the family, no attempt has been made to identify specifically the specimens here cited.

Drosera peltata Smith in Willd. Sp. Pl. **1** (1797) 1546; Diels l. c. 111.

With the preceding locally rare (No. 6207).

Previously known in the Philippines only from northern Luzon.

British India to central China, and Japan, through Malaya to Australia and Tasmania.

SAXAFRAGACEÆ.

HYDRANGEA Linn.

Hydrangea lobbii Maxim. Mem. Acad. Petersb. VII. **10** (1867) 15.

In an old clearing at 750 m. alt. (No. 5491); also in forests at 1,800 m. alt. (No. 5731), the latter referred here with some doubt.

Previously known only from the mountains of northern and central Luzon and from Panay.

Endemic in the Philippines.

PITTOSPORACEÆ.

PITTOSPORUM Banks.

Pittosporum resiniferum Hemsl. in Kew Bull. (1904) 344.

In forests at 150 m. alt. (No. 5609); found also by *Merritt* in June, 1906, at 1,700 m. alt. (Nos. 4421, 4436).

Previously known only from the mountains of central and northern Luzon; reported with doubt from Celebes by *Koorders*.

The habit of this species is very interesting, it being pseudoepiphytic, quite similar to most species of *Ficus* of the section *Urostigma*.

Pittosporum odoratum Merr. in Govt. Lab. Publ. **35** (1906) 16.

In forests at 700 m. alt. (No. 5654).

Previously known only from the mountains of central and northern Luzon.

CUNONIACEÆ.

WEINMANNIA Linn.

Weinmannia hutchinsonii Merrill n. sp.

Arbor ad 10 m. alta; ramulis fuscis, teretis, lenticellatis, glabris, junioribus pubescentibus; foliis imparipinnatis, 4 ad 5-jugatis; foliolis subsessilibus, lanceolatis, acuminatis, 3 ad 7 cm. longis, 0.7 ad 1.5 cm. latis, grosse crenato-serratis; racemis numerosis, fasciculatis terminalibus vel axillaribus, pubescentibus; floribus pedicellatis, 4-meris.

A tree about 10 m. high. Branches terete, glabrous, brown, lenticellate, somewhat compressed below the nodes, rather slender, the younger parts somewhat pubescent. Leaves opposite, unequally pinnate, about 13 cm. long, the common rachis 5 to 7 cm. long, pubescent; leaflets 4 to 5 pairs, lanceolate, coriaceous, glabrous, 3 to 7 cm. long, 0.7 to 1.5 cm. wide, the apex long, bluntly acuminate, the base inequilateral, acute or acuminate, the margins coarsely crenate-serrate; nerves about 12 on each side of the midrib, slender, the reticulations numerous; petiolules wanting. Inflorescence terminal and axiliary of many racemes, 10 cm. long or less, pubescent. Flowers white, 4-merous, pedicellate, the pedicels 1.5 mm. long, pubescent. Sepals 1 mm. long, acute, pubescent. Petals elliptical, rounded, 1.5 mm. long, very slightly pubescent. Filaments 4 mm. long, glabrous; anthers 0.5 mm. long. Ovary 2-celled, pilose.

In forests at 700 m. alt. (No. 5753).

The second species of the genus to be found in the Philippines, a third, apparently undescribed one, closely allied to the above is represented in our herbarium from Lake Lanao, Mindanao, coll. *Clemens*.

ROSACEÆ.

RUBUS Linn.

Rubus rolfei Vid. Phan. Cuming. Philip. (1885) 171.

In open heaths and thickets at 2,400 to 2,550 m. alt. (No. 5715).

Previously known only from the mountains of Luzon and from Mount Canlaon, Negros.

Rubus moluccanus Linn. Sp. Pl. (1753) 1197.

In forests at 1,800 m. alt. (No. 5595); also collected by *Merritt*, in forests at 900 m. alt. (No. 4362).

Throughout the Philippines at higher altitudes.

British India to southern China and Malaya.

Rubus fraxinifolius Poir., is abundant below 1,500 m. alt.

PHOTINIA Lindl.

Photinia luzonensis Merr. in Govt. Lab. Publ. 17 (1904) 18.

In thickets bordering an open heath at 2,400 m. alt. (No. 6205).

Previously known only from Mount Mariveles, Luzon.

LEGUMINOSÆ.

DESMODIUM DC.

Desmodium ormocarpoides DC. Prodr. 2 (1825) 327.

In open slough along the Alag River at 100 m. alt. (No. 6223).

Previously known in the Philippines from Luzon.

British India to Java.

Desmodium capitatum (Burm.) DC. l. c. 336.

Near Subaan in open grass lands, 10 m. or less above sea level (No. 6224).

Widely distributed in the Philippines.

British India to Malaya.

PITHECOLOBIUM Mart.

Pithecolobium prainianum Merr. in Philip. Journ. Sci. 1 (1906) Suppl. 61.

In forests at 300 m. alt. (No. 5702).

Previously known only from the mountains of Luzon.

RUTACEÆ.

EVODIA Forst.

Evodia reticulata Merrill n. sp.

Frutex 2 ad 3 metralis, foliis trifoliatis, rariter unifoliatis, oppositis vel suboppositis, foliolis 5 ad 9 cm. longis, coriaceis, nitidis, supra glabris, subtus ad nervos fulvo-pubescentibus, dense reticulatis, obovatis vel elliptico-ovatis, apice obtusis truncatis vel retusis, paniculis axillaribus, ad 7 cm. longis, dense fulvo-pubescentibus; flores 4-meri, 3 mm. longi, in ramulis ultimi ordinis congesti.

A shrub 1 to 3 m. high. Branches stout, the older ones nearly glabrous, light gray, the younger ones densely fulvous-pubescent. Leaves opposite or subopposite, trifoliate, rarely unifoliate, the petioles 1.5 to 2 cm. long, at first densely pubescent, becoming glabrous and rugose; leaflets coriaceous, shining, densely closely reticulate, glabrous above, fulvous-pubescent on the nerves and midrib beneath, obovate to elliptical ovate, the base subacute or obtuse, that of the lateral ones somewhat inequilateral, the apex obtuse, truncate or somewhat retuse, 5 to 9 cm. long, 3 to 5.5 cm. wide; nerves prominent, 9 to 11 on each side of the midrib; petiolule of the middle leaflet about 1 cm. long, of the lateral ones 2 to 3 mm. long, pubescent. Panicles in the upper axils, densely fulvous-pubescent, about 7 cm. long, many flowered; pedicels 1 to 2 mm. long, the bracts and bracteoles minute. Flowers white, crowded at the apices of the ultimate branches, 4-merous, 3 mm. long. Sepals broadly ovate, acute very slightly pubescent, less than 1 mm. long. Petals 4, elliptical ovate, acute, 2.5 mm. long, glabrous. Ovary glabrous, 4-lobed; style about 1.5 mm. long; stigma capitate.

In open heaths and in thickets at 2,400 m. alt. (No. 5711).

A species recognizable by its coriaceous leaves which are densely rather prominently reticulate and pubescent on the nerves and midrib beneath.

POLYGALACEÆ.

POLYGALA Linn.

Polygala venenosa Juss. ex Poir, in Lam. Encycl. 5 (1804) 493.

In forests at about 1,200 m. alt. (No. 6166). Found here also by *Merritt* at 900 m. alt. (No. 4351) June, 1906.

Known in the Philippines from Negros and from Mount Apo, Mindanao. Java, Sumatra and from near Mount Kinabalu, North Borneo.

EUPHORBIACEÆ.

ANTIDESMA Linn.

Antidesma leptocladum Tul. Ann. Sc. Nat. III. 15 (1851) 199.

In forests at 1,800 m. alt. (No. 5717).

Widely distributed in the Philippines. Endemic?

DAPHNIPHYLLUM Blume.

Daphniphyllum glaucescens Blume Bijdr. (1826) 1153.

In forests at 700 m. alt. (No. 5658).

The same form has been collected in Benguet Province, Luzon, by *Elmer* (No. 6290). The genus is new to the Philippines.

British India to southern China, Corea, etc., south to Java.

CLAOXYLON Juss.

Claoxylon sp.

In forests along the Alag River at 160 m. alt. (No. 6519). The same form collected by *Merritt*, 250 to 900 m. alt. (Nos. 4329, 4364, 4332), material in poor condition.

Claoxylon sp.

In forests at 1,800 m. alt. (No. 5668), fruit and pistillate flowers only.

MACARANGA Thouars.

Macaranga hispida (Bl.) Muell. Arg. in DC. Prodr. **15**[2] (1862) 990.

In forests at 700 m. alt. (No. 5646).

Widely distributed in the Philippines.

Moluccas.

HOMALANTHUS Juss.

Homalanthus populneus (Geisel.) Pax in Engl. and Prantl Nat. Pflanzenfam. **3**[5] (1890) 96.

In forests at 2,200 m. alt. (No. 4452 *Merritt*) June, 1906.

Widely distributed in the Philippines.

Ceylon to Java.

Homalanthus populneus (Geisel.) Pax in Engl. and Prantl Nat. Pflanzen-

In an old clearing at 900 m. alt. (No. 5593).

Widely distributed in the Philippines. Endemic.

CELASTRACEÆ.

EVONYMUS Linn.

Evonymus javanicus Blume Bijdr. (1826) 1146.

In forests at 800 m. alt. (No. 5659).

Widely distributed in the Philippines.

Malaya.

ICACINACEÆ.

STEMONURUS Blume.

Stemonurus sp.

In forests at 250 m. alt. (No. 4327 *Merritt*) June, 1906. A single, rather imperfect specimen, apparently representing an undescribed species.

AQUIFOLIACEÆ.

ILEX Linn.

Ilex fletcheri Merrill n. sp.

Ramis teretis glabris griseis, ramulis angulatis minute puberulis; foliis oblongo-ellipticis vel lanceolato-ellipticis, acuminatis, basi acutis, integris, 1.5 ad 3.5 cm. longis, nitidis, coriaceis, venis obscuris; cymis axillaribus, brevibus, paucifloribus, corollae tubo breve, segmentis oblongis, circiter 2 mm. longis, 1 mm. latis, obtusis.

A shrub 3 m. high or less, glabrous throughout, except the slightly puberulent branchlets and infloresence. Branches terete, gray, glabrous, branchlets angular, slender, dark reddish brown, minutely puberulent. Leaves oblong-elliptical to lanceolate-elliptical, 1.5 to 3.5 cm. long, 0.5 to 1.8 cm. wide, coriaceous, glabrous, shining, paler beneath, the base acute, the apex broadly acuminate or acute and minutely apiculate; nerves very obscure, nearly obsolete, the margins entire; petioles 1 to 3 mm. long. Cymes axillary, few-flowered, 5 mm. long, or less, or the inflorescence reduced to a few-flowered fascicle, puberulent the pedicels about 2 mm. long. Calyx about 1 mm. long, the teeth not prominent. Corolla tube very short, the lobes 4, oblong, about 2 mm. long, 1 mm. wide, obtuse. Filaments 1 mm. long; anthers ovoid, 0.5 mm. long. Fruit red, globose, glabrous, smooth, 2.5 mm. in diameter.

In the mossy forest on exposed ridges at 1,800 m. alt. (Nos. 5716, 5755); also collected by *Merritt* at 2,200 m. alt., in June, 1906 (Nos. 4475, 4448).

A species characterized by its small entire leaves, puberulent branchlets and inflorescence and few-flowered cymes or fascicles, the flowers on the younger branches rarely solitary. Named in honor of Mr. *Horace L. Fletcher*, who accompanied the members of the Halcon expedition to Calapan.

ACERACEÆ.

ACER Linn.

Acer philippinum Merr. in Govt. Lab. Publ. **35** (1906) 36.

This species was abundant in forests at 600 to 1,700 m. alt., but was not collected, all the specimens observed being with leaves only. The species was described from specimens collected on Mount Mariveles, Luzon, and has since been collected on Mount Data, Luzon, by the author.

Endemic in the Philippines.

BALSAMINACEÆ.

IMPATIENS Linn.

Impatiens sp.

In forests at 1.300 m. alt. (No. 4391 *Merritt*) June, 1906. Material very imperfect, but different from any other species of the genus known to me.

RHAMNACEÆ.

ALPHITONIA Reiss.

Alphitonia excelsa Reiss. ex Endl. Gen. (1836–50) 1098. *A. moluccana* T. & B.

In forests at 700 m. alt. (No. 5532). The fresh bark has a strong odor of oil of wintergreen.

Widely distributed in the Philippines.

Borneo to Polynesia and Australia.

VITACEÆ.

LEEA Linn.

Leea aculeata Blume Bijdr. (1825) 197.

In forests along the Alag River at 100 m. alt. (No. 5605).

This specimen agrees well with *Blume's* short description and moreover with *Blanco's* description of *Ticoria aculeata* [21] which *Blanco* himself later reduced to *Leea aculeata* Blume. It also agrees well with specimens in our herbarium collected on Mount Arayat by *Bolster*, a topotype of *Blanco's* species.

Widely distributed in the Philippines.

Malaya.

ELAEOCARPACEÆ.

ELAEOCARPUS Linn.

Elaeocarpus argenteus Merr. in Govt. Lab. Publ. **29** (1905) 26.

On ridges in forests at 2,200 m. alt. (No. 4462 *Merritt*) June, 1906.

Previously known only from Mount Santo Tomas, Luzon.

Elaeocarpus pendulus Merr. l. c.

On ridges in forests at 1,800 m. alt. (No. 5727), flowers immature; on exposed ridges at 2,600 m. alt. (No. 6204), fruit.

Previously known only from Mount Santo Tomas and Mount Mariveles, Luzon.

Elaeocarpus merrittii Merrill, n. sp. § *Monocera*.

Ramulis tenuis, foliis ovatis vel oblongo-ovatis, acuminatis, 6 ad 9 cm. longis, longe petiolatis, crenato-serratis, subtus in axillis venarum glandulosis, racemis axillaribus, usque ad 10 cm. longis, paucifloribus; flores 9 mm. longi, 5-meri; petala dense sericea, apice fimbriata; stamina 20; drupa 1 ad 1.3 mm. longa, ellipsoidea vel oblongo-ovoidea, 1-sperma.

A tree 10 m. high or less, nearly glabrous. Branches slender, terete, dark brownish red when dry, the younger ones slightly deciduously pubescent. Leaves submembranous, ovate or oblong-ovate, 6 to 9 cm. long, 2.5 to 4 cm. wide, the base rounded, the apex rather long acuminate, the margins slightly crenate-serrate, slightly appressed-pubescent beneath and along the midrib above, the axils of the veins beneath with prominent glands; nerves about 6 on each side of the midrib, prominent beneath;

[21] Flora Filip. (1837), 83.

petioles 2 to 2.5 cm. long, slender, glabrous or nearly so. Racemes axillary or from axils of fallen leaves, few-flowered, 10 cm. long or less, glabrous or slightly pubescent. Flowers white, the pedicels 1 cm. long, densely appressed sericeous. Sepals 5, lanceolate, acuminate, 8 to 9 mm. long, 2.5 mm. wide, appressed silvery pubescent outside. Petals oblong, equaling the sepals, very densely silvery appressed, sericeous throughout, the apex cleft into 5 to 7 linear *laciniae* 2 mm. long. Stamens 20; filaments 2.5 mm. long; anthers linear 3 mm. long, one cell with a short mucro less than 0.5 mm. long. Ovary ovoid, densely sericeous, 3-celled; style 5 mm. long. Fruit elliptical or oblong-ovoid 1 to 1.3 cm. long, dark blue when mature, 1-celled, 1-seeded, the pericarp slightly fleshy.

In forests at 900 m. alt. (No. 5582) type, also (No. 5616) from the same altitude and at 1,600 m. alt. (No. 4427 *Merritt*), June, 1906.

Elaeocarpus sp.

In forests at 1,300 and 1,600 m. alt. (Nos. 4387, 4409 *Merritt*) June, 1906. The specimens are with fruit only and apparently represent an undescribed species.

Elaeocarpus sp.

In forests at 700 m. alt. (No. 6148). Specimen with fruit only, but quite distinct from any of the genera at present represented in our herbarium.

TILIACEÆ.

HALCONIA Merrill, n. gen.

Bracteolae 6. Sepala 4, crassa, valvata. Petala 5. Stamina ∞, libera; antherae ovatae, versatiles. Ovarium 2-loculare, loculis ∞-ovulatis; stylus nullus. Capsula 2-locularis, dissepimento contrarie compressa, coriacea, apice apiculata, loculicide 2-valvis; semina ignota. Arbor. Folia subintegra penninervia et basi trinervia. Cymae axillares.

Halconia involucrata Merrill n. sp.

A tree about 8 m. high, the branches terete, brownish gray, glabrous, the younger parts stellate-lepidote pubescent. Leaves oblong, subcoriaceous, 7 to 10 cm. long, 3.5 to 5 cm. wide, slightly stellate pubescent above, beneath pale and densely minutely lepidote and with scattered stellate hairs, the margins subentire or slightly crenate, the apex acute or rounded, the base rounded or slightly cordate, strongly 3-nerved, the lateral nerves very prominent beneath, ascending, including the basal ones 4 to 5 on each side of the midrib; petioles 1 to 2.5 cm. long. Cymes axillary 14 cm. long or less, few-flowered, the flowers in groups of threes at the ends of the ultimate branchlets, each group subtended by 6, oblong, stellate-lepidote bracteoles. Sepals 4, free, elliptical-oblong, ultimately nearly 1 cm. long, densely lepidote-stellate. Petals 4, oblong, densely lepidote-stellate, about 6 mm. long, the apex truncate and obscurely 3-toothed. Stamens indefinite, free, the filaments 1.5 mm. long or less, anthers about 0.3 mm. long. Ovary ovoid, densely hirsute, 2-celled, each cell many ovuled. Capsules flattened at right angles to the dissepiment, suborbicular in outline, truncate and apiculate at the apex, about 1.5 cm. long, 2 cm. wide, coriaceous, glabrous, at least in age, 2-valved, 2-celled, dehiscing to the base. Seeds unknown.

In forests at 700 m. alt. (No. 5527).

I place the genus here proposed in the *Tilieæ*, between *Graeffia* Seem., and *Trichospermum* Blume, differing from both in its 4-merous flowers and from the latter in the presence of bracteoles.

DILLENIACEÆ.

SAURAUIA Willd.

Saurauia latibracteata Choisy in Zoll. Syst. Verz. Ind. Archip. (1854–55) 148; Merr. in Govt. Lab. Publ. **35** (1906) 41.

In forests 100 to 700 m. alt. (Nos. 5690, 5528).

Widely distributed in the Philippines. Endemic.

Saurauia elegans (Choisy) F.-Vill. Nov. App. (1880) 19; Merr. l. c. 42.

In forested ravines at 700 m. alt. (No. 5655).

Previously known only from the mountains of northern and central Luzon.

Saurauia philippinensis Merrill n. sp.

Subglabra, foliis oblongo-ovatis vel oblongo-lanceolatis, glabris, 10 ad 18 cm. longis, acuminatis; pedunculi axillares, fasciculati vel solitarii, 1-, raro 2-flori; ovarium globosum, pilosum, 3-stylum.

A shrub or small tree 7 m. high or less, nearly glabrous throughout. Branches slender, brownish, lenticellate, glabrous or nearly so. Leaves oblong-ovate to oblong-lanceolate or lanceolate, rarely somewhat oblanceolate glabrous, chartaceous, 10 to 18 cm. long, 3 to 5 cm. wide, the apex rather sharply acuminate, the base acute or acuminate, the margins rather sharply serrate; nerves 12 to 14 on each side of the midrib; petioles 1.5 to 3 cm. long. Flowers white, the peduncles solitary or fascicled in the axils of the leaves and in the axils of fallen leaves on the larger branches, the peduncles slender, somewhat strigose, 2 cm. long or less, with a small bracteole at about the middle. 1, rarely 2-flowered. Sepals elliptical to elliptical ovate, acute or rounded, 4 mm. long, the margins slightly ciliate. Petals glabrous, 7 mm. long, 4 to 5 mm. wide, cleft at the apex. Stamens 20; filaments 2 mm. long, anthers about 2.2 mm. long. Ovary somewhat pilose; styles 3, 4 mm. long, slightly united below.

In forests at 700 m. alt. (No. 5529); from the same locality I refer here No. 5633, alt. 200 m., and also No. 4394 *Merritt*, June, 1906, alt. 1,300 m. In Mindanao the species is represented by No. 4693 *Mea[illegible]s* and *Hutchinson* [illegible] Malindang, May, 1906, alt. 1,400 m., and in Basil[illegible] 4011 *Hut[illegible]* [illegible]bruary, 1906, alt. 540 m.

THEACEÆ.

THEA Linn.

Thea sp.

In forests at 350 m. alt. (No. 4328 *Merritt*) June, 1906. Material very imperfect and with fruits only, the identity of the genus therefore not certain.

TERNSTROEMIA Nutt.

Ternstroemia sp.?

In forests at 1,600 m. alt. (No. 4473 *Merritt*) June, 1906. Material very scanty and with fruits only.

ADINANDRA Jack.

Adinandra sp. near *A. dumosa.*

In forests 1,600 to 2,200 m. alt. (Nos. 4410, 4453 *Merritt*) June, 1906.

Adinandra sp. near *A. luzonica* Merr.

In forests at 500 m. alt. (No. 4344 *Merritt*) June, 1906. Material very scanty and with undeveloped flowers only.

Adinandra sp.?

In open heaths at 2,400 m. alt. (No. 5745).

EURYA Thunb.

Eurya japonica Thunb. Fl. Jap. (1784) 191. *t. 25.*

In forests 1,400 to 1,800 m. alt. (Nos. 5671, 6188); also collected by *Merritt* at 1,600 m. alt. (No. 4431).

On many of the higher mountains of the Philippines.

Corea and Japan, China, central Asia, through Malaya to the Fiji Islands.

Eurya acuminata Wall., var. **euprista** Dyer in Hook. f. Fl. Brit. Ind. 1 (1872) 285.

In forests at and below 300 m. alt. (Nos. 5749, 6146).

On many Philippine mountains.

British India to Malaya and the Fiji Islands.

RHIZOPHORACEÆ.

GYNOTROCHES Blume.

Gynotroches axillaris Blume Bijdr. (1825) 219.

In forests at 200 m. alt. (No. 4323 *Merritt*) June, 1906.

Widely distributed in the Philippines, but nowhere abundant.

Malayan Peninsula and Archipelago.

MYRTACEÆ.

MEARNSIA Merrill n. gen.

Calycis tubus anguste campanulatus, ovario adnatus; limbi segmenta 4, persistentia. Petala 4, calycis lobis longiora, patentia. Stamina 8, 1-seriata, libera, filamentis elongatis, filiformibus; antherae versatiles, loculis parallelis longitudinaliter dehiscentibus. Ovarium in fundo calycis inferum, 2-loculare; stylus filiformis, stigmate parvo; ovula in loculis

∞, multo-seriata. Capsula in calyce persistente inclusa ad apicem dehiscens, 2-loculare. Semina ∞, anguste-oblonga. Arbor. Folia opposita, pennivenia. Flores in pedunculos laterales cymosi vel racemosi.

Mearnsia halconensis Merrill n. sp.

Arbor ad 10 m. alta; foliis oppositis, oblongo-lanceolatis, coriaceis, acuminatis, glabris, ad 6 cm. longis; cymis lateralibus, paucifloribus, staminibus longe exsertis.

A tree about 10 m. high, nearly glabrous throughout. Branches terete, rough, gray or brownish, the younger branchlets brownish, glabrous, obscurely 4-angled, the growing tips appressed-pubescent. Leaves opposite, oblong-lanceolate, coriaceous, glabrous, 5 to 6 cm. long, 1.5 to 2 cm. wide, entire, the margins sightly recurved, cartilaginous, the apex sharply acuminate, the base acute, glandular-punctate beneath; lateral nerves numerous, ascending, not very distinct; petioles stout 1 to 2 mm. long. Inflorescence from the branches below the leaves, the short few-flowered cymes or racemes solitary or fascicled, the rachis appressed-pubescent, 5 mm. long or less, the bracts deciduous, narrowly ovate, 2 mm. long, the pedicels 1 mm. long or nearly obsolete. Calyx tube narrowly campanulate, sparingly appressed pubescent, about 3 mm. long, the lobes 4, broadly ovate, obtuse or acute, 1.5 mm. long, persistent. Petals 4, free, deciduous, glabrous, red, orbicular, 3.5 mm. in diameter, apex broad, rounded, base narrow. Stamens 8, 1-seriate, free, exserted, glabrous; filaments red, nearly 2 cm. long; anthers ovoid, 0.6 mm. long, versatile, 2-celled. Ovary inferior, 2-celled, the ovules many-seriate; style slender, exserted, slightly exceeding the stamens. Capsule about 7 mm. long, ovoid, slightly compressed, 4-ridged, coriaceous, crowned by the persistent calyx tube and teeth, 2-celled, many seeded, dehiscing by a single slit at the apex only and inside the persistent calyx tube. Seeds narrowly oblong, about 2 mm. long.

On exposed ridges at 1,400 m. alt. (No. 5792.)

The genus here proposed is apparently related to *Backhousia* Hook. et Harv., and to *Metrosideros* Banks, but appears to me to be very distinct from both and from all other described genera in this family. It is dedicated to Maj. Edgar A. Mears, surgeon, United States Army, with whom the author made the ascent of Mount Halcon.

LEPTOSPERMUM Forst.

Leptospermum amboinense Blume Bijdr. (1826) 1100.

In open heaths and thickets at and above 2,400 m. alt. (Nos. 5746, 5747).

On most of the higher mountains of the Philippines.

Malacca through Malaya to Australia.

EUGENIA Linn.

Two species of *Eugenia* are represented in *Merritt's* material, from forests below 900 m. alt., but the specimens are too fragmentary for accurate determination.

GUTTIFEREÆ.

GARCINIA Linn.

Garcinia binucao (Blanco) Choisy Guttif. Ind. 34; Vesque in DC. Monog. Phan. 8 (1893) 454.

In forests at 175 m. alt. (No. 4322 *Merritt*) June, 1906.

DIPTEROCARPACEÆ.

SHOREA Roxb.

Shorea squamata (Turcz.) Benth. & Hook. f. Gen. Pl. 1 (1862) 193.

In forests at 700 m. alt. (No. 5751).

Widely distributed in the Philippines.

Borneo.

Dipterocarpus grandiflorus Blanco, and *Shorea guiso* Blume were both abundant from near Subaan to the Alag River ascending to an altitude of about 400 m.

BEGONIACEÆ.

BEGONIA Linn.

Begonia incisa A. DC. in Ann. Sc. Nat. IV. 11 (1859) 129; Podr. 15[1] (1864) 321.

In forested ravines at 700 m. alt. (No. 5685).

Widely distributed in the Philippines; endemic.

Begonia pseudolateralis Warb. in Perk. Frag. Fl. Philip. (1904) 51.

Along the Alag River at 200 m. alt. (No. 6154).

Previously known from Luzon and Mindoro.

Begonia sp.

In forests at 1,200 to 1,800 m. alt. (Nos. 5515, 5607).

Begonia sp.

In forests at 1,700 m. (No. 6135).

MELASTOMATACEÆ.

MELASTOMA Burm.

Melastoma polyanthum Blume in Flora 2 (1831) 481; Cogn. in DC. Monog. Phan. 7 (1891) 354.

In forests at 1,600 m. alt. (No. 4397).

Widely distributed in the Philippines.

British India to Malaya and northern Australia.

SARCOPYRAMIS Wall.

Sarcopyramis sp.

In mossy forests 900 to 2,200 m. alt. (No. 5793); also Nos. 4386, 4461 *Merritt*, June, 1906.

This genus is generally considered to be monotypic, the only species recognized by most botanists being *S. nepalensis* Wall, extending from the Himalayan region to Java and Sumatra. The species here in question is also found on Mounts Data and Santo Tomas, northern Luzon, and in the intermediate tablelands (Nos. 4608, 4491 and 4809 *Merrill*) and also on Mount Apo, Mindanao

(*Copeland*). Dr. *C. B. Robinson*, of the New York Botanical Garden, has indicated the Philippine form, *in lit.*, as a new species, but as his description has not as yet been published, I do not consider myself free here to publish his specific name.

SONERILA Roxb.

Sonerila woodii Merrill, n. sp.

Caulis erectus vel adscendens, simplex, sparse glanduloso-setulosis; foliis consimilis, ovato-oblongis, acuminatis, basi oblique subrotundatis vel subacutis, margine setuloso-serratis; flores 3-meri; calyx ad 6 mm. longus; petala pallide violacea, usque ad 10 mm. longa; antherae lanceolatae, ad 6 mm. longae.

Erect, unbranched, 10 to 30 cm. high, the stems and petioles glandular-setose or hirsute, terete. Leaves opposite, subequal, membranous, oblong-ovate, 5 to 10 cm. long, 2 to 3 cm. wide, the apex acuminate, the base inequilateral, rounded or subacute, the margins sharply serrulate, very slightly puberulent and with few scattered setose hairs; petioles 1.5 to 3.5 cm. long, glandular-pubescent; nerves about 3 on each side of the midrib, ascending, distinct, reddish, the reticulations lax. Flowers pink, nearly 2 cm. in diameter when spread. Calyx somewhat cylindrical, 6 mm. long, slightly glandular-pubescent, the three teeth small. Petals 3, 10 mm. long, 5.5 mm. wide, oblong-ovate, abruptly acuminate, the base contracted, inequilateral. Stamens 3; filaments 5 mm. long; anthers lanceolate, 6 mm. long. Style about 10 mm. long. Capsule obconical, trigonous, 6 to 7 mm. long.

In forests 900 to 1,300 m. alt. (No. 5794); also collected by *Merritt* in June, 1906 (No. 4352). Dedicated to Maj. Gen. *Leonard Wood*, through whom the exploration of Mount Halcon was made possible.

MEDINILLA Gaudich.

Medinilla myrtiformis Triana in Trans. Linn. Soc. **28** (1871) 86; Cogn. in DC. Monog. Phan. **7** (1891) 583.

In forests at 700 m. alt. (No. 5682); also collected by *Merritt* at 900 m. alt. in June, 1906 (No. 4356).

On most of the higher mountains of the Philippines.

Amboina.

Medinilla ramiflora Merr. in Govt. Lab. Publ. **29** (1905) 35.

In forests 1,300 to 1,500 m. alt., frequently pseudoepiphytic (No. 5724); also collected by *Merritt* in June, 1906 (No. 4388).

Medinilla astronioides Triana l. c. 88; Merr. l. c. 37.

In forested ravines at 700 m. alt. (No. 6149).

Previously known only from Luzon.

Medinilla merrittii Merrill, n. sp.

Ramis teretiusculis; foliis elliptico-ovatis, glabris, coriaceis, oppositis, petiolatis, 7-nervis, nervulis transversalibus validis; cymis axillaribus, paucifloribus; floribus 5-meris; calyx campanulato, 1 cm. longo.

Scandent, glabrous throughout; branches light gray, terete, the ultimate ones brownish, slender. Leaves opposite, elliptical-ovate, abruptly short caudate-acuminate, the base broad, rounded, coriaceous, 11 to 15 cm. long, 6 to 9 cm. wide; nerves 7, prominent, the transverse nervules distinct; petioles 2.5 cm. long. Cymes solitary or fascicled on the branches below the leaves, the peduncle about 1 cm. long, bearing 3 subsessile flowers. Flowers about 2 cm. long, 5-merous. Calyx campanulate, truncate, nearly 1 cm. long, 0.8 cm. in diameter. Petals 2.5 cm. long, 1.4 cm. wide. Stamens 10.

In forests at 450 m. alt. (No. 4336 *Merritt*) June, 1906.

Medinilla verticillata Merr. in Govt. Lab. Publ. **29** (1905) 34.
In forests at 700 m. alt. (No. 5060).
Previously known only from northern Luzon.

Medinilla magnifica Lindl. in Paxt. Flower Gard. **1** (1850) 55. *t. 12.*
In forests at 500 m. alt. (No. 4341 *Merritt*) June, 1906.
Rather widely distributed in the Philippines but not abundant. Endemic.

Medinilla involucrata Merr. in Govt. Lab. Publ. **35** (1906) 51.
In forests at 250 m. alt. (No. 5634); also collected by *Merritt* at about the same altitude in June, 1906 (No. 4331).
Previously known from Mindoro and Mindanao.

Medinilla cordata Merr. in Govt. Lab. Publ. **29** (1905) 37.
On exposed ridges at 2,400 m. alt. (No. 5757); also collected by *Merritt* at 2,150 m. alt. (No. 4445). Specimens not typical, and may prove to be distinct from the Luzon plant.
Previously known from Luzon only.

Medinilla halconensis Merrill n. sp.

Ramis ramulis petiolis pedunculis bracteis calycibus foliisque dense plumoso-stellato-tomentosis; ramulis teretibus; foliis petiolatis, elliptico-ovatis, acutis vel acuminatis, nervis 5, lateralibus oppositis, cymis terminalibus, bracteis et bracteolis persistentibus; floribus 5-meris.

Branches slender, terete, densely stellate-plumose tomentose. Leaves submembranous, elliptical ovate, 4 to 9 cm. long, 2 to 4.5 cm. wide, opposite, above nearly glabrous, beneath densely stellate-tomentose, the base acute, the apex short acuminate or acute; nerves 5, petioles densely stellate-tomentose, 1 to 1.5 cm. long. Inflorescence a terminal 3 to 5-flowered cyme, 3 cm. long or less, densely stellate-tomentose throughout; bracts linear to lanceolate, 1 to 1.5 cm. long; bracteoles elliptical-ovate, white, acute, 2 cm. long, 1.3 cm. wide, densely stellate-tomentose. Calyx about 8 mm. long, tubular, inflated in fruit, the limb persistent, obscurely 5-toothed, stellate-pubescent.

Scandent on tree trunks in forests at 1,050 m. alt. (No. 5642); also collected by *Merritt* at 900 m. alt., June, 1906 (No. 4366).

Medinilla microphylla Merrill n. sp.

Frutex scandens; ramis ramulis petiolis bracteis calycibus foliisque praecipue ad nervos stellato-tomentosis; ramis teretibus; foliis oppositis, oblongo-ovatis, 5-nervis, 2 ad 4.5 cm. longis; cymis terminalibus, pauci-floribus, bracteis roseis persistentibus; floribus 4-meris.

Scandent along tree trunks, the stems and branches terete, slender, the younger parts densely stellate-plumose-tomentose. Leaves submembranous, oblong-ovate, 2 to 4.5 cm. long, 1 to 2 cm. wide, nearly glabrous above, beneath densely stellate-plumose-tomentose on the nerves, base acute, apex short-acuminate; nerves 5; petioles 0.5 mm. long, or less. Inflorescence terminal, usually reduced to a single flower subtended by about 6 pink bracts. Bracts narrowly elliptical-ovate, base and apex acute, 3-nerved, stellate-pubescent, 1.8 cm. long, 1 cm. wide. Fruit subglobose, somewhat stellate-pubescent, nearly 1 cm. long, the calyx persistent, 6 mm. long, the tube 2 mm., the lobes ovate, acute, 4 mm. long.

In forests at 900 m. alt. (No. 5599).

Medinilla sp.

In forests at 1,800 m. alt. (No. 5667), specimens in fruit only.

Medinilla sp.

In forests at 900 m. alt. (No. 4368 *Merritt*) June, 1906, material in poor condition.

Medinilla sp.

In forests at 1,300 m. alt. (No. 4382 *Merritt*) June, 1906. The last three species enumerated are all distinct from each other and different from any of the species represented at present in our herbarium, but the specimens are incomplete and are accordingly not described here.

ASTRONIA Blume.

Astronia meyeri Merr. in Govt. Lab. Publ. **35** (1906) 51.

In forests at 550 m. alt. (No. 4347 *Merritt*) June, 1906.

Previously known from Mount Mariveles, Luzon.

MEMECYLON Linn.

Memecylon preslianum Triana in Trans. Linn. Soc. **28** (1871) 157.

In forests at 100 m. alt. (No. 5604).

Widely distributed in the Philippines. Endemic.

HALORRHAGACEÆ.

HALORRHAGIS Forst.

Halorrhagis halconensis Merrill n. sp.

Planta robusta suffruticosa erecta ad 60 cm. alta; foliis oppositis vel ad ramos juniores verticillatis, elliptico-ovatis, ad 3 cm. longis, 1 ad 1.5 cm. latis; inflorescentiae ad apices caulis dense racemoso-paniculatae; flores hermaphroditi 3 ad 3.5 mm. longi in bractearum axillis solitarii.

Robust, erect, sligtly branched, about 60 cm. high, the stems terete, scabrid, the older ones reddish brown, the younger parts rather densely clothed with long brittle white hairs. Leaves elliptical-ovate 2 to 3 cm. long, 1 to 1.5 cm. wide, those on the older parts opposite, those on the younger branches in whorls of four, crowded, rigid, coriaceous, scabrid, sessile or short-petioled, the base rounded, the apex acute, the margins strongly acuminate-denticulate, both surfaces with scattered coarse white hairs. Inflorescence terminal, crowded, consisting of many simple racemes forming a terminal panicle up to 10 cm. in length and 4 or 5 cm. in diameter, the individual racemes 4 cm. long or less, ascending, the lower ones subtended by leaves. Bracts narrowly lanceolate, 3 mm. long, 1-flowered, the bracteoles acicular, 1 mm. long; pedicels about 1 mm. long, strigose. Calyx tube 4-angled, narrowly ovate, slightly strigose, not rugose nor pellucid-punctate but slightly strigose-hispid, the lobes erect, lanceolate, about 1.5 mm. long, acuminate, glabrous. Petals boat-shaped, slightly aculeate-hispid on the keel, 2.2 mm. long. Stamens 8; anthers 1.5 to 1.7 mm. long.

In open heaths at 2,400 m. alt. (No. 5700).

The fourth species of this characteristic Australian genus to be found in the Philippines, readily recognizable by its crowded, opposite and verticillate, very large leaves (for the genus), and terminal racemoso-paniculate inflorescence.

Halorrhagis micrantha (Thunb.) R. Br. ex Sieb. et Zucc. Fl. Jap. Nat. 1 (1843) 25; Merr. in Philip. Journ. Sci. (1906) Suppl. 1: 216.

With the preceding (No. 5787).

Known in the Philippines from Canlaon Volcano, Negros, and Mount Apo, Mindanao.

Bengal to Japan, Malaya, Australia and New Zealand.

ARALIACEÆ.

BOERLAGIODENDRON Harms.

Boerlagiodendron trilobatum Merrill, n. sp.

Frutex glaber 2 ad 5 m. altus; ramulis tenuis, lenticellatis; foliis glabris, submembranaceis, longe petiolatis, ad 20 cm. longis, profunde 3-lobatis, lobis acuminatis, sinuato-serratis; umbellis terminalibus, multifloribus; ovario 5-lobato.

A slender shrub, simple or sparingly branched, nearly glabrous, 2 to 5 m. high. Branches slender, terete, lenticellate. Leaves alternate, submembranous, glabrous, paler beneath, shining, about 20 cm. long, 3-lobed, the sinus reaching nearly ½ to the base of the leaf, narrow, the lobes narrowly oblong, sharply acuminate, irregularly rather coarsely sinuate-serrate, the base broadly acute, 5-nerved; petioles 13 cm. long with a prominent stipule at the base. Inflorescence terminal, short pedunculate, umbellate, about 10 cm. in diameter, the primary peduncles about 15, 2 cm. long, bearing a small umbel at the apex and usually 2 secondary peduncles 1.5 to 2 cm. long, these bearing a dense globose head of small white flowers. Calyx tubular 3 mm. long, truncate. Petals 5, valvate, oblong, acute, 3 mm. long. Stamens 5; filaments 2 mm. long; anthers 1.8 mm. long. Ovary 5-celled. Fruit dimorphous, of the primary umbels white, fleshy, globose, not ridged, 3-celled, 4 to 5 mm. in diameter, the seeds aborted; of the secondary umbels oblong-ovoid, 7 mm. long, strongly 5-ridged, 5-celled, fertile.

In forests at about 150 m. alt. (No. 5620). A closely related form is represented by No. 669 *Ahern* from Surigao, Mindanao.

SCHEFFLERA Forst.

Schefflera foetida Merrill n. sp. § *Heptapleurum*.

Scandens, glabra; foliis 6-foliatis; foliolis ovatis vel oblongo-ovatis, acuminatis, coriaceis, 4 ad 6 cm. longis; paniculis terminalibus, ramulis divergentibus; ovario 4-rariter 5-loculare.

Glabrous throughout, scandent up to 7 m., all parts when crushed with a very rank odor; branches gray, the tips very dark colored when dry, terete. Leaves alternate, the petioles 5 to 10 cm. long, digitately 5-foliate; leaflets ovate to oblong-ovate, coriaceous, shining, 4 to 6 cm. long, 2 to 3.5 cm. wide, acuminate, the base rounded or subacute, sometimes inequilateral, the midrib prominent, the lateral nerves very faint; petiolules 1 to 3 cm. long. Panicles terminal, about 12 cm. long, short-peduncled, branched along the rachis, the branches few, spreading, the primary ones (in anthesis) sometimes 10 cm. long, the umbels racemosely disposed their peduncles 1.5 to 2 cm. long, each umbel 6 to 9-flowered, the pedicels 5 to 7 mm. long. Calyx broadly funnel shaped, truncate, about 1.2 mm. long. Petals united, forming an apiculate calyptra which falls as a whole, 2 mm. long. Stamens 4, very rarely 5; filaments 4 mm. long; anthers ovoid, 1.2 mm. long. Ovary 4-celled, the top conical; stigmas 4.

On forested ridges at 1,800 m. alt. (No. 5762). No. 4423 *Merritt*, from an altitude of 1,600 m. is apparently the same, the flower buds being very immature. No. 5678 of my own collection appears also to be a very large diffuse form of the same species but it is with fruits only and has much larger more acuminate leaves and very much larger panicles than the type.

Schefflera sp.

In forests at 900 m. alt. (No. 5696) Undeterminable, with fruits only.

Schefflera insularum (*Heptapleurum insularum* Seem.) is abundant along the Alag and Binabay Rivers, and has previously been collected on the Baco River near the base of Halcon.

ARTHROPHYLLUM Blume.

Arthrophyllum sp.

In forests at 700 m. alt. (No. 5597), flowers immature, apparently representing an undescribed species.

ARALIA Linn.

Aralia glauca Merrill n. sp.

Arbor ad 10 m. alta, ramis ramulis foliisque inermibus; foliis bipinnatis 40 ad 50 cm. longis, pinnis 5-jugatis 20 ad 30 cm. longis; foliolis 4 ad 7-jugatis, oblongo-ovatis, acuminatis 6 ad 11 cm. longis, subtus pallidis; paniculis terminalibus 50 cm. vel ultra longis et latis, ramis elongatis ad 40 cm. longis, ramulis racemose dispositis, 4 ad 6 cm. longis; ovario 5-loculare.

A spineless tree about 10 m. high, the branches thickened, the ultimate ones 1.5 to 2 cm. in diameter, lenticellate, striate when dry, glabrous or nearly so, the large bipinnate leaves crowded at the apices of the branches, the inflorescence of several large spreading terminal panicles. Leaves about 50 cm. long, the pinnæ 5-jugate, opposite, the rachis glabrous terete, enlarged at the nodes which are prominently jointed, the petiole 20 to 25 cm. long, glabrous; pinnæ 20 to 30 cm. long, the lowest pair shorter; leaflets 4 to 7-jugate, opposite, oblong-ovate to oblong-lanceolate, glabrous, subcoriaceous, dull, the lower surface glaucous, 6 to 11 cm. long, 2 to 4 cm. wide, the base rounded or cordate, gradually narrowed above to the rather prominently acuminate apex, the margins distantly irregularly sinuate-crenate, the teeth frequently apiculate, nerves 5 to 6 on each side of the midrib, rather prominent beneath; petiolules 1 to 4 mm. long. Panicles at least 50 cm. long and about as wide, several from the apex of the same branch, the branches racemosely disposed, about 40 cm. long, the rachis thick, about 20 cm. long, slightly hirsute, branches slightly ferruginous-pubescent, the branchlets and pedicels rather densely so; branchlets racemosely disposed, slender, 4 to 6 cm. long, numerous, each supplied with many scattered lanceolate acuminate bracts or bracteoles 2 to 3 mm. long the flowers umbellately disposed, usually 2 umbels towards the apex of each branchlet and frequently solitary flowers in the axils of the upper bracts below the umbels. Umbels 15 to 20-flowered, the pedicels 5 mm. long or less. Flowers (immature) white, petals 1.5 mm. long. Calyx 1.5 mm. long, the lobes 5, broadly ovate, acute, 0.3 mm. long. Stamens 5; anthers 1 mm. long. Ovary 5-celled; styles 5, free, 1 mm. long.

In forests at 700 m. alt. (No. 6177), locally rather abundant and recognizable by being entirely unarmed.

The very spiny *Aralia hypoleuca* Presl is abundant on recent "slides" on Halcon at an altitude of about 1,800 m. It is widely distributed in the Philippines and has been reduced by *Forbes* and *Hemsley* to the very widely distributed *Aralia spinosa* Linn.

UMBELLIFERÆ.

DIDISCUS DC.

Didiscus saniculaefolius (Stapf) *Trachymene saniculaefolia* Stapf in Trans. Linn. Soc. Bot. II. 4 (1894) 167; Hook. Icon. IV. 4 (1895) *pl. 2308.*

In open heaths at 2,400 m. alt. (No. 6174), locally abundant, but found only in the open heaths associated with *Gleichenia, Lycopodium, Dipteris, Vaccinium, Leptospermum, Isachne, Drosera, Patersonia,* etc.

While the above specimens differ slightly from the description and figure given by *Stapf,* still I can detect no constant characters by which the Philippine form can be distinguished from the Bornean, and without comparison with the type material I do not care to describe the Halcon plant as a new species. The genus is new to the Philippine flora, and its discovery is another link in the chain of evidence regarding possible previous land connections with Borneo on the one hand, and with Australia on the other. *Didiscus saniculaefolius* was previously known only from Mount Kinabalu, North Borneo, being related to the Australian *D. humilis.* Of the genus, about fourteen species are recognized, twelve in Australia, one in New Caledonia and one in Mindoro and Borneo; of the genus *Trachymene,* in which the above species was placed by *Stapf,* about twelve species are recognized, all the others confined to Australia.

CLETHRACEÆ.

CLETHRA Linn.

Clethra lancifolia Turcz. in Bull. Soc. Nat. Mosc. 36[2] (1863) 231.

In forests at 700 m. alt. (No. 5575); also collected by *Merritt* in June, 1906 at 2,200 m. alt. (No. 4455).

Previously known only from the mountains of Luzon and Negros.

ERICACEÆ.

RHODODENDRON Linn.

Rhododendron quadrasianum Vid. Rev. Pl. Vasc. Philip. (1886) 170.

Terrestrial and epiphytic, mossy ridge-forests at 1,350 m. alt. (No. 6158); also collected by *Merritt* in forests at 1,600 m. alt. (No. 4455).

On most of the higher mountains of the Philippines from northern Luzon to southern Mindanao. Endemic in the Philippines.

Rhododendron rosmarinifolium Vid. l. c. 172.

In open heaths at 2,400 m. alt. (No. 5736).

Previously known only from the mountains of northern Luzon; the Halcon specimens not quite typical, the leaves shorter and broader than in specimens from northern Luzon. This is probably the form reported from "Dulangao" (a spur of Halcon) by *Rendle,* as the Bornean *Rhododendron cuneifolium* Stapf, and appears to be rather intermediate between *R. rosmarinifolium* and *R. cuneifolium,* but nearer the former.

GAULTHERIA Linn.

Gaultheria cumingiana Vid. Phan. Cuming. Philip. (1885) 184.

Subscandent in ridge forests 1,800 to 2,200 m. alt. (No. 5725).

Previously known only from the mountains of Luzon and Formosa.

DIPLYCOSIA Blume.

Diplycosia merrittii Merrill, n. sp.

Frutex pseudoepiphyticus scandens, inflorescentiis excepti, glabra; foliis coriaceis oblongo vel elliptico-ovatis, rariter oblongo-lanceolatis, breviter acuminatis, 6. ad 10 cm. longis, basi acutis nervis utrinque 2 vel 3, subtus prominentibus; flores axillares, fasciculati; corolla ovoidea, 6 ad 7 mm. longa.

A scandent pseudoepiphytic shrub often 6 m. high, glabrous except the inflorescence. Branches gray or brown, terete, the younger ones angular. Leaves coriaceous, oblong-ovate to elliptical-ovate, rarely oblong-lanceolate, 6 to 10 cm. long, 2 to 5 cm. wide, the base acute, the apex slightly acuminate, the margins obscurely denticulate, recurved, paler and glandular punctate beneath; nerves 2 to 3 on each side of the midrib, ascending, impressed above, rather prominent, the reticulations nearly obsolete; petioles stout, 4 to 8 mm. long, rugose. Flowers pink, fascicled, 2 to 8 in each axil, the pedicels slender, slightly pubescent, 1 to 1.5 cm. long, the apical bracts two, orbicular-ovate, 1.3 mm. long. Calyx very slightly pubescent, 3.5 mm. long, the lobes ovate or narrowly ovate, acute, 2 mm. long. Corolla ovoid, narrowed below, 6 to 7 mm. long, the lobes 5, ovate, broadly acuminate, 2 mm. long, reflexed. Stamens 10, glabrous; filaments 3 mm. long; anthers oblong, 1.5 mm. long. Ovary glabrous, style 2 mm. long. Fruit ovoid or subglobose, soft, fleshy, 1 cm. in diameter, black when mature.

In ridge forests at 1,400 m. alt. (No. 5670) (type), very abundant, the fruit edible but nearly tasteless; also collected by *Merritt* in June, 1906, at an altitude of from 1,600 to 1,700 m. (Nos. 4413, 4415, 4437). The same form has been collected in Palawan, Victoria Peak (666 *Foxworthy*) March, 1906.

Of the Philippine species of this genus, apparently most closely related to DIPLYCOSIA LUZONICA (A. Gray) (*Gaultheria luzonica* A. Gray), from Mount Banajao and Mount Santo Tomas, Luzon. I have before me a single leaf from the type of *Gray's* species, kindly supplied by Dr. *J. N. Rose* of the U. S. National Museum. No. 5932 *Elmer* from Mount Santo Tomas seems to match it exactly. The species proposed above differs from this in its larger, differently shaped leaves, much longer pedicels and more prominently nerved leaves, the venation in the two species being quite different. I have seen no flowers of *Gaultheria luzonica*.

VACCINIUM Linn.

Vaccinium mindorense Rendle in Journ. Bot. **34** (1896) 355.

Epiphytic, mossy ridge forests 1,950 to 2,200 m., and terrestrial above 2,200 m. alt. (No. 5676); also collected by *Merritt* at 1,600 m. in June, 1906 (No. 4414).

The type of the species was from Mount Dulangan, a spur of Halcon. It is also apparently represented by specimens from Mount Madiaas, Panay (*A. E. Yoder*), April, 1905, and from Mount Apo, Mindanao (*Copeland*), October, 1904.

Vaccinium banksii Merr. in Govt. Lab. Publ. **35** (1905) 54.

In an open heath at 2,400 m. alt. (No. 5506).

Previously known only from Canlaon Volcano, Negros.

Vaccinium villarii Vid. Rev. Pl. Vasc. Filip. (1886) 166.

With the preceding species (No. 5502).

Extending from the high table lands of northern Luzon to Mount Apo, Mindanao. Endemic in the Philippines.

Vaccinium hutchinsonii Merrill n. sp.

Epiphyticum, glabrum, foliis late elliptico-ovatis, abrupte subcaudato-acuminatis, coriaceis, 8 ad 11 cm. longis, 5.5 ad 7 cm. latis, basi acutis; flores usque ad 17 mm. longi, rubri, in racemos bracteatos axillares dispositi, filamentis pause setoso-pilosis.

A scandent epiphytic or pseudoepiphytic shrub about 5 m. high. Branches glabrous, light gray or brown, the younger ones somewhat angular. Leaves broadly elliptical-ovate, coriaceous, glabrous, the base acute or acuminate, the apex abruptly subcaudate-acuminate, shining, entire, 8 to 11 cm. long, 5.5 to 7 cm. wide; nerves about 7 on each side of the midrib, mostly basal, ascending, distinct, the reticulations distinct; petioles 1 to 1.5 cm. long. Racemes axillary, glabrous, 8 to 14 cm. long, the bracts oblong-lanceolate, reddish, membranous, deciduous, glabrous, acuminate, 2 cm. long, 5 mm. wide; pedicels rather distant, solitary in the axil of each bract, about 1.5 cm. long. Calyx globose, rugose, 3 to 4 mm. in diameter, the teeth 5, triangular-ovate, acute, 1 mm. long. Corolla red, tubular-campanulate, glabrous, 14 mm. long, gradually wider above. Stamens 10; filaments 6 to 7 mm. long, with few stiff hairs below; anthers narrowly oblong, 5 to 6 mm. long, the terminal tubes half the length of the anthers. Ovary glabrous; style glabrous, 15 mm. long.

Epiphytic or pseudoepiphytic in mossy ridge forests at 2,000 m. alt. (No. 5524).

Most closely related to *Vaccinium barandanum* Vid., from northern Luzon, differing in its much broader, relatively shorter and differently shaped more numerously veined leaves, shorter flowers and slightly setose-pilose filaments. Named in honor of *W. I. Hutchinson* of the Philippine Forestry Bureau, my companion in the ascent of Halcon.

Vaccinium halconense Merrill n. sp.

Scandens, epiphyticum; foliis oblongo-elliptico-ovatis vel obovatis, acutis vel breviter acuminatis, basi acutis, 6 ad 9 cm. longis, coriaceis; racemis axillaribus, rhachidibus pedicellis fructibusque ferrugineo-pilosis.

A scandent shrub or subarborescent, 5 to 10 m. high, epiphytic or pseudoepiphytic. Branches reddish brown, glabrous, terete, the growing tips slightly pubescent. Leaves coriaceous, oblong-elliptical-ovate or somewhat obovate, 6 to 9 cm. long, 2 to 3.5 cm. wide, glandular-punctate beneath and paler than above, slightly shining, the base acute, the apex acute or shortly acuminate; nerves about 3 on each side of the midrib, mostly basal, ascending, not very distinct; petioles stout, 5 mm. long or less, glabrous or slightly pubescent. Racemes axillary, 5 to 7 cm. long, the rachis, pedicels and fruits ferruginous-pilose [illegible] densely so; pedicels about 1 cm. long. Fruit [illegible] about 8 [illegible]ameter.

On exposed ridges, epiphytic on *Podocarpus*, at 1,350 m. alt. (No. 5665); also collected by *Merritt* in June, 1906, at 1,600 m. alt. (No. 4422).

A species distinguishable from all other Philippine representatives of the genus known to me by its pilose racemes and fruits.

Vaccinium pyriforme Merrill n. sp.

Epiphyticum, glabrum, scandens; foliis elliptico-oblongis vel anguste elliptico-obovatis, integris, obtusis, 1.5 ad 2 cm. longis, ad 5 mm. latis, glabris, coriaceis; racemis axillaribus, paucifloribus, 1.5 cm. longis; fructibus pyriformibus.

A slender scandent epiphyte, glabrous throughout. Stems slender, reddish brown, angular. Leaves elliptical-oblong or narrowly elliptical-obovate, the apex obtuse, the base acute, 1.5 to 2 cm. long, about 5 mm. wide, coriaceous, shining, pale when dry, entire, the nerves few, indistinct; petioles 1 to 2 mm. long. Racemes axillary, few flowered, 1.5 cm. long, the rachis about 1 cm. long, the pedicels 5 mm. long. Flowers unknown. Fruit pyriform, glabrous, about 4 mm. long, the apex subtruncate and somewhat pubescent inside the persistent obscure calyx teeth.

Epiphytic in forests at 1,600 m. alt. (No. 4424 *Merritt*) June, 1906. A species characterized by its small entire leaves, axillary racemes and pyriform fruit.

Vaccinium whitfordii Merrill n. sp.

Frutex glaber; foliis coriaceis, anguste obovatis vel elliptico-obovatis, basi acutis, apice obtusis, obscure crenatis, usque ad 1 cm. longis; flores axillares, solitarii, rubri, ad 8 mm. longi; filamentis pilosis.

An erect shrub 0.7 to 3 m. high, terrestrial, or sometimes epiphytic, nearly glabrous throughout. Branches slender, gray or brown, angular, the younger ones somewhat puberulent. Leaves 1 cm. long or less, narrowly obovate or elliptical-obovate, coriaceous, glabrous, the apex obtuse, the base acute, the margins somewhat crenate especially above; nerves obsolete or nearly so; petioles about 1 mm. long. Flowers axillary, solitary, the pedicels slightly pubescent, 2 to 3 mm. long. Calyx 3.5 mm. long, the tube ovoid, the lobes spreading, narrowly ovate, glabrous, 1 mm. long. Corolla narrowly urceolate, red, glabrous, 7 to 8 mm. long, 4 mm. wide below, narrowed above and 2 mm. wide below the mouth, the lobes 5, ovate reflexed, acute, 1 mm. long. Stamens 10; filaments 3 to 4 mm. long, thickened below, pilose; anthers oblong, 1.5 mm. long. Style thick, 8 mm. long. Fruit subglobose, or ovoid, glabrous, 5 mm. in diameter.

On open heaths at 2,400 m. alt. (No. 5798), a shrub about 70 cm. high. Also found in the District of Lepanto, Luzon, at 1,500 m. alt. (No. 5741 *Klemme*) November, 1906, a shrub up to 3 m. in height, and on Mount Silay, Negros (No. 1534 *Withford*) May, 1906, epiphytic in the latter place.

A species characterized by its small crenate leaves and solitary axillary flowers.

Vaccinium sp.

Epiphytic in the mossy forest at 1,800 m. alt. (6133). Sterile material, apparently representing an undescribed species.

MYRSINACEÆ.

ARDISIA Swartz.

Ardisia elmeri Mez in Philip. Journ. Sci. **1** (1906) Suppl. 273.

In forests at 1,800 m. alt. (No. 6138); also collected by *Merritt* in June, 1906, in forests 2,100 to 2,200 m. alt. (Nos. 4444, 4457).

Previously known only from northern Luzon.

Ardisia racemoso-panniculata Mez l. c. 271.

In forests at 450 m. alt. (No. 4334 *Merritt*), June, 1906.

Previously known only from Mount Apo, Mindanao. *Merritt's* specimen is not quite typical and is very fragmentary, but I consider it referable to this species.

Ardisia saligna Mez in Engler's Pflanzenreich **9** (1902) 143.

In forests at 300 m. alt. (No. 5567).

Previously known from Luzon and Polillo.

Ardisia boissieri A. DC.; Mez l. c. 129.

In forests at 1,450 m. alt. (No. 5669); also collected by *Merritt* at about 1,000 m. alt. (Nos. 4371, 4355).

Endemic in the Philippines and frequently confused with *A. humilis.* The specimens cited above are all with fruit and accordingly the identification must be considered as somewhat doubtful.

Ardisia serrata (Cav.) Pers. Syn. **1** (1805) 233; Mez l. c. 137.

In forests at 1,800 m. alt. (Nos. 5675, 5732, 6145); also collected by *Merritt* in June, 1906, at 1,300 m. alt. (No. 4372).

Widely distributed in the Philippines.

Borneo.

Ardisia serrata (Cav.) Pers., var. **brevipetiolata** Merrill n. var.

Foliis breviter (3 mm.) petiolatis, basi anguste rotundato-cordatis.

In forests at 550 m. alt. (No. 4346 *Merritt*) June, 1906. The type is No. 4049 *Merrill* from the Baco River, near the base of Halcon, March, 1905. More abundant and better material may prove this form to be worthy of specific rank.

Ardisia sp.?

An undershrub less than 1 m. high, in forests at 250 m. alt. (No. 5743), specimens in fruit only, and possibly not this genus.

LABISIA Lindl.

Labisia pumila (Blume) F.-Vill. Nov. App. (1883) 123; Mez in Engler's Pflanzenreich **9** (1902) 171 ("Benth. et Hook.") var. **genuina** Mez l. c.

In forests at 450 m. alt. (No. 4335 *Merritt*) June, 1906.

A monotypic genus extending from Cochin China and the Malayan Peninsula to Java, Sumatra and Borneo, the variety *genuina* in Java, Penang, Singapore and Cochin China.

Reported from Luzon by *F.-Villar*, but not found in the Philippines by any other botanists or collector until discovered by *Merritt.*

DISCOCALYX Mez.

Discocalyx sp.

In forests at 1,400 m. alt. (No. 5608), specimens with immature fruit only, apparently undescribed.

Discocalyx sp.

In forests at 1,800 m. alt. (No. 5508).

RAPANEA Aubl.

Rapanea retusa Merrill n. sp.

Frutex glaber ad 3 m. alta; foliis oblongo-oblanceolatis, coriaceis, apice retusis, basi cuneatis, 2 ad 5 cm. longis, subtus valde glanduloso-punctatis; flores 4-meri, 2 mm. longi, fasciculati; petalis ovatis, acutis, punctatis usque ad ¼ connatis.

An erect much branched shrub about 3 m. high, glabrous throughout. Branches brown or gray, the younger ones glandular-punctate. Leaves oblong-oblanceolate, to narrowly elliptical-oblanceolate, 2 to 5 cm. long, 0.5 to 1.5 cm. wide, coriaceous, shining above, both surfaces glandular-punctate, the lower one more prominently so, margins entire, the apex retuse, the base cuneate; petioles 3 to 8 mm. long; nerves and reticulations obscure. Flowers fasciculate in the leaf-axils, usually about 5 in a fascicle, the pedicels glabrous, glandular-punctate, 3 to 4 mm. long. Calyx lobes 4, ovate, acute, nearly 1 mm. long, glandular-punctate. Corolla 2 mm. long, the lobes 4, narrowly-ovate, acute, glandular-punctate, united for the lower ¼. Anthers suborbicular-ovate, about 1.2 mm. long. Fruit globose, about 3 mm. in diameter, slightly glandular-punctate or nearly epunctate, crowned by the style which is apparently sessile and coarsely lobed.

In open heaths at 2,400 m. alt. (Nos. 5734, 5735); both specimens in fruit; also collected by *Merritt* in July, 1906, at from 1,600 to 2,200 m. alt. (Nos. 4426, 4449), both specimens with staminate flowers.

Apparently a distinct species, characterized by its 4-merous flowers, and oblong-oblanceolate, retuse, glabrous, glandular-punctate leaves, but in the absence of pistillate flowers I am not sure of its affinity, but it appears to belong in the group with *Rapanea myrtillina*, *M. platystigma*, etc., this group being developed in New Zealand, Australia, etc., with a single species extending as far north as New Guinea.

EMBELIA Burm.

Embelia halconensis Merrill n. sp. (§ *Pattara?*)

Frutex vel arbor erecta, glabra; foliis oblongo-ovatis, integris, 5 ad 7 cm. longis; racemis axillaribus, solitariis, 3 ad 4 cm. longis, basi squamis imbricatis destitutis; flores 5 et 6-meri, petalis basi breviter connatis; filamentis quam petalis brevioribus.

A shrub or tree glabrous throughout, reaching a height of 10 m. Branches dark gray, slender, lenticellate. Leaves oblong-ovate, entire, the apex broadly rather obscurely acuminate, the base acute, 5 to 7 cm. long, 1.5 to 3 cm. wide, coriaceous, somewhat shining, glandular-punctate beneath; nerves 5 to 6 on each side of the midrib, obscure; petioles 6 to 8 mm. long. Racemes axillary, solitary, the basal bracts wanting, 3 to 4 cm. long, few flowered, the pedicels 4 to 5 mm. long, each subtended by small basal bract, the flowers white. Sepals 6, rarely 5, ovate, acute, about 1 mm. long. Petals 6, rarely 5, symmetrical oblong-ovate, obtuse, 2 to 2.5 mm. long, epunctate, united for the lower 0.5 mm. Filaments 1 mm. long; anthers broadly ovoid, 0.5 mm. long. Ovary rudimentary in staminate flowers, ovoid, glabrous.

In ridge-forests at 1,800 m. alt. (No. 5771).

Rather an anomalous species for this genus because of its usually 6-merous flowers and with its petals manifestly united below. Careful dissection of many flowers shows them to be mostly 6-merous, but sometimes 5-merous on the same branches and even in the same racemes.

PRIMULACEÆ.

LYSIMACHIA (Tourn.) Linn.

Lysimachia ramosa Wall. Cat. (1828) n. 1490; Knuth in Engler's Pflanzenreich. **22** (1905) 271.

In forests at 2,000 m. alt. (No. 4443 *Merritt*) June, 1906.

Previously known in the Philippines only from the mountains of northern Luzon.

Himalayan region to Java and the northern Philippines.

SAPOTACEÆ.

PALAQUIUM Blanco.

Palaquium sp. aff. *P. luzoniensi* Vid.

In forests along the Alag River at 100 m. alt. (No. 5767).

Previously known only from Luzon and Mindoro.

Palaquium sp. aff. *P. luzoniensi* Vid.

In forests at 1,100 m. alt., fragmentary imperfect material from fallen branches of a large tree.

SYMPLOCACEÆ.

SYMPLOCOS Linn.

Symplocos adenophylla Wall. Cat. (1828) No. 4427; Brand in Engl. Pflanzenreich **6** (1901) 48.

In exposed ridge-thickets at 2,450 m. alt. (No. 5752); also collected by *Merritt* in June, 1906, at from 1,500 to 2,200 m. alt. (Nos. 4406, 4428, 4440, 4447).

Specimens of the above were sent to Dr. *A. Brand*, who has identified them as above. No specimens with mature flowers were collected, only with immature buds and mature fruits, the fruit being slightly longer than in the type specimens. The species is new to the Philippines.

Penang, Singapore, Banca and North Borneo.

APOCYNACEÆ.

ALYXIA R. Br.

Alyxia monilifera Vid. Rev. Pl. Vasc. Filip. (1886) 182.

In ridge thickets at 2,500 m. alt. (No. 5713).

Previously known only from Mount Mariveles and Mount Banajao, Luzon.

ASCLEPIADACEÆ.

Seven species of *Dischidia* and *Hoya* are represented in the material collected by the author on Halcon, but as much of our material of these genera is at present in the hands of Dr. *Schlechter* for identification, no attempt is here made to determine the species.

VERBENACEÆ.

CALLICARPA Linn.

Callicarpa caudata Maxim. in Bull. Acad. Pétersb. **31** (1887) 76.

In forests at 800 m. alt. (No. 5556).

On the higher mountains from northern Luzon to southern Mindanao. Endemic.

The Halcon specimen has pure white fruits, while specimens from northern Luzon have purple fruits.

CLERODENDRON Linn.

Clerodendron sp.

In forests at 1,800 m. alt. (No. 5516).

Apparently an undescribed species, but without flowers, the persistent calyx and bracts purplish. The same species is represented by No. 5713 *Klemme* from Balbalasan, District of Lepanto, Luzon, alt. 1,600 m., also without flowers.

LABIATÆ.

SCUTELLARIA Linn.

Scutellaria luzonica Rolfe in Journ. Linn. Soc. Bot. **21** (1884) 315.

On ledges along the Binabay River at 200 m. alt. (No. 5640), a form with much larger leaves than the type.

Luzon and Formosa.

GOMPHOSTEMMA Wall.

Gomphostemma philippinarum Benth. in DC. Prodr. **12** (1848) 551.

In old clearings at 900 m. alt. (No. 5581).

Throughout the Philippines at higher altitudes. Endemic.

SOLANACEÆ.

SOLANUM Linn.

Solanum parasiticum Blume Bijdr. (1826) 697; Prain ex King in Journ. As. Soc. Beng. **74**[2] (1905) 330.

Epiphytic in forests at 100 m. alt. (No. 6157).

This is the form that has been reported from the Philippines as *Solanum blumei* Nees (873 *Cuming*), but judging from the descriptions it is nearer *S. parasiticum* Bl. I refer here the following specimens: PHILIPPINES (837 *Cuming*). MINDANAO, Davao (329 *Copeland*); Lake Lanao, Camp Keithley (428 *Clemens*).

Malayan Peninsula, Java and Sumatra.

Solanum nigrum Linn. Sp. Pl. (1753) 329.
In old clearings at 900 m. alt. (No. 5571).
Widely distributed in the Philippines; a weed.
Tropical and temperate regions generally.

SCROPHULARIACEÆ.

VANDELLIA Linn.

Vandellia grandiflora Merr. in Philip. Journ. Sci. 1 (1906) Suppl. 237.
In forests at 1,500 m. alt. (No. 4401 *Merritt*) June, 1906.
Previously known only from the highlands of northern Luzon.

TORENIA Linn.

Torenia polygonoides Benth. Scroph. Ind. (1835) 39.
In an old clearing at 1,050 m. alt. (No. 5495).
Widely distributed in the Philippines but nowhere abundant.
British India to the Malayan Peninsula and Borneo.

GESNERIACEÆ.

TRICHOSPORUM Don.

Trichosporum philippinense (Clarke) O. Ktz. Rev. Gen. Pl. (1891) 478.
In forests at 1,300 m. alt. (No. 6141), ascending to 2,200 m.; also collected by *Merritt* in June, 1906, at 1,300 m. alt. (No. 4379).
Previously known only from Luzon.

Trichosporum rubrum Merr. in Philip. Journ. Sci. 1 (1906) Suppl. 227.
In forests at 1,400 m. alt. (No. 5769); also collected by *Merritt* in June, 1906, at 2,200 m. alt. (No. 4450).
Previously known only from northern Luzon, with a closely related if not identical form from Canlaon Volcano, Negros.

DICHROTRICHUM Reinw.

Dichrotrichum chorisepalum Clarke in DC. Monog. Phan. 5 (1883) 53.
In forests at 1,800 m. alt. (No. 6142).
Previously known from the mountains of Luzon, Negros and Mindanao.

CYRTANDRA Forst.

Cyrtandra cumingii Clarke in DC. Monog. Phan. 5 (1883) 263.
In forests at 1,500 m. alt. (No. 5578).
Widely distributed in the Philippines. Endemic.

Cyrtandra parvifolia Merrill n. sp.

Ramis gracilibus, glabris, junioribus plus minus ferrugineo-hirsutis; foliis oppositis, lanceolatis vel oblongo-lanceolatis, basi acutis, plus minus repando-crenatis vel subintegris, 2 ad 4.5 cm. longis; pedicellis axillaribus, solitariis, elongatis, medio bibracteolatis bracteolis minutis; calyce persistente; corolla circa 1.5 cm. longa.

A slender shrub 1 to 3 m. high. Branches light gray or brownish, glabrous, slender, terete, the younger ones more or less ferruginous-hirsute. Leaves opposite, lanceolate to oblong-lanceolate, 2 to 4.5 cm.

long, 0.5 to 1.5 cm. wide, the base inequilateral, acute, the apex acute or somewhat acuminate, the tip blunt, the margins slightly repand-crenate or subentire, submembranous, glabrous above, paler and somewhat ferruginous-hirsute on the midrib beneath or quite glabrous; nerves 4 to 5 on each side of the midrib; petioles 1 cm. long or less, ferruginous-hirsute. Flowers axillary, solitary, long-pedicelled, about 1.5 cm. long, the pedicels sparingly hirsute, 1 to 2 cm. long, slender, bibracteolate at about the middle, the bracts narrow, 2 mm. long or less. Calyx glabrous or nearly so, the tube broad, about 3 mm. long, the teeth about 4 mm. long, broadened at the base, narrowed abruptly and linear-lanceolate above, persistent and slightly accrescent in fruit. Corolla about 1.5 cm. long, glabrous, the lobes narrowly-ovate, obtuse, 5 to 6 mm. long. Samens 2; anthers broad, about 1.2 mm. long. Style slightly hirsute. Fruits ovoid, fleshy, dark purple, glabrous, about 5 mm. long.

In forests at 1,800 m. alt. (Nos. 5718, 5777). The same species, but with shorter petioles and somewhat narrower leaves has been collected on Mount Malindang, Mindanao, at an altitude of 1,700 m. (No. 4753 *Mearns and Hutchinson*) May, 1906. A sterile specimen from Canlaon Volcano, Negros, collected by *Banks* in March, 1902, with more strongly sinuate leaves is probably referable here.

Cyrtandra sp.

In forests at 900 m. alt. (No. 4350 *Merritt*) June, 1906.

A characteristic, apparently undescribed species, with very long petioles, but the material rather imperfect.

Cyrtandra sp.

In forests at 1,400 m. alt. (No. 5770), an undershrub 1 to 1.5 m. high.

LENTIBULARIACEÆ.

UTRICULARIA Linn.

Utricularia orbiculata Wall. Cat. (1828) No. 1500.

On seepy slopes, open heath at 2,400 m. alt. (No. 6168); flowers pale purple. Not previously reported from the Philippines.

Southeastern Asia, through the Malayan Peninsula to Mount Kinabalu, North Borneo.

Utricularia sp.

On rocks along the Alag River at 150 m. alt. (No. 5547). Possibly referable to the preceding, but the material very imperfect.

ACANTHACEÆ.

JUSTICIA Linn.

Justicia luzonensis C. B. Clarke in Govt. Lab. Publ. 35 (1905) 91.

In damp shaded ravines along the Alag River at 150 m. alt. (No. 5622). Previously collected on the Baco River, near the base of Mount Halcon by *Merrill* (No. 1778), April, 1903, and by *McGregor* (No. 156) March, 1905.

Known only from Luzon and Mindoro.

ERANTHEMUM Linn.

Eranthemum curtatum C. B. Clarke in Govt. Lab. Publ. **35** (1905) 89.

In thickets near the Alag River at 100 m. alt. (No. 6153). Previously collected on the Baco River, near the base of Mount Halcon by *Merrill* (No. 1779) April, 1903, and by *McGregor* (No. 144) March, 1905.

Known only from Luzon, Mindoro and Ticao.

STROBILANTHES Blume.

Strobilanthes halconensis Merrill n. sp.

Subglabrus; foliis oppositis, inaequalibus, usque ad 19 cm. longis, 8 cm. latis, longe subcaudato-acuminatis; spicis 5 ad 8 cm. longis; bracteis aculeatis, 5 mm. longis, in paribus distantibus; corolla 2 cm. longa; filamentis pilis longis ornatis.

Erect or ascending, much branched, glabrous except the somewhat aculeate bracts and sepals, 1 to 2 m. high. Leaves opposite, unequal, 5 to 19 cm. long, 1.5 to 8 cm. wide, ovate-lanceolate, the apex rather slenderly subcaudate-acuminate, the base acute, the margins subentire or obscurely crenate-dentate; nerves about 6 on each side of the midrib; petioles 2 cm. long or less. Spikes many, axillary, solitary, 5 to 8 cm. long; bracts in pairs, rather distant, ovate, obtuse, 5 mm. long, aculeate-hispid. Calyx segments 5 mm. long, oblong, obtuse, aculeate at the apices. Corolla white, 2 cm. long; stamens 4; filaments clothed with stout, brittle, jointed hairs.

In thickets bordering the forest at an altitude of 900 m. (No. 5586). Also collected by *Merritt* (No. 4370), at 920 m. alt. in June, 1906.

Perhaps as closely related to *Strobilanthes merrillii* Clarke, as to any other Philippine species, but distinct.

STAUROGYNE Wall.

Staurogyne debilis (Andres) C. B. Clarke in herb. *Ebermaiera debilis* Andres in Journ. Linn. Soc. Bot. **9** (1867) 452, in nota; Vidal Rev. Pl. Vasc. Filip. (1886) 203. *Ebermaiera elongata* Nees in DC. Prodr. **11** (1847) 721 var. β only. *Erythracanthus elongatus* Nees l. c. 78, var. β only.

On ledges along the Binabay River at 200 m. alt. (No. 5554).

Luzon, Negros and Mindanao.

RUBIACEÆ.

HEDYOTIS Linn.

Hedyotis hispida Retz. Obs. **4** (1779–91) 23.

In an old clearing at 100 m. alt. (No. 5694).

Previously recorded from the Philippines only by *F.-Villar,* Nov. App. (1883) 107, also represented in our herbarium by specimens from Rizal Province, Luzon, 1108 *Ramos; 3312 Ahern's collector.*

British India to southern China, the Malayan peninsula and archipelago.

Hedyotis elmeri Merr. in Philip. Journ. Sci. **1** (1906) Suppl. 127.

No. 4381 *Merritt*, June, 1906, altitude not given.

Endemic in the Philippines.

Hedyotis congesta R. Br. in Wall. Cat. No. 844.
In an old clearing at 700 m. alt. (No. 5531).
Widely distributed in the Philippines.
Malayan peninsula and archipelago.

Hedyotis eucapitata Merrill n. sp.

Frutex vel suffrutex 0.6 ad 1.4 m. altus, ramis ramulisque gracilibus aut crassiusculis, puberulis, foliis oblongo-ovatis vel lanceolato-ovatis, acutis vel acuminatis, 3 ad 7 cm. longis, pubescentibus, nervis 3 ad 5 utrinque, stipulis liberis, 4 ad 5 mm. longis, fimbriatis, pubescentibus; inflorescentiae axillares, pedunculo 2 ad 4 cm. longo, puberulo; floribus capitato-congestis, bracteis foliaceis plus minusve involucratis.

Suffrutescent or woody, erect, much branched, 0.6 to 1.4 m. high. Branches slender or somewhat thickened, brown, pubescent, becoming glabrous. Leaves membranous, oblong-ovate to ovate-lanceolate, 3 to 7 cm. long, 1 to 2.5 cm. wide, the apex acute or sharp-acuminate, the base acute, both surfaces pubescent with weak scattered hairs, dull; nerves 3 to 5 on each side of the midrib, ascending, distinct; petioles pubescent, 0.5 to 1.5 cm. long; stipules pubescent, laciniate, the laciniæ setiform, pubescent, 4 to 7 mm. long. Inflorescence axillary, the peduncles pubescent, 2 to 4 cm. long, slender, each bearing a single terminal head of sessile flowers, 1 cm. in diameter or less, the bracts foliaceous, forming an involucre, the bracts and flowers more or less hispid-pubescent, the pedicels 1 mm. long or less. Calyx tube about 1 mm. long, the lobes 4, linear or linear-lanceolate, 2.5 mm. long, the mature capsule 2 to 2.5 mm. long.

In forests at 1,800 m. alt. (No. 5726); also collected by *Merritt* at 1,600 m. alt. (Nos. 4417, 4430).

A species evidently closely related to *Hedyotis macrostegia* Stapf, from Mount Kinabalu, North Borneo, differing from that species in its pubescent, fewer nerved leaves, much shorter stipules, and other characters. It is distinguished from all other Philippine species of the genus by its long peduncled capitate solitary axillary inflorescence.

Hedyotis whiteheadii Merrill n. sp.

Frutex 2 ad 2.6 m. altus, ramis ramulisque crassiusculis tetragonis, glabris, foliis rigide coriaceis, ovatis, acutis, glabris, nitidis, 1.5 ad 3 cm. longis, basi late rotundatis aut subtruncatis; inflorescentiae axillares, pauciflorae, cymosae; corolla alba, 6 mm. longa.

An erect branched shrub glabrous throughout, 2 to 2.6 m. high. Branches stout, tetragonous, green or brown, smooth. Leaves ovate, usually broadly so, 1.5 to 3 cm. long, 1 to 1.8 cm. wide, the base broad, rounded or subtruncate, gradually narrowed above to the acute apex, rigid coriaceous, shining, the margins often recurved; nerves 3 to 4 on each side of the midrib, not prominent; petioles stout, 2 mm. long or less; stipules 3 to 4 mm. long, usually trifid. Inflorescence axillary, few flowered, cymose, the peduncles 1.5 cm. long or less, the branches few, the bracts and bracteoles foliaceous, the former oblong-ovate, 0.8 mm. long, the latter oblong, about 3 mm. long; pedicels 2 mm. long. Calyx tube ovoid, 2 mm. long, the lobes 4, narrowly oblong, blunt, 1.5 mm. long. Corolla 6 mm. long, the tube broadened above, 3 mm. long, the lobes somewhat ciliate on the margins, 3 mm. long, oblong. Filaments 1.5 mm. long; anthers 1.4 mm. long. Style included 2 mm. long.

In dense thickets on exposed ridges at 2,500 m. alt. (No. 5783).

Named in honor of *John Whitehead* who first attempted to ascend Halcon.

Hedyotis montana Merrill n. sp.

Frutex 1 m. altus, ramis ramulisque teretis vel obscure tetragonis, dense puberulis, foliis rigide coriaceis, oblongo-ovatis, acuminatis, glabris, nitidis, 3 ad 4 cm. longis, dense confertis, basi acutis; inflorescentiae axillares et terminales, pauciflorae, cymosae; corolla purpurea, 11 mm. longa.

An undershrub about 1 m. high. Branches stout, terete or obscurely tetragonous, brown, rather densely puberulous. Leaves oblong-ovate, rigid, coriaceous, shining, 3 to 4 cm. long, 1.5 cm. wide or less, the margins recurved, the apex rather prominently acuminate, the base acute; nerves 3 to 4 on each side of the midrib, prominent beneath, ascending; petioles 1.5 to 3 mm. long; stipules short, trifid. Cymes axillary and terminal, all borne near the apices of the branchlets appearing like an interrupted terminal inflorescence, the peduncles puberulent, 1 cm. long or less, each bearing about 6 congested flowers, the bracts foliaceous, narrowly ovate. 4 mm. long, the bracteoles smaller; pedicels 0.5 to 1.5 mm. long. Calyx tube ovoid, 1.5 mm. long, the lobes 4, narrowly oblong, blunt, 2 mm. long. Corolla purple, 11 mm. long, the tube about 8 mm. long, the lobes 3 to 4 mm. long, narrowly oblong, blunt, slightly ciliate on the margins. Filaments 0.5 mm. long; anthers 2 mm. long. Style slightly exserted, 9 mm. long.

On an open heath at 2,400 m. alt. (No. 5782).

Closely related to the preceding species, differing in its puberulent branches, longer narrower leaves which are acuminate at the apex and acute at the base and with prominent nerves, and much larger purple flowers.

OPHIORRHIZA Linn.

Ophiorrhiza venosa Merrill n. sp.

Herba vel suffruticosa, simplex vel pauciramosa, usque ad 60 cm. alta, foliis longe petiolatis, membranaceis, oblongo-ellipticis, basi et apice acuminatis, glabris, nervis 15 ad 20 utrinque, prominentibus, petiolo 3 ad 5 cm. longo; stipulis acuminato-lanceolatis, 3 mm. longis; cymae terminales, ferrugineo-puberulae.

An erect herbaceous or suffrutescent plant, simple or slightly branched, about 60 cm. high, glabrous except the inflorescence. Stems stout. Leaves oblong-elliptical, 13 to 20 cm. long, 4.5 to 7 cm. wide, membranous, green above, pale beneath, somewhat shining, glabrous throughout, base and apex acuminate; nerves 15 to 20 on both sides of the midrib, very prominent and brownish beneath, spreading, anastomosing near the margins; petiole 3 to 5 cm. long, the lamina somewhat decurrent-acuminate; stipules lanceolate-acuminate, about 3 mm. long. Cymes terminal, ferruginous puberulent, the peduncle 4 to 5 cm. long, the branches 4 cm. long or shorter. Calyx 2.5 to 3 mm. long, the teeth 5, short, acute. Corolla white, 8 mm. long, the tube cylindrical, slightly inflated below, 6 mm. long, the lobes 5, elliptical-oblong, obtuse, 2 mm. long.

In humid forests at 150 m. alt. (No. 5628).

A species characterized by its glabrous, long petioled leaves, the nerves subparallel, very numerous and prominent.

Ophiorrhiza oblongifolia DC. Prodr. 4 (1830) 415.

In forests at 1,200 m. alt. (No. 5496), small leaved form; at about the same altitude (No. 4385 *Merritt*) June, 1906, a large leaved form.

Widely distributed in the Philippines. Endemic.

ARGOSTEMMA Wall.

Argostemma solaniflorum Elmer Leaflets Philip. Bot. 1 (1906) 2.

In forests at 1,350 m. alt. (Nos. 6105, 6186); also collected by *Merritt* (Nos. 4470, 4390).

Previously known only from northern Luzon.

UNCARIA Schreb.

Uncaria philippinensis Elmer Leaflets Philip. Bot. (1906) 38.

In thickets at 700 m. alt. (No. 5530).

Previously known from Luzon and Mindoro.

NAUCLEA Linn.

Nauclea sp.

In forests at 450 m. alt. (4337 *Merritt*) June, 1906. Undeterminable, the material being very fragmentary and in poor condition.

MUSSAENDA Linn.

Mussaenda anisophylla Vidal Phanerog. Cuming. Philip. (1885) 178.

In thickets at 250 m. alt. (No. 4330 *Merritt*) June, 1906.

Apparently widely distributed in the Philippines. Endemic.

UROPHYLLUM Wall.

Urophyllum glabrum Jack ex Roxb. Fl. Ind. ed. Carey, 2: 186.

In forests at 300 m. alt. (No. 5603).

From Burma through the Malayan Peninsula to the Malayan Archipelago.

Urophyllum bataanense Elmer Leaflets Philip. Bot. (1906) 40; Merr. in Philip. Journ. Sci. 1 (1906) Suppl. 129.

In forests at 1,500 m. alt. (Nos. 6144, 6179); also collected by *Merritt* as low as 450 m. alt. (Nos. 4318, 4339, 4389).

Previously known only from Luzon.

Urophyllum sp.

In forests at 400 m. alt. (No. 5580). Material imperfect, mature fruits only.

Urophyllum sp.?

In forests at 900 m. alt. (No. 5573). A form represented in our herbarium by several specimens from Mindanao, but unfortunately no flowers are available.

RANDIA Houst.

Randia sp.?

In forests at 1,800 m. alt. (No. 5522). Specimens with immature fruits.

IXORA Linn.

Ixora sp.

In forests at 550 m. alt. (No. 5569). A very characteristic species with pure white fruits and setiform much elongated stipules, apparently undescribed, but the specimens are without flowers.

PSYCHOTRIA Linn.

Psychotria sarmentosa Blume Bijdr. (1826) 964.

Mount Halcon, without data (No. 5683).

Previously known in the Philippines from Luzon.

British India to Malaya.

Psychotria tacpo (Blanco) Rolfe in Journ. Linn. Soc. Bot. **21** (1884) 312.

Without data (No. 6150).

Widely distributed in the Philippines. Endemic.

Psychotria diffusa Merrill in Philip. Journ. Sci. **1** (1906) Suppl. 134.

In forests, 900 to 1,800 m. alt. (Nos. 6176, 6170); also No. 4435 *Merritt*, June, 1906, alt. 1,700 m.

Previously known only from Luzon.

In addition to the above species of the genus, no less that six others are represented by the following numbers, from Halcon, all of them differing from the material at present in our herbarium. Unfortunately all the specimens are with fruit only, and accordingly no attempt is here made to describe them. Nos. 4324, 4349, 4365, 4396, 4456 *Merritt;* Nos. 6131, 6159, 5576 *Merrill.*

LASIANTHUS Jack.

Lasianthus copelandi Elmer Leaflets Philip. Bot. (1906) 10.

In forests at 300 m. alt. (No. 5778).

Previously known only from Negros, a species very closely related to *L. appressus* Hook. f., of the Malayan Peninsula.

Lasianthus obliquinervis Merr. in Philip. Journ. Sci. **1** (1906) Suppl. 136.

In forests at 1,400 m. alt. (No. 6189).

Previously known from Luzon and Negros.

Lasianthus tashiroi Matsum. in Tokyo Bot. Mag. **15:** 37.

In forests at 1,400 m. alt. (Nos. 5739, 5776).

I have based the identification of the above numbers largely on a Formosan specimen, No. 1301 *Kawakami*, so determined by *Hayata*.

Formosa.

HYDNOPHYTUM Jack.

Hydnophytum formicarium Jack in Trans. Linn. Soc. 14 (1823) 124.

Epiphytic, in forests along the Alag River below 100 m. alt. (No. 6182).

Malayan Peninsula, Cochin China, Sumatra and Borneo.

Hydnophytum nitidum Merrill n. sp.

Tuber diametro ad 25 cm.; cuales ramique lignescentes teretes vel ramuli juniores leviter compressi, foliis oblongo-ellipticis, coriaceis, nitidis, obtiusis, 3 ad 6 cm. longis, subsessilibus; flores breviter tubulosi, sessiles, ad articulationes fasciculati; corolla 3.5 mm. longa.

Tuber about 25 cm. in diameter, glabrous; the stems glabrous, brown or gray, terete, the younger branchlets reddish-brown and slightly compressed, 60 to 80 cm. long, branched. Leaves oblong-elliptical, 3 to 6 cm. long, 1 to 2.5 cm. wide, coriaceous, glabrous, the upper surface shining, the lower dull, margins slightly recurved, the apex rounded, the base subacute or rounded, the midrib prominent, the lateral nerves three or four on each side of the midrib, obscure or nearly obsolete; petiole very short or wanting. Flowers fasciculate, sessile at the nodes, few, white. Calyx truncate, glabrous. Corolla 3.5 mm. long, the tube cylindrical 2 mm. long, barbulate with tufts or hairs at the throat between the insertion of the stamens, the lobes 4, narrowly ovate, acute, 1.5 mm. long. Stamens 4, filaments wanting; anthers 0.8 mm. long. Style 3 mm. long, slightly cleft at the apex.

Epiphytic in the mossy forest at 1,400 m. alt. (No. 6181); also collected by *Merritt* at an altitude of about 800 m. (No. 4358).

A species characterized by its coriaceous shining elliptical-oblong leaves and small flowers, apparently belonging in the group with *H. formicarum* Jack, following *Beccari's* classification.[13]

NERTERA Banks & Soland.

Nertera depressa Banks & Soland. ex Gaertn. Fruct. 1 (1788) 124 *t. 26*.

On bowlders along shaded streams at 700 m. alt. (No. 5614); also collected by *Merritt* on exposed ridges at 2,250 m. alt. (No. 4459).

On many of the higher mountains of the Philippines.

Widely distributed in Malaya, Australia and South America.

CAPRIFOLIACEÆ.

SAMBUCUS Linn.

Sambucus javanica Reinw. ex Blume Bijdr. (1826) 657.

In old clearings at 900 m. alt. (No. 5572).

Widely distributed in the Philippines.

British India to Japan and Malaya.

[13] Malesia 2 (1884–85), 123–175.

CAMPANULACEÆ.

PENTAPHRAGMA Wall.

Pentaphragma philippinensis Merril n. sp.

Foliis amplis, membranaceis ovatis vel oblongo-ovatis, acuminatis, basi acutis inaequilateralibus, 20 ad 30 cm. longis, 10 ad 22 cm. latis, supra glabris, subtus plus minus tomentellis; floribus ad 4 cm. longis 5-meris; calycis lobis ovatis, 1.5 ad 2 cm. longis, tubo 5-angulato.

An erect unbranched suffrutescent herb 1 m. high or less, the stems thick, glabrous or nearly so, yellowish when dry. Leaves ovate or oblong-ovate, membranous, 20 to 30 cm. long, 10 to 22 cm. wide, glabrous above, beneath paler and somewhat tomentose, the margins rather finely crenate-dentate, the apex acuminate, base inequilateral, acute; nerves 5 to 6 on each side of the midrib, prominent, ascending, the reticulations lax; petioles 5 to 10 cm. long. Racemes axillary, few or many flowered, not unilateral, the peduncles short, the bracts membranous about 2 cm. long. Flowers white or greenish white when fresh, yellowish when dry, 5-merous, the pedicels 2 to 3.5 cm. long. Calyx tube 1.5 to 2 cm. long, oblong, narrowed below, 5-angled, glabrous, the lobes 5, ovate, acute, two nearly 2 cm. long and 1.3 cm. wide, three 1.5 cm. long and 0.7 mm. wide. Corolla lobes 5, glabrous, equal, oblong-ovate, acute, about 1 cm. long, 0.5 cm. wide, the tube short. Stamens 5, filaments 2 mm. long, the anthers about the same length. Ovary 5-celled, ovules very numerous; style 5 to 6 mm. long; stigma oblong-ovoid, 5-ridged.

In forests along the Alag River at 100 m. alt. (No. 6136), ascending to 1,500 m. alt.; also collected by *Merritt*, in June, 1906 (No. 4333). In addition to the above specimen, the following are referable here, all from Mindanao: Province of Misamis, Mount Malindang (4702 *Mearns & Hutchinson*) May, 1906; Province of Surigao (354 *Bolster*) May, 1906; Lake Lanao, Camp Keithley (229 Mrs. *Clemens*) February, 1906.

Apparently most closely related to *P. macrophylla* Oliv., from New Guinea, differing from that species, as described in its smaller leaves, longer bracts and 5-angled, not terete, calyx tube.

The genus is new to the Philippines, the known species being *P. begoniaefolium* Wall., from Burma and the Malayan Peninsula, *P. scortechinii* King & Gamb., and *P. ridleyi* King & Gamb., from the Malayan Peninsula and Singapore, *P. aurantiaca* Stapf, from Mount Kinabalu, North Borneo, *P. macrophylla* Oliv., from New Guinea and *P. grandiflorum* Kurz from the Moluccas.

COMPOSITÆ.

MIKANIA Willd.

Mikania scandens (Linn.) Willd. Sp. Pl. 3 (1800) 1743. *Willughbaeya scandens* O. Kunze.

In thickets at 100 m. alt. (No. 5699).

Throughout the Philippines.

Cosmopolitan in the Tropics.

PLUCHEA Cass.

Pluchea scabrida DC. Prodr. **5** (1836) 453.
In an old clearing at 300 m. alt. (No. 5565).
Luzon to Mindanao. Endemic in the Philippines.

DICHROCEPHALA DC.

Dichrocephala latifolia DC. Prodr. **5** (1836) 372.
In an old clearing at 300 m. alt. (No. 5584).
Previously known in the Philippines only from Luzon.
Tropical Africa, to China and Japan.

LAGENOPHORA Cass.

Lagenophora billardieri Cass. Dict. Sc. Nat. **25** (1826) 111.
In forests at 2,000 m. alt. (No. 4442 *Merritt*) June, 1906.
Previously known in the Philippines only from the mountains of northern and central Luzon.
British India to Japan, Malaya and northern Australia.

SENECIO Linn.

Senecio mindoroensis Elm. Leaflets Philip. Bot. **1** (1906) 155.
In an old clearing at 300 m. alt. (No. 5570); also collected by *Merritt* (No. 4402) in June, 1906, at 1,500 m. alt., the latter specimens not typical.
Luzon to Mindanao. Endemic in the Philippines.

BIDENS Linn.

Bidens pilosa Linn. Sp. Pl. ed. 2 (1763) 832.
In an old clearing at 300 m. alt. (No. 5566).
Widely distributed in the Philippines.
Temperate and tropical regions of the World.

AINSLIAEA DC.

Ainsliaea reflexa Merr. in Philip. Journ. Sci. **1** (1906) Suppl. 242.
In mossy ridge forests at 2,400 m. alt. (No. 5781).
Previously known only from the mountains of Luzon.

LACTUCA Linn.

Lactuca thunbergiana (A. Gray) Maxim. in Bull. Acad. Pétersb. **19** (1874) 530.
In crevices of bowlders and ledges along the Alag River 100 to 300 m. alt. (No. 6143).
From northern Luzon to southern Mindanao, but usually at much greater altitudes.
Japan, southern China and Formosa.

CREPIS Linn.

Crepis japonica (DC.) Benth. Fl. Hongk. (1861) 194.
In an old clearing at 300 m. alt. (No. 5583).
Japan, southern China and India through Malaya to northern Australia.
Widely distributed in the Philippines.

ORCHIDACEÆ HALCONENSES: AN ENUMERATION OF THE ORCHIDS COLLECTED ON OR NEAR MOUNT HALCON, MINDORO, CHIEFLY BY ELMER D. MERRILL.

By Oakes Ames, A. M., F. L. S.

Most of the orchirds enumerated in this paper were collected by *Elmer D. Merrill* in Mindoro at high altitudes on Mount Halcon. In addition, those species have been included here which were found near Mount Halcon during April and May, 1905, by *R. C. McGregor*, and in June, 1906, by *M. L. Merritt*. Among the species are many known to come from Java, Sumatra and Borneo; and, aside from the new ones, which constitute about half of the orchids in this paper, there are several interesting additions to the Philippine flora.

I have adopted the sequence of genera proposed by *Pfitzer* in *Engler & Prantl's* "Die natürlichen Pflanzenfamilien" and have arranged the species alphabetically under their respective genera. The notes relative to the geographical distribution of species that occur outside of the Philippine Islands have been made from lists and floras; therefore they are not definitive, as closely allied species are not infrequently confused by authors, and their ranges consequently exaggerated.

The types of the new species herein described are in the herbarium of the Bureau of Science at Manila. Usually cotypes are to be found in my own herbarium.

NEUWIEDIA Bl.

N. veratrifolia Bl. in Hoev. & De Vriese, Tijdschr. 1: 142 (1834).

Terrestrial, ridge forest, at 2,300 ft. alt. on Mount Halcon, flowers yellow, Nov. 8, 1906, *Merrill* (No. 5681).

APOSTASIA Bl.

Apostasia Wallichii R. Br. Wall. Cat. 4448 (1828).

Terrestrial in ridge forest at 3,000 ft. alt. on Mount Halcon, flowers yellowish (old), probably white when young, Nov. 10, 1906, *Merrill* (No. 5521). Terrestrial in humid forest on ridge at about 1,000 ft. alt. on Mount Halcon, flowers odorless, yellowish white, Nov. 2, 1906, *Merrill* (No. 5639).

Nepal, Assam, Khasia, Perak, Ceylon, Penang, Sumatra, Borneo, Java, New Guinea. Doubtfully ascribed to Luzon, P. I.

HABENARIA Willd.

Habenaria (§ Seticaudae) alagensis Ames sp. nov.

Plants about 6 dm. high, leafy at the base. Leaves oblong-lanceolate, acuminate, acute, about 2 dm. long, about 3 cm. wide. Floral bracts lanceolate, acuminate, very acute, 1.5–2 cm. long, shorter than the ovary. Ovary 3 cm. long, narrowed above into an elongated, slender neck. Flowers greenish, odorless, in an elongated loose raceme. Lateral sepals somewhat elliptical, deflexed, 1 cm. long. Upper sepal strongly concave, 1.4 cm. long. Petals 1.4 cm. long, 1 mm. wide, simple, linear above the dilated base, where they are 2.5 mm. wide and anteriorly protuberant. Labellum about 1.2 cm. long simple, strongly deflexed beyond the middle, caudate-tipped. Spur about 2 cm. long, very slender above the middle, strongly incurved and dilated-clavate toward the apex, resembling somewhat the abdomen of an inchneumon fly (*Ophion purgatus*). Anther canals much exceeding the stigmatic processes.

Terrestrial in humid forest at 400 ft. alt. along the Alag River, Nov. 5, 1906, *Merrill* (No. 5803).

Habenaria angustata (Bl.) O. K. Rev. Gen. Pl. 2: 664 (1891).—*Mecosa angustata* Bl. Bijdr. 404, fig. 1 (1825).

Terrestrial in mossy forest at 8,000 ft. alt. on Mount Halcon, Nov. 20, 1906, *Merrill* (No. 5802).

The material on which my determination is based is not very satisfactory for a sure diagnosis.

Java, Borneo.

Habenaria (§ Mecosa) halconensis Ames sp. nov.

Plant about 3 dm. high. Leaves much reduced, cordate-ovate to lanceolate, acute about 4 cm. long, 2–3 cm. wide, passing above into clasping lanceolate, acute bracts. Floral bracts exceeding the ovaries, lanceolate, acute, about 1 cm. long. Flowers green, in a loose, erect raceme. Lateral sepals linear-oblong, acute, about 1 cm. long, 2.5–3 mm. wide. Upper sepal broadly ovate, obtuse, 9 mm. long, about 6 mm. wide. Petals simple, lanceolate, falcate, obtuse, 8 mm. long, 4.5 mm. wide at base. Labellum simple, linear-oblong, tapering gradually to the obtuse tip, 11 mm. long, 3 mm. wide at base, 1 mm. wide near the tip. Spur nearly straight, slightly exceeding the labellum in length, 12 mm. long. (Uppermost flowers slightly smaller in all their parts than the lowermost.)

Terrestrial in open heath, at 8,000 ft. alt. on Mount Halcon, Nov. 20, 1906, *Merrill* (No. 5835).

The material from which the above description is drawn is somewhat scrappy and not well provided with foliage. The flowers are much like those of *Habenaria angustata* O. K., but considerably larger and different in detail.

CRYPTOSTYLIS R. Br.

Cryptostylis arachnites (Bl.) Hasskarl Cat. Bog. 48 (1844); Reichb. f. Bonpl. **5:** 36 (1857).—*Zosterostylis arachnites* Bl. Bijdr. 419, *fig. 32.* (1825).

Terrestrial in forest at 3,000 ft. alt. on Mount Halcon, petioles and leaves mottled. Nov. 10, 1906, *Merrill* (Nos. 5478, 5488).

Ceylon, S. India, Malay Peninsula, Java, Borneo.

GALEOLA Lour.

Galeola Hydra Reichb. f. Xen. Orch. **2:** 77 (1862).—*Galeola Kuhlii* Reichb. f. Xen. Orch. **2:** 78.—*Erythrorchis Kuhlii* Reichb. f. *loc. cit.* t. 119.

Near the Baco River, saprophyte on dead tree trunk, Apr. 11, 1903, *Merrill* (No. 1811).—Baco River, Apr.–May, 1905, *R. C. McGregor* (No. 225).

"Whole plant yellowish, leafless. This species grew on a dead stump near the edge of an old clearing, in a very humid forest apparently rooting in the ground at the base of the stump. The Baco Valley is a very broad, mostly heavily timbered plain not much above the sea level (perhaps 10 m.), 6 or 8 miles inland. It is a very humid locality owing to the proximity of Mount Halcon to the north and west."—*Field-notes by the collector.*

India, Tenasserim, Penang, Perak, Singapore, Malacca, Java and Sumatra.

APHYLLORCHIS Bl.

Aphyllorchis pallida Bl. Bijdr. *fig.* 77 (1826), Mus. Bot. Lugd.-Bat. **1:** 30.

Terrestrial in humid forest on ridges at 1,000 ft. alt. along the Binabay River, Nov. 2, 1906, *Merrill* (No. 5810).—Terrestrial in ridge forest, at 2,800 ft. alt. on Mount Halcon, Nov. 10, 1906, *Merrill* (No. 5811).

Java.

VRYDAGZYNEA Bl.

Vrydagzynea albida Bl. Fl. Jav. Orch. 62, *t. 19, f.* 2. (1858).—*Etaeria albida* Bl. Bijdr. 410 (1825).

Terrestrial in forest at 850 ft. alt. along the Binabay River, flowers white, Nov. 4, 1906, *Merrill* (No. 6125).

The material on which the above determination is based differs in several minor details from the specimens preserved at Leiden, which constitute *Blume's* type, but is not sufficiently different for separate treatment at this time.

Java, Sumatra.

CYSTORCHIS Bl.

Cystorchis aphylla Ridl. Journ. Linn. Soc. **32:** 400 (1896).

Terrestrial on forested ridge at about 1,000 ft. alt. on Mount Halcon, plant white, bracts pinkish, Nov. 8, 1906, *Merrill* (No. 5662).—Sapropyte, ridges in forest, about 1,000 ft. alt. along the Binabay River, whole plant salmon pink, base and tips of floral segments paler, Nov. 2, 1906, *Merrill* (No. 5797).

Malay Peninsula and Java.

HERPYSMA Lindl.

Herpysma Merillii Ames sp. nov.

Plants 1–3 dm. high. Rhizome creeping. Leaves about 5, ovate-lanceolate, shortly acuminate, acute, rounded at the base, 5–7 cm. long, about 3 cm. wide, passing into slender petioles. Base of the petioles scarious, sheathing the stem. Peduncle sparsely pubescent, with about 3 lanceolate, scarious, about 1 cm. long bracts below the loose, racemose inflorescence. Floral bracts linear-lanceolate, acute, scarious, about 1 cm. long. Flowers white. Lateral sepals linear-oblong, about 1.2 cm. long, 2 mm. wide, concave, lightly carinate, cucullate with several hairs at the tip. Upper sepal similar to the laterals, broader, adhering lightly to the petals. Petals 1.2 cm. long, spathulate, obtuse, linear below the middle, free from each other at the base, cohering above the middle by their inner margins. Labellum adhering to the column, produced at base into a rather slender bilobed-tipped spur which protrudes between the lateral sepals; free portion narrow, a little dilated beyond the column, then 4-lobed; proximal lobes divaricate, oblong, obtuse, 1.5 mm. long, 1 mm. wide, separated from the distal lobes by a short 1 mm. long claw or isthmus; distal lobes divaricate, 1.5 mm. long, about 2 mm. wide, margin irregular; on the disc two thin longitudinal lamellæ, free at the obliquely truncate apex. Two wart-like calli are situated in the spur near its base, on the dorsal wall. From tip of spur to apex of labellum 1.5 cm.

Terrestrial in damp ravine, by small stream on Mount Halcon, Nov. 9, 1906, *Merrill* (No. 5836).

ZEUXINE Lindl.

Zeuxine luzonensis Ames Orchidaceæ, fasc. 2 (1907) *ined.*

Terrestrial in forest at 2,800–4,000 ft. alt. on Mount Halcon, flowers white with faint odor, sepals purplish green or brownish, Nov. 13, 1906, *Merrill* (No. 5841).

CHEIROSTYLIS Bl.

Cheirostylis octodactyla Ames sp. nov.

Related to *C. Griffithii* Lindl. Plants rather stout in relation to their height, 4–8 cm. tall, few-flowered. Leaves ovate-lanceolate, acute, 0.7–2 cm. long, 5–11 mm. wide, 5–6 mm. apart on the stem. Petioles short, sheathing at base. Flowers one or two, white, 1 cm. long, standing at right angles to the erect stem. Lateral sepals united nearly to the apex, lanceolate, subacute, 8 mm. long. Upper sepal oblong-lanceolate, acute, somewhat dilated near the base, 8 mm. long. Petals lightly adhering to the upper sepal, linear-spathulate, about 8 mm. long, 2 mm. wide near the tip. Labellum linear-oblong, about 9 mm. long to the tip of the slightly dilated 8-fingered apex, 2.5 mm. wide at base; on each side a row of 7–8 setæ. Digitate divisions of the apex 3 mm. long.

Terrestrial in very dense mossy thicket on ridge at 8,200 ft. alt. on Mount Halcon, Nov. 22, 1906, *Merrill* (No. 5834).

GOODYERA R. Br.

Goodyera sp.

Flowers too much withered for analysis.

Leaves slightly mottled, *i. e.*, the veins and veinlets white, above, inflorescence old, Mount Halcon, Nov. 16, 1906, *Merrill* (No. 5512).

MYRMECHIS Bl.

Myrmechis gracilis Bl. Fl. Jav. Orch. 64, *t. 21, f. 2* (1858).—*Anoectochilus gracilis* Bl. Bijdr. 413 (1825).

Mount Halcon, June 15–27, 1906, *M. L. Merritt* (Nos. 4432, 4460).

Java, Japan.

HAEMARIA Lindl.

Haemaria Merrillii Ames sp. nov.

Plants erect or ascending, up to 3 dm. high, leafy at the base. Leaves 5–7, ovate-lanceolate, acute, 1.5–3 cm. long, 1–1.5 cm. wide. Peduncle pubescent, provided with about 5 closely appressed, lanceolate, scarious 1–2 cm. long bracts. Flowers whitish, in a rather dense, short raceme. Floral bracts exceeding the ovaries, narrowly lanceolate, margin ciliate. Lateral sepals triangular-lanceolate, obtuse, 1-nerved, about 6 mm. long, slightly exceeding 2 mm. in width at the base. Upper sepal adhering lightly to the petals, lanceolate, 1-nerved, about 6 mm. long, slightly exceeding 2 mm. in width. Labellum about 5 mm. long, broadly unguiculate from a short, round, saccate base; margins of the claw erect (involute?) above; the claw is dilated into a transversely oblong, retuse, mucronate lamina with a crenate or bluntly-toothed margin; claw about 3 mm. long; lamina 4 mm. wide, 2 mm. long. Within the sac 2 roundish sessile calli are situated.

Terrestrial in mossy forest at 4,300 ft. alt. on Mount Halcon, flowers white, odorless, calyx greenish, Nov. 13, 1906, *Merrill* (Nos. 5840, *type*, and 5819).

HYLOPHILA Lindl.

Hylophila rubra Ames Orchidaceæ, fasc. 2 (1907) *ined.*

At 4,300 ft. alt. on Mount Halcon, flowers brownish red, June 15–27, 1906, *M. L. Merritt* (No. 4378).

TROPIDIA Lindl.

Tropidia mindorensis Ames sp. nov.

Plants about 4 dm. tall, slender, graceful, branching. Stems about 2 mm. thick, clothed by the sheathing bases of the numerous, alternate, linear-lanceolate, acuminate, acute, 3–5-nerved, 10–15 cm. long, 1–2 cm. wide leaves. Racemes leaf-opposed, short, few-flowered, not exceeding 3 cm. long. Peduncle clothed with imbricating, nervose bracts. Pedicels comparatively stout, about 5 mm. long. Flowers yellowish white, tinged with green, odorless, about 1.2 cm. long. Lateral sepals about 1.2 cm.

long, 3 mm. wide near the base, oblong-lanceolate, acute. Upper sepal about 9 mm. long, ovate-lanceolate, acute. Petals narrowly lanceolate, slightly carinate dorsally along the median nerve, 7 mm. long, about 2.5 mm. wide. Labellum 7 mm. long, apex strongly deflexed, subacute, saccate; lamina somewhat rhombic in outline, intramarginally bicarinate.

In humid forest, terrestrial, at 900 ft. alt. along the Binabay River, Nov. 2, 1906, *Merrill* (No. 5552).

In habit *T. mindorensis* recalls *T. graminea* Bl., differing from our species mainly in its smaller, terminal raceme and flowers. The measurements given above for the height of the plant apply to specimens which appear to have been broken. Specimens with roots were not collected by Mr. *Merrill.*

NEPHELAPHYLLUM Bl.

Nephelaphyllum mindorense Ames. sp. nov.

Closely allied to *N. pulchrum* Bl.

Plants about 2 dm. tall. Rhizome creeping, slender, rooting at intervals. Stems purple. Leaves with the under surface uniformly dark purple, upper surface mottled, ovate-lanceolate, acuminate, acute, 8–10 cm. long, 3–5.5 cm. wide near the base. Petioles relatively slender, about 3 cm. long. Peduncles exceeding the leaves, clothed with several scarious, tubular, acute sheaths. Inflorescence loosely few-flowered. Bracts about 1 cm. long, linear, acute, scarious, somewhat shorter than the pedicels of the white flowers. Lateral sepals linear-acute, 1-nerved, 9 mm. long, 1.5 mm. wide. Upper sepal similar and equal to the laterals. Petals oblong, acute, slightly broader above than below the middle, 1-nerved, about 8 mm. long, 3 mm. wide. Labellum suborbicular, entire 9–10 mm. long, 9–10 mm. wide, with 3 prominent converging lamellæ near the apex, which pass basally into the main nerves of the hairy disc. Spur blunt, inflated, 4–5 mm. long.

Terrestrial in humid forest at about 900 ft. alt. along the Binabay River, Nov. 2, 1906, *Merrill* (No. 5623).

CHRYSOGLOSSUM Bl.

Chrysoglossum villosum Bl. Bijdr. 338, *f.* 7 (1825).

Terrestrial, flower odorless, or nearly so, petals twisted, dark purple in the middle, Binabay River, Nov. 2, 1906, *Merrill* (No. 5838).—Mount Halcon, June 15–27, 1906, *M. L. Merritt* (No. 4398).

The specimens on which the above determination is based agree too well with *C. villosum* Bl. to be separated from it specifically.

Java, Borneo, Perak.

DENDROCHILUM Bl.

Dendrochilum (§ Platyclinis) arachnites Reichb. f. Gard. Chron. n. s. 17: 256 (1882).

Epiphyte at 6,800 ft. alt. on Mount Halcon, flowers pale greenish, odorless, Nov., 1906, *Merrill* (No. 5511).

Endemic in the Philippines.

Dendrochilum (§ **Acoridium**) **bicallosum** Ames Orchidaceæ, fasc. 2 (1907) *ined.*

Epiphyte in ridge forest at 2,500 ft. alt. on Mount Halcon, flowers brownish red, Nov. 28, 1906, *Merrill* (No. 5812).

Var. **minor** Ames Orchidaceæ, fasc. 2 (1907) *ined.*

On ledge in ridge forest at about 1,500 ft. alt. on Mount Halcon, flowers pale salmon color, odorless, Nov. 8, 1906, *Merrill* (No. 5663).

Dendrochilum (§ **Acoridium**) **exile** Ames Orchidaceæ, fasc. 2 (1907) *ined.*

Epiphyte at 6,800 ft. alt. on Mount Halcon, flowers greenish yellow, odorless, Nov. 15, 1906, *Merrill* (No. 5721).

Related to *D. tenellum* Ames and *D. Williamsii* Ames, from which it differs in its much shorter leaves and very different labellum.

Dendrochilum (§ **Platyclinis**) **glumaceum** Lindl. Bot. Reg. 1841, Misc. p. 23.

Flowers white, slightly fragrant, Mount Halcon, June 15–27, 1906, *M. L. Merrit* (No. 4373).

Dendrochilum (§ **Acoridium**) **hastatum** Ames Orchidaceæ, fasc. 2 (1907) *ined.*

On bases of trees, flowers purplish, at 6,800 ft. alt. on Mount Halcon, Nov., 1906, *Merrill* (No. 5759).—Near same locality, June 15–27, 1906, *M. L. Merritt* (No. 4411).

Dendrochilum hastatum is clearly distinguished from all other species known to be natives of the Philippines by its hastate labellum.

Dendrochilum (§ **Acoridium**) **Hutchinsonianum** Ames Orchidaceæ, fasc. 2 (1907) *ined.*

In dense mossy thickets, on exposed ridge, both terrestrial and epiphytic at 8,000 ft. alt. on Mount Halcon, flowers flesh colored, odorless, Nov., 1906, *Merrill* (No. 5813).

D. Hutchinsonianum is related to *D. pumilum* Reichb. f. from which it differs in its larger proportions and flowers.

Dendrochilum (§ **Platyclinis**) **magnum** Reichb. f. Walp. Ann. **6**: 240 (1861).

Epiphyte at 6,000 ft. alt. on Mount Halcon, flowers pale brownish, odorless. Nov. 15, 1906, *Merrill* (No. 5730).—Epiphyte in mossy forest on Mount Halcon, Nov., 1906, *Merrill* (No. 5611).

Although the labellum is not 3-lobed in any of the specimens examined, I am of the opinion that my determination of the Mount Halcon material, while provisional, is the only safe one, waiting an examination of the type of *D. magnum*. From a sketch of the labellum in Lindley's herbarium at Kew it would seem highly probable that the 3-lobed character was the result of malformation or injury.

Distribution obscure.

Dendrochilum (§ **Acoridium**) **mindorense** Ames Orchidaceæ, fasc. 2 (1907) *ined.*

Epiphyte with greenish yellow flowers, Mount Halcon, Nov. 15, 1906, *Merrill* (No. 5729).—In same locality, Nov., 1906, *Merrill* (No. 5795).

Allied to *D. recurvum* Ames.

Dendrochilum (§ **Acoridium**) **pumilum** Reichb. f. Bonpl. **3**: 222 (1855).

Mount Halcon, Nov., 1906, *Merrill* (No. 6196).

Endemic in the Philippines.

Dendrochilum (§ Acoridium) recurvum Ames Orchidaceæ, fasc. 2 (1907) *incd.*—*Acoridium recurvum* Ames Proc. Biol. Soc. Wash. 19: 148 (1906).

Epiphyte in ridge forest at 8,400 ft. alt. on Mount Halcon, flowers yellowish, Nov., 1906, *Merrill* (No. 5831).—On mossy trees, at 8,200 ft. on Mount Halcon, Nov., 1906, *Merrill* (No. 5509).

Dendrochilum (§ Acoridium) tenellum Ames Ochidaceæ, fasc. 2 (1907) *incd.*—*Acoridium tenellum* Nees & Meyen Nov. Act. Nat. Cur. 19 (suppl. 1): 131 (1843).—*Dendrochilum junceum* Reichb. f. Bonpl. 3: 222 (1855).

Epiphyte in mossy forest, on ridge, at 4,300 ft. alt. on Mount Halcon, flowers whitish, Nov., 1906, *Merrill* (No. 5839).—Near same locality, Nov., 1906, *Merrill* (No. 5720).

Dendrochilum (§ Eudendrochilum) Woodianum Ames Orchidaceæ, fasc. 2 (1907) *incd.*

Epiphyte at 6,800–8,000 ft. alt. on Mount Halcon. flowers dark red, odorless, Nov. 15, 1906, *Merrill* (No. 5816).

This very interesting species belongs to the section *Eudendrochilum*, characterized by a lateral inflorescence produced on a leafless shoot. The floral structure is mainly that of *Acoridium*. The species is named in honor of Maj. Gen. Leonard Wood, through whose interest the expedition to Mount Halcon was undertaken.

Dendrochilum (§ Acoridium) sp.

In habit similar to *D. oliganthum* Ames. The immature flower shoots and withered remains of the persistent perianth insufficient for a sure diagnosis.

Mount Halcon, Nov., 1906, *Merrill* (No. 5540).

MICROSTYLIS Nutt.

Microstylis (§ Eumicrostylis) alagensis Ames sp. nov.

Plant about 25 cm. tall, slender, leafy at the base. Leaves lanceolate to ovate-lanceolate, acute, about 7 cm. long (sometimes those nearest the base much reduced), about 3 cm. wide. Flowers somewhat congested near the summit of the elongated peduncle. Bracts linear-lanceolate. Lowermost flowers purple, those along the middle of the spike yellowish. Lateral sepals suborbicular, 2 mm. long. Upper sepal elliptic-oblong, obtuse, slightly longer than the laterals. Petals linear-oblong, 2 mm. long. Labellum simple, broadly crescentiform, 2 mm. long from its tip to base of column, with a callus near the base; auricles (which form the horns of the crescent) triangular, acute, or subobtuse, distant from each other.

Terrestrial in humid forest at 400 ft. alt. along the Alag River, flowers odorless, Nov. 5, 1906, *Merrill* (No. 5801).—In forests along the Alag River, Nov., 1906, *Merrill* (No. 5807) type.

Microstylis (§ Eumicrostylis) binabayensis Ames sp. nov.

Plant about 3 dm. tall, leafy at the base. Leaves ovate-lanceolate, acuminate, acute, or subobtuse, 6–9 cm. long, 3.5–8 cm. wide, broadest near the base. Petioles about 2 cm. long, those of the lower leaves sheathing with their bases the bases of the petioles of the leaves above. Peduncle about 2 dm. long, rather stout. Bracts linear, deflexed, about 5 mm. long, shorter than the pedicels of the large yellow odorless flowers.

Pedicels of the lowermost flowers about 1 cm. long, very slender. Lateral sepals elliptic to suborbicular, relatively small, about 3 mm. long, very obtuse. Upper sepal oblong, obtuse, convex, 4.5–5 mm. long. Petals linear, about 4 mm. long. Labellum auriculate; auricles (or lateral lobes) dolabriform, obtuse, 5.5 mm. long, about 3 mm. wide; middle lobe 4 mm. wide, oblong; 6.5 mm. long, from the rounded tip to base of column; monocallose at the base.

Terrestrial in forests at 1,000 ft. alt. along the Binabay River, Dec. 4, 1906, *Merrill* (No. 5804).

Microstylis (§ Eumicrostylis) dentata Ames sp. nov.

Leafy plants with elongated strict racemes of small flowers. Leaves 5 or 6, lanceolate, very acute, about 10 cm. long, 2–3.5 cm. wide, bases imbricating, sheathing the stem. Peduncle elongated, 2–5 dm. long, often copiously bracteate. Bracts linear, deflexed, about 1 cm. long. Lateral sepals elliptic, very obtuse, 2 mm. long, about 1.5 mm. wide. Upper sepal similar to the laterals. Petals elliptic-oblong, very obtuse, about 2 mm. long. Labellum 3-lobed, auriculate; auricles obtuse, 1 mm. long; middle-lobe oblong, bifid, 1 mm. long; on the anterior margin of each lateral lobe 3 acute teeth are situated; at the base of the labellum under the column is a minute callus.

In forests at about 3,000 ft. alt. along the Alag River, Nov., 1906, *Merrill* (No. 5806) type.—In forests on Mount Halcon, Nov., 1906, *Merrill* (No. 5808).

Microstylis (§ Eumicrostylis) Hutchinsoniana Ames sp. nov.

A species well characterized by the large, rotund, overlapping auricles of the 3-lobed labellum. Plants about 2 dm. tall. Leaves ovate, acuminate, 4–6 cm. long, about 2.5 cm. wide, petiolate. Petioles 2 cm. long. Peduncle graceful, exceeding the leaves. Bracts linear, acute, the lowermost, 6 mm. long. Flowers comparatively large, about 5 mm. across, pale purple, odorless (greenish when dry). Pedicels very slender, 5–7 mm. long. Lateral sepals elliptical, very obtuse, 3 mm. long, margin very strongly revolute. Upper sepal 3.5 mm. long, somewhat narrower in relation to its length than the lateral sepals. Petals linear, obtuse, about 3 mm. long. Labellum 3-lobed, lobes subequal; middle lobe broadly oblong, rounded at the tip, very obtuse, 3.5 mm. long from tip to base of column, about as broad as long; lateral lobes rotund or subreniform, overlapping behind the column, 4 mm. long, 3 mm. wide.

Terrestrial in forest at 2,300 ft. alt. on Mount Halcon, Nov. 9, 1906, *Merrill* (No. 5809).

This species in named in honor of Mr. W. I. Hutchinson, who was a member of the Mount Halcon Expedition.

Microstylis (§ Commelinoides) Merrillii Ames sp. nov.

Distantly related to *M. commelinifolia* Zoll., from which it is to be distinguished by the very different, larger leaves and dissimilar flowers. Plants creeping, rooting at intervals. Roots long, slender, few. Rhizome with several tubular bracts. Leaves numerous, those near the rhizome often much reduced (1 cm. long, 5 mm. wide), those nearest the inflorescence ovate-lanceolate, acute, comparatively large (2–4 cm. long, about 1.5 cm. wide). Peduncle graceful, ascending or erect, about 7 cm. long, provided with deflexed, linear, acute, about 4 mm. long bracts. Inflorescence loosely few-flowered. Pedicels slender, about 2 mm. long. Lateral sepals oblong-lanceolate, obtuse, 3 mm. long. Upper sepal similar to the laterals, but somewhat narrower. Petals linear, 3 mm. long. Labellum entire, bluntly sagittate, the auricles oblong, falcate, about 2 mm. long.

Terrestrial at base of cliff in dense forest, at 5,800 ft. alt. on Mount Halcon, Nov. 15, 1906, *Merrill* (No. 5820).

Microstylis (§ Eumicrostylis) quadridentata Ames sp. nov.

Flowers similar in structure to those of *M. oculata* Reichb. f. Plants about 1.5 dm. tall. Leaves about 6, narrowly lanceolate, very acute, about 7 cm. long, 6 mm. wide, dilated into a sheating base below the short petiole. Peduncle graceful, exceeding the leaves. Bracts linear, acute, the lowermost 4–5 mm. long, exceeding the short pedicels. Flowers purplish, in a rather strict raceme. Lateral sepals elliptic-ovate, very obtuse, rounded at the tip, 2 mm. long. Upper sepal similar to the laterals. Petals linear-oblong, about 2 mm. long, obtuse. Labellum auriculate, auricles triangular obtuse, less than 1 mm. long; blade of the labellum about 2 mm. long with a minute callus at base, shortly cleft at the apex, bidentate on each side at about the middle; distal tooth of each pair 1 mm. long, narrower and longer than the basal teeth.

Terrestrial in damp shaded ravine at 3,000 ft. alt. on Mount Halcon, *Merrill* (No. 5805). Terrestrial in forest at about 2,500 ft. alt. on Mount Halcon, Nov. 10, 1906, *Merrill* (No. 5818) type.

CESTICHIS Pfitzer.

Cestichis disticha (Thou.) Pfitzer in Engler & Prantl's Pflanzenfamilien 2. pt. 6, p. 131 (1888). *Malaxis disticha* Thouars Orch. Iles Afr. *t.* 88 (1882).—*Liparis disticha* Lindl. Bot. Reg. sub *t.* 882 (1825).

On ledge in ridge forest at 1,800 ft. alt. on Mount Halcon, flowers odorless, reddish (yellow when dry), Nov. 8, 1906, *Merrill* (No. 5643).—Epiphyte in humid forest at 650 ft. alt. along the Binabay River, Nov., 1906, *Merrill* (No. 5638).

The material on which my determination is based has the spathulate petals and very characteristic labellum of the figure in *Thouars's* Flore des Iles Australes de L'Afrique and agrees with *C. disticha* from Ceylon. The Mount Halcon plants are very unlike the *Liparis disticha* Lindl. of *Cuming's* Philippine orchids and should not be confused with it. The Mount Halcon plants are the only specimens of *C. disticha* which I have seen from the Philippines. In habit *C. gracilis* Ames might readily be mistaken for it, but the floral parts at once distinguish it. The *Cuming* plant in *Lindley's* herbarium at Kew, determined by *Lindley* as *Liparis disticha*, is (?) *Cestichis Merrillii* Ames. Lindley's *Liparis gregaria*, if I am not mistaken, is referable to the present species.

Mauritius, Bourbon, Ceylon. (Distribution uncertain as several species wrongly referred to *C. disticha* by authors make an examination of material necessary in giving range.)

Cestichis (§ Laxiflorae) halconensis Ames sp. nov.

A very distinct bifoliate species, about 3 dm. high. Pseudobulbs about 1.5 cm. long, somewhat cylindrical, covered by 4 or 5 distichous, acute sheaths when immature. Leaves oblong-lanceolate, very acute, 15–20 cm. or more long, about 3 cm. wide, contracted into a winged petiole. Peduncle graceful, strongly bialate, exceeding the leaves. Lowermost bracts elongated, linear-acute, 1–1.5 cm. long, those of the inflorescence about half as long as the slender pedicels. Pedicels of the lowermost flowers 1.5 cm. long. Inflorescence loosely many-flowered. Flowers grass-green, turning yellow with age. Lateral sepals oblong, very obtuse, 5 mm. long, 2 mm. wide. Upper sepal similar to the laterals. Petals linear, 5 mm. long, about 1 mm. wide. Labellum 5.5 mm. long, suborbicular from an oblong-cuneate base; distal margin crenulate and obscurely blunt-mucronate; in the middle of the claw is a fleshy subcucullate callus. Column 3 mm. long, rather slender, strongly arcuate near the summit.

Terrestrial in ridge forest at 1,200–2,200 ft. alt. on Mount Halcon, Nov. 8, 1906, *Merrill* (No. 5799).

Cestichis Merrillii Ames Orchidaceæ, fasc. 1, p. 11, *t. 3* (1905).

Epiphyte in mossy forest at 2,800 ft. alt. on Mount Halcon, flowers brownish-yellow, Nov. 12, 1906, *Merrill* (No. 5617).

The Mount Halcon plants have smaller flowers than the type.

Cestichis philippinensis Ames Orchidaceæ, fasc. 1, p. 7, *t. 2* (1905).

Epiphyte at 6,800 ft. alt. on Mount Halcon, flowers brownish-yellow, Nov., 1906, *Merrill* (No. 5764).

The lanceolate, acute labellum of this species is very characteristic.

OBERONIA Lindl.

Oberonia McGregorii Ames sp. nov.

Closely allied with *O. ciliolata* Hook f. Plants caulescent, when in flower about 12 cm. high from base of stem to tip of the densely flowered cylindrical spike. Leaves distichous, obliquely spreading from below the middle, ensiform, about 4 cm. long, acute. Peduncle relatively stout, bracteate, minutely scurfy pubescent. Bracts linear-lanceolate, acuminate, acute, ciliate-pubescent, about 2 mm. long, exceeding the pedicels of the flowers. Ovaries finely pubescent. Lateral sepals ovate-lanceolate, or triangular-lanceolate, acute, ciliolate 0.75 mm. long. Upper sepal nearly elliptical, subobtuse, 0.75 mm. long, ciliolate. Petals linear-oblong, rounded at the tip, much shorter and narrower than the sepals, minutely ciliolate. Labellum pandurate (or oblong, constricted at the middle), coarsely several-toothed at the dilated tip, 0.75 mm. long, slightly auriculate at base.

Only one specimen seen, found on prostrate tree, Balete, Baco River, April 23, 1905, *R. C. McGregor* (No. 291).

Oberonia mindorensis Ames sp. nov.

Allied to *O. aporaphylla* Reichb. f. Plants caulescent, 1.5–4 dm. or more tall from base of stem to tip of elongated, slender, densely flowered spike. Leaves distichous, 5–10 cm. or more long, acute or subobtuse, obliquely ascending. Upper half or two-thirds free. Spike 1–2.5 dm. long, about 5 mm. in diameter, somewhat scurfy pubescent. Bracts linear, about 2 mm. long. Flowers minute, greenish. Lateral sepals elliptic ovate, or ovate, 1 mm. long. Upper sepal similar to the laterals. Petals linear, obtuse, 0.75 mm. long. Labellum 1 mm. long, 3-lobed; middle lobe emarginate, about 1 mm. wide; lateral lobes not very conspicuous, prolonged slightly behind the column. In general outline the labellum is subpanduriform, emarginate, or equally 4-lobed.

Epiphyte on trees at 1,200 ft. alt. along the Alag River, Nov., 1906, *Merrill* (No. 5613).

PODOCHILUS Bl.

Podochilus cornutus (Bl.) Schlechter Mem. Herb. Bois. no. 21, p. 34 (1900).—*Appendicula cornuta* Bl. Bijdr. 302 (1825).

On rotten log in forest at 600 ft. alt. along the Binabay River, Nov., 1906, *Merrill* (No. 5843).—Near same locality on tree fern, in humid forest, Nov., 1906, *Merrill* (No. 5541).

India, Malay Peninsula, Singapore, Java, Borneo, China and Luzon, P. I.

Podochilus pendulus (Bl.) Schlechter Mem. Herb. Bois. no. 21, p. 48 (1900).—*Appendicula pendula* Bl. Bijdr. 298 (1825).

Epiphyte in forest at 2,400 ft. alt. on Mount Halcon, flowers yellowish, Nov. 10, 1906, *Merrill* (No. 5858).—On bowlder in forest, Mount Halcon, Nov. 28, 1906, *Merrill* (No. 5697).

Malay Peninsula, Java, New Guinea and the Philippines.

?Podochilus philippinensis Schlechter Mem. Herb. Bois. no. 21, p. 49 (1900).

Growing on a fallen tree near the Baco River, Apr. 22, 1905, *R. C. McGregor* (No. 290).

My determination, while reasonably sure, is unfortunately based on a single, imperfect specimen.

?Podochilus reflexus (Bl.) Schlechter Mem. Herb. Bois. no. 21, p. 31 (1900).—*Appendicula reflexa* Bl. Bijdr. 301 (1825).

Epiphytic on trees along the Alag River, Nov., 1906, *Merrill* (No. 5842).

Material inadequate for a sure diagnosis, as the flowers are withered, the labellum is imperfect, and the specimens for the most part are in fruit.

Malay Peninsula, Java and Borneo.

Podochilus xytriophorus (Reichb. f.) Schlechter Mem. Herb. Bois. no. 21, p. 47 (1900).—*Appendicula xytriophora* Reichb. f. Seem. Fl. Vit. 299 (1868).

Epiphytic at 500 ft. alt. along the Alag River, flowers odorless, greenish, labellum white, turning yellowish, throat purple, Nov. 6, 1906, *Merrill* (No. 5817).

Malay Peninsula, Borneo and the Philippines.

AGROSTOPHYLLUM Bl.

Agrostophyllum Merrillii Ames sp. nov.

Allied to *A. saccatum* Ridl. Plants stout, about 10 dm. tall, somewhat similar in habit and general conformation to *A. longifolium* Reichb. f. but with different flowers, the petals being broader. Leaves oblong, about 2 dm. long, about 2 cm. wide, rounded at base, tapering very gradually to the bilobed apex. Sheaths formed by the persistent bases of the leaves, smooth, about 7 cm. long. Inflorescence capitate, dense, about 2.5 cm. in diameter (3 cm. when pressed for the herbarium). Spikelets 3–4-flowered. Flowers white. Lateral sepals oblong-lanceolate, acute, 5 mm. long, 2–2.5 mm. wide. Upper sepal oblong, acute, 5 mm. long, about 2 mm. wide. Petals ovate-lanceolate, 5 mm. long, about 2 mm. wide. Labellum 5 mm. long; hypochil saccate, with blunt, rounded, erect lateral lobes, which are connected in front by a transverse plate or callus; epichil very broadly ovate, subacute, 3 mm. long, 4.5 mm. wide, rather fleshy. Column 3.5 mm. long, bent at the middle, above the bend provided with a blunt, fleshy, 0.5 mm. long process. Fruit about 9 mm. long.

Epiphyte in forest at 3,000 ft. alt. on Mount Halcon, Nov. 27, 1906, *Merrill* (No. 5844).

CERATOSTYLIS Bl.

Ceratostylis ramosa Ames and Rolfe Orchidaceæ, fasc. 2 (1907) *ined.*

Plants somewhat more slender than the type.

Mount Halcon, June 15–27, 1906, *M. L. Merritt* (No. 4418).

Ceratostylis subulata Bl. Bijdr. 306 (1825).—*Ceratostylis gracilis* Reichb. f. Xen. Orch. **2:** 92, t. 127. not Bl.

Epiphyte on exposed ridge at 4,500 ft. alt. on Mount Halcon, flowers dark purple, Nov. 14, 1906, *Merrill* (No. 5766).

British India, Assam, Malay Peninsula, Borneo, Sumatra, Java and the Philippines.

Phaius Lour.

Phaius halconensis Ames sp. nov.

Plants graceful, about 3 dm. high. Leaves 3, lànceolate-acuminate, acute, tapering to both ends, about 14 cm. long, 1.5–2 cm. wide. Scape rather graceful, sparsely pubescent, about 3 dm. long, clothed at intervals with loose, tubular sheaths. Floral bracts caducous. Flowers odorless, white, turning yellowish, about 8 mm. long, pubescent externally and sparsely so internally on the sepals and petals. Lateral sepals lanceolate, acute, 9–10 mm. long, 4 mm. wide near the base. Upper sepal similar to the laterals. Petals lanceolate, 3-nerved, 9 mm. long, slightly exceeding 3 mm. in width. Labellum 7 mm. long, 3-lobed; disc pubescent, bicarinate to the base of the middle-lobe, 6–7 mm. wide across the middle when flattened; lateral lobes ovate, obtuse, ragged-fimbriate on the anterior margin; middle lobe subcuneate, truncate, dentate, 3.25 mm.

wide at the tip, 2 mm. long, margin irregular or dentate. Column stout, 5 mm. long.

Terrestrial on steep, wooded, damp slopes at about 6,400 ft. alt. on Mount Halcon, Nov. 16, 1906, *Merrill* (No. 5513).

Phalus mindorensis Ames sp. nov.

Plants about 3 dm. high, branching. Stems slender. Leaves 3–4, lanceolate, acute, 10–18 cm. long, 2.5–4 cm. wide. Bracts tubular, scarious. Peduncle slender, pubescent, with several tubular, inflated, obtuse bracts. Raceme loosely flowered. Flowers about 10 in number, yellowish (buds white). Floral bracts caducous. Ovary densely pubescent. Lateral sepals lanceolate, acute, 7 mm. long, 2.5 mm. wide. Upper sepal ovate-lanceolate, acute, 7–8 mm. long, about 3 mm. wide. Petals ovate acuminate, acute, 7 mm. long, 2.5 mm. wide near the middle. Labellum 6 mm. long, 3-lobed; lateral lobes about 1 mm. long, 1 mm. wide, oblong-truncate, margin with several irregular, short, blunt teeth; middle lobe flabelliform, retuse, apiculate, 1.75 mm. long, 3–5 mm. wide; disc pubescent, bicarinate, the carinæ extending to the base of the middle lobe; when spread out the labellum is 7 mm. wide between the tips of the lateral lobes. Column relatively stout, about 3 mm. long.

Epiphyte in mossy forest at 4,300 ft. alt. on Mount Halcon, Nov. 13, 1906. *Merrill* (No. 5612).

P. mindorensis is closely allied to *P. halconensis* but differs from it in its smaller flowers, different labellum and broader leaves.

Phalus sp.

This appears to be a new species but the material is in bad condition and rather scanty.

Terrestrial in humid forest at 700 ft. alt. along the Binabay River, flowers odorless, sepals green outside, brown-purple inside, labellum yellowish, Nov. 3, 1906, *Merrill* (No. 5800).

CALANTHE R. Br.

Calanthe angustifolia (Bl.) Lindl. Orch. Pl. 251 (1833).—*Ambyglottis angustifolia* Bl. Bijdr. 369 (1825).—*Calanthe phajoides* Reichb. f. Bonpl. **5:** 37 (1857).

Terrestrial at 6,500 ft. alt. on Mount Halcon, flowers white, odorless, Nov. 15, 1906, *Merrill* (No. 5677).—At 7,250 ft. alt. on Mount Halcon, June 15–27, 1906, *M. L. Merritt* (No. 4454).

Java, Sumatra, and Malay Peninsula.

Calanthe halconensis Ames sp. nov.

In habit similar to *C. pulchra* Lindl. Flowers pale straw yellow, or nearly white, labellum riddish-yellow, spur straight. Leaves oblong-lanceolate, acuminate, acute, about 4 dm. long, about 7 cm. wide, petiole about 2 dm. long. Scape with several ample bracteate sheaths, about 5 cm. long. Inflorescence rather densely many-flowered, nearly 2 dm. long, about 4 cm. in diameter. Floral bracts fugacious. Pedicels slender, wiry, 7–10 mm. long. Lateral sepals subfalcate, acute, 11 mm. long, 3.5 wide. Upper sepal oblong-lanceolate, acute, 11.5 mm. long. Petals oblong-oblanceolate, somewhat acuminate, acute, 1 cm. long, 4.5 mm. wide above the middle. Labellum quadrate, apiculate; plate 4 mm. long, about 4 mm. wide, sometimes constricted or obscurely lobed near the middle; obscurely if at all tricallose in the throat on the middle nerves. Spur glabrous, nearly straight, dilated slightly near the tip, 8 mm. long.

Terrestrial in forest at 2,300 ft. alt. on Mount Halcon, Nov. 9, 1906, *Merrill* (No. 5480).

Calanthe McGregorii Ames sp. nov.

Leaves lanceolate, acuminate, acute, about 3 dm. long, about 8 cm. wide, long-petioled. Petioles about 2.5 dm. long. Scape sparsely pubescent about 9 dm. long, with several tubular, acute sheaths. Inflorescence rather strict. Bracts 5–10 mm. long, lanceolate, acute, persistent. Flowers white, with a yellow spot in the throat, numerous, small. Pedicels slender, elongated, about 2 cm. long. Lateral sepals oblong, acute, about 6 mm. long, 3 mm. wide. Upper sepal elliptic-lanceolate. Petals linear-oblong, acute, somewhat dilated near the middle, 6.5 mm. long, about 2 mm. wide. Labellum 3-lobed; middle lobe deeply cleft, the divisions about equal to the lateral lobes; lateral lobes 5 mm. long, 3.5 mm. wide, oblong, obtuse, broader at the apex than below the middle; middle lobe 7 mm. long divisions divaricate, spathulate, 3.5 mm. long, 2.5–3 mm. wide at the apex. On the disc near the column several papillæ are situated. Spur straight, glabrous, 8 mm. long.

Balete, along the Baco River, March 31, 1905, *R. C. McGregor* (No. 177).

Calanthe mindorensis Ames sp. nov.

Leaves long-petiolate, petioles about 18 cm. long; lamina oblong-lanceolate to elliptic-lanceolate, acuminate, acute, about 3 dm. long, about 1 dm. wide. Scape 4 dm. or more long, sparsely pubescent, provided with several oblong, closely appressed, acute bracts. Floral bracts broadly ovate, acute, about 1 cm. long. Pedicels about 2 cm. long. Flowers very large, odorless, pale violet. Lateral sepals oblanceolate, falcate, acute, 1.5 cm. long, 7 mm. wide above the middle. Upper sepal lanceolate, 1.5 cm. long, acute. Petals spathulate-oblanceolate, 1.5 cm. long, 5.5–6 mm. wide above the middle. Labellum 3-lobed; middle lobe 12 mm. long, 16 mm. wide, cuneate-obcordate or flabelliform, deeply cleft; lateral lobes 1 cm. long, 6 mm. wide, oblong, obliquely truncate; on the disc near the colunm is a 3-lobed or 3-plaited callus, beyond which are several rows of erect papilæ. Spur curved, glabrous, about 1.5 cm. long.

Terrestrial in ridge forest at 2,300 ft. alt. on Mount Halcon, Nov. 9, 1906, *Merrill* (No. 5525).

Calanthe pulchra (Bl.) Lindl. Och. Pl. 250 (1833).—*Ambyglottis pulchra* Bl. Bijdr. 371 (1825).—*Calanthe curculigoides* Lindl. Wall. Cat. 7340 (1828).

Terrestrial, border of forest at 2,300 ft. on Mount Halcon, Nov. 8, 1906, *Merrill* (No. 5850).—Terrestrial along the Alag River at 400 ft. alt., Nov. 5, 1906. *Merrill* (No. 5849).

Calanthe pulchra has yellow flowers.

Java, Sumatra, Singapore and Malay Peninsula, Malacca, Penang.

Calanthe triplicatis (Willem.) Ames Orchidaceæ, fasc. 2 (1907) *ined.*—*Orchis triplicatis* Willem. in Usteri Ann. Bot. **18**: 52 (1796).—*Calanthe veratrifolia* R. Br. Bot. Reg. sub *t. 573* (1821).—*C. furcata* Batem. Bot. Reg. 1838, Misc. p. 28.

Baco River, March, 1905, *Merrill* (No. 4065).

After a careful study of the rich collections of *Calanthe* in the herbarium of the Bureau of Science I am quite convinced that *C. furcata* is untenable as a distinct species. Although the average specimens from the Philippines are smaller than *C. triplicatis* they do not exhibit any structural characters which plainly separate them from it. The lobes of the labellum are very variable and the spur is frequently simple, with no indication of a furcate tip. Lindley considered the larger size of the lateral lobes of the labellum of *C. furcata* the chief distinction by which to separate it from *C. veratrifolia*, but this distinction fails absolutely if applied to a large series of specimens. Cuming's No. 2064 in the British Museum herbarium has large apical lobes on the labellum and leaves fully 3 dm. long by 6 cm. wide.

Japan to New South Wales, and from S. India to the Fiji Islands, New Guinea, Java, Cochin China, Ceylon, Formosa, Liu Kiu and Borneo. Very common in the Philippines.

PLOCOGLOTTIS Bl.

Plocoglottis Copelandii Ames sp. nov.—*P. acuminata* Ames Orchid. fasc. 1, p. 82 (1905), *not* Blume.

Rhizome creeping, sheathed with scarious, tubular bracts which persist as elongated fibers. Leaves elliptic-lanceolate, acuminate, acute, about 14 cm. long, 4.5–6.5 cm wide. Petioles 3–4 cm. long rather slender. Scape about 3 dm. long, graceful, pubescent, with several sheathing, tubular, obliquely truncate, acute bracts at intervals. Inflorescence an elongated, rather loose raceme of yellowish, red-brown spotted, odorless flowers. Rachis of the raceme rather densely pubescent. Floral bracts triangular-lanceolate, acute, about 4 mm. long, shorter than the pubescent pedicels of the flowers. Pedicels and ovary together 1.5 cm. long. Lateral sepals linear-lanceolate, acute, 1.5 cm. long, about 4 mm. wide. Upper sepal similar to the laterals but somewhat shorter. Petals linear, gradually tapering from the base to the acute apex, 11 mm. long. Labellum about 6 mm. long, 5 mm. wide at the tip, oblong-cuneate, convex, with a long, acuminate, circinate tip which is invisible from above, being concealed by the apical margin. Column stout, 6 mm. long.

The type from Gimogan River, Negros, where it was collected on January 5, 1904, by *E. B. Copeland* (No. 134).—Growing in leaf mold in forests along the Baco River, April and May, 1905, *R. C. McGregor* (No. 308).—Terrestrial in humid forest at 700 ft. alt. along the Binabay River, flowers yellowish with purple spots, Nov. 3, 1906, *Merrill* (No. 5624).

A careful study of the material collected by *Merrill* and *McGregor* in Mindoro leads me to believe that it is conspecific with the specimens collected on the Island of Negros by *Copland* and is not the same as *Plocoglottis acuminata* Bl., although a closely allied species.

Plocoglottis mindorensis Ames sp. nov.

Allied to *P. javanica* Bl. Plants 6–8 dm. or more tall, very slender, graceful, not much thickened at the base. Petioles about 2 dm. long. Leaf narrowly oblong-lanceolate, acuminate, very acute, tapering at both ends, about 3 dm. long, 2.5–5 cm. wide, smooth. Scape about 8 dm. long, pubescent, densely so near the summit, clothed at intervals with tubular, lanceolate, pointed sheaths. Bracts of the inflorescence rather rigid, triangular-lanceolate, acute, somewhat concave, pubescent, about 6 mm. long. Flowers purplish, rather numerous in loose racemes. Pedicels slender, pubescent, about 1 cm. long. Sepals externally pubescent, the laterals falcate-lanceolate, about 12 mm. long, 6 mm. wide near the middle. Upper sepal oblong-lanceolate, 1.3 cm. long, about 5 mm. wide. Petals linear, falcate, obtuse, 11 mm. long, 2–2.5 mm. wide. Labellum about 6 mm. long, cuneate-quadrate, about 9 mm. wide near the tip; in the middle of the anterior margin is a deflexed, triangular, 1 mm. long tooth.

Terrestrial in humid forest at 650–900 ft. alt. along the Binabay River, Nov. 3, 1906, *Merrill* (No. 5837).

SPATHOGLOTTIS Bl.

Spathoglottis aurea Lindl. Journ. Hort. Soc. **5**: 34 (1850); Gard. Chron. n. s. **4**: 92, f. 9 (1888).

At 4,250 ft. alt. on Mount Halcon, flowers yellow, June 15–27, 1906, *M. L. Merritt* (No. 4375).

Apical lobe of the labellum bilobed. The material on which my determination is based is not very satisfactory.

Java, Sumatra, Borneo and Malay Peninsula.

Spathoglottis plicata Bl. Bijdr. 400, 401 (1825).

Flowers light purple, odorless, plants growing in damp soil among bowlders along stream, Alag River, Nov. 6, 1906, *Merrill* (Nos. 5621 5815).

Malay Peninsula, Java, Sumatra, Borneo, Celebes, Moluccas, New Guinea, Solomon and Fiji Islands. Very common in the Philippines.

DENDROBIUM Sw.

Dendrobium (§ Grastidium) alagensis Ames sp. nov.

Allied to *D. salaccense* (Bl.) Lindl. Plants about 6 dm. tall. Stems 3–4 mm. in diameter, very graceful, comparatively slender, clothed by the tubular, sheathing bases of the numerous leaves. Leaves linear-lanceolate, acute, about 1.5 cm. apart, about 14 cm. long, about 1 cm. wide, very unequally bilobed at the apex. Flowers 2, pale straw-yellow, greenish tinged, leaf-opposed. Pedicels elongated, 1 cm. long, very slender. Lateral sepal oblong-lanceolate, subacute, 11 mm. long, 7 mm. wide at base. Upper sepal oblong-lanceolate, 11 mm. long, 3.5 mm. wide, subobtuse. Petals oblong, obtuse, 8–9 mm. long, about 3 mm. wide. Labellum 9 mm. long, with several raised, longitudinal nerves, 3-lobed, cuneate at base; lateral lobes comparatively small, triangular, acute or subobtuse, 1 mm. long, 0.75 mm. wide at base; middle lobe suborbicular, 3.5 mm. long, 3.75 mm. wide; from the tip of the lateral lobes to the base of the labellum 5.5 mm.

Epiphyte on trees along the Alag River, Dec. 2, 1906, *Merrill* (No. 5846).

Dendrobium (§ Virgatae) polytrichum Ames Orchidaceæ, fasc. 2 (1907) *ined.*

Epiphyte in humid forest at 800 ft. alt. along the Alag River, flowers white, fragrant, Nov. 5, 1906, *Merrill* (No. 5630).

In habit similar to *D. setifolium* Ridl., having subulate leaves. The flowers, however, are well characterized by the 3-lobed labellum of which the apical lobe is provided along the margin with an elongated, copious fringe.

Dendrobium Victoriae-Reginae Loher Gard. Chron. ser. 3, **21**: 399 (1897), var. **exile** Ames var. nov.

Stems ramose, graceful, slender, 1.5–3 mm. thick. Leaves linear-lanceolate, 6 cm. long, 9 mm. wide below the middle, acuminate, bilobed at the tip, lobes acute. Lateral sepals 2 cm. long, 6 mm. wide. Labellum 2.2 cm. long, 9 mm. wide near the apex. Similar to the type but much more slender and graceful throughout. Distinguished mainly by the more slender stems.

Epiphyte in dense, wet, mossy forest, at 8,000 ft. alt. on Mount Halcon, flowers odorless, bluish purple, petals white at base, Nov. 20, 1906, *Merrill* (No. 5503) type.—At 6,500 ft. alt. on Mount Halcon, flowers purplish, June 15–27, 1906, *M. L. Merritt* (No. 4438).

Endemic in the Philippine Islands.

Dendrobium sp.

Two plants of the section *Aporum* without flowers. Indeterminable.

Growing on fallen trees along the Baco River, April–May, 1906, *R. C. McGregor* (No. 286).

Dendrobium sp.

Material not sufficient for description.

Epiphyte at 5,000 ft. alt. on Mount Halcon, Nov. 26, 1906, *Merrill* (No. 5577).

The paucity of *Dendrobium* species in the collections from Mount Halcon and its neighborhood is noteworthy.

ERIA Lindl.

Eria aëridostachya Reichb. f. ex Lindl. Journ. Linn. Soc. **3:** 48 (1859); Reichb. f. in Seem. Fl. Vit. 301.

Epiphyte at about 6,400 ft. alt. on Mount Halcon, flowers brown-purple, odorless, Nov. 15, 1906, *Merrill* (No. 5518).

Philippines and Fiji Islands; also ascribed to Java and the Malay Peninsula. Several closely allied species are likely to have been confused in the literature of distribution under the name *E. aëridostachya*. Lindley's plant and Reichenbach's Seemann plant agree.

Eria (§ Trichotosia) binabayensis Ames sp. nov.

Allied to *E. oligantha* Hook. f. Plants rather stout, 4–5 dm. tall. Stems about 5 mm. in diameter. Leaves oblong-lanceolate, tapering gradually toward the point, about 1 dm. long, 1–2 cm. wide, rigid, coriaceous, pubescent. Inflorescence leaf-opposed, clothed with dense, cinnabar-red tomentum. Bracts about 1 cm. long. Raceme short, 3 cm. long, probably becoming longer as the flowers develop. Lateral sepals triangular-lanceolate, densely tomentose externally, 1 cm. long, about 8 mm. wide at base. Upper sepal oblong, narrower than the laterals. Petals linear-spathulate, subobtuse, 7–8.5 mm. long, 2 mm. wide above the middle. Labellum 1 cm. long, narrowly cuneate at base, then gradually dilated to within 2 mm. of the tip, where it is constricted and about 4 mm. wide; above the constriction it is again dilated into a transversely oblong, 6–7 mm. wide plate; disc bicarinate.

Epiphyte in humid forest along the Binabay River, flowers nearly flesh colored, the narrow petals white, labellum with yellow-purple spots, Nov. 3, 1906, *Merrill* (No. 5661).

It is highly probable that the racemes of the specimens examined had not attained their full length when collected. At maturity they most likely resemble the racemes of such closely allied species as *E. vulpina*, *E. ferox*, and *E. velutina*.

Eria (§ Hymeneria) compacta Ames. sp. nov.

Roots elongated, much branched. Pseudobulbs approximate, elongated-pyriform, about 5 cm. long, about 1 cm. in diameter at base, bifoliate. Leaves rigid, coriaceous, oblong-lanceolate, obtuse, 4.5–5.5 cm. long, 10–14 mm. wide. Inflorescence about 3 cm. long. Bracts ovate-lanceolate, 5 mm. long, about 2 mm. wide, acute, equaling or exceeding the pedicels of the flowers. Lateral sepals triangular-lanceolate, acute, 6 mm. long, 2.5 mm. wide at base. Upper sepal lanceolate, slightly broader than the laterals. Petals lanceolate, acute or subacute, about 6 mm. long, 2 mm. wide, 3-nerved. Labellum ovate-lanceolate, subobtuse, cordate at base, 4 mm. long, 1.75 mm. wide near the base.

At 5,250 ft. alt. on Mount Halcon, June 15–27, 1906, *M. L. Merritt* (No. 4420). The type consists of a single specimen and is unfortunately not in good condition. The flowers are not very numerous, and the only raceme seen may not be wholly characteristic.

Eria cymbiformis J. J. Smith Rec. Trav. Bot. Neerland. 1: 152 *with fig.* (1904).

Epiphyte in ravine forest at 4,000 ft. alt. on Mount Halcon, flowers white, with faint odor, Nov. 26, 1906, *Merrill* (No. 4847).

Sumatra.

Eria (§ Trichotosia) halconensis Ames sp. nov.

Plant comparatively slender, about 5 dm. tall. Stems about 7 mm. in diameter near the base, tapering gradually upwards. Leaves linear-lanceolate, acuminate, acute, pubescent, about 1 dm. long, 7–11 mm. wide. Racemes leaf-opposed, shorter than the leaves, about 5 cm. long, somewhat flexuose, densely covered with reddish yellow hairs. Bracts broadly ovate or suborbicular, 4–6 mm. long, hairy, abruptly acuminate. Lateral sepals triangular, externally hairy, subacute, 7 mm. long about 4 mm. wide at base. Upper sepal oblong, externally hairy. Petals linear, subspathulate, tapering to a subacute or subobtuse apex, 6 mm. long, 1.5 mm. wide. Labellum 7 mm. long, linear-cuneate at the base, dilated above, then constricted within 2 mm. of the tip, 3-lobed; lateral lobes minute, obtuse, formed by the constriction; middle lobe subquadrate, apiculate, 3–4 mm. wide; disc with a prominent mid-nerve. Mentum about 3 mm. long.

Epiphyte on exposed ridge at 4,500 ft. alt. on Mount Halcon, flower pink-purple, Nov. 10, 1906, *Merrill* (No. 5742).—Terrestrial on banks in mossy forest, at 6,000 ft. alt. on Mount Halcon, flowers pink-purple, Nov. 15, 1906, *Merrill* (No. 5510) type.

This is a rather graceful species, the slender stems sometimes attaining 9 dm. in length and less than 1 cm. in thickness at base.

Eria (§ Hymeneria) Hutchinsoniana Ames sp. nov.

Allied to *E. tenuifolia* Ridl. Rhizomes woody, about 5 mm. in diameter. Pseudobulbs 5 cm.–1 dm. apart, abbreviated, 1.5–3 cm. long, clothed with scarious sheaths, leafy at the summit. Leaves linear-lanceolate, about 1 dm. long, about 8 mm. wide, acute, tapering gradually toward both ends. Inflorescence few-flowered, about 6 cm. long, near the summit of the pseudobulbs. Flowers white and purple. Peduncle, pedicels, and ovaries covered with rufous or dark yellowish hairs. Bracts lanceolate, acute, about 4 mm. long, about 1.5 mm. wide. Lateral sepals triangular-lanceolate, acute, slightly protuberant anteriorly at base near the apex of the column foot, about 8 mm. long, 2.5 mm. wide near the middle, broader below. Upper sepal lanceolate, acute, about 9 mm. long, narrower than the laterals. Petals linear-lanceolate, or linear-oblong, subobtuse, 3-nerved, about 8 mm. long, 2 mm. wide near the middle. Labellum ecallose, smooth, 5 mm. long, basal half with the sides erect (conduplicate when dry), apical half oblong, rounded at the apex, 2 mm. wide.

Terrestrial in mossy ridge forest at 7,000 ft. alt. on Mount Halcon, very abundant but only one plant found in flower, flowers white, base of tube purple within, Nov. 16, 1906, *Merrill* (No. 5514).

Eria (§ Convolutae) Merrillii sp. nov.

Pseudobulbs about 1 dm. long, very stout, compressed, 3 cm. or more in diameter, diphyllous (sometimes bearing more than 2 leaves). Leaves oblong-lanceolate, about 3 dm. long (up to 6 dm.), 4–7 cm. wide. Peduncle comparatively stout, arising from near the summit of a pseudobulb (erect? or drooping?), bearing numerous very large, nearly white, somewhat purple-tinged flowers in a dense, elongated raceme. Raceme about 3 dm. long. Bracts triangular-lanceolate, acute, 1.5–2 cm. long, about 4 mm. wide at base. Ovary very strongly winged, distantly resembling an auger on account of the spiral turnings of the wings. Lateral sepals 1.5 cm. long, linear-falcate, tapering to an acute apex from a 6 mm. broad base. Upper sepal linear, 1.9 cm. long, 3 mm. wide, tapering gradually to an obtuse tip. Petals similar to the lateral sepals, 1.6 cm. long, about 4 mm. wide at base. Labellum 11–11.5 mm. long, 3-lobed; lateral lobes comparatively small, curved, about 1 mm. long, 1 mm. wide, obtuse, 5.5 mm. from the base of the labellum; middle lobe 7 mm. long, 2.5 mm. wide, oblong, acute. Through the disc of the labellum extend 5 prominent nerves or carinæ.

Epiphyte at 1,250 ft. alt. along the Alag River, flowers very fragrant, with odor of the swamp Habenaria of the eastern United States, Nov. 12, 1907, *Merrill* (No. 5519).

This robust *Eria* is allied closely to *E. rugosa* Lindl., *E. striolata* Reichb. f., *E. fragrans* Reichb. f., and *E. cochleata* Lindl. The strongly developed, spirally twisted wings of the ovary are very curious and quite distinctive of this species and *E. cochleata*.

Eria (§ Hymeneria) Merrittii Ames sp. nov.

Pseudobulbs rather stout, cylindrical from a stout rhizome, 7–10 cm. long, about 5 mm. thick, 2–3-leaved at the summit. Leaves linear-lanceolate, about 12 cm. long, about 1.5 cm. wide, acuminate, acute. Peduncle short, about 4 cm. long, breaking forth from the upper part of the pseudobulbs. Bracts lanceolate, acute. Flowers white. Lateral sepals linear-lanceolate, acute, about 6–7 mm. long, 1.5 mm. wide at base. Upper sepal linear-lanceolate, about equal to the laterals. Petals narrowly lanceolate, acute, about 6.5 mm. long, 1.5 mm. wide, 3-nerved, acute. Labellum lanceolate, ecallose, smooth, 4 mm. long, 1.5 mm. wide at base.

At 3,950 ft. alt. on Mount Halcon, June 15–27, 1906, *M. L. Merritt* (No. 4357).

Unfortunately the material on which the above description is based is not in good condition.

Eria vulpina Reichb. f. Bonpl. **3**: 222 (1855).

Epiphyte in forest at 2,300 ft. alt. on Mount Halcon, flowers purplish, bracts yellowish red, Nov. 10, 1906, *Merrill* (No. 5501).

Endemic in the Philippines. Originally collected by H. Cuming on the island of Bohol.

Eria (§ Hymeneria) Woodiana Ames sp. nov.

Allied to *E. ovata* Lindl. Stems about 18 cm. long, leafy at the summit, closely sheathed. Leaves oblong-lanceolate, up to 3 dm. long, 3–4.5 cm. wide, acute. Inflorescence racemose, much shorter than the leaves. Flowers pale yellow. Peduncles about 1 dm. long, floriferous nearly to the base, breaking forth from the leafy summit of the pseudobulbs. Bracts ligulate, acute, about 5 mm. long. Lateral sepals oblong-lanceolate, acute, 6 mm. long, 2 mm. wide. Upper sepal narrowly lanceolate, 6.5 mm. long. Petals narrowly lanceolate, subacute, 5.5 mm. long, 1.5 mm. wide Labellum 3-nerved, about 4 mm. long, orbicular at base, contracted at about the middle into the oblong, obtuse, apical half, monocallose at base in front of the claw.

Epiphyte in ridge forest at 3,000 ft. alt. on Mount Halcon, No. 10, 1906, *Merrill* (No. 5490).

PHREATIA Lindl.

Phreatia sulcata (Bl.) J. J. Smith Orch. Java 505 (1905).—*Dendrolirium sulcatum* Bl. Bijdr. 347 (1825).—*Eria sulcata* Lindl. Orch. Pl. 69 (1830).

Epiphyte in ridge forest at 6,000 ft. alt. on Mount Halcon, flowers white, with faint odor, Nov. 15, 1906, *Merrill* (No. 5765).

Java and Sumatra.

Phreatia prorepens Reichb. f. Otia Bot. Hamb. 54 (1878).

Epiphyte in ridge forest at about 5,800 ft. alt. on Mount Halcon, flowers white, fragrant, Nov. 15, 1906, *Merrill* (Nos. 5758, 5814, 6190).

This very rare species was originally collected by the expedition commanded by Captain *Wilkes* which visited the Philippines between the years 1838 and 1842. One of the specimens collected by this expedition is preserved in the Gray Herbarium and bears the name *Eria* (*Phreatia*) *prorepens* in *H. G. Reichenbach's* hand. The material on which my determination is based agrees with this specimen in essential details.

Endemic in the Philippines.

BULBOPHYLLUM Thouars.

Bulbophyllum adenopetalum Lindl. Bot. Reg. 1842, Misc. p. 85.

Epiphyte at 3,550 ft. alt. on Mount Halcon, flowers odorless, nearly white, slightly straw-colored, Nov. 10, 1906, *Merrill* (No. 5684).

The petals of the specimens collected by Mr. *Merrill* are rather lanceolate and acute than spathulate as in *Lindley's* colored sketch of the Singapore plants on which is based the description of *B. adenopetalum*. The Mount Halcon specimens agree very well in habit with *Cuming's* Philippine specimens in *Lindley's* Herbarium determined as *B. adenopetalum* by *Lindley*. Unfortunately I have no record of the shape of the petals of *Lindley's* specimens aside from a copy of his drawing of the Singapore plant which he received from Messrs. *Loddiges* in 1842. This drawing shows a flower very similar to that of the specimens in question, if the petals are excepted. Notwithstanding the discrepancy that exists between the petals of *Lindley's* drawing and the Mount Halcon plant I refer the material collected by Mr. *Merrill* to *B. adenopetalum*.

B. adenopetalum belongs to a section of *Bulbophyllum* which appears to be very well represented in the Philippines, mainly characterized by the absence of well-developed pseudobulbs. This section is in great need of careful study and revision. To it belongs *B. dasypetalum* Rolfe.

Singapore and the Philippines.

Bulbophyllum (§ Monanthaparva) alagense Ames sp. nov.

Rhizome creeping, slender. Pseudobulbs approximate to each other or sometimes 2 cm. apart, pyriform, 4–6 mm. long, narrowed above, 3–4 mm. in diameter at base. Leaves ovate, apiculate, 1.2–2 cm. long, 4–9 mm. wide. Apicule 0.75 mm. long, awn-like. Scape very slender, filiform, exceeding the pseudobulbs, 8–11 mm. long, sheathed at base by a tubular, truncate, 1–2 mm. long bract. Flower solitary, pale yellow, nearly white. Pedicel slender, graceful, subtended by a loose, tubular, obliquely truncate, apiculate, 2 mm. long bract, which is dilated above. Sepals triangular-lanceolate, 5–7 mm. long, caudate-tipped, the tails about 4 mm. long. Petals minute, about 2 mm. long, spathulate, acute. Labellum about 1.5 mm. long, strongly curved, 3-lobed; lateral lobes erect, half-round, when spread out forming an orbicular plate, 1 mm. long, 1 mm. wide; middle lobe fleshy, oblong, obtuse, about 1 mm. long. Column minute with blunt wings.

On mossy branches overhanging the water along the Alag River, at 1,250 ft. alt., Nov. 12, 1906, *Merrill* (No. 5494).

Bulbophyllum dasypetalum Rolfe in Ames Orchidaceæ, fasc. 1, p. 98, *with fig.* (1905).

Epiphyte at 3,550 ft. alt. on Mount Halcon, flowers odorless, pale yellow, Nov. 10, 1906, *Merrill* (No. 5649).—Same locality, Nov. 15, 1906, *Merrill* (No. 6129).—Same locality on exposed ridge at 4,500 ft. alt., Nov. 14, 1906, *Merrill* (No. 5719).

Type from Mount Mariveles, Province of Bataan, Luzon.

Bulbophyllum (§ Monanthaparva) halconense Ames sp. nov.

Rhizome thread-like, less than 1 mm. thick. Pseudobulbs 1–3 cm. apart, round-pyriform when mature, rugose when dry, about 5 mm. long, about 4 mm. in diameter at base. Leaves narrowly elliptic-oblong, very fleshy (not apiculate), 1.5–2.5 cm. long, 4–6 mm. wide, acute, contracted into a very slender petiole. Scape exceeding the pseudobulbs, filiform, 1.5–3 cm. long, sheathed at base. Flower solitary, relatively large. Pedicel filiform, subtended by a tubular, obliquely truncate, apiculate bract dilated at its mouth. Lateral sepals narrowly lanceolate, caudate-tipped, 12–14 mm. long, about 3 mm. wide near the base. Upper sepal similar to the laterals, 12–14 mm. long, caudate-tipped. Petals ovate, acute, 5 mm. long, 2.5 mm. wide. Labellum lanceolate from a cordate base, acute, dilated at the middle, 3.5–4 mm. long, 2 mm. wide. (From dried speciments it appears to have been strongly convex in life.) Column short with a minute tooth in front at about the middle.

On trees in ridge forest, flowers dark purple, at 4,500 ft. alt. on Mount Halcon, Nov. 16, 1906, *Merrill* (No. 5832). On the same sheet with the type is a small species of the *Monanthaparva* section with much smaller flowers and shorter scapes.

Bulbophyllum (§ Racemosae) Merrittii Ames sp. nov.

Near *B. cylindraceum* Lindl. Roots copious. Rhizome creeping. Pseudobulbs small, about 1 cm. long, about 5 mm. thick, when mature surrounded by the elongated fibrous remains of sheathing bracts. Leaves coriaceous, 7–16 cm. long, 2–3.2 cm. wide, rounded at the tip, gradually tapering into the comparatively slender, about 3 cm. long petiole. Scape slender, exceeding the leaves, provided with several distant, closely appressed bracts. Bracts of the inflorescence minute, triangular, 1 mm. long, acute. Flowers in a dense, slender, 4–5 cm. long raceme. Lateral sepals strongly deflexed, ovate-falcate, subobtuse, 3-nerved, 2.5–3 mm. long, 1.25 mm. wide near the middle. Upper sepal ovate-lanceolate, subobtuse, 2.75–3 mm. long, about 1.5 mm. wide at base. Petals oblong, obtuse, 1-nerved, 1.5 mm. long, 0.75 mm. wide. Labellum lingulate, 1.5 mm. long, 1 mm. wide, rounded-obtuse at the tip, very fleshy with two callus-like thickenings at base. Column minute.

At 1,475 ft. alt. on Mount Halcon, June 15–27, 1906, *M. L. Merritt* (No. 4338).

Bulbophyllum (§ Monanthaparva) mindorense Ames sp. nov.

Rhizome inconspicuous, concealed by the depressed, 3–4 mm. long pseudobulbs, which form a continuous, sometimes branching chain, which is closely appressed to the bark of trees on which the species is epiphytic. Leaves lanceolate, acute, 6–8 mm. long, up to 3 mm. wide, shortly petiolate. Scapes filiform, up to 4 cm. long. Flower with a very long pedicel, apparently wihout a subtending bract, the pedicel being fully 5 mm. long. Floral bract tubular, obliquely truncate. Flower straw-yellow, relatively large. Lateral sepals narrowly lanceolate, acute, 3-nerved, 8 mm. long, 2 mm. wide below the middle, margin minutely ciliolate. Upper sepal similar to the laterals, about equally long. Petals linear-oblong, subspathulate, acute, 1-nerved, 3 mm. long, about 0.75 mm. wide. Labellum linear-lanceolate, 3.5 mm. long, about 1 mm. wide. Column 1 mm. long, with a tooth or protuberance in front near the base or below the middle.

Epiphyte in deep shaded ravine at 3,000 ft. alt. on Mount Halcon, Nov. 27, 1906, *Merrill* (No. 5796).

Pseudobulbs as in *B. cernuum* (Bl.) Lindl.

Bulbophyllum (§ Monanthaparva) pleurothalloides Ames sp. nov.

Rhizome obscure. Pseudobulbs 3 mm. long, much depressed, forming a chain. Leaves about 1 cm. long, 2–4 mm. wide, oblanceolate to spathulate, obtuse, minutely apiculate, contracted below into a slender petiole. Scapes filiform, elongated, exceeding the leaves, 3 cm. long, with scarious sheaths at base, and with a tubular obliquely truncate bract subtending the solitary, minute, brownish-yellow flower. Lateral sepals lanceolate, acute, about 4 mm. long, 2.5 mm. wide, 3-nerved. Upper sepal similar and equal to the laterals. Petals lanceolate, 2 mm. long, 0.75 mm. wide. Labellum lanceolate-cordate, obtuse, or subsagittate, 2 mm. long, 1 mm. wide at base, about 0.5 mm. wide near the tip. Column 1 mm. long.

On mossy trunks of trees at 4,500 ft. alt. on Mount Halcon, Nov. 14, 1906, *Merrill* (No. 6128).

In habit allied to *B. cernuum* (Bl.) Lindl. but with very different leaves. The general aspect of the plant when in flower recalls some species of Pleurothallis.

Bulbophyllum vagans Ames and Rolfe Orchidaceæ, fasc. 2 (1907) *ined.*

Epiphyte on mossy tree trunks in ridge forest at 6,500 ft. alt. on Mount Halcon, flowers odorless, straw colored, Nov. 21, 1906, *Merrill* (No. 6217).

Endemic in the Philippines.

Bulbophyllum vagans var. **angustun** Ames var. nov.

Differs from the type in its linear-lanceolate, acuminate, acute, 9–13 cm. long, 1–1.5 cm. wide leaves.

Epiphyte in mossy forest on tree trunks at 650 ft. alt. on Mount Halcon, flowers greenish yellow, Nov. 21, 1906, *Merrill* (No. 6218).

This variety is very remarkable in that it agrees with *B. vagans* in all respects except foliage. The leaves of *B. vagans* are oblong-ovate, abruptly rounded at base and subobtuse, 4–7 cm. long, 3 cm. wide.

Bulbophyllum sp.

At 2,950 ft. alt. on Mount Halcon, June 15–27, 1906, *M. L. Merritt* (No. 4360).

A single specimen insufficient for description.

Bulbophyllum sp.

Balete, Baco River, flowers yellowish, April–May, 1905, *R. C. McGregor* (No. 329).

A single specimen found on a fallen tree trunk.

Bulbophyllum sp.

Epiphyte on exposed ridge at 4,500 ft. alt. on Mount Halcon, Nov. 14, 1906, *Merrill* (No. 5833).

A single specimen insufficient for description.

THELASIS Bl.

Thelasis carinata Bl. Bijdr. 386 (1825).

Epiphyte in humid forest at 500 ft. alt. along the Alag River, flowers odorless, brownish, except white tips of petals, Nov. 7, 1906, *Merrill* (No. 5679).

Java, Sumatra and Borneo.

PHALAENOPSIS Bl.

Phalaenopsis Aphrodite Reichb. f. Hamb. Gartenz. **18:** 35 (1862).

Epiphyte in forest at 300 ft. alt. along the Alag River, flower odorless, pure white, except for the purple-striped and spotted labellum, Dec. 3, 1906, *Merrill* (No. 5845).

Very nearly allied to, if not merely a form of, *P. amabile* Bl.

Philippines and Formosa.

SARCANTHUS Lindl.

?Sarcanthus striolatus Reichb. f. Gard. Chron. n. s. **18:** 168 (1882).

On large bowlders in Alag River, at 500 ft. alt., flowers greenish, labellum white, turning yellowish, Nov. 6, 1906, *Merrill* (No. 5517).

Endemic in the Philippines.

SACCOLABIUM Bl.

?Saccolabium compressum Lindl. Bot. Reg. 1840, misc. p. 9.

Epiphyte in mossy forest at 6,000 ft. alt. on Mount Halcon, Nov. 18, 1906, *Merrill* (No. 5564).

Although the material on which the above determination is based is in fruit, it agrees well with *S. compressum* Lindl. in habit and general aspect.

Endemic in the Philippines.

ANGRAECUM Thouars.

Angraecum philippinense Ames sp. nov.

Plants 3–6 cm. high. Roots very fleshy. Leaves elliptic-oblong, obtuse, 2–5.5 cm. long, 0.6–1.4 cm. wide, on contracted stems. Peduncles fleshy, stout, conspicuously winged, few-flowered, about 4 cm. long. Bracts rigid, fleshy, 5 mm. long, conduplicate, triangular, acute. Pedicels elongated, about 2.5 cm. long, including the ovary. Flowers large, white, odorless. Lateral sepals elliptic, rounded and very obtuse at the apex, about 2.2 cm. long, about 1.5 cm. wide. Upper sepal similar to the petals, cuneate at base, about 2.2 cm. long, 14–15 mm. wide. Petals broadly spathulate, about 2.2 cm. long, 1.5 cm. wide, very obtuse. Labellum 3-lobed; middle lobe oblong, rounded at the tip, 9 mm. long, about 7.5 mm. wide; lateral lobes somewhat similar to the middle lobe, but shorter, 4–5 mm. long, 6.5 mm. wide at base. Spur slender, 3.5 cm. long. Column about 7 mm. long, rather stout.

Epiphyte at about 2,500 ft. alt. on forested slopes of Mount Halcon, Nov. 28, 1906, *Merrill* (No. 5698).

THRIXSPERMUM Lour.

Thrixspermum McGregorii Ames sp. nov.

Stem about 5 mm. in diameter. Leaves oblong, falcate, acute or subacute, about 12 cm. long, about 1.5 cm. wide. Scapes very slender, almost filiform, shorter than the leaves, about 6 cm. long, naked, bearing a short raceme of minute flowers. Floral bracts minute, triangular, acute. Lateral sepals ovate, subacute, 1.5 mm. long, less than 1 mm. wide near the middle. Upper sepal strongly concave, narrower than the laterals, 1.5 mm. long. Petals somewhat oblanceolate or spathulate, obtuse, slightly exceeding 1 mm. in length, 0.5 mm. wide. Labellum about 1 mm. long, 3-lobed; lateral lobes semi-rotund, erect, 0.5 mm. long and about 0.5 mm. wide; middle lobe with a minute lobule on each side, one in front of each lateral lobe, otherwise fleshy, minute, blunt. Flowers when spread out 3 mm. across.

Growing on fallen tree near Balete, Baco River, April 23, 1905, *R. C. McGregor* (No. 288).

This is a very small-flowered species of a most puzzling group, closely allied to *Dendrocolla Zollingeri* Reichb. f.

Thrixspermum sp.

Binabay River, Nov., 1906, *Merrill* (No. 5539). A single specimen.

Thrixspermum sp.

Mount Halcon, Nov. 1906, *Merrill* (No. 5691). Specimen in fruit.

MUSCI HALCONENSES.

By V. F. BROTHERUS.
(*Helsingfors, Finland.*)

The mosses below enumerated were collected by *Elmer D. Merrill* in his ascent of Mount Halcon, Mindoro, in November, 1906. Thirty-one species are represented in the collection, of which four are described as new.

SPHAGNALES.

Sphagnum Junghuhnianum Doz. et Molk.
In open heaths at 2,400 m. alt. (No. 5707), sterile specimens.
Area: Khasia, Sikkim, Java, Celebes, Batjan and the Philippines.

Sphagnum sericeum C. Müll.
On cliffs in forests at 1,970 m. alt. (No. 6161 ex p.), sterile specimens.
Area: Java, Sumatra.

BRYALES.

DICRANACEÆ.

Dicranoloma Blumei (Nees) Ren.
On trees, forested ridges at 1,800 m. alt. (No. 6192 ex p.), sterile.
Area: Ceylon, Java, Luzon and New Guinea.

Pilopogon Blumei (Doz. et Molk.) Broth.
Terrestrial in open heaths at 2,400 m. alt. (No. 6111).
Area: British India and Ceylon to Japan, Luzon, Malaya and Polynesia.

LEUCOBRYACEÆ.

Leucobryum subsanctum Broth. n. sp.

Dioicum; robustum, caespitosum, caespitibus laxis, albescentibus, nitidiusculis; *caulis* usque ad 6 cm. altus dense foliosus, simplex vel furcatus; *folia* sicca laxe imbricata, humida eecto-patentia, haud subsecunda, e basi ovali, valde concava ovato-lanceolata, acutissima vel apiculo terminata, 5–6 mm. longa, basi circ. 1.7 mm. lata, dorso laevia, marginibus superne late involutis, integerrimis, limbo inferne 3–4 seriato, superne sensim angustiore, cellulis alaribus numerosis, lamina bistratosa; *bractae perichaetii* minutae, internae e basi oblonga, vaginante subito lanceolato-subulatae; *seta* circ. 1.5 cm. alta, tenuissima, sicca flexuosula, rubra, apice leviter scabriuscula; *theca* erecta, minuta, breviter oblonga, strumulosa, sicca indistincte plicatula, nitidiuscula, fuscidula; *operculum* e basi conica subulatum; *calyptra* cucullata, trifida.

LUZON, Province of Bataan, summit of Mount Mariveles (Nos. 3540, 3549 *Merrill*) on trees. MINDORO, Mount Halcon, on prostrate logs at 1,350 m. alt. (No. 6208 ex p.); on cliffs at 1,970 m. alt. (No. 6161 ex p.).

Species ob folia cellulis alaribus numerosis praedita cum *L. snacto* (Brid.) Hamp. comparanda, sed foliis erectis, acutissimis, dorso leavibus nec non theca erecta, haud arcuata calyptraque trifida optime diversa.

Leucobryum sanctum (Brid.) Hamp.
On logs along the Binabay River, 180 to 240 m. alt. (Nos. 5636, 5601).
Area: Nepal, Malacca, Malaya, Philippines and Polynesia.

Schizomitrium apiculatum Doz. et Molk.
On trees at 1,800 m. alt. (No. 6192 ex p.).
Area: Java, Sumatra, Borneo.

Schizomitrium Nieuwenhuisii Fleisch.
On trees at 1,360 m. alt. (No. 6209).
Area: Borneo.

Leucophanes candidum (Hornsch.) Lindb.
On bowlders along the Alag River at 150 m. alt. (No. 5632).
Area: Ceylon, Malacca, Malaya, New Guinea and Samoa.

ORTHOTRICHACEÆ.

Macromitrium fasciculare Mett.
On trees at 1,360 m. alt. (No. 5709).
Area: Ceylon and Java.

Macromitrium Reinwardtii Schw.
On branches of trees, exposed ridges at 2,360 m. alt. (No. 6198).
Area: Java, Borneo, Celebes, Luzon, Tasmania and Tahiti.

Macromitrium (Goniostoma) mindorense Broth. n. sp.

Dioicum robustum, caespitosum, caespitibus densiusculis, rufescentibus, nitidis; *caulis* elongatus, repens, per totam longitudinem fuscoradiculosus, densiuscula ramosus, ramis erectis, strictis, vix ultra 1.5 cm. longis, dense foliosis, obtusis; *folia ramea* sicca laxe adpressa, apice patentia, flexuosulo humida horrida recurvo-patula, carinato-convaca, e basi oblongo-ovali plicata lanceolato-subulata, aristata, circ. 0.5 mm. longa, basi circ. 1 mm. lata, marginibus erectis, subintegris vel superne minutissime denticulatis, nervo rufescente, in aristam tenuam, hyalinam excedente, cellulis pellucidis rhombeis, incrassatis, lumine anguste elliptico, basilaribus elongatis, lumine angustissimo, papillis altis praeditis; *bractae perichaetii* foliis subsimiles, nervo longius excedente, cellulis omnibus elongatis; *seta* vix ultra 1 cm. alta, tenuissima, sicca dextrorsum torta, fuscescenti-rubra, scaberula; *theca* erecta, ovalis, microstoma ore plicatula, fuscescenti-rubra, laevissima; *peristomium* simplex; *exostomii* dentes brevissimi, albidi, truncati, valde papillosi. Caetera ignota.

On trees, 2,000 to 2,350 m. alt. (Nos. 5559, 6165).

Species pulcherrima, *M. cuspidato* Hamp. habitu similis, sed notis supra allatis longe diversa.

Macromitrium Blumei Nees.

In exposed ridge-thickets at 2,540 m. alt. (No. 5505).

Area: Sumatra, Java, Borneo, Celebes and Luzon.

MNIACEÆ.

Mnium rostratum Schard.

On rocks in forests at 1820 m. alt. (No. 5740).

Area: Widely distributed throughout the temperate, tropical and subtropical parts of the World.

RHIZOGONIACEÆ.

Rhizogonium spiniforme (L.) Bruch.

On trees in forests 900 to 1,060 m. alt. (Nos. 5534, 5591).

Area: Widely distributed throughout the tropical and subtropical regions of the World.

POLYTRICHACEÆ.

Pogonatum macrophyllum Doz. et Molk.

Terrestrial in forests at 1,820 m. alt. (No. 5761).

Area: Sumatra, Java, Batjan and Mindanao.

SPIRIDENTACEÆ.

Spiridens Reinwardtii Nees.

On trees at 1,360 m. alt. (No. 5763).

Area: Java, Celebes, Tidor, Batjan, Mindanao, Luzon and New Guinea.

NECKERACEÆ.

Pterobryella longifrons (C. Müll.) C. Müll.

On trees in forests at 1,820 m. alt. (No. 6183).

Area: Luzon.

ENTODONTACEÆ.

Symphyodon Merrillii Broth. n. sp.

Dioicus; robustiusculus, mollis, lutescens, nitidus; *cualis* primarius filiformis, repens, secundarius 2–3 cm. altus densiuscula foliosus, complanatulus, obtusus vel sensim subflagelliformiter attenuatus, pinnatim ramosus, ramis patentibus, brevibus, strictis, obtusis; *folia* erecto-patentia, concava, oblonga, breviter acuminata, acuta, marginibus erectis, inaequaliter serrulatis, nervis binis, luteis, brevibus, inaequalibus, cellulis angustissimis, sublaevibus, basilaribus infimis abbreviatis; *bracteae perichaetii* erectae, internae e basi oblonga, vaginante lanceolato-subulatae, superne serrulatae; *seta* usque ad 3 cm. alta, flexuosula, rubra, inferne laevis, superne scaberrima; *theca* erecta, oblongo-elliptica, fuscescenti-rubra, dense spinulosa annulus angustus; *peristomium* duplex; *exostomii* dentes lanceolato-subulati lutei minutissime papillosi, laméllati; *endostomium* sordide lutescens, minutissime papillosum, corona basilaris humilis, processus breves, angusti, carinati; *spori* 0.012–0.015 mm., minutissime papillosi; *operculum* e basi conica recte subulatum. *Calyptra* ignota.

On trees at 1,800 m. alt. (No. 6193).

Species *S. Perrottetii* Mont. affinis.

RHACOPILACEÆ.

Rhacopilum spectabile Reinw. et Hornsch.
On wet rocks along the Alag River at 180 m. alt. (No. 5641).
Area: Sumatra, Java, Mindanao, New Guinea and Polynesia.

LESKEACEÆ.

Thuidium plumulosum (Doz. et Molk.) Bryol. jav.
On shaded rocks along the Alag River at 90 m. alt. (No. 5689).
Area: Ceylon, Malaya, Philippines, New Guinea and Polynesia.

SEMATOPHYLLACEÆ.

Warburgiella cupressinoides C. Müll.
On trees at 1,820 m. alt. (Nos. 5596 and 6192 ex p.).
Area: Mindanao.

Trichosteleum hamatum (Doz. et Molk.) Jaeg.
(No. 5535 ex p.).
Area: Malaya, Mindanao and Polynesia.

Sematophyllum subulatum (Hamp.) Jaeg.
(No. 5535 ex p.), sterile.
Area: Sumatra, Java, Celebes, Luzon and Mindanao.

Acanthocladium Korthalsii (C. Müll.).
On prostrate logs and on trees at 1,360 m. alt. (Nos. 6281 ex p., 6193 ex p.).
Area: Malacca and Java.

Acanthocladium Prionodontella Broth.
Terrestrial in forests at 1,300 m. alt. (No. 5499).
Area: Mindanao.

Acanthocladium lancifolium (Harv.) Broth.
On logs along the Binabay River 180 to 300 m. alt. (Nos. 5602, 5637).
Area: Nepal, Malacca, Malaya and Luzon.

STEREODONTACEÆ.

Isopterygium albescens (Schwaegr.) Jaeg.
On wet rotten logs in forests at 900 m. alt. (No. 5587).
Area: British India to Japan and Malaya.

Ectropothecium Meyenianum (Hamp.) Jaeg.
On wet rocks in shaded ravine at 90 m. alt. (No. 5548).
Area: Luzon.

HYPNACEÆ.

Rhynchostegium mindorense Broth. n. sp.

Autoicum; tenellum, caespitosum, caespitibus laxiusculis, depressis, mollibus, sordide viridibus, haud nitidis; *caulis* elongatus, repens, per totam longitudinem fusco-radiculosus, laxe foliosus, dense pinnatim ramosus, ramis erectiusculis, vix ultra 1 cm. longis, laxiuscule foliosis, complanatis, obtusis: *folia* sicca contracta, humida patentia, planiuscula, caulina minuta, plerumque destructa, ramea ovalia vel ovato-ovalia, obtusa, circ. 1 mm. longa et circ. 0.5 mm. lata, marginibus infima basi tantum recurvis, inferne minute, superne argute serrulatis, nervo crassiusculo, longe infra apicem folii evanido, cellulis anguste rhomboideis, pellucidis, utriculo primordiali repletis, minutissime papillosis, alaribus sat numerosis, quadratis; *bracteae perichaetii* suberectae, e basi ovata sensim lanceolato-subulatae, superne serrulatae; *seta* vix 1 cm. lata, tenuissima, rubrar scaberula; *theca* asymmetrica, cernua, ovalis, sicca deoperculata sub ore vix constricta, leptodermis, lutea; *annulus* latus, longe persistens; *peristomium* duplex; *exostomii* dentes lanceolata, rufescentes, dense striolati; *apice* hyalini, papillosi, dense et alte lamellati; *endostomium* hyalinum minute papillosum, corona basilaris alta, plicata, processus dentium longitudinis, carinati, rimosi, cilia brevia; *spori* inaequales, 0.015 mm. vel 0.025–0.030 mm., lutescenti-virides, minutissime papillosi; *operculum* e basi convexa breviter subulatum.

On damp shaded bowlders along the Alag River at 150 m. alt. (No. 5546).

Species *Rh. menadensi* (Bryol. jav.) affinis, sed statura multo minore, foliis obtusis, superne argute serrulatis facillime dignoscenda.

HYPNODENDRACEÆ.

Hypnodendron Reinwardtii (Hornsch.) Lindb.
On dead trees in forests at 1,350 m. alt. (No. 5737), forma *breviseta.*
Area: Malaya and Polynesia.

Mniodendron divaricatum (Reinw. et Hornsch.) Lindb.
On prostrate logs in forests at 1,820 m. alt. (No. 6185).
Area: Malaya.

INDEX TO PHILIPPINE BOTANICAL LITERATURE, II.

By ELMER D. MERRILL.
(*From the botanical section of the Biological Laboratory, Bureau of Science.*)

Anonymous. Decades Kewensis, Decas XLII. (*Kew Bull.* (1906) pp. 200–205.)
One Philippine species. *Peracarpa luzonica* Rolfe, is described on page 201, from Northern Luzon, the only other known species of the genus, *P. carnosa* Hook. f. et Th., extending from Northern India to Yunnan.

Boorsma, W. G. Ueber philippinische Pfeilgifte. (*Bull. l'Inst. Bot. Buitenzorg* **6** (1900) pp. 14–18.)
A consideration of *Lunasia amara* Blanco and *Lophopetalum toxicum* Loher. in connection with the use of the bark as a source of arrow poison.

Ceron, S. Catálogo de las plantas del herbario recolectado por el personal de la suprimida comisión de la flora forestal. (Manila (1892), pp. 1–231, plate 1.)
A catalogue of a portion of the plants collected in the Philippines by *Vidal*, including those enumerated by the latter in his Revision de plantas vasculares Filipinas (1886), and some collected after the publication of that work. A number of genera and species are credited to the Philippines for the first time and one new species is described, *Calophyllum vidalii F.-Villar,* l. c. 229, with plate, *C. cuneatum* Vidal and *C. rolfei* Vidal being cited as synonyms. It is not entirely clear who is the author of the entire work, but *Ceron's* name, then "Inspector general de Montes" for the Philippines appears on page 5 at the end of the introduction. The enumeration of species is however apparently the work of *Regino García.*

Chevalier, Auguste. Monographie des Myricacées. (Thèses presentées á la faculté des sciences de Paris (1901) pp. 1–257, plates 9, reprint from *Mém. Soc. Sci. Nat. Cherbong* **32** (1901.)
Three genera, *Gale, Comptonia* and *Myrica* are recognized, the latter containing 51 species and many varieties, represented in the Philippines by the endemic *Myrica vidaliana* Rolfe. Other species of the genus have since been found in the archipelago.

Chodat, Robert. Polygalaceae novae vel parum cognitae, V. (*Bull. Herb. Boiss.*, **4** (1906) pp. 233–237.)
Securidaca philippinensis is described as new.

Chodat, Robert. Conspectus systematicus Generis Xanthophylli. (*Bull. Herb. Boiss. XX,* **4** (1906) pp. 254–264.)
Xanthophyllum bracteatum, X. philippinense and *X. robustum* are described from Philippine material, the first two endemic, the last extending from the Philippines to Borneo and Malaca.

Chodat, Robert. Monographia Polygalacearum. (*Mém. Soc. Phys. et Hist. Nat. Genève* (1903) pp. 1–500, plates 35.)

Of the genus *Polygala* 404 species are recognized, but one *P. warburgii* Chod., Philippines and New Caledonia being credited to the archipelago (*P. telephioides* of Philippine authors, non Willd.). Several other species are however found in the archipelago, *P. venenosa* Juss., *P. chinensis* L., *P. polifolia* Presl, *P. luzoniensis* Merr., and *P. septemnervia* Merr.

Elmer, A. D. E. Leaflets on Philippine Botany (1906–07) pp. 1–208. Of the above work 9 articles have appeared, as follows:

Article 1, April 8, 1906, Philippine Rubiaceae, by *A. D. E. Elmer*, pp. 1–41. Of this family 149 species representing 42 genera are enumerated, of which several genera are reported from the Philippines for the first time, *Amaracarpus*, *Chasalia*, *Coelospermum*, *Galium*, *Mussaendiopsis*, and *Tricalysia*, and the following 45 species are described as new: *Amaracarpus longifolius*, *Argostemma solaniflora*, *Coelospermum ahernianum*, *Gardenia whitfordii*, *G. merrillii*, *G. elliptica*, *G. acutifolia*, *Ixora sparsiflora*, *I. bibracteata*, *Lasianthus hispidus*, *L. copelandi*, *L. bordeni*, *L. culionensis*, *Mussaenda grandiflora*, *M. benguetensis*, *Mussaendiopsis multiflora*, *Nertera dentata*, *Nauclea vidalii*, *Oldenlandia apoensis*, *O. yoderi*, *O. benguetense*, *O. banksii*, *O. ciliata*, *Ophiorrhiza biflora*, *Psychotria longipedicellata*, *P. bataanensis*, *P. subsessiliflora*, *P. rubiginosa*, *P. banahaensis*, *P. pinnatinervia*, *P. barnesii*, *Randia mindorensis*, *R. samalensis*, *R. uncaria*, *R. umbellata*, *R. fasciculiflora*, *Sarcocephalus ovatus*, *Tricalysia tinagaoense*, *Timonius attenuatus*, *T. benguetensis*, *T. quadrasii*, *T. obovatus*, *Uncaria philippinensis*, *Urophyllum sablanense* and *U. bataanense*. Many other species are credited to the Philippines for the first time but without citation of specimens representing them, the inference being that these species are represented in the herbarium of the Bureau of Science. In some genera, *Plectronia*, *Stylocoryne*, etc., new combinations are made without references to previously described species. Keys are given to the species under each genus, but no keys to the genera. According to the date of issue this work antedates Supplement I to the *Philippine Journal of Science*, on pages 126–137 of which some of the species mentioned above are also published. Careful work will be necessary to correlate these species. No attempt is made to enumerate all the species of the family credited to the archipelago by various authors, and no synonomy is given.

Article 2, April 10, 1906, pp. 1–21 (42–62). A Fascicle of Benguet Figs, by *A. D. E. Elmer*. Twenty-eight species are listed, of which the following are described as new: *Ficus fastigiata*, *F. irisana*, *F. cucaudata*, *F. confusa*, *F. umbrina*, *F. longipedunculata* (Merr.) Elm., *F. magnifica*, *F. rudis arborea*, *F. subintegra* (Merr.) Elm., *F. repandifolius*, and *F. integrifolia*.

Article 3, April 12, 1906, pp. 63–73. Additional New Species of Rubiaceae, by *A. D. E. Elmer*. The following 14 species are described as new: *Argostemma quadripetiolata*, *Oldenlandia filifolia*, *Psychotria subalpina*, *P. puloense*, *P. ellipticifolia*, *Gardenia morindaefolia*, *Ophiorrhiza pubescens*, *Ixora meyeri*, *I. leytensis*, *Tricalysia purpureum*, *Urophyllum banahaense*, *U. lucbanense*, *Timonius arboreus* and *Lasianthus morus*. As with article 1, this paper antedates Supplement I to the *Philippine Journal of Science*, according to the date of issue, where some of the above species are also published, not always however based on the same material.

Article 4, April 15, 1906, pp. 74–77. Pandans of East Leyte, by *A. D. E.*

Elmer. *Pandanus radicans* Blanco is redescribed, and *P. paloensis* and *P. muricatus* are proposed as new.

Article 5, July 26, 1906, pp. 78–79. A New Polypodium and Two Varieties, by *E. B. Copeland.* *Polypodium* (*Phymatodes*) *monstrosum* Copel. is described, and the two varieties, *leucophlebium* and *integriore.*

Article 6, August 1, pp. 78 bis–82. New Pandanaceae from Mount Banahao, by *A. D. E. Elmer.* *Freycinetia monocephala, Pandanus banahaensis* and *P. utilissimus* are described as new.

Article 7, August 16, 1906, pp. 83–186. Manual of the Philippine Compositae, by *A. D. E. Elmer.* In this paper 60 genera and 103 species are credited to the Philippines, including introduced and cultivated species. *Ethulia, Centipeda, Epaltes, Anaphalis, Erechtites,* and *Chrysogonum* are reported from the Philippines for the first time, and the following species are described as new: *Vernonia lenticellata, V. benguetensis* sub *B. Vialis* D. C., *Eupatorium toppingianum, E. sambucifolium, Blumea laxiflora, Gnaphalium oblancifolium, Aster luzonensis, Senecio benguetense, S. confusus, S. rubiginosus, S. mindorensis,* and *Chrysogonum philippinense.* New names appear in *Gynura, G. vidaliana* for *G. purpurascens* Vid., non DC., and *G. latifolium* (*Crassocephalum latifolium* Moore). An attempt was made to account for all the species credited to the Philippines by various authors, important synonomy is given, keys to the tribes, genera and species, and short descriptions of all the genera and species admitted.

Article 8, December 10, 1906, pp. 187–205. A Fascicle of East Leyte Figs, by *A. D. E. Elmer.* Twenty-six species of *Ficus* are numerated, of which the following are described as new: *Ficus johnsoni, F. benguetensis leytensis, F. fiskei, F. guyeri, F. carpenteriana, F. satterthwaitei, F. cassidyana, F. ruficaulis paloense* and *F. latsoni.*

Article 9, April 11, 1907, 207–208. A new Trigonostemum, by *Otto Stapf.* *Trigonostemum philippinense* Stapf is described, the genus being new to the Philippines.

Forbes, Francis Blackwell, and **Hemsley, William Botting.** An Enumeration of all the Plants known from China Proper, Formosa, Hainan, Corea and the Luchu Archipelago, and the Island of Hongkong, together with their Distribution and Synonomy. (*Journ. Linn. Soc. Bot.* **23** 1886–1888) pp. 1–521, plates 14: **24** (1889–1899) pp. 1–592, plates 10: **36** (1903–1905) pp. 1–686, with an Historical Note, Index, and List of Genera and Species discovered in China since the publication of the various parts of the "Enumeration.")

In the above work 8,271 species, of which 4,230 are endemic or not known to occur outside of the Chinese Empire are enumerated, but Sir *William T. Thistleton-Dyer* considers that the most moderate estimate can not put the whole flora as containing less than 12,000 species. Very many of the species enumerated extend to the Philippines, especially to northern Luzon, and the work is quite essential to the student of the Philippine flora.

Giesenhagen, K. Die Farngattung Niphobolus (1901, pp. 1–223, figures 20).

Fifty species are recognized, of which the following are credited to the Philippines: *Niphobolus splendens* (Hook.) Giesenh., endemic, *N. sticticus* Kze., British India and Ceylon to south China and Luzon, *N. nummularifolius* J. Sm., British India to Malaya, *N. lanuginosus* Giesenh., endemic, *N. samarensis* Giesenh., endemic, and *N. adnascens* Klf., south China to Malaya and Samoa. Several other species have since been found in the archipelago.

Gray, Asa. Characters of New or Obscure Species of Plants of the Monopetalous Orders in the Collection of the United States South Pacific Exploring Expedition under Captain Charles Wilkes, U. S. N., with Occasional Remarks, etc. (*Proc. Am. Acad.* **5** (1862.)

On page 324 a single species from the Philippines is described, *Gaultheria* (*Diplycosia*) *luzonica* = *Diplycosia luzonica* (A. Gray) Merr.

Harms, H. Anomopanax Harms, Eine im Herbar des Mus. Bot. Hort. Bogoriensis entdeckte neue Araliaceen-Gattung. (*Ann. Jard. Bot. Buitenzorg* II. **4** (1904) pp. 13–16.)

The new araliaceous genus *Anomopanax* is described, with three species, two, *A. celebicus* and *A. warburgii* from Celebes, the third, *A. philippinensis*, from Mindanao.

Hasskarl, J. K. Ueber einige neue Pflanzen der Philippinen aus der Cumingschen Sammlung. (*Flora*, **38** (1865) pp. 401–403.)

Three species are described: *Anredera cumingii* Hassk. (= *A. scandens* Moq.), *Symphorema glabrum* Hassk. (= *S. luzonicum* (Blanco) F.-Vill.) and *Tribulus macranthus* Hassk. (= *T. cistoides* L.).

Hemsley, W. Botting. Revision of the Synonomy of the Species of Aleurites. (*Kew Bull.* (1906) pp. 119–121.)

Four species of *Aleurites* are considered in connection with a preceding article on the source of Chinese wood-oil, *A. cordata* R. Br., Japan to Formosa and south China, *A. fordii* Hemsl., China, *A. triloba* Forst. (*A. moluccana* (L.) Willd.) Malaya and Polynesia and naturalized in many other tropical countries, and *A. trisperma* Blanco. The last two are common and widely distributed in the Philippines, the latter being endemic.

Laguna y Villanueva, Maximo. Apuntes sobre un nuevo roble (Q. jordanae) de la flora de Filipinas (1875) pp. 1–8, with plate.

In this work, all the species of *Quercus* then known from the Philippines are enumerated, and on page 7 *Quercus jordanae* is described, with a plate showing a branch and fruit, natural size.

Massee, George. Revision of the Genus Hemileia Berk. (*Kew Bull.* (1906) pp. 35–42, with one plate.)

Four species are recognized, of which one, *H. vastatrix* Berk. & Broome, the cause of the devastating coffee-leaf disease, is credited to the Philippines, on leaves of *Coffea arabica* L., and *C. liberica* Hiern. (It is abundant on the leaves of the former throughout the Philippines, and has practically killed the coffee industry in the Archipelago.)

Maxon, William R. A New Name for Kaulfussia Blume, a Genus of Marattaceous Ferns. (*Proc. Biol. Soc. Wash.* **18** (1905) pp. 239–240.)

The new generic name *Christensenia* is proposed, *Kaulfussia* Blume being invalidated by earlier use of the same name by *Dennstedt* and *Nees* in the *Polygalaceæ* and *Compositæ*. A single species is recognized, *Christensenia æsculifolia* (Blume) Maxon. The genus is represented in the Philippines by a distinct species, *C. cumingiana* Christ.

Moore, Spencer le M. Alabastra Diversa, Part XII. (*Journ. Bot.* **43** (1905) pp. 137–150.)

Among various species described from different parts of the World are three from the Philippines, *Aster philippinensis* from northern Luzon, and *Pogostemon philippinensis* from Luzon and Panay, and *Crassocephalum latifolium* from Negros.

Müller, J. Nouvelle espèce de Loranthus (L. mirabilis Van Huerck at Muell. Arg.) provenant des îles Philippines. (*Verhandl. Schweiz. Naturff. Gesellsch.* **55** (1872), pp. 47–48.)

A single species of *Loranthus* described, based on *Cuming's* No. 1966 from the Philippines, the species having been overlooked by the authors of Index Kewensis and by *Van Tieghem*. (See *Merrill* in *Philip. Journ. Sci.* **1** (1906) Suppl. 187.)

Rendle, A. B. New Philippine Plants. (*Journ. Bot.* **34** (1896) pp. 355–358.)

A short paper based on collections made by *John Whitehead* in the highlands on northern Luzon and in Mindoro, Mount "Dulangau" (correctly Dulangan), a spur of Mount Halcon. *Podocarpus falciformis* Parl., *Phyllocladus hypophylla* Hook. f., *Dacrydium elatum* Wall., *Cephalotaxus mannii* Hook. f., *Burmannia longifolia* Becc., *Platyclinis latifolia* Hemsl., *Litsaea villosa* Blume, *Gaultheria borneensis* Stapf, *Rhododendron cuneifolium* Stapf and *Strobilanthes penstemonoides* T. Andr., are reported from the Philippines for the first time, and *Vaccinium mindorense*, *Rhododendron lussoniense*, *R. whiteheadi*, *R. subsessile*, *Microstylis mindorensis* and *Zeuxine whiteheadi* are described as new. Of the above *Cephalotaxus mannii* Hook. f., can be excluded, the identification having been made from sterile material, and Mr. *Rendle* informs us that he now considers the plant to be *Taxus baccatus* subsp. *wallichiana* (Zucc.) Pilger.

Schlechter, R. Neue Orchidaceen der flora des Monsun-gebietes. (*Bull. Herb. Boiss.* II. **6** (1906) pp. 295–310.)

Among other species described from the monsoon region is *Platyclinis microchila* Schltr., p. 302 "Kultiviert in einem Garten in Sandakan, in Britisch Borneo; soll von Manilla importiert sein."

Stapf, Otto. The Oil-grasss of India and Ceylon. (*Kew Bull.* (1906) pp. 297–363, plates 1.)

An exhaustive account of the grasses yielding the products known as Citronella oil, Lemon-grass oil, Vetiver, etc., with a consideration of the species including full synonomy, at least two of the species considered extending to the Philippines.

Stein, B. Leptospermum (Glaphyria) annae Stein. (*Gartenflora* **34** (1885) pp. 66–68, plate 1184.)

The above species described and figured, the type from Mount Apo, Mindanao.

Stein, B. Rhododendron kochii Stein. (*Gartenflora* **34** (1885) pp. 193–195, plate 1195.)

The above species described and figured, the type from Mount Apo, Mindanao.

THE PHILIPPINE

JOURNAL OF SCIENCE

C. BOTANY

VOL. II — OCTOBER, 1907 — No. 5

PHILIPPINE WOODS.

By FRED W. FOXWORTHY.

(*From the botanical section of the Biological Laboratory, Bureau of Science.*)

CONTENTS.

I. INTRODUCTION.

Much misinformation is current as to the names and characteristics of our native woods. A wood is often variously designated in the same or in different provinces and again, several different kinds are frequently found under an identical name, for example molave (*Vitex* spp.) has more than forty different names in the Archipelago, and this multiplicity of names for the same wood naturally results in confusion which is very much increased when, as often happens, the same name applies to different woods in different localities. This makes it very easy for the unscrupulous dealer to substitute a poor quality for a better. There is evident need of some quick and sure way of identifying the woods needed for furniture, construction, and other purposes, and therefore it has seemed desirable to prepare a brief guide and description of those which are found in commercial quantities in the Manila market. This has been a task of some difficulty, because of their large number and the unsteady and uncertain supply of any one species at any given time.

There are about sixty-five commercial woods furnished by about one hundred species which are nearly always to be found in Manila, and in addition, there are several times as many which may occasionally be brought here in small quantities, so that the resulting complication is considerable. It follows that the chances for error are very great; so that this paper at best can be only preliminary to the more complete work indicated by the title.

PREVIOUS WORK.

But little has been done in the way of careful study of the native woods; the literature is as follows:

VIDAL Y SOLER (D. DOMINGO).—Manual del Maderero en Filipinas (1877), and other works by the same author.

Scattered notes by other Spanish authors.

FOREMAN (JOHN).—The Philippine Islands. London (1899), 2d edition, 367–373; (1906), 3d edition, 312–317.

This author gives notes on some of the best-known commercial woods.

AHERN, GEORGE P.—Important Philippine Woods. Manila (1901).

This is a compilation of notes from previous writers. This book brought together what had been written of the Philippine woods before 1901.

GARDNER, R.—Mechanical Tests, Properties, and Uses of Thirty Philippine Woods. Manila, *For. Bur. Bull.* (1906), 4, *2d edition* (Aug., 1907).

WHITFORD, H. N.—A preliminary Check-list of the Philippine Commercial Timbers. Manila, *For. Bur. Bull.* (1907), 7. (In press.)

The last two publications are most useful at the present time and they have been quoted extensively in this paper.

SCOPE AND METHODS OF THE PRESENT WORK.

In this paper the attempt has been made to give: 1. A general and technical discussion of wood. 2. A key to the common commercial woods. 3. Short notes on the structure, appearance, common names, range, and usefulness of individual species. 4. A very complete index.

Botanical material has furnished the starting point in correlating the name and wood which should go together; the botanical determination being made from herbarium material taken from the same tree as is the wood specimen; when the scientific name has been fixed and the structure studied, the wood is compared with commercial material until the latter can be determined definitely under its different names. Sections, whenever necessary, and as many as were necessary, have been made to determine doubtful points of structure.

The usefulness of this paper should consist in the ready classification of the commoner native woods; in the better understanding of their uses; in the finding of new applications for them and in discovering the relationships existing between the woods of the Philippine Islands and those of the rest of the world.

In addition to the ones already mentioned, the following sources of information have been used:

Roth and Fernow. Timber. Bul. Bur. of Forestry, U. S. Dept. of Agriculture (1895), **10**.

Gamble, J. S. A Manual of Indian Timbers. London (1902).

Janssonius, H. H. Mikrographie des Holzes. Leiden (1906).

Each of the American foresters of the Philippine Forest Service has aided the writer with material and observations. Special acknowledgments, however, are due Dr. H. N. Whitford and H. M. Curran, of the Bureau of Forestry, for their constantly helpful observations and the large amount of material furnished by them for the study of different woods. The field notes of Mr. J. R. Hillsman, of the Bureau of Internal Revenue, have also been of service.

II. GENERAL DISCUSSION.

1. STRUCTURE.

(a) GROSS MORPHOLOGY OF WOOD.

CLASSES OF WOOD.—All woody plants may be grouped according to their stem structure and botanical relationships as *Pteridophytes, Monocotyledons* (*Endogens*), and *Exogens.*

Pteridophytes.—The hard tissue is scattered in large, irregular bundles through the stem; the latter is uneven, being made up of soft and very hard material. *Tree ferns* are included in this class; they do not come into the market, but the trunks of certain species are used locally in Benguet and elsewhere in northern Luzon as posts for houses.

Monocotyledons or Endogens.—The wood is composed of scattered, small bundles of hard, woody tissue, the interspaces being filled with soft tissue. This group includes the *bamboos, palms, pandans,* etc.

Bamboos.—No work on the woods of the Philippine Islands would be complete without some mention being made of the bamboos which furnish so large a part of the structural materials of the Archipelago. Several different species are used, but they all agree in having the peculiar monocotyledonous structure already described, modified by the stem being hollow and jointed. They also contain a considerable proportion of silica.

The *palms* do not have jointed stems and are not hollow, but the central part of the stem is usually very soft and brittle. From the outer part, which is very hard and which will take a high polish, canes, bows, and other articles are made. Palma brava (*Livistona* spp.) and the coconut palm (*Cocos nucifera* L.) are the ones most used. Some palm stems are also suitable for the manufacture of small ornamental pillars, where the top and bottom are not exposed to the air, and where the defective nature of the inner part of the stem is not displayed. Palms are also to some extent used for flooring and for corner posts of houses.

The *bejucos* and *rattans* (*Calamus, Daemonorops*) also belong in this group, but as they occur in such small dimensions they are not considered in this paper.

The *pandans* or *screw pines* (*Pandanus* spp.) are widely distributed throughout the Archipelago. They are from a number of different species of the genus *Pandanus.*

The outer part of the stem of the *Pandanus* is usually very hard. I do not know of its commercial use here, but in some Pacific islands it furnishes an ornamental wood similar in texture, but inferior in finish, to that of the coconut (*Cocos nucifera* L.) and palma brava (*Livistona* spp.).

Exogens.—The remainder of our woody plants may be grouped together as *Exogens;* that is, the stem consists of a woody cylinder which grows in diameter by the addition of concentric layers about the wood already formed; there are two great groups; the *Gymnosperms,* or *Conifers,* and the *Angiosperms,* or broad-leaved plants. These may be distinguished as follows:

Conifers.—Wood, except in the first layer about the pith, containing no vessels; that is, nonporous; exceedingly regular in structure. There are a number of *Conifers* native to the Islands, but they are scattered in small patches or in almost inaccessible places on the mountains. The only native *Conifer* that is cut at all for timber is the Benguet pine (*Pinus insularis* Endl.) and it scarcely comes into the Manila market at all. However, a large amount of coniferous wood is imported: nearly all of this is California redwood or Oregon pine, although an occasional piece of coniferous timber from Australia, Japan, or China is encountered.

Angiosperms.—The remaining group, the broad-leaved trees, furnishes practically all of the Philippine wood found in the lumber yards, and further discussion will apply to woods of this group unless otherwise indicated.

PARTS OF THE STEM.

Pith, wood, and bark.—In examining the end of a log, three distinct areas are seen; namely, a small, central portion, the pith, made up of soft tissue; an outside, more or less corky covering, the bark, for purposes of protection; and, the wood, which is the hard tissue making up the greater part of the log and extending from bark to pith.

The pith is usually of very small diameter; it is rarely, as in Malapapaya (*Polyscias nodosa* Seem.), greater than one centimeter. This fact is of importance because the pith is an element of weakness in the wood.

Sapwood and heartwood.—The outer part of the log is often of a much lighter color, less in specific gravity and much softer than the center. The distinct, central part of the log is known as the heartwood and this outer portion is termed the sapwood. Many woods do not show any heartwood. The relative amount of sap- and heart-wood is very variable according to the individual tree, the age and the part of the tree from which it is taken.

Pith-rays.—Radiating from the pith to the bark are connecting lines of soft tissue, the medullary or pith-rays. These are among the most important characteristics to be observed in the structure of a wood, since they have an intimate connection with both the strength and beauty. They differ in size in different woods, being very large and distinct in some, as for example in teluto (*Pterocymbium tinctorium* Merr.), catmon

(*Dillenia* spp.), etc., and, in others, so small as to be invisible without the aid of a magnifying glass, as in acle (*Pithecolobium acle* (Blco.) Vid.), betis (*Illipe betis* (Blco.) Merr.), camagon (*Diospyros* spp.), or banaba (*Lagerstroemia speciosa* (L.) Pers.). The pith-rays may be all of the same size in the same tree, or there may be some large ones (the primary pith-rays) running from pith to bark, and some finer (secondary rays) starting beyond the pith. Compound pith-rays, where several are crowded together, may also occur. Pith-rays may take either a crooked or a straight course from the pith to the bark, but if curved, they usually are not abruptly so. The height of pith-rays is variable; they may be so short as scarcely to appear to have this dimension, or again it may be quite appreciable.

Growth rings.—The wood is formed in layers about the pith; and these may be formed only during certain seasons, the tree resting the remainder of the time. Where this is the case, each period of growth produces a ring about the pith. These rings are found in nearly all woods of temperate regions and in some of those of the Tropics. Where but one of them is formed during the year, it is called an annual ring, but manifestly, this name is not suitable for use with our woods, since we do not know whether one year sees the growth of one or of several. Consequently, the term annular, or seasonal growth rings has seemed preferable and will be used in this paper.

Seasonal rings seem to be characteristic of some of our woods only. It seems probable that the same species may have them when grown under one set of conditions and not under different ones. It also appears that many trees exhibit rings of seasonal growth when they are young but not afterwards. We have begun, in coöperation with the Bureau of Forestry, a series of observations on the manner and rate of formation of growth rings, but it will necessarily be some years before any safe general conclusions can be reached.

Distinct seasonal rings seem to be of constant occurrence in narra (*Pterocarpus* spp.), banaba (*Lagerstroemia speciosa* (L.) Pers.), calantas (*Toona* spp.), ipil (*Intsia* spp.), supa (*Sindora supa* Merr.), molave (*Vitex* spp.), and several other woods, but there seems to be a considerably greater number where they are not so.

False seasonal rings.—A number of woods show distinct, concentric lines bearing a strong, superficial resemblance to seasonal rings. These false rings may be caused by lines of soft tissue, as is the case in dita (*Alstonia scholaris* R. Br.) and palo maria (*Calophyllum* spp.), where they are so close together as to make it unlikely that they will often be mistaken. Lines of whitish resin-canals often give the appearance of seasonal rings in lauan (*Shorea* spp.), apitong (*Dipterocarpus* spp.), yacal (*Hopea* spp.), guijo (*Shorea guiso* Bl.), etc. These may readily be distinguished from the true seasonal rings by their irregularity of occurrence and by the fact that they usually fade out before completely encircling the log, and where they are numerous some of them can usually be seen to do this even in a small piece.

Vessels.—Fine, tubular passageways are found in all of these woods; in observing the end view of the log they appear as pores or sieve-like openings.

Concentric lines of soft tissue are found in some woods. These may be fine or coarse, wavy, broken, or straight. They are of very constant occurrence and serve clearly to delimit certain groups. The size, number, and arrangement of the vessels as well as the relation of the soft tissue and vessels to each other and to the pith-rays is very important.

PLANES OF SECTION.

Each wood should be observed in the following three planes of section:

Cross section.—Any section directly across the stem at right angles to the direction of growth; in this the pith-rays appear as long lines from pith to bark.

Tangential section or slab cut.—Any longitudinal section parallel to the bark and at right angles to the pith-rays. This is the one used in making ordinary, cheap planking, and it shows what is known as the cat-faced or bastard grain. In this plane of section the vessels appear as long lines through the wood and the pith-rays are seen in end view.

Radial section.—Any longitudinal section parallel to the pith-rays. Here the pith-rays appear as flat, expanded surfaces and the vessels as long lines; the timber so cut is known as quartered or rift-sawed, and has the beautiful silver grain which is familiar to most users of wood. This is the best method of cutting to secure the maximum of beauty and strength, but the tangential cut is much the easier to make, as it necessitates less handling and involves less waste; however, it gives an inferior timber.

GRAIN.

This is the figure presented by the structure of the wood. It is fine or coarse, straight or crooked, according as the elements of the wood are coarse or fine, crowded or loosely put together, straight or twisted. The best grain of the wood is brought out by careful attention to the cutting. The occurrence of a knot or branch, an irregularity in the trunk or root, or some local imperfection in the wood, may produce a regional modification of the grain, causing what is known as curly, or bird's-eye grain, or burl. Specimens showing the latter are at times very pretty and are much prized for certain classes of furniture. One of the best-known modifications of the grain is found in the large buttresses or buttress roots of some of our trees; some of these are of sufficient size to furnish single-piece table tops. Narra (*Pterocarpus* spp.) is probably the most widely known for this purpose, but we have a number of different trees showing this habit. Tindalo (*Pahudia rhomboidea* Prain), palo maria (*Calophyllum* spp.), tanguile (*Shorea polysperma* (Blco.) Merr.), calantas (*Toona* spp.) may be mentioned among the trees showing the fancy burl or bird's-eye grain.

Spiral grain.—A tree in growing often takes a spiral direction as indicated by the twistings of the bark; this gives the grain a spiral twist and the wood, in splitting, shows a series of flutings. A moderately pronounced spiral or twisted grain is evident in a number of our woods which show a resistance to smoothing in planing and working. When planed in one direction, portions of the surface are smoothed and certain others are roughened, and when the operation is reversed, the smooth surface becomes roughened as the rough surface is smoothed. This irregularity of grain is often noticed in amuguis (*Koordersiodendron pinnatum* Engl.), lauan (*Shorea* spp.), guijo (*Shorea guiso* Bl.) and mayapis (*Anisoptera* spp.).

(b) MINUTE ANATOMY.

Elements.—The elements making up wood are, vessels or tracheæ, tracheids, wood-fibers, pith-ray cells, and wood parenchyma cells.

Tracheæ, vessels, or *pores* are long tubes extending through the wood for some distance. Their size, arrangement in rows or scattering, and their relation to other elements are of great importance in the classification of woods. Large vessels are found in calantas (*Toona* spp.), lauan (*Shorea* spp.), and batitinan (*Lagerstroemia batitinan* Vid.); very small ones in bolongeta (*Diospyros* spp.), calamansanay and mancono (*Xanthostemon verdugonianus* Naves).

Wood-fibers.—These are long and slender, thick-walled cells, containing lignin in their walls. Their abundance and the thickness of their walls is usually sufficient to account for the weight and hardness of the wood.

Tracheids.—These are elongated, tapering cells, not so thick-walled as the wood-fibers, of relatively greater diameter, with walls more pitted and shorter.

Pith-ray cells.—These are short, prismatic, thin-walled cells containing starch grains, resin, or other deposits; they are nearly always with their long axes horizontal.

Wood parenchyma.—This is formed by thin-walled, prismatic cells, with starch or other inclusions. The cells are scattered with more or less regularity through the wood; the long axes being vertical. The wood parenchyma in some woods is arranged in fine, concentric or wavy, broken lines. These are usually of a lighter color than the surrounding tissue.

Pith-rays.—These are usually made up of unlignified cells and extend in a radial direction.

Resin-canals.—These are passages lined with thin-walled cells which secrete a resin which is often found exuding from the cells into the central passage, or completely filling it. Resin-canals are found in but few of our woods; for example, Benguet pine (*Pinus insularis* Endl.), lauan (*Shorea* spp.), apitong (*Dipterocarpus* spp.), yacal (*Hopea* spp.), tanguile (*Shorea polysperma* (Blco.) Merr.), guijo (*Shorea guiso* Bl.), mangachapuy (*Hopea acminata* Merr.), mayapis (*Anisoptera* spp.).

Deposits in vessels, etc.—The nature and color of the deposits in the vessels of certain woods is a distinctive character. Thus ipil (*Intsia* spp.) is distinguished by the sulphur-yellow deposits in its vessels; acle (*Pithecolobium acle* (Blco.) Vid.) and catmon (*Dillenia* spp.) by white ones; lumbayao, calantas (*Toona* spp.), and duguan (*Knema* and *Myristica* spp.) by red deposits; ebony, camagon (*Diospyros* spp.), and bolongeta by the very dense, black deposits in all of the wood elements of the heartwood; palo maria (*Calophyllum* spp.), betis (*Illipe betis* (Blco.) Merr.), and bansalaguin (*Mimusops elengi L.*) have pale-yellowish deposits in the vessels.

2. PHYSICAL AND CHEMICAL PROPERTIES OF WOOD.

PHYSICAL PROPERTIES.

Color.—The heart- and the sap-wood are often very widely different in color. Usually, the former is very much darker than the latter and the line of demarcation between the two is often very distinct. In some cases, such as agoho (*Casuarina equisetifolia* Forst.) the heart is only different in degree from the sap, being only a few shades darker in color and showing a gradual change from sap- to heart-wood. In other instances there is no heartwood, the color being the same throughout; examples are dita (*Alstonia scholaris* R. Br.) and lanete (*Wrightia* spp.).

There is usually some range of color within a species, but still not so much as to prevent the recognition of the characteristic color. However, in some species there is the greatest latitude of variability. In narra (*Pterocarpus* spp.), for instance, three colors of wood, respectively known as white, yellow, and red narra, seem to be obtained from the same species.

Color may be due to deposits in vessels, parenchyma and pith-ray cells, or to the presence of some pigment in all the elements of the wood. In calantas (*Toona* spp.), the elements all contain a certain amount of pigment and there is also the red-colored substance in the vessels. The black color of camagon (*Diospyros* spp.), bolongeta (*Diospyros* spp.), and ebony (*Maba buxifolia* Pers. and *Diospyros* spp.) is caused by a compound of tannic acid which fills all the elements of the heartwood.

Odor.—Certain woods are recognizable by their disagreeable odor, as, for example, cupang (*Parkia roxburghii* Don.) and Eugenia sp. Calantas (*Toona* spp.) has an odor resembling that of cedar; narra (*Pterocarpus* spp.), a sweetish cedary, and teak (*Tectona grandis* L. f.) a distinctly aromatic odor. Others of our woods have their peculiar odors, which, though fainter and difficult of description, are yet distinctive.

Taste.—A number of our woods may be recognized by their bitter taste; among these are anubing (*Artocarpus* spp.), batino (*Alstonia macrophylla* Wall.), betis (*Illipe betis* (Blco.) Merr.), bansalaguin (*Mimusops elengi* L.),. dita (*Alstonia scholaris* R. Br.), and yacal (*Hopea* spp.).

Weight and specific gravity.—We have quite a large number of heavy woods, although perhaps not so large a proportion as is found in some other tropical countries. I have classified our woods as very heavy, heavy, moderately heavy, and light, following the classification used by Gardner.[1] We have many woods which when green will sink in water, but the number of these which has a greater specific gravity than water when dry is relatively small. The following table gives a list of Philippine and American commercial woods, with their weight and specific gravity so far as known.

The heavy woods which are italicized frequently come into the "very heavy" class.

Comparative weights of Philippine and American woods.

PHILIPPINE WOODS.

Very heavy. Sp. gr., 0.90 or more. Weight.—Metric system, 900 kilos or more per cu. m.; English system, 56 lbs. or more per cu. ft.; Spanish system, 42 lbs. or more per cu. ft.	Heavy. Sp. gr., 0.70–0.90. Weight.—Metric system, 700–900 kilos per cu. m.; English system, 44–56 lbs. per cu. ft.; Spanish system, 32–42 lbs. per cu. ft.	Moderately heavy. Sp. gr., 0.50–0.70. Weight.—Metric system, 500–700 kilos per cu. m.; English system, 31–44 lbs. per cu. ft.; Spanish system, 23–32 lbs. per cu. ft.	Light. Sp. gr., 0.50 or less. Weight.—Metric system, 500 kilos or less per cu. m.; English system, 31 lbs. or less per cu. ft.; Spanish system, 23 lbs. or less per cu. ft.
Mancono.[a]	*Duñgon.*[a]	Narra.[a]	Lauan.[a]
Duñgon-late.[b]	*Ipil.*[a]	Acle.[a]	Baticulin.
Ebony.	Molave.[a]	Teak.[b]	Calantas.[a]
Camagon.	*Yacal.*[a]	Guijo.[a]	Mayapis.[a]
Bolongeta.	Tindalo.[a]	Apitong.[a]	Red lauan.[a]
	Betis.[a]	Amuguis.[a]	Dita.[b]
	Bansalaguin.[a]	Palo maria.[a]	Cupang.[a]
	Supa.[a]	Banaba.	Teluto.
	Macaasin.[a]	Anubing.	Malapapaya.
	Batitinan.[a]	Bancal.[b]	
	Aranga.[a]	Tamayuan.	
	Sasalit.[a]	Sacat.[a]	
	Liusin.[a]	Malasantol.[a]	
	Tucan-calao.	Balacat.[a]	
	Alupag.	Malugay.[a]	
	Catmon.[a]	Banuyo.[a]	
	Agoho.[a]	Tanguile.[a]	
	Calamansanay.	Lanete.	
	Mangachapuy.[a]	Duguan.	
	Batete.	Santol.[b]	
	Lanotan.[a]	Nato.	
		Dalinsi.	
		Calumpit.	
		Talisay.[b]	
		Baliñhasay.	
		Lumbayao.[a]	
		Batino.	

[a] The specific gravity of these woods was obtained from tests made in Manila.
[b] These woods were grouped by data found in Gamble's Manual of Indian Timbers.

[1] *Bur. For. Bull.*, Manila (1907), **4**, 51.

Comparative weights of Philippine and American woods—Continued.

AMERICAN WOODS.[a]

Very heavy.	Heavy.	Moderately heavy.	Light.
	Hickory.	Ash.	White cedar.
	White oak.	White elm.	White pine.
	Red oak.	Sweet gum.	White Spruce.
	Persimmon.	Hard pine.	Bald cypress.
	Osage orange.	Cherry.	Red cedar.
	Black locust.	Birch.	Hemlock.
	Hackberry.	Maple.	Redwood.
	Blue beech.	Walnut.	Oregon pine.
		Sour gum.	Basswood.
		Coffee tree.	Chestnut.
		Honey locust.	Butternut.
		Tamarack.	Tulip.
		Douglas spruce.	Catalpa.
		Western hemlock.	Buckeye.
		Soft maple.	Poplar.
		Sycamore.	Willow.
		Sassafras.	
		Mulberry.	

Resonance.—We have no commercial wood in the Islands which is suitable for making good sounding boards. Imported coniferous wood is usually used for this purpose in guitars and other stringed instruments of local manufacture, the backs and sides of the instruments being made of lanotan (*Bombycidendron campylosiphon* (Tcz.) F. Vill.), lanca (*Artocarpus integrifolia* L. f.) or other even-grained ornamental woods.

Moisture content, shrinkage, seasoning.—Wood is much heavier when green than when dry, because of the large amount of water which it contains; air-dry it still holds 8 to 10 per cent of moisture and even when it is kiln-dried there is usually some water left in it. It is exceedingly hygroscopic; a piece which has been very thoroughly dried will, if placed in a moist place, take up enough water to equalize its moisture content with that of the surrounding air. This capacity for taking up water is responsible for the swelling and warping of timber. The loss of water from the wood causes shrinkage and where this is uneven, checking.

Seasoning.—The process by which water is gradually removed from wood is known as seasoning. In seasoning, certain chemical and physical changes take place which render the wood stronger, more durable, and usually harder and heavier. The nature of these changes is rather imperfectly understood, but it seems probable that certain materials contained in the pith-ray and wood parenchyma cells become changed into tannins, resins, and other substances which have a preservative and strengthening effect. When properly seasoned a wood is always stronger

[a] The classification of American woods was taken from Roth's Bull. Timber., U. S. Bur. of Forestry (1895), 10.

than it is when unseasoned. There may be several kinds of seasoning, as follows:

Natural seasoning taking place in the tree.—This results in the formation of heartwood by the means already indicated. A loss of water occurs simultaneously with the chemical change taking place, and the deposit of certain substances in the cells more than counterbalances the loss in weight, so that the heartwood is specifically heavier, although lower in moisture content, than the sapwood. This change from sap- to heart-wood is very important in considering the value of a timber. Sapwood seems incapable of equaling heartwood, no matter how carefully it may be handled after leaving the tree.

Artificial seasoning.—*In the standing tree:* In some cases, as for instance in the teak forests in India, the tree is girdled and then left on the stump for a year or more before being cut. It is claimed that the disadvantages of this method are that the resulting wood is more brittle than if it is seasoned in the usual way, and moreover, during the process it is more exposed to the attacks of burrowing insects. To offset this there is the advantage of rapid seasoning, with but little checking. For some species this is probably the best method.

In the log: Material left to season in the log usually becomes noticeably checked. Rapid seasoning is most safely accomplished in pieces of small dimensions.

By air-drying: The greater part of our material is air-dried—that is, seasoned by standing in piles of lumber exposed to the air. If properly piled, the process will proceed at a fairly rapid rate and the checking will be very slight. The pile should be so arranged that the air can reach the wood from all sides.

By kiln-drying: This is accomplished by means of a controlled supply of artificial heat. Kiln-drying is resorted to whenever it is desired to reduce the percentage of moisture below that of air-dry wood or whenever especially rapid seasoning is required. If the operation is carefully performed, the wood is seasoned with a minimum amount of checking; it is made stronger and is less liable to decay. Of course, kiln-dried wood will take up moisture from the air, but it will not absorb it in as great quantity or as rapidly as the air-dried material; therefore, it actually remains drier than wood which has been seasoned in the air. The best results are obtained by prolonged and careful air-drying, followed by kiln-drying. If properly handled, wood is always improved by being kiln-dried. Unfortunately, the process is not as much practiced with the native woods as it should be.

Seasoning in fluids: Timbers sometimes are submerged in sea water for years before being dried, additional strength and durability apparently being given to them. For many years this has been the process with oak used for shipbuilding in England. Of course this method of seasoning can only be employed where the material can be so submerged as to be free from teredo attack. Timbers occasionally are encountered which have had a part of their seasoning in fresh water or in the mud at the bottom of fresh-water streams or lakes, an example being the swamp cypress logs which are raised from the mud of rivers and bayous in the southern United States, after having lain there for many years.

Small pieces of woods for certain purposes are *seasoned in oil or other fluids.* All these methods of submerging woods during seasoning have the very great advantage that the process is thereby made a very gradual and uniform one, checking being reduced to a minimum. However, these methods are suited only to special cases.

Heating power.—This varies with the content of carbon and contained resins, oils, etc. Our best firewoods are usually those with very thick, dense cell-walls.

CHEMICAL PROPERTIES.

It is not the intention here to treat of the chemical composition of wood or of its behavior under any but the simplest reagents. However, there are a few very simple tests which may aid in the determination of particular woods and these are included here:

Molave (*Vitex* spp.) turns to a bright greenish-yellow when treated with an alkaline solution; narra (*Pterocarpus* spp.) gives a fluorescent, blue color to water; betis (*Illipe betis* (Blco.) Merr.) or bansalaguin (*Mimusops elengi* L.) will form a lather if the surface of the wood be rubbed with water or saliva; calumpit (*Terminalia edutis* Blco.), dalinsi (*T. pellucida* Presl.) sacat (*T. nitens* Presl.), or talisay (*T. catappa* L.) will color water a dirty, straw-yellow; catmon (*Dillenia* spp.) causes water to become pale red.

Others of our woods will doubtless also be found to give distinct reactions with simple reagents.[3]

3. DURABILITY AND DECAY.

Fungi and bacteria.—These grow abundantly in warm and moist situations. Wood which is partly submerged, or in contact with the ground, is most subject to the attack of these organisms, a continual supply of moisture favoring their development. Piling, railroad ties, and portions of buildings in contact with the ground give the best illustrations of destruction by these means, but while wood is always liable to damage from these causes, they are not the most serious considerations in this climate.

Beetles.—Woods frequently are encountered which are completely riddled by the burrows of wood-boring beetles; these extend in all directions and very perceptibly weaken the wood. The presence of beetles is indicated by the open burrows or by fine wood dust pushed out from them. It is said that no woods are entirely immune from beetle attack. In the very hard woods, however, it is usually only the sapwood which is affected. Logs left in the forest or piled with beetle-eaten material are most subject to attack. Certain woods, such as dita (*Alstonia scholaris* R. Br.) and lanete (*Wrightia* spp.), are particularly liable to be damaged in this way. No entirely satisfactory means of preventing beetle attacks has as yet been found.

Anay or "white ants."—Termites, very generally known in the Islands as *anay*, destroy a great many of the softer woods, completely hollowing them out until only a shell is left.

Shipworm or teredo.—This is the most serious enemy to piling, boat keels, and other wooden articles which are immersed in sea water. The work of these small animals constitutes so serious a nuisance as to render any but a very few of our very hard woods useless for piling.

[3] A discussion of the mechanical properties of wood is given in Timber, Bull. 10, U. S. Bur. For., and for a discussion of the mechanical properties of Philippine woods the reader is referred to Gardner, Bull. For. Bur., Manila (1906), 4, (1907), 2d edition.

IMMUNITY FROM ATTACK.

Hardness.—A few woods, such as mancono (*Xanthostemon verdugonianus* Naves) for piling and molave (*Vitex* spp.) for house construction, seem to be immune from attack because of their hardness. As a rule the hardwoods are very much freer from insect and teredo attack than are the soft kinds.

Taste and odor.—It seems probable that some woods may be safe from insects because of a taste or odor which is not agreeable to the invaders. This is supposed to be the case with calantas (*Toona* spp.).

IMPREGNATION AND OTHER ARTIFICIAL MEANS OF DEFENSE AGAINST INSECTS AND TEREDO.

Creosoting.—Creosoting, in the very few cases in which it has been tried in the Islands, has been very satisfactory, but it can not as yet come into common use because of the present prohibitive cost of creosote in Manila.

Impregnation with mineral salts may prove effective, if some way can be found of precipitating the salts in the wood so that they will not leech out under the action of this moist climate.

Painting a wood has proved effective so long as the painted surface does not become cracked.

There is further need of experiment to determine what Philippine woods are most immune to insect and teredo attack, and what are the best artificial means of defense.

USES OF PHILIPPINE WOODS.

While complete tests have not been carried out for any Philippine woods, certain of them have been found to be particularly well fitted for especial uses, and the effort is here made to group the woods of commerce according to their use.

1. *In places exposed to salt water and teredo attack.*—For piling: Liusin, betis, aranga, mancono, banaba, batitinan, bolongeta, duñgon, duñgon-late, mangachapuy, molave, and yacal are used; but the first four mentioned give the best satisfaction.

In addition to these there is agoho, which by its great hardness and its normal, tapering shape seems to be well fitted for piling. It seems not yet to have been tried for that purpose.

For ship and boat building: Teak, usually of the first importance as a shipbuilding wood, is of small importance in the Philippines because of its very restricted occurrence. It is obtainable only in small quantities.

For keels and other parts of ships exposed to salt water: Aranga, banaba, bansalaguin, betis, duñgon, guijo, liusin, molave, narra, palo maria, and yacal are used.

For small boats, bancas, cascoes, etc., a large number of different woods are employed, among which are: Apitong, amuguis, bancal, banuyo, calantas, white lauan, lumbayao, malasantol, malugay, mangachapuy, and tanguile.

2. *In places where the wood is in contact with the ground.*—For corner posts of houses (*harigues*): Molave, ipil, acle, agoho, alupag, anubing, aranga, banaba, bansalaguin, banuyo, batitinan, betis, calamansanay, duñgon, duñgon-late,

liusin, macaasin, manono, mangachapuy, narra, palo maria, sasalit, supa, tamayuan, tucan-calao, yacal.

For railroad ties: Molave, ipil, acle, betis, aranga, duñgon, yacal, tindalo, sasalit, supa, anubing, banaba, bolongeta, agoho. In addition to these, the following have been recommended by the Forestry Bureau as worth testing: Tong, dao (*Dracontomelum* sp.), apitong, amuguis, banuyo, malaruhat (*Eugenia* sp.), palo maria.

For paving blocks: Molave is the only native wood which is known to be satisfactory as a paving block. Several of the woods used for railroad ties should be tried for this purpose.

3. *For use as construction timbers.*—For heavy framing and general high-grade construction: Acle, agoho, alupag, aranga, banaba, bansalaguin, batitinan, betis, catmon, duñgon, duñgon-late, ipil, liusin, macaasin, mangachapuy, molave, narra, palo maria, sasalit, supa, tamayuan, tucan-calao, yacal.

For medium-grade construction: Anubing, lumbayao, guijo, malasantol, malugay, lanotan, calamansanay, banuyo, batete, apitong, amuguis, tanguile.

For light or temporary construction: Balacat, balinhasay, bancal, batino, cakantas, calumpit, cupang, dugnan, dalinsi, dita, lanete, white lauan, red lauan, malapapaya, marapis, nato, sacat, santol, talisay.

4. *For use in making furniture and ornaments.*—For the better grades of furniture there are used: Tindalo, acle, palo maria, catmon, teak, supa, ipil, narra, calamansanay, banuyo.

Cheap furniture is made of guijo, bancal, apitong, calumpit, dalinsi, sacat, talisay, dita, santol, baticulin, batete, malugay. Tanguile, red and white lauan, apitong and lumbayao make cheap furniture of excellent quality.

Besides the above-mentioned woods the following are used in cabinet making: Anubing, aranga, banaba, bansalaguin, camagon, bolongeta, ebony, lanete, lanutan, macaasin, tucan-calao, yacal, narra. Lanete, molave, and santol are among the woods used for wood carving.

III. KEY TO PHILIPPINE COMMERCIAL WOODS.

It has been the effort to make this key cover all the woods in the first three groups on which the Bureau of Internal Revenue collects a tax; and in addition, such of the fourth-group woods as are commonly found on the Manila market. The one wood of the first three groups which has not been included is malacadios. Specimens of wood bearing this name have been received from several different provinces. These specimens represent three or four widely different species. Repeated effort has failed to discover here a wood bearing the name of "malacadios," although a number of dealers have said that it is occasionally found in the Manila market. We must, for the present, leave this wood out of our consideration.

EXPLANATION OF THE TERMS USED IN THE KEY.

Nonporous and porous woods. (See p. 354.)

Seasonal rings (see p. 355).—These are defined by a line at either margin or by the greater density of structure in the outer part of the ring.

Ring porous.—With one or more rows of large vessels in the early part of the ring; the later part of the ring having the vessels smaller and scattered.

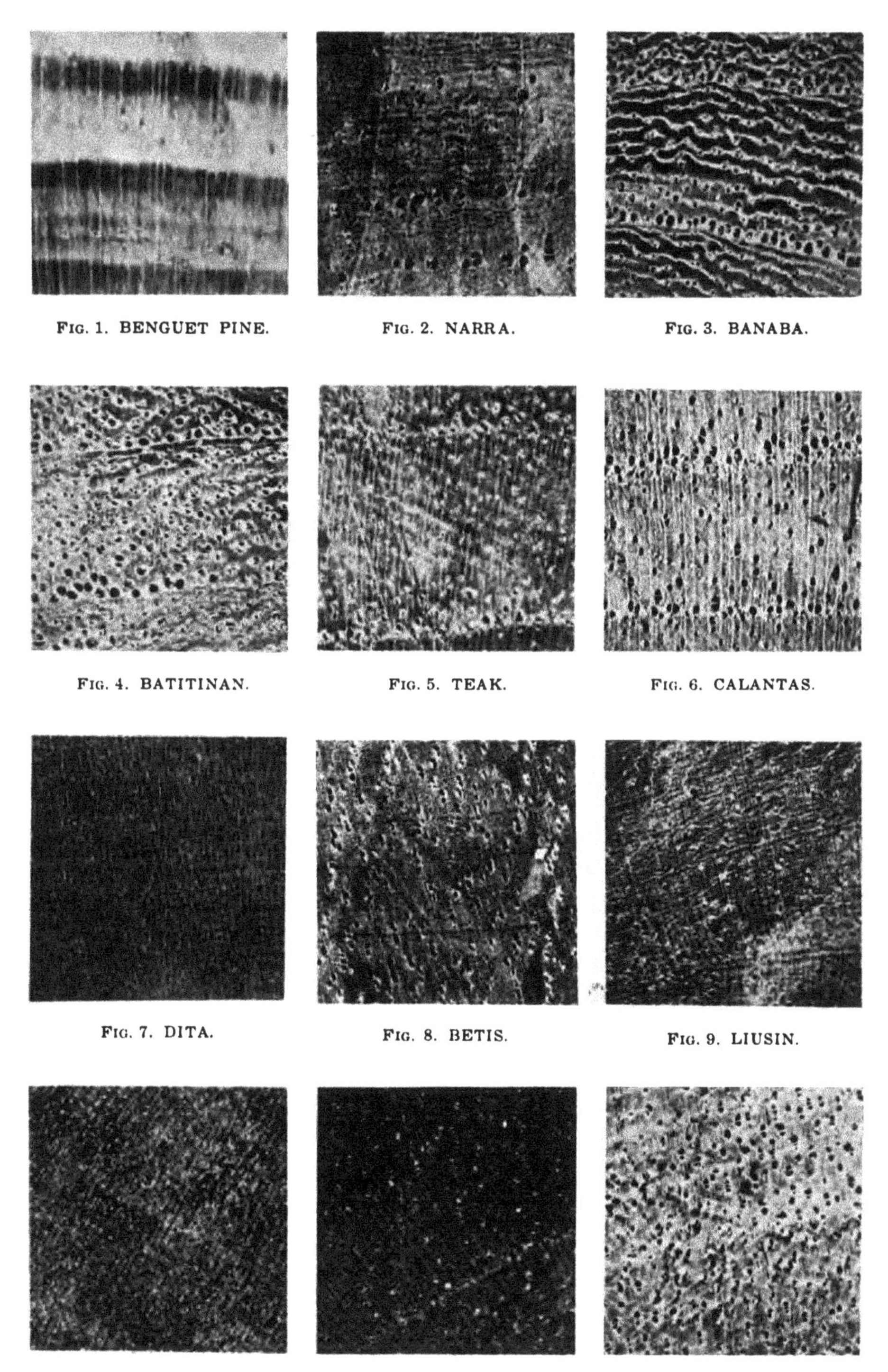

FIG. 1. BENGUET PINE.

FIG. 2. NARRA.

FIG. 3. BANABA.

FIG. 4. BATITINAN.

FIG. 5. TEAK.

FIG. 6. CALANTAS.

FIG. 7. DITA.

FIG. 8. BETIS.

FIG. 9. LIUSIN.

FIG. 10. TAMAYUAN.

FIG. 11. DUÑGON.

FIG. 12. BALACAT.

Diffuse porous.—With the vessels of approximately uniform size and scattered through the ring.

Wherever the rings have seemed to be doubtfully distinct, the wood has been included in both sections of the key. This will explain why a number of woods are found more than once.

Unless especially otherwise indicated, the color credited to a given wood is that of the heartwood.

In using this key, no special equipment is needed beside a sharp jack-knife and a small lens capable of magnifying as much as five diameters. The end of the block used should be carefully smoothed with the knife, so that the different structural features may be seen with the lens. Ordinarily, the cross section is the only one used, as it shows nearly all of the points which it is necessary to observe.

The figures used to illustrate this key have been made with a uniform magnification of about five diameters.

Key to Philippine commercial woods.

a. NONPOROUS WOODS .... BENGUET PINE (see fig. 1).
OREGON PINE.
CALIFORNIA REDWOOD.

aa. POROUS WOODS.
- *b*. SEASONAL RINGS DISTINCT.
  - *c*. Ring porous.
    - *d*. Wood parenchyma in wavy, tangential lines.
      - *e*. Tangential section showing faint, parallel, transverse lines; distinct, sweetish, cedary odor; wood coloring water a fluorescent blue. NARRA (see fig. 2).
      - *ee*. Not as above.
        - *f*. Pale to dark red .... BANABA (see fig. 3).
        - *ff*. Brownish to dark brownish; vessels containing glistening dark-colored deposits .... BATITINAN (see fig. 4).
    - *dd*. Parenchyma lines wanting.
      - *e*. Moderately heavy to heavy and hard; yellow to very dark brown; strong, spicy odor .... TEAK (see fig. 5).
      - *ee*. Light and soft; red; cedary odor .... CALANTAS (see fig. 6).
  - *cc*. Diffuse porous.
    - *d*. Wood parenchyma prominent.
      - *e*. In concentric lines.
        - *f*. Pith-rays very small.
          - *g*. Wood soft and white, with bitter taste .... DITA (see fig. 7).
          - *gg*. Not white.
            - *h*. Wood forming a lather when rubbed with water or saliva; vessels in oblique radial lines.
              - *i*. Very hard and heavy; very dark red .... BANSALAGUIN.
              - *ii*. Lighter in weight and color; not quite so hard. BETIS (see fig. 8).
            - *hh*. Wood not forming a lather when rubbed.
              - *i*. Vessels medium size, scattered, not numerous; wood parenchyma lines very distinct .... LIUSIN (see fig. 9).

ii. Vessels small, very numerous, in short radial rows; parenchyma in very fine, broken lines.. TAMAYUAN (see fig. 10).

ff. Pith-rays of medium size; very hard and heavy; reddish brown to chocolate color; parenchyma lines very fine.
DUÑGON (see fig. 11).
DUÑGON-LATE.

ee. Clustered about large vessels.

f. Wood white; not heavy.

g. Pith-rays small; wood parenchyma forming a very narrow fringe about the vessels .... BALACAT (see fig. 12).

gg. Pith-rays medium size; wood parenchyma forming a very distinct fringe about the vessels; very bad odor when first cut.
CUPANG (see fig. 13).

ff. Wood not white; heavier.

g. Wood dark brown.

h. Seasonal rings rather narrow; whitish deposits in vessels.
ACLE (see fig. 14).

hh. Seasonal rings broad; vessels without whitish deposits.
BANUYO (see fig. 15).

gg. With yellow or reddish tinge.

h. Dark red; parenchyma sometimes forming lines between vessels .... TUCAN-CALAO (see fig. 16).

hh. Parenchyma only about vessels.

i. Yellow to dark reddish-brown; sulphur-yellow deposits in vessels .... IPIL (see fig. 17).

ii. Yellowish-red; no yellow deposits in vessels.
TINDALO (see fig. 18).

dd. Wood parenchyma not prominent.

e. Wood white, or grayish.

f. Gray with faint greenish-yellow tinge; very hard and heavy; vessels small and scattered; wood turning bright greenish-yellow when treated with an alkaline solution...... MOLAVE (see fig. 19).

ff. White; moderately heavy and moderately hard; vessels very small, in short radial lines.

g. Very even grained; tasteless .... LANETE (see fig. 20).

gg. Less even grained; with bitter taste............ BATINO (see fig. 21).

ee. Not white or grayish.

f. Vessels clogged with white substance; heartwood brownish or greenish or greenish-yellow.. ANUBING (see fig. 22).

ff. Vessels not clogged with whitish deposits.

g. Seasonal ring marked off by a distinct line.

h. Vessels exuding an oil whenever cut across; heavy and hard.

i. Yellowish-brown .... SUPA (see fig. 23).

ii. Reddish-brown; all surfaces stained by the oil.
BATETE (see fig. 24).

hh. Vessels not exuding an oil; pink or pale reddish.

i. Edge of seasonal ring a fine line; hard and moderately heavy .... MALUGAY (see fig. 25).

ii. Edge of seasonal ring a coarser line; wood softer and lighter; pith-rays more distinct...... DUGUAN (see fig. 26).

gg. Seasonal rings not marked off by a distinct line.

h. Wood purple; with parallel transverse markings in longitudinal sections .... LANOTAN (see fig. 27).

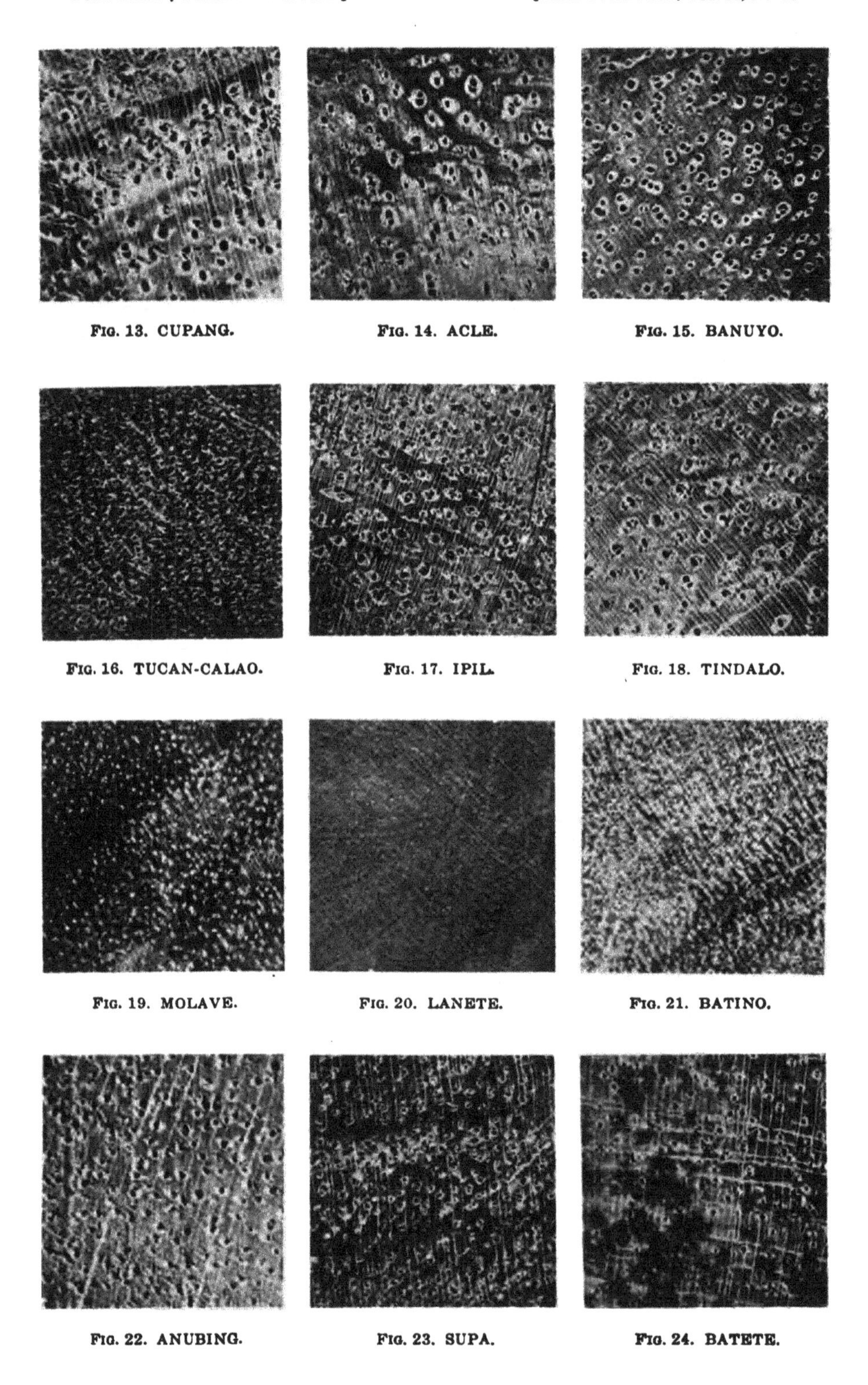

FIG. 13. CUPANG. FIG. 14. ACLE. FIG. 15. BANUYO.

FIG. 16. TUCAN-CALAO. FIG. 17. IPIL. FIG. 18. TINDALO.

FIG. 19. MOLAVE. FIG. 20. LANETE. FIG. 21. BATINO.

FIG. 22. ANUBING. FIG. 23. SUPA. FIG. 24. BATETE.

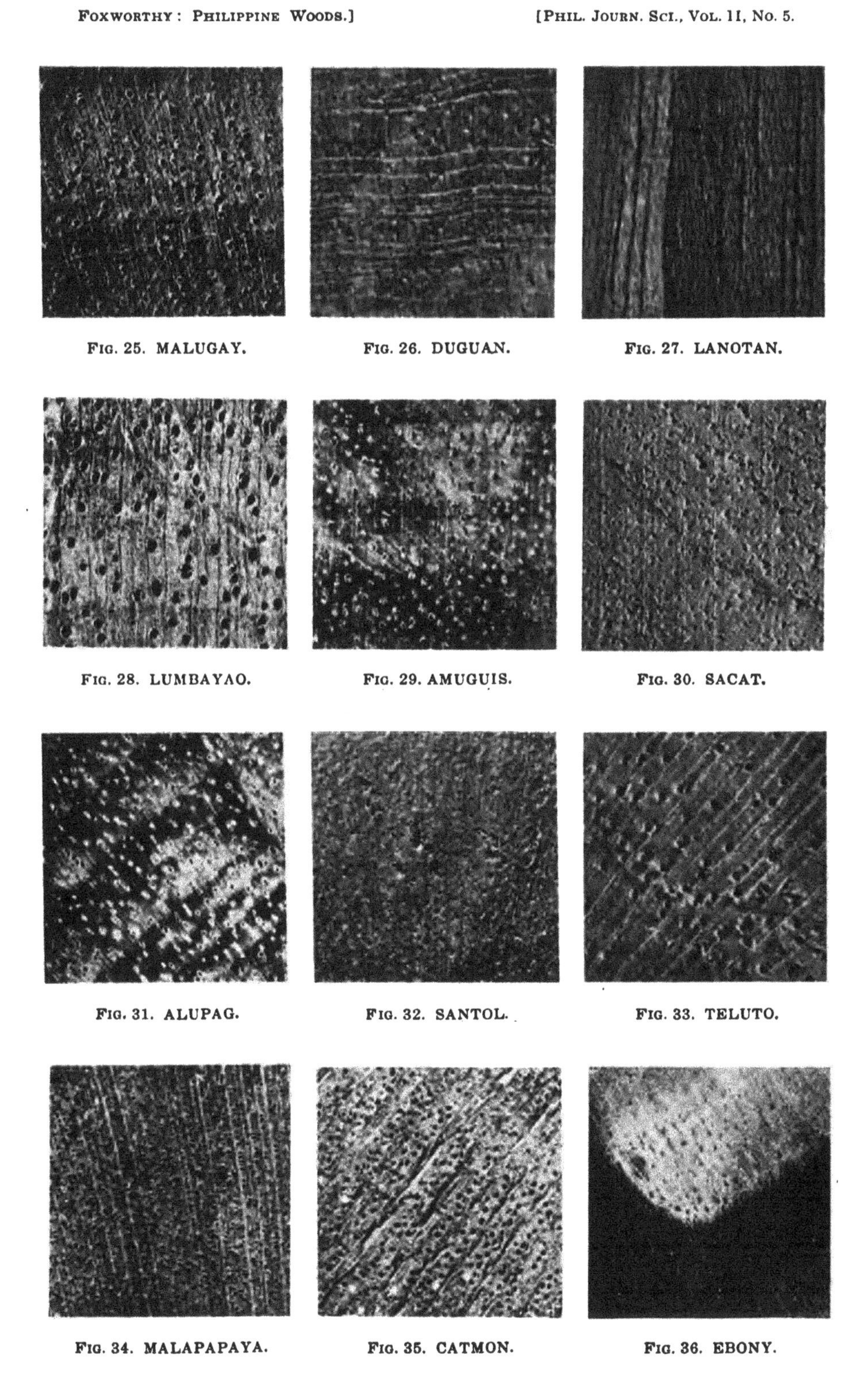

FIG. 25. MALUGAY.

FIG. 26. DUGUAN.

FIG. 27. LANOTAN.

FIG. 28. LUMBAYAO.

FIG. 29. AMUGUIS.

FIG. 30. SACAT.

FIG. 31. ALUPAG.

FIG. 32. SANTOL.

FIG. 33. TELUTO.

FIG. 34. MALAPAPAYA.

FIG. 35. CATMON.

FIG. 36. EBONY.

hh. Wood not purple; without transverse markings.

i. Dark red.

j. Vessels large, scattered, with red deposits.

LUMBAYAO (see fig. 28).

jj. Vessels medium size, scattered, without red deposits.

AMUGUIS (see fig. 29).

ii. Not dark red.

j. Wood coloring water a pale straw-yellow; moderately heavy and moderately hard; vessels of medium size, often in more or less wavy lines in outer part of seasonal ring .... CALUMPIT.

DALINSI.

SACAT (see fig. 30).

TALISAY.

jj. Wood not coloring water straw-yellow; vessels scattered, medium size to small; pale reddish.

k. Very hard and heavy .... ALUPAG (see fig. 31).

kk. Soft and moderately heavy; faint, camphor-like odor when first cut........ SANTOL (see fig. 32).

bb. SEASONAL RINGS NOT DISTINCT.

c. Pith-rays broad or very broad.

d. Wood soft and light; whitish.

e. Tangential section showing numerous fine, parallel, transverse lines.

PELUTO (see fig. 33).

ee. Tangential section without transverse lines.

MALAPAPAYA (see fig. 34).

dd. Wood hard and heavy; reddish or purplish; vessels with white deposits .... CATMON (see fig. 35).

cc. Pith-rays not broad.

d. Wood parenchyma in concentric lines or clustered about vessels.

e. Wood parenchyma in regular concentric lines.

f. Heartwood black or with black streaks; very heavy and very hard.

BOLONGETA.

CAMAGON.

EBONY (see fig. 36).

ff. Heartwood not black.

g. Pith-rays of medium size; wood very hard and heavy; heartwood chocolate color; parenchyma lines very fine.

DUÑGON (see fig. 11).

DUÑGON-LATE.

gg. Pith-rays fine and indistinct.

h. Vessels medium size or small, in irregular radial rows, with occasional yellowish deposits.

i. Very hard and heavy; forming lather when rubbed with water or saliva.

j. Dark red and very hard........ BANSALAGUIN.

jj. Slightly lighter in color and weight.. BETIS (see fig. 8).

ii. Moderately heavy and moderately hard; not forming a lather with water or saliva........ NATO (see fig. 37).

hh. Vessels medium size to large; scattered; without yellowish deposits; heavy to very heavy and very hard.

i. Wood dark brownish or reddish-brown. AGOHO (see fig. 38).

ii. Pale reddish .... LIUSIN (see fig. 9).

ee. Wood parenchyma not in regular concentric lines.
f. In very irregular wavy lines.
g. Connecting the vessels; wood heavy and hard; yellowish to dark reddish-brown or chocolate color............ MACAASIN (see fig. 39).
gg. Not connecting the vessels.
h. Vessels of medium size, in irregular, branching lines and containing pale yellow deposits .... PALO MARIA (see fig. 40).
hh. Vessels medium sized, scattered; wood pale reddish; moderately heavy and moderately hard........ DUGUAN (see fig. 26).
hhh. Vessels very small, in short radial rows; parenchyma lines very minute .... TAMAYUAN (see fig. 10).
ff. Wood parenchyma clustered about the vessels.
g. Vessels large, scattered.
h. Wood brown .... ACLE (see fig. 14).
hh. Wood white, with bad odor......................... CUPANG (see fig. 13).
gg. Vessels smaller, in irregular wavy lines; wood with reddish tinge .... TUCAN-CALAO (see fig. 16).
dd. Wood parenchyma, if present, without regular arrangement.
e. Vessels very small to medium size.
f. White; vessels very small.
g. With bitter taste .... BATINO (see fig. 21).
gg. Without bitter taste; finer grained............. LANETE (see fig. 20).
ff. Not white.
g. Dark brown; very hard and very heavy; vessels very small.
h. With slight purplish tinge; oil exuding from freshly cut vessels; vessels very small................ MANCONO (see fig. 41).
hh. With faint greenish-yellow tinge............ SASALIT (see fig. 42).
gg. Yellow; vessels small to medium size.
h. With pinkish tinge; vessels few and scattered.
CALAMANSANAY (see. fig. 43).
hh. Without pinkish tinge; vessels numerous, small or medium size; wood with greasy feel .... BANCAL (see fig. 44).
ggg. Pale, or dark red.
h. Pith-rays numerous, crowded, edged with white.
ARANGA (see fig. 45).
hh. Not as above.
i. Wood heavy and very hard; pale red.... ALUPAG (see fig. 31).
ii. Moderately heavy; not very hard.
j. Grain crooked, not smoothing readily with the plane.
k. Dark red .... AMUGUIS (see fig. 29).
kk. Lighter in color and weight......................... BALINHASAY.
jj. Grain straight, working well.
k. Soft; pale red to red; with faint camphor-like odor when first cut .... SANTOL (see fig. 32).
kk. Moderately hard; heartwood brownish-red; faintly disagreeable odor when first cut.
MALASANTOL (see fig. 46).
ee. Vessels medium size to large.
f. Without resinous deposits or resinous odor; wood white, with greenish tinge; pith-rays white and prominent.
BATICULIN (see fig. 47).

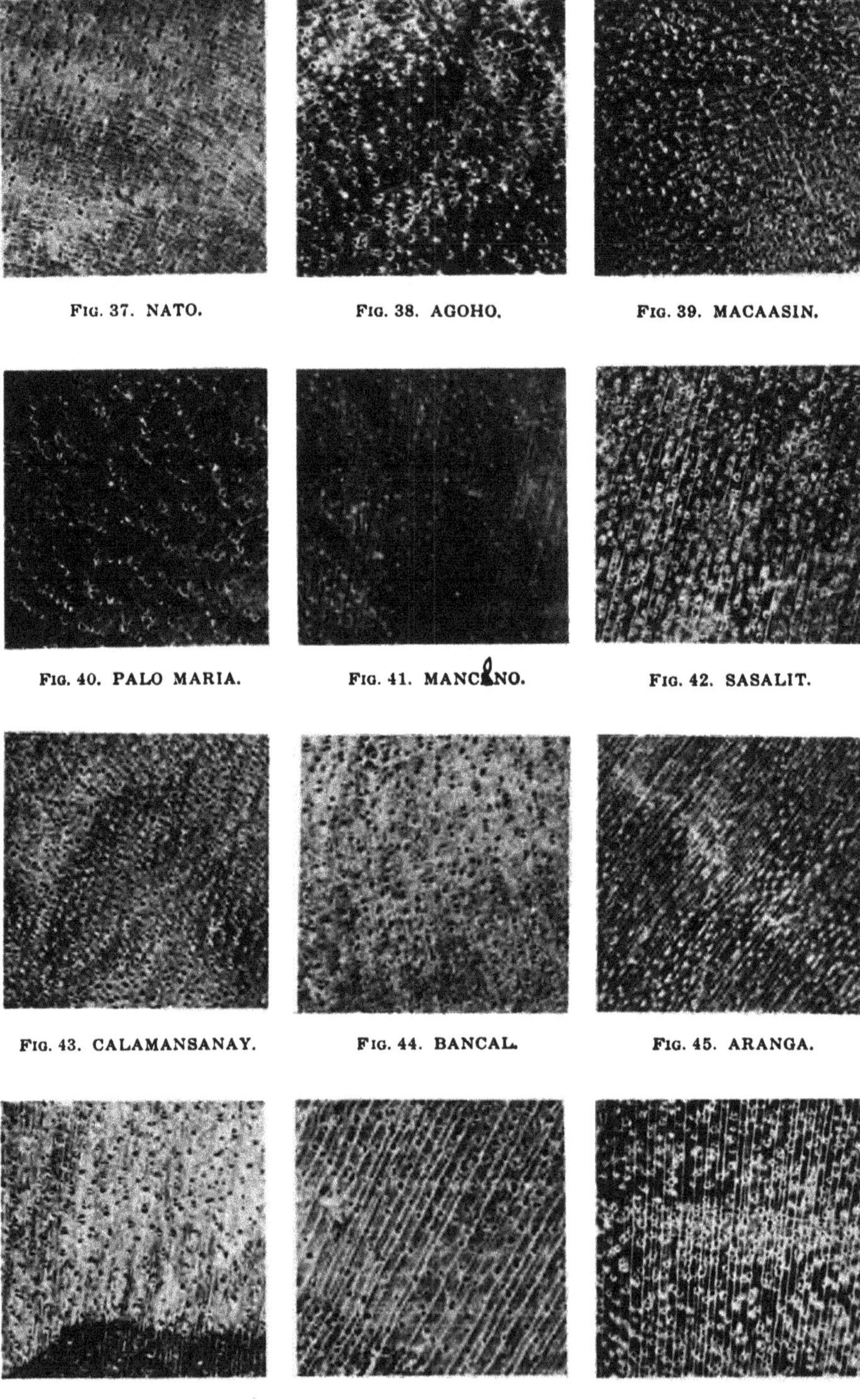

FIG. 37. NATO.

FIG. 38. AGOHO.

FIG. 39. MACAASIN.

FIG. 40. PALO MARIA.

FIG. 41. MANCONO.

FIG. 42. SASALIT.

FIG. 43. CALAMANSANAY.

FIG. 44. BANCAL.

FIG. 45. ARANGA.

FIG. 46. MALASANTOL.

FIG. 47. BATICULIN.

FIG. 48. MANGACHAPUY.

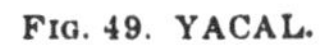

FIG. 49. YACAL.

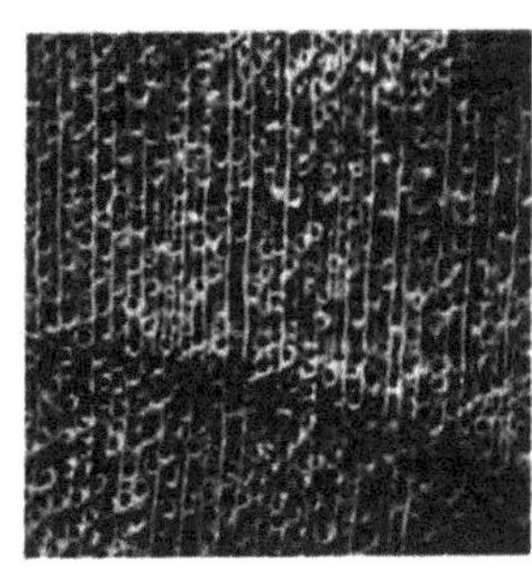

FIG. 50. MAYAPIS.

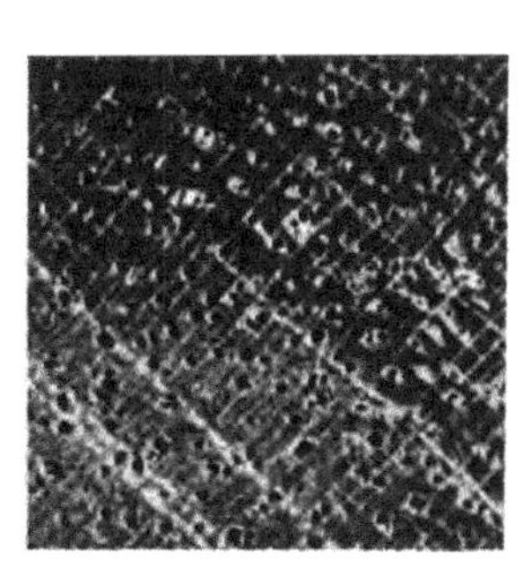

FIG. 51. WHITE LAUAN.

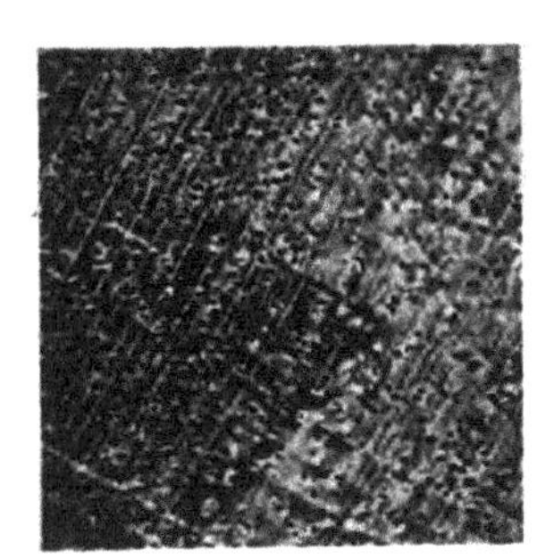

FIG. 52. TANGUILE.

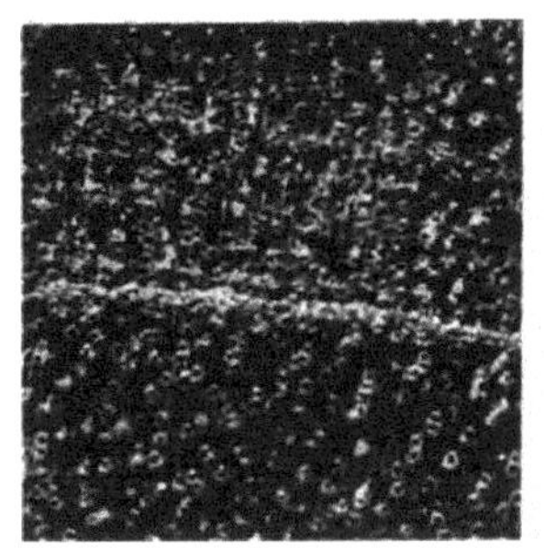

FIG. 53. GUIJO.

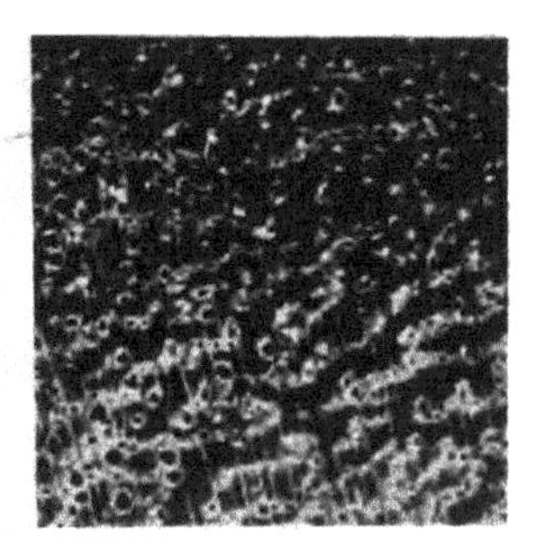

FIG. 54. APITONG.

FIG. 55. RED LAUAN.

ff. With resinous deposits and, sometimes, resinous odor.
  g. Wood without reddish tinge.
    h. Very hard and heavy.
      i. Straw color or almost white, when fresh. MANGACHAPUY (see fig. 48).
      ii. Light to dark brown .... YACAL (see fig. 49).
    hh. Soft and light.
      i. White to grayish, uneven grain .... MAYAPIS (see fig. 50).
      ii. Grayish; very soft and light; comparatively straight grain. WHITE LAUAN (see fig. 51).
  gg. With reddish tinge.
    h. Vessels of medium size.
      i. Rather dark and coarse grained, shining, brittle. TANGUILE (see fig. 52).
      ii. Lighter color, grayish-red, dull; heavier and harder; finer grain .... GUIJO (see fig. 53).
    hh. Vessels large and numerous.
      i. Wood moderately heavy and hard; dark, often with purplish tinge .... APITONG (see fig. 54).
      ii. Soft and light; pale red .... RED LAUAN (see fig. 55).

VI. NOTES ON SPECIES.[4]

ACLE.

Pithecolobium acle (Blco.) Vid. Fam. **LEGUMINOSÆ.**
(*Mimosa acle* Blanco.)

Acle (T.); anaguep (I., T.); languip, quitaquita, tabalangi (V.).

Philippines. Widely distributed through the Archipelago. Said closely to resemble the pynkadoo (*Xylia dolabriformis* Benth.) of India.

Moderately heavy and hard. Sp. gr. 0.610 to 0.693.

Fine grained, durable and seasons well.

Sapwood whitish; heartwood dark brown, like English walnut. Decided peppery odor, causing sneezing when planed or sawed.

Acle chips soaked in water, color the water a dark brown and give it a slightly aromatic odor. The coloring matter is also soluble in alcohol.

Uses.—Acle is one of the most satisfactory woods for fine furniture and cabinetmaking. It is used in first-class construction of various sorts. It is said to be difficult to burn it. It is known to be used for railroad ties; posts; bridges; chairs; desks; doors; floors; planks; siding of houses; naval construction; sides of guitars.

[4] In the following species notes, most of the common names have been taken from Dr. H. N. Whitford's Check List. The notes on the uses of individual woods have, many of them, been taken from data on file in the Bureau of Forestry. The following abbreviations are used to designate the dialect or the region in which a particular name is used:

B.	Bicol.	Il. Sur.	Ilocos Sur.	Pang.	Pangasinan
Cag.	Cagayan.	Neg.	Negrito.	T.	Tagalog.
I.	Igorot.	N. Luz.	Northern Luzon.	V.	Visayan.
Il.	Ilocano.	Pam.	Pampanga.	Z.	Zambales.

Structure.—Pith-rays very fine, indistinct in the heartwood because of the dark color. Vessels moderately large, scattered, in more or less wavy lines, surrounded by a fringe of wood parenchyma and containing whitish deposits.

Banuyo is the wood most often substituted for acle. It may be distinguished from acle by its more distinct and wider annual rings, its lighter color, the absence of whitish deposits in the vessels and by its greater softness.

Bull. For. Bur. Manila (1906) **4**, 59; 2d ed. (1907), 63. Ahern, l. c., 16–18.

AGOHO.

Casuarina equisetifolia Forst. Fam. **CASUARINACEÆ.**

Agoho (T., Il. Sur.); agoso (Pamp., T.); aguso (Z.); antong (N. Vis.); aro (Il.); ayo (V.); caro (Il.); karamutan (Moro); malabohoc (V.). Foreign names: Aru (New Guinea); beefwood (Australia, Ceylon, India); cassowary tree (India); filaro (Santa Lucia); horsetail pine (Trinidad); ironwood; ru (India); swamp oak, she oak, or botany oak (Australia); whistling pine (Jamaica).

From Queensland northward throughout the Eastern Tropics.

Common along the seacoast from the extreme northern to the extreme southern parts of the Archipelago, growing in the white sand of the beach, also back along the streams into the mountains up to 1,000 meters altitude.

Other species of *Casuarina* occur but have not yet become of commercial importance in the Archipelago. Their wood may be distinguished from that of *C. equisetifolia* by the much larger pith-rays.

A very hard wood, heavy to very heavy. Specific gravity, 0.704 to 0.942.

Very durable, but little used because of the difficulty of working.

Sapwood very light brown, becoming gradually darker in the formation of heartwood, with no sharp line of demarcation between the two. Grain fine and straight. No seasonal rings.

Uses.—Agoho is an extremely fine firewood. It makes an intensely hot fire, too hot for some purposes. The wood is used for general construction work; posts; boards; charcoal; railroad ties. It is said to furnish excellent piling in some of the countries where it is grown. It is not known to be used for piling in the Philippines; but its great hardness and durability would seem to indicate fitness for such work. It should also be tried for paving blocks.

Structure.—Pith-rays usually very fine and indistinct; but occasional compound rays are found. Vessels medium size, scattered or in irregular branching or obscurely radial lines. Wood parenchyma in fine, concentric, wavy lines.

The wood most likely to be confused with agoho is liusin, which is useful for the same things and which is lighter in color, with larger and more widely scattered vessels.

Bull. For. Bur. Manila (1907), **4**, 68.

ALUPAG.

Euphoria cinerea Radlk. Fam. **SAPINDACEÆ.**

(*Euphoria litchi* Blanco. *Sapindus cinereus* Turcz.)

It is probable that other closely related species in the genus *Nephelium* furnish some of this wood.

Alipay, alupag (T.) ; alupag ama, alupe (Pamp.) ; apalong (N. Luz.) ; bulala (B.) ; dulit (Neg.) ; halupag (Tayabas).

Philippines.

Heavy and very hard.

Seasonal rings distinct. Diffuse porous. Sapwood very pale reddish ; heartwood very slightly darker. Fine and straight grained.

Uses.—House construction ; flooring ; posts ; ax handles.

Structure.—Pith-rays very fine and indistinct. Vessels small and scattered, often two, three or four in a row, separated only by cross partitions.

Malugay is the wood most like alupag. It may be distinguished from alupag by its lighter weight, larger vessels and the distinct line at the end of the seasonal ring.

AMUGUIS.

Koordersiodendron pinnatum ~~Engl.~~ (Blco.) Merr. Fam. **ANACARDIACEÆ.**

(*Odina multijuga* Vid. *Helicteres pinnata* Blanco. *Crytocarpa quinquestila* Blanco. *Odina speciosa* Blume.)

Ambogues (V.) ; ampopo (C.) ; bancalari (I.) ; bancochasi (Il.) ; calumanog (V.) ; carugcog (B.) ; dangila (T.) ; laco-laco (V.) ; marsantog (N. Luz.) ; palo-santo (T.) ; sambulauan (V.) ; twi (Neg.).

Philippines, Celebes and New Guinea.

Very widely distributed in the Islands.

Moderately heavy and hard. Sp. gr. 0.690.

Sapwood pale red ; heartwood dark red, both sap and heartwood often with irregular, lead-colored areas ; black spots often scattered through the wood. Seasonal rings sometimes distinct, sometimes not. Diffuse porous. Rather fine, but not straight grained, not smoothing readily under the plane. Said to have a slightly disagreeable odor when freshly worked.

Uses.—Ordinary construction and cabinet work ; posts ; rafters ; flooring ; furniture ; inside partitions of houses ; naval construction ; planks ;

superstructure of ships; framing; carriage making; recommended for railroad ties.

Structure.—Pith-rays fine and indistinct. Vessels small and scattered.

Balinhasay is the wood which is most like amuguis in structure. It is, however, distinctly lighter in color and weight.

Other woods of red color are often substituted for amuguis. One local merchant has been known to have for sale at one time as many as *eight* different woods under this one name.

Bull. For. Bur. Manila (1906) **4**, 59; 2d ed. (1907) **4**, 62. Ahern, l. c., 25–26.

ANUBING.

Artocarpus cumingii Trec. Fam. **ARTOCARPACEÆ.**
(*Artocarpus ovata* Blanco.)

It is quite probable that some of the other species of *Artocarpus* furnish some of this wood.

Anabiong (V.); anobion (Pamp.); bayuco (V.); ubien (I.).

Philippines.

Moderately hard and moderately heavy. Straight grained.

Seasonal rings distinct. Sapwood white; heartwood greenish-yellow to brownish. Very disagreeable taste and slightly disagreeable odor, when fresh.

Uses.—House construction; posts; rafters; railroad ties.

Structure.—Pith-rays medium size, distinct, white or lighter than the surrounding wood. Vessels medium size, scattered, filled with white, glistening, milky deposit and surrounded by a very thin fringe of wood parenchyma.

The wood known as cubi is practically the same as anubing. It may come from another species of *Artocarpus*, but is not distinguished. Antipolo (*Artocarpus incisa* L. f.) is very much like anubing, but softer. The wood known as lanca or nunca comes from *Artocarpus integrifolia* L. f. This wood is softer and more even in texture than anubing and is used for the backs and sides of guitars and other stringed instruments.

The only other wood which might be mistaken for anubing is baticulin; this is whiter, softer and without the white deposits in the vessels.

Ahern, l. c. 21, 22.

APITONG.

Dipterocarpus grandiflorus Blanco. Fam. **DIPTEROCARPACEÆ.**
(*Mocanera grandiflora* Blanco.)

Philippines, Borneo, Malay Peninsula.

Dipterocarpus sp.

Anahaon (B.); apitong (T.); canunyao (Il.); duco (N. Luz.).

PANAO.

Dipterocarpus vernicifluus Blanco.
(*Mocanera vernicifluna* Blanco.)

Philippines.

Balao (T.); bulay (Batangas); camiling (Neg.); liga, Malapaho, panao (T.); panantulen (Pang., Il.).

HAGACHAC.

Dipterocarpus lasiopodus Perk.

Philippines.

These three woods pass for the same thing. They are widely distributed and furnish much of the timber of the Islands.

Moderately heavy and moderately hard.

Sapwood light colored; heartwood dark, with reddish or purplish tinge. Grain straight, but coarse. With more or less distinct resinous odor.

Uses.—Ordinary construction; shipbuilding; bancas; cascoes; planks for ships; ship bottoms and sides; piles; interior finish; rafters; sills; recommended for railroad ties.

Structure.—Pith-rays medium size, distinct. Vessels large and scattered. Wood parenchyma scattered, sometimes abundant. Resin-canals often very distinct. Wood with considerable quantities of resin exuding when fresh.

Bull. For. Bur. Manila (1906), **4**, 51; 2d ed. (1907), **4**, 53–54. Ahern, l. c., 22–24.

Sp. gr., of apitong, 0.620.

ARANGA.

Homalium luzoniense F. Vill. Fam. **FLACOURTIACEÆ.**
H. Panayanum F. Vill.
H. villarianum Vid.

All Philippine species.

Ampupuyot (V., Il.); cuela, laing (Rizal); puyot (V.).

Wood very hard and heavy. Sp. gr. 0.863.

Pale reddish. Fine grained. No seasonal rings.

Uses.—Piling; high grade construction; boat building; cabinetwork; flooring; posts; paddles for beating clothes (palo-palo); rafters; railroad ties. A strong and durable wood, which resists the teredo.

Structure.—Pith-rays fine, very numerous and closely packed bending outward to pass vessels, appearing to be margined with white and taking up a very large part of the wood. Vessels small, in short radial rows.

Bull. For. Bur. Manila (1906) **4**, 61; 2d ed. (1907) **4**, 65. Ahern, l. c. 24–5.

BALACAT.

Zizyphus zonulatus Blanco. Fam. **RHAMNACEÆ.**
(*Rhamnus zonulatus* Blanco. *Zizyphus arborea* Merr.)

Aggue (Cag.); aligamon (Il.); balacat (Pamp.); bigaa (Rizal); danlic (Tayabas); ligaa (T.).

Philippines.

Probably the other species of *Zizyphus* furnish some of the wood known as balacat.

Closely related to the Indian species of *Zizpyhus*.

A moderately heavy soft wood. Sp. gr. 0.517.

White to light brown. Coarse and straight grained. Seasonal rings distinct. Diffuse porous. Heartwood slightly darker, but scarcely distinct from the sap. Faintly unpleasant odor.

Used in light and temporary construction.

Structure.—Pith-rays small but distinct. Vessels of medium size, in very short, radial rows and sometimes in wavy lines connected by wood parenchyma, with clear and glistening deposits in the vessels.

Bull. For. Bur. Manila (1906), **4,** 67; 2d ed. (1907), **4,** 60.

BALACBACAN.

See TANGUILE.

BALINHASAY.

Buchanania florida Schauer var. **arborescens** Engl. Fam. **ANACARDIACEÆ.**

Philippines. Frequent but not abundant, usually in small dimensions.

Moderately heavy and moderately hard.

Pale reddish. Rather fine but not straight grained.

Uses.—Light or temporary construction; not durable.

Structure as in amuguis, but slightly lighter and coarser.

BANABA.

Lagerstroemia speciosa Pers. Fam. **LYTHRACEÆ.**
(*Munchausia speciosa* Blco.)

Agaro, alagaa (Cag.); banaba (B., T.); danioura (N. Luz.); macabalo (Pang.); mittla (Pamp.); panao (T.); soglogan (Neg.); tabangao (Il.).

Throughout the Eastern Tropics. Said to be the same as the jarul wood of India.

Moderately heavy and hard.

Sapwood very light to pinkish; heartwood dark reddish brown. Seasonal rings distinct. Ring porous. Straight grained.

Uses.—A high grade construction timber, used for boat construction; interior partitions and finish; planks; rafters; sills; wharves; piling; furniture; carabao yokes; barrels; railroad ties.

Structure.—Pith-rays very fine. Vessels, one large row in the inner part of each ring, smaller vessels in more or less broken lines in outer part of ring. Dark-colored, glistening deposits in the vessels. Wood parenchyma in wide, wavy, tangential lines in outer part of each ring, connecting and surrounding the vessels.

Ahern, l. c., pp. 26–28.

BANCAL.

Sarcocephalus cordatus Miq. Fam. **RUBIACEÆ.**

(*Nauclea glaberrima* Blanco. *Nauclea lutea* Blanco.)

Bancal (T.); bulala (Il.); cabag (V.); nababalos.

Malacca and Ceylon, India, through the Malay Archipelago to north Australia. Common throughout the Archipelago on low lands near the coast.

Soft and moderately heavy. Sp. gr. about 0.550.

Sapwood light-yellow; heartwood darker yellow, no very sharp line of demarcation between the two. Seasonal rings distinct. Diffuse porous. Wood with a decidedly greasy feeling.

Uses.—Small boats; partitions; posts; rafters; flooring; ceilings; chairs; desks; barrel staves; paddles for beating clothes (palo-palo); tubs.

Structure.—Pith-rays small. Vessels of medium size, usually subdivided, in rough radial rows between the numerous fine pith-rays, which bend around them. Vessels often clogged with whitish deposits.

Ahern, l. c., 28–30.

BANSALAGUIN.

Mimusops elengi L. Fam. **SAPOTACEÆ.**

Bansalaguin, cabique (T.); duyogduyog (V.); pasac (T.); talipopo (V.).

Eastern Tropics. The same as the "bullet tree" or "horse-flesh wood" of India.

Very hard and heavy to very heavy. Sp. gr. 0.850 to 0.900.

Sapwood, pale-reddish; heartwood dark-red. Very fine grained. Seasonal rings present or absent. Diffuse porous. Bitter taste. Rubbing with water or saliva produces a lather.

Uses.—A first-class construction timber, used for posts; shipbuilding; treenails in shipbuilding; keels; marlin spikes; belaying pins; spokes and handles of ship's wheel; tool handles; turnery.

Structure.—Pith-rays very fine and indistinct. Vessels very small, in oblique, radial lines, with yellowish deposits. Wood parenchyma in discontinuous, wavy, tangential lines.

Exceedingly like betis but has finer grain, darker color, greater weight and hardness.

Bull. For. Bur. Manila (1906), **4**, 60; 2d ed. (1907) **4**, 64. Ahern, l. c., 30–31.

BANUYO.

Fam. LEGUMINOSÆ.

The exact botanical position of Banuyo is not known; but, from its structure, it is a legume.

Banuyo (T.); hamago (B.); magtululung (Il.); manglati, malatigue.

Masbate, Negros Occidental, Ambos Camarines.

Moderately heavy and moderately hard. Sp. gr. 0.525.

Golden brown, with fine grain, similar to Acle. Seasonal rings distinct. Diffuse porous.

Uses.—Fine furniture and cabinetwork; light construction; flooring; interior finish; siding; bancas; outriggers; telegraph poles; recommended for railroad ties.

Structure.—Pith-rays fine and indistinct. Vessels of medium size, scattered. Wood parenchyma grouped about the vessels.

Differs from acle in being coarser grained, softer, lighter in color and weight and in the absence of the whitish deposits in the vessels.

Bull. For. Bur. Manila (1906), **4**, 62 2d ed. (1907), **4**, 65–6.

BATETE.

The botanical position of batete is not known.

Moderately heavy and hard.

Seasonal rings distinct. Diffuse porous. Very full of oil. Colors water brown, with purplish tinge.

Uses.—Furniture; flooring; interior finish; siding.

Structure.—Pith-rays small and indistinct. Vessels few, of medium size, scattered. End of seasonal ring marked by a distinct line. Oil exuding from freshly cut vessels and darkening the surrounding wood.

Very much like supa in structure, but sufficiently distinct owing to the darker color and greater amount of oil.

BATICULIN.

Litsea perottetii F. Vill. Fam. LAURACEÆ.

(*Litsea obtusata* F. Vill. and *Olax baticulin* Blanco also refer to this wood.)

Aban, anago (Il.); ansohan (C.); bacan (C.); baticuling, diraan (T.); indang (V.); marang (T.).

Philippines.

Moderately hard and light.

Straight and coarse grained. No distinct sap- and heart-wood. White or greenish-yellow. Said to have a pleasant odor when first cut. No seasonal rings.

Uses.—Light or ordinary construction; partitions; ceilings; boxes; foundry molding; writing desks; sculpture; wood carving.

Structure.—Pith-rays medium size, very distinct, almost white. Vessels grouped 2 to 3 together and then arranged in wavy, tangential lines. Wood parenchyma a small amount surrounding the vessels.

Ahern, l. c., 31–33.

BATINO.

Alstonia macrophylla Wall. Fam. **APOCYNACEÆ.**

(*Alstonia batino* Blanco. *Echites trifida* Blanco.)

Batino (T.); tañgitan, tanguilan (V.).

British India and Malaya.

Moderately heavy and moderately hard.

Whitish or pale yellowish. Grain close and tolerably straight. Bitter taste. Seasonal rings distinct. Diffuse porous.

Uses.—Building; posts; rafters; boards.

Structure.—Pith-rays fine but distinct and numerous. Vessels small in radial rows of 3 to 7. Wood parenchyma scattered.

Ahern, l. c., 33–34.

BATITINAN.

Lagerstroemia batitinan Vid. Fam. **LYTHRACEÆ.**

(*Lagerstroemia hexaptera* Vid.)

Bingas (Il.); boticalag (Pang.); batitian (T., Pamp.); batitinan (T.); bugaron, dumate, lasila (N. Luz.); lumasi, lumati, larila (Il); magatululung (Il.); manglate, miao (V.); nathubo, saguimsim (V.); tinaan (B.).

Philippines.

Hard and heavy. Sp. gr. 0.786.

Sapwood light; heartwood dark brown. Fine, straight grain. Seasonal rings distinct. Ring porous.

Uses.—General construction; posts; sleepers; flooring; joists; planks; rafters; keelsons of ships; masts; piles; telegraph poles.

Structure.—Pith-rays fine and indistinct. Vessels, a large row in the inner part of each ring, outer vessels of ring smaller, in more or less wavy lines. Black, glistening deposits in vessels.

Wood parenchyma in wavy lines surrounding vessels.

Bull. For. Bur. Manila (1906), **4**, 61; 2d ed. (1907), **4**, 65. Ahern, l. c., 34–35.

BETIS.

Illipe betis (Blco.) Merr. Fam. **SAPOTACEÆ.**

(*Azaola betis* Blanco. *Payena betis* F. Vill.)

Bacayao (Il., Pang., T.); baniti (B.); banicac (V.); betis (T., Pamp.); pagpagan (Cag.); pailan (T.); pianga (Cag., Il. Sur).

Philippines.

Very hard and heavy. Sp. gr. 0.773.

Dull, dark reddish. Bitter taste. Clear and straight grained.

Seasonal rings present or absent. Diffuse porous. Wood making a lather when rubbed with water or saliva.

Uses.—High grade construction; shipbuilding; keels of ships; sternposts; wharves; piling; posts; doors; rafters; railroad ties.

Structure.—Pith-rays very minute. Vessels small, in irregular, oblique radial rows, with yellowish deposits. Wood parenchyma in numerous minute tangential lines.

Differs from bansalaguin in being softer and of lighter weight and color.

Bull. For. Bur. Manila (1906), **4**, 60; 2d ed. (1907), **4**, 63. Ahern, l. c., 36–37.

BITANHOL.

See Palo Maria.

BOLONGETA.

Diospyros pilosanthera Blanco. Fam. **EBENACEÆ.**

Amara, amaga (V.); alintatao (T.); apopuyot (Cag.); ata-ata (V.); batolinao (Cag.); barlis, bolonguita (T.); dalondong (V.); calohadia, caloyanang, galariga (T.); tamil (Yacan); malatalang (T.); tapilac (Moro).

Philippines.

Very heavy and very hard.

Sapwood light pink; heartwood black or streaked. Grain close and straight.

Uses.—Fine furniture; inside finish; gun carriages; railroad ties; also, reported as being used for piling.

Structure.—Pith-rays very minute. Vessels very small and scattered in small, radial lines. Wood parenchyma in numerous, very fine tangential lines. In the black heart nearly all details of structure are obscured by the black substance that fills all the elements.

For discussion of this wood in connection with camagon and ebony, see ebony.

CALAMANSANAY.

Nauclea sp. Fam. **RUBIACEÆ.**

Flacourtia inermis Roxb. Fam. **FLACOURTIACEÆ.**

Malayan region.

Terminalia calamansanai Rolfe. Fam. **COMBRETACEÆ.**

(*Gimbernatia calamansanai* Blanco. *Terminalia bialata* Vid.)

Philippines.

Vidal thought it a species of *Diospyros*, Fam. Ebenaceæ.

The *Terminalia* species has been considered the source of this wood, but I think it more likely that it is a *Nauclea*. The only examples of calamansanay which we have as herbarium material are a *Nauclea* from

Masbate and a *Flacourtia* from Zambales. The structure described by Vidal for this wood would seem to be more that of one of the two just mentioned than of a *Terminalia*. The structure here described is that of *Nauclea* material, but the *Flacourtia* is very much like it.

Bayabo (Il.); bancalauag (V.); bancalauan (T.); bisal (Pang.); calamansanai (T.); camansac (Z.); calamansauan (T.); calumagon (B.); himbabalut (V.); lisac (T.); lumanog (V.); magatalay (I.); magobinlod, maytalisay (V.); malacalumpit (B.); magatolay (N. Luz.); subo-subo (Z).

Heavy and hard. Yellow, with a rose tint. Close and straight grained. No seasonal rings.

Uses.—Flooring; masts of boats; beams in interior construction; furniture; ordinary construction; posts for houses or in contact with the ground; siding; shipbuilding; telegraph poles; window sills.

Structure.—Pith-rays very fine. Vessels very small and scattered.

In *Flacourtia* the vessels contain a reddish substance.

Ahern, l. c., 39–41.

CALANTAS.

Toona spp. Fam. **MELLACEÆ.**

This includes the Philippine material which has been credited to *Cedrela Toona* Roxb. and *Cedrela odorata* Blanco.

Alam (Mindoro); balongcamit (B.); calantas (T.); cantiñgen (Il.); dampia (I.); danigga (Cag.); danupia (Cag.); lanigpa (V.); porac (Il.); taratara (T.); saggued (Palanan, Isabela); sandana (V.).

Light and very soft. Sp. gr. 0.438.

Sapwood a narrow rim of very pale reddish color; heartwood darker. Coarse and straight grained; occasionally a tree is found with a good burl. The burl is always in demand for furniture.

Distinct, cedary odor. Seasonal rings distinct. Ring porous.

Uses.—Cigar boxes; small boats; furniture; cabinetmaking; pattern making; carving; bancas; ceiling; doors; partitions; sides of guitars.

This is the best and in fact nearly the only wood used for cigar boxes in the Islands.

Structure.—Pith-rays fine but distinct. Vessels large, in spring wood, gradually becoming smaller toward outer part of ring. Wood parenchyma not prominent. Vessels with reddish deposits.

This wood is much like lumbayao, and red lauan. It may be distinguished from the former by its very light weight and the more distinct seasonal rings, and from the latter by the seasonal rings and the cedary odor.

Bull. For. Bur. Manila (1906), **4,** 58; 2d ed. (1907), **4,** 61.

Ahern, l. c., 38–39.

Sapwood narrow, very pale reddish; heartwood dull, dark reddish. No seasonal rings. Coarse and twisted grain. Wood staining water a pale reddish color.

Uses.—Framing of native houses; heavy construction; furniture.

Structure.—Pith-rays broad to very broad, not running straight, appearing as broad and crooked lines in the tangential section. Vessels medium size, scattered, containing white deposits.

CUPANG.

Parkia coxburghii G. Don. Fam. **LEGUMINOSÆ.**
(*Mimosa peregrina* Blanco.)

Baguen, balay-oac (Il.); butaric (N. Luz.); cupang (Il., T.).

India, Indo-China, and Malaya. Widely distributed throughout the Archipelago in lowland forests.

Light and soft. Sp. gr. 0.285.

Almost white in color when first cut, but discolors rapidly. With strong, disagreeable odor when first cut. Seasonal rings sometimes distinct, sometimes not. Diffuse porous.

Uses.—Light or temporary construction; matches; shoes; paper pulp; not durable.

Structure.—Pith-rays of medium size, white, distinct. Vessels of medium size, numerous and scattered, not conspicuously clogged with secretions, usually surrounded by a white fringe of wood parenchyma cells.

Bull. For. Bur. Manila (1907), **4**, 2d ed., p. 69.

DUGUAN.

Myristica spp. Fam. **MYRISTICACEÆ.**
Knema heterophylla Warb.

Tambalao (Z.); duguan (T.).

Light and soft. Very faint reddish. Seasonal rings distinct. Grain fine and straight. Diffuse porous. Sapwood slightly lighter than the heart and with blood-red sap. Wood soft and spongy.

Uses.—Light or temporary construction.

Structure.—Pith-rays fine and indistinct. Vessels small and scattered. Wood parenchyma a more or less coarse line in outer part of ring.

DALINSI.

Terminalia pellucida Presl. Fam. **COMBRETACEÆ.**

Philippines.

Moderately heavy and moderately hard. Coarse and straight grained. Seasonal rings distinct. Diffuse porous. Brownish, with greenish tinge. Wood coloring water a pale yellow.

Uses.—House construction; interiors of houses.

Structure.—Pith-rays fine and indistinct. Vessels medium size, numerous, scattered. In the outer part of the ring the vessels are arranged in more or less regular, wavy lines and are sometimes connected by wood parenchyma.

(See also note under "Talisay.")

DITA.

Alstonia scholaris (L.) R. Br. Fam. **APOCYNACEÆ.**
(*Echites scholaris* L.)

Andarayan (Cag.); dalipaoyan (Il.); polay (Pang.); tanitan (V.).

Tropical Africa to Australia. The "chatwan" of India.

Light and soft. Sp. gr. about 0.45.

White, with pleasant odor. No distinct sap and heartwood. Fine and twisted grain. Very bitter taste. Seasonal rings distinct. Diffuse porous.

Uses.—Light construction and cabinetwork—easily worked; shoes; ceilings; furniture; matches; musical instruments; paper pulp; rafters; resin boxes.

Structure.—Pith-rays small but distinct. Vessels few, widely scattered, and of medium size. Wood parenchyma in prominent concentric lines.

Ahern, l. c., 46–47.

DUNGON.

Tarrietia sylvatica (Vid.) Merr. Fam. **STERCULIACEÆ.**
(*Heritiera silvatica* Vid.)

Bingas (Il.); dungon (T., V.); palonapin (Z.).

Philippines.

Very hard and heavy. Small amount of pinkish or brownish sapwood; heartwood chocolate brown. Sp. gr. 0.852.

Fine, crossed grain. No distinct seasonal rings. Slightly disagreeable odor when fresh.

Uses.—High-class construction work; naval construction work; piling; posts; anchors; beams; boat ribs; bridge construction; buildings; cogwheels; inner support to keel construction of ships; hoists; keels of ships; oars; pillars; sills; stemposts; hubs of wheels; railroad ties.

Structure.—Pith-rays medium size, prominent, and dark colored. Pith-ray cells with oblique ends and many cells filled with a reddish brown substance. Vessels medium size, scattered, frequently with dark-colored deposits. Wood parenchyma in very fine and very numerous broken concentric lines.

Bull. For. Bur. Manila (1906), **4**, 56; 2d ed. (1907), **4**, 59.

Ahern, l. c., 47–48.

DUÑGON-LATE.

Heritiera littoralis Dry. Fam. **STERCULIACEÆ.**

(*Helicteres apetala* Blanco. *Sterculia cymbiformis* Blanco.)

Dungon-dungon (V.); duñgon-late (T.); malarungon (T.).

Tropics of the Old World.

This is the same species as the sundri of Burma.

Very hard and heavy, "63–75 lbs. per cu. ft."

Widely scattered in mangrove swamps throughout the Archipelago.

Color as in duñgon, but said to have a larger amount of sapwood.

Uses.—Same as for duñgon.

Structure.—In all respects as in duñgon.

Uses.—Canoes; outriggers; firewood; house posts; joists; presses; shipbuilding; telegraph poles and posts for small houses; wheel hubs; same uses as duñgon, if found in large sizes; small sizes used as ribs for small boats.

Ahern, l. c., 49–50.

EBONY.

Maba buxifolia Pers. Fam. **EBENACEÆ.**

Diospyros spp.

Ata-ata (V.); balatinao (Il.); caloyanan (Pamp.); ebano (Span.); galarigal (T.); luyong (T.); malatalang (T.); tanguintin (Surigao).

Tropical Africa to Australia.

Very hard and very heavy. Sapwood ashy; heartwood black, of even color. Fine and straight grained. No distinct seasonal rings.

Uses.—Cabinetwork; canes; desks; furniture; frames; inlaid work; shipbuilding; saber handles.

Structure.—Pith-rays very small and indistinct. Vessels very small and scattered. Wood parenchyma in numerous, very fine, concentric lines. All the elements of the heartwood full of a very dense, black substance.

Ebony, in the commerce of the world, is a heavy, hard black wood with the characteristic structure of the family *Ebenaceæ*. There are, in the Philippine Islands belonging to the ebony group, woods known by the names of bolongeta, camagon, and ebano or ebony. The first of these, bolongeta, is furnished by *Diospyros pilosanthera* Blanco, and probably by other species. *Diospyros pilosanthera* is a fair-sized tree in the ridge forests. It is of wide distribution in the Archipelago and may be said to be fairly abundant. Its wood is pink or red, streaked with black, the black streaks being very much denser and harder than the rest of the wood.

Camagon is the name applied to the wood of *Diospyros discolor* Willd. and other *Diospyros* species, whenever the sapwood is grayish or mottled and the heartwood black, with brownish streaks.

By true ebony, in the Philippines, is usually meant the wood of *Maba buxifolia* Pers., a small tree with white, or light-colored, sapwood and clear, black heart.

The pure-black portion of any one of these three woods is true ebony. It is formed from the sapwood by the deposit in all the wood elements of a very dense, black substance, which causes the wood to become very heavy, very hard, and very brittle. It will be seen from this that there are several species of trees in the Islands that furnish true ebony. They are, none of them, of large dimensions, but they are of frequent occurrence and rather wide distribution.

Ahern, l. c., pp. 50–51.

GUIJO.

Shorea guiso (Blanco) Blume. Fam. **DIPTEROCARPACEÆ.**

(*Mocanera guiso* Blanco. *Dipterocarpus guiso* Blanco.)

Betic (Laguna); carucat, catapang (Nueva Viscaya); guijo, guijo-bitic (T.); niquet (Il.); saray (Il., N. Luz.); yamban, yate, zilan (Il.).

Very closely related to the "sal" wood of India.

Philippines.

Moderately heavy and hard. Sp. gr. 0.688.

Ashy-red. No distinct seasonal rings. Fairly straight-grained, warping badly. Sap and heartwood distinct; sapwood very light colored. An appearance as of seasonal rings often caused by lines of resin canals. Wood with faint, resinous odor.

Uses.—General construction; shipbuilding; carriage wheels, frames, and shafts; flooring and girders of houses; inferior furniture; beams; beams of ships; booms of ships; bridges; casks and barrels; decking; docks; hubs; keels of ships; masts; naval construction; outrigger supports; oars and paddles; partitions; plows; posts; rafters; side planking of ships; telegraph poles; wheel rims; wheel spokes.

Structure.—Pith-rays small. Vessels medium size to small and scattered. Wood parenchyma scanty and scattered. Resin canals in incomplete concentric lines, superficially resembling seasonal rings and whitish in color, because of the contained resin.

Guijo and the finer grades of apitong are so much alike that it is practically impossible to tell them apart. Ordinarily, however, the apitong is much coarser grained. For many purposes, it is probable that, contrary to common opinion, apitong is the better wood.

Bull. For. Bur. Manila (1906), **4**, 52; 2d ed. (1907), **4**, 55. Ahern, l. c., 52–53.

IPIL.

Intsia bijuga (Colebr.) O. Kuntze. Fam. **LEGUMINOSÆ.**

(*Afzelia bijuga* A. Gray. *Eperua decandra* Blanco.)

Intsia Acuminata Merr.

Ipil (T., V.); obien (Pang.).

Intsia bijuga is distributed throughout the Eastern Tropics. It is one of the most important Bornean woods. *I. acuminata* is known only from the Philippines.

Heavy and very hard. Sp. gr. 0.758 to 0.909.

Distinct sap- and heart-wood. Sapwood whitish or light yellow; heartwood dark reddish-brown. Grain straight. Seasonal rings distinct. Diffuse porous.

Uses.—General high-class construction; posts of houses; sleepers; electric-light poles; furniture; cabinetmaking; paving blocks; beams; bridge construction; doors; joists; keels; pillars; rafters; sleepers; sternposts; window frames; railroad ties.

Structure.—Pith-rays small but distinct. Vessels of medium size, scattered. Wood parenchyma clustered about the vessels. Vessels frequently containing sulphur-colored deposits.

Bull. For. Bur. Manila (1906), **4**, 55–6; 2d ed. (1907), **4**, 59. Ahern, l. c., 54–57.

LANETE.

Wrightia laniti (Blanco) Merr. Fam. **APOCYNACEÆ.**

(*Anasser laniti* Blanco. *Wrightia ovata* A. DC..)

Alanote (T.); bantolinao (B.); tanghas, tiguig (V.).

Philippines.

Moderately heavy and moderately hard.

Seasonal rings present or absent. Diffuse porous. Fine and cross grained. White or pale yellowish. No distinct sap and heartwood. Disagreeable odor when fresh.

Uses.—Carving and light construction; bolo scabbards; boxes; canes; cooking utensils; chairs; decoration; furniture; house construction; musical instruments; shoes; trunks; turning; wardrobes; window sills in native houses; the most-used wood for wood carving.

Structure.—Pith-rays small and short. Vessels small, in short, radial rows.

Ahern, l. c., pp. 57–59.

LANOTAN.

Bombycidendron campylosiphon (Tcz.) Warb. Fam. **MALVACEÆ.**

(*Thespesia campylosiphon* Rolfe. *Hibiscus grewiaefolius* (Hassk.) Miq. *Hibiscus vidalianus* Naves. *Hibiscus campylosiphon* Tcz.)

Philippines.

Moderately heavy to heavy, and moderately hard. Sp. gr. 0.732.

Seasonal rings distinct. Diffuse porous. Sap- and heart-wood very distinct. Sapwood light yellow or white; heartwood purple. Straight grained.

Uses.—Cabinetmaking; carriage building, shafts; flooring; ordinary construction; ordinary furniture; planks; planking for ships; shipbuilding; telegraph poles; sides of guitars and mandolins.

Structure.—Pith-rays medium size, distinct. Vessels medium size or small, scattered. Tangential section showing fine, parallel, transverse markings as in narra and teluto. Fig. 27 shows these in longitudinal section.

Ahern, l. c., pp. 59–60.

WHITE LAUAN.

Shorea contorta Vid. Fam. **DIPTEROCARPACEÆ.**
Shorea squamata Benth. and Hook. f.
Shorea malaanonan Blume.
(*Mocanera malaanonan* Blanco, *Dipterocarpus malaanonan* Blanco.)
And probably other species.

Acab-acab (Basilan); almon (V.); anteng (Il.); apnit (Il., N. Luz.); bagtican (V.); duyong (Il.); sandana (T.).

Shorea contorta and *S. malaanonan* are Philippine. *S. squamata* occurs in the Philippines and Borneo.

Light and soft. Sp. gr. 0.461.

Both heart- and sap-wood very light colored. Straight and coarse grained. No seasonal rings; but frequent false rings, caused by lines of resin canals. Resinous odor.

Uses.—Light and temporary construction; cabinetmaking; inferior furniture; small boats; bancas; canoes; cascoes; masts; paper making; planks for ships; rafts; resin; shipbuilding; carriage making; axles; door panels.

Structure.—Pith-rays medium size, distinct. Vessels medium size, scattered. Wood parenchyma scattered. Resin-canals numerous, with white resin.

Bull. For. Bur. Manila (1906), **4,** 50; 2d ed. (1907), **4,** 52.

Ahern, l. c., 60–62.

RED LAUAN.

Shorea sp. Fam. **DIPTEROCARPACEÆ.**

Red lauan (T.); mangachapuy (V.).

Moderately heavy to light, and very soft. Sp. gr. 0.542.

Reddish-brown. Coarse, but straight grained.

Sapwood and heartwood distinct; sapwood very light; heartwood red. No seasonal rings, but the whitish, concentric lines of resin canals often present much the appearance of seasonal rings.

Uses.—Light or temporary construction; inferior cabinetwork.

Structure.—Pith-rays medium size. Vessels large and scattered. Resin canals abundant, with a whitish appearance, because of the contained resin.

Bull. For. Bur. Manila (1906), **4,** 62; 2d ed. (1907), **4,** 53.

LIUSIN.

Parinarium griffithianum Benth. Fam. **ROSACEÆ.**
(*Parinarium salicifolium* Miq. and *Grymania salicifolia* Presl. are names that are found applied to liusin by the earlier Philippine writers.)

Cubel (Rizal); culitingan (Pamp.); malapuyao (Tayabas); sampinit, tapgas (Guimaras).

India and Malaya.

Very hard and heavy. Sp. gr. 0.710.

White and reddish in color. Fine and straight grained.

Uses.—Piling; wharf construction; shipbuilding; keels of ships.

Structure.—Pith-rays fine and indistinct. Vessels medium size or large, scattered, with some indication of special groupings. Wood parenchyma in concentric lines.

Bull. For. Bur. Manila (1906), **4**, 64; 2d ed. (1907), **4**, 67.

LUMBAYAO.

Lumabayao is unknown botanically. It comes into the market only from Mindanao.

Soft and moderately heavy. Sp. gr. 0.550.

Reddish-brown. Seasonal rings distinct. Ring porous. Coarse and straight grained.

Uses.—General construction; house construction; bancas.

Structure.—Pith-rays small. Vessels of medium size, scattered, the larger ones in the inner part of the ring. Vessels with bright-red deposits.

Bull. For. Bur. Manila (1906), **4**, 64; 2d ed. (1907), **4**, 68.

MACAASIN.

Eugenia spp. Fam. **MYRTACEÆ.**

Macaasin (T.); binolo (B.).

Heavy and hard. Sp. gr. 0.705.

Purplish brown, occasionally with yellowish or greenish tinge. Fine grained. No distinct seasonal rings.

Uses.—General construction; cabinetwork; beams; flooring; furniture; interior construction; joists; naval construction; planks; rafters; ship rudders; telegraph poles; posts; window sills; tool handles; washbowls.

Structure.—Pith-rays fine and indistinct. Vessels of medium size, scattered. Wood parenchyma in wavy, broken lines, connecting the vessels.

Macaasin is taken to include the wood of any of the different species of *Eugenia*, and it may also include wood of some other closely related Myrtaceous plants. It is quite likely that, as we become better acquainted with the members of this family, we will be able to recognize several distinct woods now grouped together under this name.

Bull. For. Bur. Manila (1906), **4**, 57; 2d ed. (1907), **4**, 61.

Ahern, l. c., 62–64.

MALAPAPAYA.

Polyscias nodosa Seem. Fam. **ARALIACEÆ.**

(*Aralia pendula* Blanco.)

Bias-bias, bonglin (T., V.); tucud lañgit (Bataan).

Malaya.

Light and soft. Whitish. No distinct sap and heartwood. Seasonal rings usually not distinct. Straight grained.

Uses.—Matches; light construction; paper making;

Structure.—Pith-rays medium size to broad. Vessels medium size and scattered.

MALASANTOL.

Sandoricum vidalii Merr. Fam. **MELIACEÆ.**
(*Sandoricum harmsianum* Perk.)

Cateban (Pamp.); malasantol (T.); panantolen (Il.).

Philippines.

Moderately heavy and moderately hard. Sp. gr. 0.640.

Sapwood white, or pinkish; heartwood, brownish-red. Straight and coarse grained. Disagreeable odor. No distinct, seasonal rings.

Uses.—General construction; bancas; roof timbers;

Structure.—Pith-rays small and indistinct. Vessels small and scattered.

Bull. For. Bur. Manila (1906), **4,** 56; 2d ed. (1907), **4,** 60.

MALUGAY.

Pometia pinnata Forst. Fam. **SAPINDACEÆ.**

This was first identified from sterile material as a species of *Dracontomelum*. When fertile material was secured, it was found to be a *Pometia*.

Malay Peninsula, Malay Archipelago, and New Guinea.

Thus far, known in these Islands only from Mindoro.

Moderately heavy and moderately hard. Sp. gr. 0.658.

Pale red. Fine and straight-grained. Seasonal rings distinct. Diffuse porous.

Uses.—General construction; cabinetwork; interior finish; ribs and planking for small boats. A very tough wood.

Structure.—Pith-rays small and indistinct. Vessels medium size to small and scattered. End of seasonal ring marked by a definite line.

Bull. For. Bur. Manila (1906), **4,** 63; 2d ed. (1907), **4,** 66.

MANCONO.

Xanthostemon verdugonianus Naves. Fam. **MYRTACEÆ.**

Mancono (T., V.); palo de hierro (V.); tamulauan (V.).

Philippines.

Very heavy and very hard, the hardest and heaviest of any known Philippine wood. Sp. gr. 1.2363.

Sapwood, pale reddish; heartwood, dark brown, with purplish tinge. Fine and straight grained. No distinct seasonal rings.

Uses.—Posts; pillars; piling; bearings for machinery; a possible substitute for lignum vitæ.

Structure.—Pith-rays very fine and indistinct. Vessels very small and scattered. Wood parenchyma sparingly present. Oil exuding from some of the freshly cut vessels.

Ahern, l. c., 65–66.

MANGACHAPUY.

Hopea acuminata Merr. Fam. **DIPTEROCARPACEÆ.**

Bacuog (Pang.); banacao (Il.); caliot (Pang.); dalindingan (Bataan); mangachapuy (T.).

Philippines.

Hard and heavy. Sp. gr. 0.726.

Light yellowish or whitish; heart slightly darker than the sap. Wood darkening rather rapidly on exposure to the air. Straight and coarse grained. No seasonal rings. Faintly resinous odor.

Uses.—Barrotos; beams in interior construction; ceilings; decks; flooring; house construction; inside partitions of houses; masts; naval construction; outrigger supports; piles; planks; sacayanes; waterways.

Structure.—Pith-rays medium size, distinct. Vessels medium size, scattered. Resin canals present, containing whitish resin.

Bull. For. Bur. Manila, 2d ed. (1907), **4**, 68–69.

Ahern, l. c., 67–68.

MAYAPIS.

Anisoptera vidaliana Brandis. Fam. **DIPTEROCARPACEÆ.**

(*Mocanera mayapis* Blanco. *Dipterocarpus mayapis* Blanco.)

Anisoptera thurifera (Blanco) Blume.

(*Mocanera thurifera* Blanco, *Dipterocarpus thurifer* Blanco), which has sometimes been credited with furnishing white lauan is one of the probable sources of mayapis.

Apu (Cag.); dagang (Rizal); mayapis (T.); paihapy (Z.); palosapis (Z., Bataan).

Philippines.

Light and soft. Sp. gr. 0.399.

White and gray; heartwood slightly darker than the sap. Coarse and crooked grained. No distinct seasonal rings. Slightly resinous odor.

Uses.—Bancas; boxes; canoes; flooring; light or temporary construction; paper making; rice mortars.

Structure.—Pith-rays medium size, distinct. Vessels medium size, scattered. Resin canals fairly numerous, with whitish resin. Wood parenchyma scanty and scattered.

Bull. For. Bur. Manila (1906), **4**, 63; 2d ed. (1907), **4**, 66.

MOLAVE.

Vitex littoralis Decne. Fam. **VERBENACEÆ.**

(*Vitex altissima* Blanco.)

Vitex pubescens Vahl.

(*Vitex latifolia* Blanco.)

Agubarao (V.); amugauan (Il.); bulaon (T., V., Pamp.); calipapa (Moro); molauin (T.); salincapa (Guimaras); tugas (V.).

Malaya.

Hard and heavy. Sp. gr. 0.778.

Heartwood pale yellow; sapwood only slightly lighter in color. Fine and usually cross-grained. Seasonal rings present. Diffuse porous. Slightly acid odor. Bitter taste. Turning greenish-yellow when treated with an alkali. Staining water a greenish-yellow color.

Uses.—Axles; beams; bridges; cabinetmaking; carabao yokes; cog-wheels; general high-grade construction; docks; doors; finishing of houses; firewood; flooring; footings in the ground; futtocks; palo (wooden club to pound rice); posts; joists; knees; piles; pillars; pinions; planks; plows; rafters; rice mortars; shipbuilding; cutwater; ships' knees; ribs; frames; siding of houses; sleepers; sternposts; sugar presses; wedges; wheel rims; wheels; undersills; paving blocks; railroad ties.

Structure.—Pith-rays fine and indistinct. Vessels small and scattered.

Bull. For. Bur. Manila (1906), **4,** 53; 2d ed. (1907), **4,** 55.

Ahern, l. c., 69–74.

NARRA.

Pterocarpus indicus Willd. Fam. **LEGUMINOSÆ.**
Pterocarpus echinatus Pers.
Pterocarpus blancoi Merr.
Pterocarpus klemmei Merr.

Apalit (Pamp.); asana (T.); narra (T., V.); odiao (Pamp.); sangque (Il., V.,) taga (N. Luz.); urian (Pamp.).

The first two species are widely distributed throughout tropical Asia, the other two are known only from the Philippines. Said to be closely related to the "padouk" or "amboyna wood" of India.

Moderately heavy and moderately hard. Sp. gr. 0.580.

Heartwood white, yellow or red; sapwood nearly white. Grain coarse and more or less twisted. Seasonal rings very distinct. Ring porous. Faint, sweet, cedary odor. Turning water a fluorescent blue.

Uses.—Bancas; bridge construction; cabinetmaking; carabao yokes; cascoes; chests; general construction; doors; dyewood; finishing of houses; floors; furniture; posts; planks; rafters; ships; siding of houses; table tops; walls; window sills; washbowls; door panels; carriage making.

Structure.—Pith-rays very small. Row of large vessels in inner part of ring and larger ones scattered in later part of season's growth. Vessels surrounded by parenchyma. Wood parenchyma in fine, more or less wavy, concentric lines. Fine, parallel, transverse lines in tangential and radial sections.

Bull. For. Bur. Manila (1906), **4,** 54; 2d ed. (1907), **4,** 57.

Ahern, l. c., 74–76.

NATO.

Palaquium spp. Fam. **SAPOTACEÆ.**

Banilac (V.); narec (V., T.); malac malac, palac palac, dolitan.

Light to moderately heavy; soft to moderately hard.

Pale-reddish; sapwood only slightly lighter than the heart. Straight-grained. No seasonal rings.

Uses.—Light construction; flooring; siding.

Structure.—Pith-rays small and indistinct. Vessels small or medium size, in short radial rows. Wood parenchyma in numerous, fine, concentric lines.

This wood is very often substituted for amuguis.

PALO MARIA.

PALO MARIA DE LA PLAYA OR BITAOG.

Calophyllum inophyllum L. Fam. **GUTTIFERÆ.**

This species is cosmopolitan in the Tropics. It is known as poon in India.

PALO MARIA DEL MONTE OR BITANHOL.

Calophyllum wallichianum Planch. et Tr.

Calophyllum spp.

Bancalan (T.); dincalan (T.); pamitaogan (V.); zarumayen (Il.).

Hard and moderately heavy. Sp. gr. 0.621.

Reddish-brown. Grain fine and crossed. No seasonal rings.

Uses.—General construction; masts; hubs; turnery; fine furniture; bridge building; carriage building; cascoes; decks of ships; flooring; futtock timbers; house posts; oars; paddles; ships' booms; bowsprits; spars; wagon shafts; keels of boats; recommended for railroad ties.

Structure.—Pith-rays fine and indistinct. Vessels of medium size, in irregular, branching, radial lines; with sulphur-yellow deposits. Wood parenchyma in a few, coarse, concentric lines and scattered.

Bitaog, or palo maria de la playa, differs from bitanhol, or palo maria del monte, in having a much more twisted and crooked grain. It is therefore the finer furniture wood. Otherwise these two woods are suitable for the same things and not distinct in structure.

Bull. For. Bur. Manila (1906), **4,** 60–61; 2d ed. (1907), **4,** 64.

Ahern, l. c., 77–78.

PANAO.

For description see APITONG. Panao is like apitong in all respects, except that it is less resinous and has less wood parenchyma. Not ordinarily distinguished from apitong.

Bull. For. Bur. Manila (1906), **4,** 52; 2d ed. (1907), **4,** 54.

Ahern, l. c., 79–80.

SACAT.

Terminalia nitens Presl. Fam. **COMBRETACEÆ.**

Philippines.

Moderately heavy and moderately hard. Sp. gr. 0.589.

Sapwood very light; heartwood gray or brownish-yellow. Coarse and straight grained. Seasonal rings distinct. Diffuse porous. Colors water a pale yellow.

Uses.—Light construction.

Structure.—Pith-rays medium size, distinct. Vessels medium size and scattered, occasionally roughly grouped in irregular wavy tangential lines.

Bull. For. Bur. Manila (1906), **4**, 55; 2d ed. (1907), **4**, 58.

(See also note under "Talisay.")

SANTOL.

Sandoricum indicum Cav. Fam. **MELIACEÆ.**

Southern Asia and Malaya. The false or wild mangosteen of India. Commonly cultivated for its fruit.

Moderately heavy and soft. "Weight about 36 lbs. per cu. ft."

Pinkish; straight grained. Aromatic odor somewhat camphor-like but faint. Seasonal rings present or absent. When soaked in water, producing a reddish tinge.

Uses.—Light framing; cabinetwork; house posts; wood carving; wooden blocks for shaping hats. This wood is exceedingly easy to work.

Structure.—Pith-rays small but distinct. Vessel small and scattered. Wood parenchyma scanty and scattered.

SASALIT.

Vitex aherniana Merr. Fam. **VERBENACEÆ.**

Gualberto (Il.); igang (Baler); dungula (V.); sasalit (T.).

Philippines.

Very hard and heavy. Sp. gr. 0.872.

Pale yellow to dark, yellowish-brown. Fine and wavy grained. Seasonal rings not present.

Uses.—Posts and general house construction; telegraph poles; railroad ties.

Structure.—Pith-rays fine but distinct. Vessels small and scattered. Wood parenchyma scanty or wanting.

Sasalit is exceedingly like molave in structure, but it is much harder and it does not turn the same bright greenish-yellow as molave, when treated with an alkali. It also appears very much like yacal; but it differs in structure by its smaller vessels and its lack of wood parenchyma. It also behaves differently when treated with an alkali.

Bull. For. Bur. Manila (1906), **4**, 63–64; 2d ed. (1907), **4**, 67.

SUPA.

Sindora supa Merr. Fam. **LEGUMINOSÆ.**
(*Sindora wallichii* Vid.)

Malapajo (V., T.); Pancalian, Paimo (Cag.); Yacal-dilao (T.).

Philippines.

Heavy and hard. Sp. gr. 0.729.

Yellow and brown. Fine and straight-grained. Seasonal rings distinct. Diffuse porous. Faint peppery odor. Colors water a dark reddish-brown.

Uses.—General construction; flooring; interior house trim; furniture; cabinetmaking; baseball bats; bridge construction; door frames; posts; joists; naval construction; pillars; door panels; railroad ties.

Structure.—Pith-rays small and distinct. Vessels medium sized, scattered, exuding oil when first cut. Wood turning much darker as it is covered by the oil. End of seasonal ring a distinct line.

Supa is said to be sometimes substituted for ipil; but the two woods are very distinct in appearance.

Bull. For. Bur. Manila (1906), **4**, 57; 2d ed. (1907), **4**, 60.

Ahern, l. c., 80–81.

TALISAY.

Terminalia catappa L. Fam. **COMBRETACEÆ.**
(*Terminalia latifolia* Blanco.)

Asiatic Tropics. "Indian almond."

Moderately heavy and moderately hard. Sp. gr. about 0.700.

Wood red, with lighter-colored sapwood. Wood coloring water a yellowish color.

Structure.—Pith-rays very fine. Vessels medium size, scattered, sometimes joined by irregular, wavy, short, concentric bands of wood parenchyma.

The different species of *Terminalia,* furnishing our woods known as calumpit, dalinsi, sacat, and talisay, are exceedingly alike in structure and are used for the same purposes. Thus far we can distinguish them only by color, and not certainly by that.

TAMAYUAN.

Strombosia philippinensis (Baill.) Vid. Fam. **OLACACEÆ.**
(*Strombosia dubia* Vid.)

Camayuan (Bataan); Tamabayan (T.).

Philippines.

Moderately heavy to heavy and hard.

Dull-yellowish to pinkish. Fine and straight grained. Seasonal rings present or absent.

Uses.—Posts; house building; joists; roofing; ax handles.

Structure.—Pith-rays small but distinct. Vessels small, scattered. Numerous very fine, irregular lines of wood parenchyma.

Ahern, l. c., 84–85.

TANGUILE.

Shorea polysperma (Blco.) Merr. Fam. **DIPTEROCARPACEÆ.**

(*Mocanera polysperma* Blco.)

(*Dipterocarpus polysperma* Blco.)

Adamuy (B.); araca (Il.); balacbacan (V.); panansogin, tanguile (T.).

Philippines.

Light and moderately hard. Sp. gr. 0.491.

Sapwood light-colored; heartwood light reddish-brown. Coarse and straight grained. No seasonal rings. Faint resinous odor.

Uses.—Bancas; boat building; boxes; canoes; medium-grade construction; furniture; house construction; shipbuilding; interior cabinetwork.

Structure.—Pith-rays medium size, distinct. Vessels medium size, scattered. Wood parenchyma scattered. Resin-canals scattered.

Differs from red lauan in being heavier and harder and in splitting more readily. It is also of slightly darker color. Tanguile does not have such large vessels as red lauan.

Bull. For. Bur. Manila (1906), **4**, 55; 2d ed. (1907), **4**, 58.

Ahern, l. c., 85.

TEAK.

Tectona grandis L. f. Fam. **VERBENACEÆ.**

Coloyate, dalondan (V.); ticla (T.); Yati (V.).

India and Malaya.

Teak is of infrequent occurrence in the Archipelago. It is known to be found in Mindanao and on the Island of Basilan. It is reported as occurring in Jolo, and in some of the other islands, but we have no herbarium specimens except from the places first mentioned and from two trees at Tanay, Rizal Province. The teak in the Manila market is probably imported.

Heavy and hard. "Weight 40 to 58 lbs. per cu. ft."

Sapwood white, usually small; heartwood dark golden-yellow, turning brown, dark brown, and finally almost black with age. Grain coarse and straight. Seasonal rings distinct; ring porous. Distinct aromatic odor.

Uses.—Shipbuilding; high-class construction; carving; furniture.

Structure.—Pith-rays moderately broad to broad. Vessels a row of large ones in the inner part of the ring, smaller scattered ones in the outer part of the ring.

Ahern, l. c., 86–87.

TELUTO.

Pterocymbium tinctorium Merr. Fam. **STERCULIACEÆ.**

Philippines.

Very light and soft.

White. No distinct sap- and heart-wood.

Uses.—Rafts; matchwood.

Structure.—Pith-rays large and distinct. Vessels medium size, scattered. Prominent parallel, transverse lines in tangential section as in narra.

TINDALO.

Pahudia rhomboidea (Blanco) Prain. Fam. **LEGUMINOSÆ.**
(*Afzelia rhomboidea* Vid. *Eperua rhomboidea* Blanco.)

Balayong (T.); ipel (Pang.); magalayao (N. Luz.); uris (Il.).

Philippines.

Heavy and hard. Sp. gr. 0.878.

Sapwood white; heartwood saffron or red, becoming very much darker with age. Fine and usually straight-grained, sometimes with bird's-eye grain. Seasonal rings distinct. Diffuse porous. Agreeable odor.

Uses.—Often considered our finest furniture wood; cabinetmaking; construction above ground; beams; chairs; desks; doors; floors; posts; interior finish of houses; joists; naval construction; planks; rafters; railing; siding; window sills; railroad ties.

Structure.—Pith-rays small but distinct. Vessels of medium size, scattered and surrounded by a fringe of wood parenchyma.

Bull. For. Bur. Manila (1906), **4**, 58; 2d ed. (1907), **4**, 62.

Ahern, l. c., 81–84.

TUCAN-CALAO.

Aglaia clarkii Merr. Fam. **MELIACEÆ.**

Tansuyot (B.); tucan-calao (T.).

Philippines.

Moderately heavy and hard. Reddish or brownish. Fine, curly grain. Seasonal rings present or absent; if present, then diffuse porous. Faintly resinous odor.

Uses.—Buildings; flooring; furniture; house construction; shipbuilding; interior finish of houses.

Structure.—Pith-rays small and distinct. Vessels of medum size, scattered. Wood parenchyma clustered about vessels.

YACAL.

Hopea plagata Vid. Fam. **DIPTEROCARPACEÆ.**

(*Mocanera plagata* Blanco. *Dipterocarpus plagatus* Blanco.)

Biti, callot (Il.); linap (V.); maraga (Neg.); maratuba (Cag.); panigayan, salapugud, taggoy (Il.).

It seems likely that one or more other species may furnish some of the wood known as yacal.

Philippines.

Heavy and hard. Sp. gr. 0.830.

Dirty-yellow to brownish. Coarse and straight grained. No seasonal rings. Slightly resinous odor.

Uses.—Beams; bridge construction; cabinetmaking; general construction; desks; doors; flooring; girders; joists; keels; naval construction; outrigger supports; piles; posts; rafters; sleepers; sternposts; walls, exterior; window sills; railroad ties; masts; rudder posts; spokes and felloes of carriage wheels.

Structure.—Pith-rays small but distinct. Vessels of medium size, scattered, frequently a slight border of wood parenchyma about the vessels. Resin canals present, frequently conspicuous.

Bull. For. Bur. Manila (1906), **4**, 53; 2d ed. (1907), **4**, 56.

Ahern, l. c., 88–90.

V. INDEX TO PHILIPPINE WOODS.

A.

B.

THE PHILIPPINE

JOURNAL OF SCIENCE

C. BOTANY

VOL. II DECEMBER, 1907 No. 6

NOTES ON THE STEERE COLLECTION OF PHILIPPINE FERNS.

By EDWIN BINGHAM COPELAND.

(From the Bureau of Education, Manila, P. I.)

By the kindness of Dr. J. B. Pollock, of the University of Michigan, I have been fortunate enough to receive fragments of practically all the ferns collected in the Philippines by Dr. Steere the determinations of which were published by Prof. Harrington.[1] Although made at Kew, a few of these determinations are doubtful, and a few clearly are wrong. The latter include some species described as new, or accredited to these Islands on the strength of these determinations alone. Some others of Harrington's species have since received other names by mistake, or his names have been misapplied.

Dryopteris aoristisora (Harr.) C. Chr. is nearly related to *D. canescens* (Bl.) C. Chr. Its sporangia are glabrous.

Nephrodium Bakeri Harrington is a relative of *D. canescens*, as Christ has pointed out;[2] but it is so peculiar that I can not follow Christ in submerging it in that aggregation. It must rather be known as **Dryopteris Bakeri** (Harrington) Copel. n. comb.

Dryopteris Luerssenі (Harrington) C. Chr. This plant is included by Christ [3] in *D. Foxii*, being the large form to which he refers, and including a majority of the specimens he cites. I have never regarded these

[1] *Journ. Linn. Soc.* (Botany), (1877), **16**, 25.
[2] *Ann. Jard. Bot.*, Buitenzorg (1898), **15**, 131.
[3] *This Journal, Bot. Sec. C.* (1907), **2**, 208.

and the small, typical *D. Foxii* as identical, and am therefore still disposed to maintain the latter species as originally diagnosed; but this may be an error. The plant called *D. Luersseni* by Christ is altogether different; it is a *Goniopteris* in affinity, in spite of its free veins.

Polystichum horizontale Pr. This is the form determined by Harrington under the general head of "*Aspidium aculeatum* Swtz."

Hemionitis Zollingeri Kurz. This is one of our most-named ferns. It is the same fern described by me as *Hemionitis gymnopteroidea* in Perkins Fragmenta (1905), **1**, 183. I have since become convinced that it is absolutely identical with *Leptochilus latifolius* (Meyen) C. Chr., and have distributed it under that name, calling attention to its synonymy, and have treated it as such in my "Ecology of San Ramon Polypodiaceæ."[4] Still more recently, Christ has described it as a type of a new genus, *Hemigramma*,[5] distinguished from *Leptochilus* by the venation and by the restriction of the sori to the veins. It was with the same opinion of the value of the sorus character that I called the plant *Hemionitis*, and that Presl[6] called it *Gymnopteris* instead of *Leptochilus*. If the plant be maintained as generically distinct, its name must be **Hemigramma latifolia** (Meyen) Copel. n. comb., this being a far older specific name than *Zollingeri*. Its known range is Malaya and the Philippines.

Asplenium lunulatum Swtz. of this collection is the fern commonly determined here as *A. tenerum* Forst. The original specimen at Ann Arbor is a single imperfect frond. A various lot of Philippine specimens is now grouped under *A. tenerum*, but I would not call Steere's, or any other of the many collections, *A. lunulatum*.

Asplenium wightianum Wall. My material of this fern, collected in Panay, is insufficient for positive identification, but it shows that the pinnæ are too inequilateral to be the species stated, which should therefore not be accredited to these Islands. The plant is almost certainly *A. vulcanicum* Bl., which is found in Negros and as far north as Mount Maquiling in Luzon.

Asplenium Steerei Harrington. This is the fern described by me as *A. laxivenum*.[7]

Stenochlaena areolaris (Harrington) Copel. nom. nov. (*Lomaria*, Harrington, I. c., p. 28). The specialized, inflexed margin (indusium), because of which Harrington described this fern as *Lomaria*, is very evident in the specimen sent me, but such a margin is not rare in *Stenochlaena*. This is very distinct from its nearest known relative, *S. palustris*.

[4] *Ibid.* (1907), **2**, 3.

[5] *Ibid.* (1907), **2**, 170.

[6] "Sori hemionitidei, nec acrostichacei adsunt inde nec Leptochilo inserendum est, quemadmodum clar. Fée autumat". Epimeliae Botanicae, Prague (1849), 150.

[7] *This Journal, Bot. Sec. C.* (1907), **2**, 132.

Monogramme paradoxa (Fée) Bedd. (*M. Junghuhnii* Hook.) is credited to Formosa by Harrington, but the label of the specimen sent me reads "Growing in tufts on trunks of trees, Philippine Islands." The determination is correct, and this species should probably be added to our known flora.

Polypodium craterisorum Harrington is unmistakably *P. celebicum* Blume. It is described in error as membranous (l. c., p. 32). Elmer (No. 7760) has collected this fern on Mount Banajao (which is Mount Majayjay). Some of the fronds are contracted below so abruptly as to suggest its relative, *P. decrescens* Christ.

Polypodium hammatisorum Harrington turns out, as I [8] have anticipated, to be identical with *P. nummularium* Mett.

Polypodium Schenkii Harrington is, as stated by Baker,[9] our common *P. obliquatum* Bl.

[8] *Ibid. Supp. IV* (1906), 1, 256.
[9] *Ann. of Bot.* (1892), 5, 467.

A REVISION OF TECTARIA WITH SPECIAL REGARD TO THE PHILIPPINE SPECIES.

By Edwin Bingham Copeland.
(*From the Bureau of Education, Manila, P. I.*)

TECTARIA Cavanilles.

A genus of Aspidieæ, derived from Dryopteris (§) Eudryopteris, characterized by the comparatively undissected fronds, and corresponding venation of reticulate veins which do not anastomose regularly in pairs, fronds more or less deltoid in form, and the fertile and sterile not exceedingly different.

It has become possible, with the accumulation in Manila of a large amount of material, large not only in the number of collections to be cited, but also in the presence of many specimens of single collections, to present this confused genus in a way that was not at all possible at the time of publication of my[1] Polypodiaceæ of the Philippines. It has also become possible to substitute for the artificial grouping of species according to the indusium a classification which for the most part may be advanced with some confidence in its naturalness. In many parts this arrangement is that adopted by Diels in Engler and Prantl's "Die Natürlichen Pflanzenfamilien."

As would be anticipated, the chief difficulty and uncertainty in the delimitation of groups of species is among the most primitive, just as in a family it is the most primitive genera which are most difficult to characterize, or even to recognize with certainty. I have already established the place of *Dryopteris* (§) *Lastraea* as among our most primitive ferns, and as the parental type of *Aspidieæ*. In this group, *Dryopteris dissecta* (Bory) O. K. represents, most nearly of known species, the probable origin of *Tectaria*. At a future date I shall show that some "*Athyria*" are so near to *D. dissecta* that their generic assignment is a matter of judgment. We have in the Philippines a number of species of *Tectaria* which, while in reality and by easy definition members of the genus *Tectaria*, are more like this parental type than they are like any of the

[1] *Publications of the Bureau of Government Laboratories*, Manila (1904), No. 28.

well-differentiated groups within *Tectaria*. In a family it is necessary, for the sake of identification, to refer these undifferentiated members to genera, as, in Polypodiaceæ, to *Dryopteris*, or *Athyrium* or *Acrophorus*, but within a genus a forced classification would serve no purpose.

The Philippine species apparently too primitive for any finer classification than as *Tectariæ*, are *T. devexa* and *T. calcarea*. *T. ambigua* is apparently isolated in the Philippines, but is a relative of *T. gigantea*. (*Aspidium* Blume, Enumeratio 159, 1828.) I regard it as also a likely relative of *Stenosemia*.

The better differentiated and understood groups are:

a. Cicutariæ.—Frond in the more generalized forms decompound; as in *T. angustius* (Christ, *Sagenia*, Bull. Herb. Boiss **6** (1906) 165) and *T. cicutaria* (Linnaeus *Polypodium*, Syst. Nat. **2**, p. 1326, 1759) of the American tropics, and *T. malayensis*. *T. latifolia* (Forst., *Polypodium*, Prod. 83, 1786) and *T. melanocaulon* belong in this group, and retain the dissected fronds. A simplification of the fronds is evident in *T. angelicifolia* (*Polypodium*, Schum. Vid. Selsk. Afh. **4** (1827) 228) and has gone further in *T. Hippocrepis* (*Polypodium*, Jacq. Collect. **3** (1789) 816), *T. apiifolia* (*Aspidium* Schkuhr 1809), and in our *T. Christii*.

b. 1. *Crenatæ*.—Simply pinnate plants with indusiate sori in regular rows parallel to main veins. The several supposed Philippine species seem safely referable to one, *T. crenata*.

b. 2. *Trifoliatæ-Polymorphæ*.—A group whose common ancestry with the preceding is probably not exceedingly remote. Fronds less cut and more ample; indusia mostly fugacious. Sori in regular rows in *Trifoliatæ; T. trifoliata* Cav., *T. subtriphylla* (*Polypodium* Hook. & Arn. Bot. Beech. Voy. 256, 1838–40), and *T. siifolia*. Sori scattered in *Polymorphæ; T. Menyanthidis* (not improbably, though not intimately, related to *T. crenata*), *T. Barberi*, *T. polymorpha*, and *T. irrigua*, and *T. Labrusca* (*Polypodium*, Hooker. Sp. Fil. **5** (1863) 73); these stand here in the probable order of their differentiation, the most ancestor-like first. The group is specialized in adaption to a moist, still habitat, as in gorges. American representatives are *T. Plumierii* (*Aspidium* Presl. Rel. Haenk. 29, 1825), a near relative of *T. trifoliata*, and *T. martinicense* (*Aspidium* Spr. Anleitung **3** (1804) 133). The *Polymorphæ* of America and the Old World have very likely been derived separately, and might well be regarded as distinct minor groups.

c. 1. *Decurrentes*.—Coarse ferns, with the rachis mostly or entirely winged, the main veins prominent and the sori in regular rows. *T. grandifolia*, *T. decurrens*. *T. draconoptera* (*Aspidium*, Eaton, Mem. Am. Acad. II, **8** (1860) 211) is an American fern intermediate between this group and the *Vastæ*.

c. 2. *Vastæ.*—Like the *Decurrentes,* but sori scattered. *T. vasta* (*Aspidium,* Blume. Enum. 142, 1828), *T. Bryanti.*

d. 1. *Pleocnemia.*—Fronds ample, at least bipinnate, veins forming a single series of areolæ, without free included veinlets, crown of caudex and bases of stipes densely covered with harsh, dark, slender paleæ.

T. leuzeana.—The inclusion of various species having nothing in common except the venation robs *Pleocnemia* of all naturalness. *Dryopteris dissidens* (Mett.) O. K. is sometimes improperly placed here.

d. 2. *Arcypteris.*—Like the preceding, but fronds usually less dissected and areolæ more numerous.

T. irregularis.

Tectaria is a genus of tropical ferns the specialization of which for the most part is in adaption to moist and windless habitats. The general process of phylogenetic differentiation has been a simplification of the frond and a correlated anastomosis of the veins. In three of the groups, this has led to some examples with simple fronds. With increasing complexity of the venation the sori become scattered. In *Pleocnemia* there is an adaption to the same environment by great development in size.

In this paper I have adopted the principles of nomenclature exemplified in Christensen's Index, believing that uniformity will most quickly be reached by general conformity with so valuable a model. According to these principles, as Christensen states in the fascicle last published, *Tectaria,* and not *Aspidium,* is the "*nomen optimum.*"

Key to Philippine species.

1. Frond ample, pinnæ broadly decurrent, and stipe winged.
   2. Frond simple.
      3. Indusium wanting; sori scattered .... 1. *T. Bryanti*
      3. Indusium present; sori in rows .... 2. *T. decurrens*
   2. Pinnæ not all connected by the broad wings .... 3. *T. grandifolia*
1. Stipe not broadly winged; fully developed frond simply pinnate, and none save the basal and apical pinnæ deeply cut.
   2. Lateral pinnæ usually not more than 2 pairs, not pinnatifid, the lowest pair forked or oblique; indusia fugacious.
      3. Not at all dimorphous; margin entire, sori scattered.
         4. Pinnæ ample, ovate with curved margins .... 4. *T. polymorpha*
         4. Pinnæ contracted, angular .... 5. *T. irrigua*
      3. Fertile frond somewhat restricted, sori in rows .... 6. *T. siifolia*
   2. Lateral pinnæ usually more numerous; or if only 1 or 2 pairs, the lowest pair not forked and the indusium persistent.
      3. Base of pinnæ round or truncate .... 7. *T. Barberi*
      3. Base of pinnæ acute.
         4. Position of sori not marked on upper surface .... 8. *T. Menyanthidis*
         4. Position of sori evident on upper surface .... 9. *T. crenata*
1. Pinnæ lobed halfway to the costa, the lowest not forked .... 10. *T. ambigua*

1. Fully developed fronds bipinnate, at least at the base.
   2. Fronds 30 cm. or less high including stipe, at least bipinnate.
      3. Sori few, fronds depauperate-looking.. 11. *T. calcarea*
      3. Sori minute and numerous; fronds ample................................ 12. *T. devexa*
   2. Fully developed fronds more than 30 cm. high.
      3. Stipe and rachises black .... 13. *T. melanocaulon*
      3. Axes not ebeneous, veins forming more than 1 series of areolæ.
         4. Indusia present.
            5. Stipe scaly, frond glabrous... 14. *T. malayensis*
            5. Stipe (unless at base) smooth, frond ciliate.................... 15. *T. Christii*
         4. Indusia absent (*Arcypteris*) .... 16. *T. irregularis*
      3. Axes hardly black; veins forming a single series of areolæ (*Pleocnemia*) .... 17. *T. leuzeana*

1. Tectaria Bryanti Copel. nom. nov.

Aspidium, Copel. Perkins Fragmenta, 175; 1905; Polypod. of the Philippines 34, 1905.

Negros, *Copeland* 82: Basilan, *Hallier.*

Aspidium Bolsteri Copel. (This Journal **I** Suppl. (1906) 252) is probably a very large plant of the same species.

Mindanao, Surigao *Bolster* 305. The *A. Bryanti* reported by Christ from *Loher's* collection on Mount Maquiling, Luzon, is almost certainly different.

Aspidium vastum Bl., a relative of *T. Bryanti,* with indusiate sori, is taken up as a Philippine fern in my Polypodiaceae of the Philippine Islands, but probably by mistake. J. Smith in his enumeration of *Cuming's* ferns calls No. 356 *Aspidium alatum* Wall., which is synonymous with *A. vastum;* but this number is cited by *Hooker* Sp. Fil. **4**: 48, under *A. pteropus* Kze. (=*T. decurrens*); and the material distributed under this number included *T. grandifolia.*

2. Tectaria decurrens (Presl) Copel. Elmer's Leaflets, 1 (1907) 234.

Aspidium Presl, Rel. Haenkeanae 28, 1825; *A. heterodon* Copel. Perkin's Fragmenta 177, 1905; *A. Copelandi* C. Chr. Index 661, 1906.

This is a most variable species, if all the plants I am including in it are really identical. It was originally stated to have an erect caudex, but more recent works have described it as having a horizontal rhizome. As a matter of fact, we find this member horizontal, oblique or erect. The margin of the frond is entire or variously lobed. Fruiting specimens with simple and entire fronds occur on Mount Maquiling. Specimens in our herbaria are:

Cuming 148, Luzon; *Bolster* 146, Cagayan (Luzon); *Whitford* 966, *Elmer* 7999, Mt. Banajao; *Merrill* 1773, 4042, 5872, *McGregor* 133, *Merritt* F. B. 6778, Mindoro; *Clemens* 378, Camp Keithley, Mindanao; *Copeland* 951, 967, Davao; *Copeland* 1603, Zamboanga.

3. **Tectaria grandifolia** (Presl.) Copel. nom. nov.

Aspidium, Presl. Epim. Bot. 64, 1849; *A. grande* J. Sm. 1841 (name only), Mett. 1858.

Except as to the indusium, which Presl, probably in error, described as orbiculate and peltate, and Mettenius described as reniform, the diagnoses of *A. grande* and *A. grandifolium* are sufficiently alike to be construed as referring to the same species. Both are based on Philippine plants collected by *Cuming*, Presl's being from Panay. No one of *Cuming's* plants in Manila is referable here, but *Merrill's* No. 5871 fits Mettenius' description very closely and agrees with Presl's wherever it differs from Mettenius'. As Presl states, it is a relative of *T. decurrens*.

(Panay, *Cuming* s. n.)[2] Mindoro, *Merrill* 5871.

4. **T. polymorpha** (Wall.) Copel. nom. nov.

Aspidium, Wallich. List No. 382, 1828; Hooker Sp. Fil. **4** (1862) 54. *A. angulatum* Christ in Bull. Herb. Boiss II **6** (1906) 1003, non J. Sm. Mett.

Luzon, Rizal, *Ramos B. S.* 2160; Los Baños and Mt. Maquiling (very common below falls), *Copeland* 2024, *Matthew, Elmer* 8329, *Topping* 634, 706; Mindoro, *Merrill* 5874; Mindanao, Davao, *Copeland* 1311, 1466.

A polymorphous species indeed. Cordate, simple, entire fronds are sometimes fertile. The sori are often elongated along their veins, and the indusia are of various shapes and fugacious. *T. irrigua* is not sharply distinguished from this species, and the two species succeeding it are near relatives. From the material I have from the locality from which Christ reports *A. angulatum*, I have no doubt that this is the plant. Christ is also in error in giving the Journal of Botany **3** as the place of publication of the name; *A. angulatum* J. Sm. was a manuscript name first published by Mettenius.

5. **T. irrigua** (J. Sm.) Copel. nom. nov.

Aspidium J. Smith, Journal of Bot. **3** (1841) 410, Presl. Epim. Bot. 62, 1849, *A. lamaoense* Copel. Perkin's Fragmenta 176, 1905.

Luzon, *Cuming* 31; Lamao Forest Reserve, *Copeland* 223, *Meyer* F. B. 2497; Pagsanjan, *Copeland* 1993; Los Baños, *Copeland* 2023, *Elmer* 8330, *Matthew, Topping* 686; Indang, Cavite, *Copeland* s. n. Known only in Luzon.

This fern grows on rocks in creek beds and on their banks where it is submerged by floods. It is a smaller fern than the preceding and much narrower in all its divisions, with comparatively straight margins forming sharp bases and apices. *T. polymorpha* would be torn to pieces and destroyed if it grew in the characteristic habitat of *T. irrigua*.

[2] Collections cited in parentheses were not seen.

6. **T. siifolia** (Willd.) Copel. nom. nov.

Polypodium, Wild. Sp. Pl. **5** (1810) 196: *Aspidium*, Mett. 1864. non Blume, 1828; *A. angulatum* Copel. Phil. Journ. Sc. **I** Suppl. (1906) 145, non J. Sm., Mett.; *A. biseriatum* Christ. Bull. Herb. Boiss. II **6** (1906) 1002.

Luzon, *Cuming* 4; Pagsanjan, *Copeland* 1985; Mt. Maquiling, *Matthew;* Mindanao, San Ramon, *Copeland* 1776; Palmas Id., *Merrill*, 5353. (Mt. Pinatubo, Luzon, *Loher*).

The other specimens cited here from our own herbarium are identical with *Cuming's* No. 4 of which our specimens are sterile. The fertile frond is sometimes almost as ample as the sterile, but more often is strongly contracted; Mettenius, *Aspidium* No. 287, says "*Folia . . . . . subdifformia.*" In spite of my declaration in this JOURNAL, **I** Suppl. (1906) 145, that the plant is without indusia, I must now state that on immature fronds they can often be seen, and are irregular in form.

7. **T. Barberi** Copel. nom. nov.

Polypodium Hooker, Sp. Fil. **5** (1864) 100. *Aspidium*, Copel. Polyp. Phil. 34, 1905.

Luzon (Laguna, Majayjay, *Loher*, teste Christ); Mindoro, *Merrill* 5875.

Merrill's specimens fit Hooker's description perfectly except that the bases of the pinnæ, while broad and unequal, are not cuneate. The peculiar venation distinguishes this from any other Philippine fern.

8. **Tectaria Menyanthidis** (Presl) Copel. nom. nov.

Aspidium Presl, Rel. Haenkeanae **I** (1825) 28; *A. repandum* J. Sm. non Willd.

Luzon, *Cuming* 183; Mindanao, Surigao, *Bolster* 263.

Cuming's plant in Herb. Bureau of Science is labeled *Aspidium repandum* in Smith's writing, of which we have abundant specimens (Cf. Presl. Epim. Bot. p. 523). Instead of being, as stated by Hooker, Sp. Fil. **4**: 58. too near *A. pachyphyllum*, this species is closely related to *T. polymorpha* and *T. irrigua*, from which it is distinguished by the long, narrow pinnæ and more persistent indusia. We have fertile fronds which are simple, trifoliate, and with 3 pairs of lateral pinnæ.

9. **Tectaria crenata** Cav. Descr. 250, 1802.

Aspidium repandum Willd. Sp. pl. 5 (1810) 216; *A. platyphyllum* Presl, Epim. Bot.; *A. pachyphyllum* Kze. Bot. Zeit. (1848) 259, at least as to Philippine plants; (?) *A. persoriferum* Copel. Perkins Fragmenta 177, 1905.

Batanes Ids., *Mearns* B. S. 3157, 3158, 3159, 3161, 3166: Luzon (*Cuming* 224); Benguet *Topping* 337, 338, *Elmer* 6171; Mt. Mayon *Mearns* B. S. 2909, 2917; Cebu (*Cuming* 339, 340), *McGregor* B. S.

1736; Mindanao (*Cuming* 290); Camp Keithley, *Clemens* 160, 251; Davao, *Copeland* 929, 1467. The herbarium of the Bureau of Science has two unnumbered specimens of Cuming's collection, labeled *Sagenia platyphylla* J. Sm. According to Christensen (Index p. 614) this is *A. cicutarium;* but he (l. c. p. 87) interprets *A. platyphyllum* Presl which Presl says = *S. platyphylla* J. Sm. as *A. repandum* Willd., which is my judgment as well. *Elmer* No. 7060, from Leyte, is also this species, in my judgment; it has the ancestral form of the genus *Hemigramma,* Christ.

This is a reasonably variable fern, in form, margin, and indusia; but the chief difficulty in treating it is not natural, but due to the jumble of names in use. Willdenow published the name *repandum* with the avowal that Cavanilles' description would fit his plant except that the latter stated the pinnæ to be alternate. They are indeed rather more often opposite, but by no means constantly so. The indusia are peltate and reniform on the same plants. The manner in which this species has been shifted between *Sagenia* and *Tectaria* illustrates the futility of trying to maintain in this group genera based on the indusium. The fertile fronds are sometimes like the sterile, but usually somewhat contracted (Cf. Sp. Fil. **4**; 57); in view of this fact, I doubt the propriety of maintaining *persoriferum* as a separate species. Cavanilles says *"fructibus numerosissimis."*

10. **Tectaria ambigua** (Presl) Copel. nom. nov.

Digrammaria Presl, Tent. Pterid. 117, 1836. *Aspidium profereoides* Christ, Phil. Journ. Sc. **2** C. (1907) 158.

Davao, *Copeland* 1467 A. This is a very distinct species whose venation and texture, and sparsely hairy sinuses and contracted fertile fronds strongly suggest *Stenosemia.* It is related to *T. gigantea* (Blume).

11. **Tectaria calcarea** (J. Sm.) Copel. nom. nov.

Sagenia J. Sm., Journ. Bot. **3** (1841) 410. *Aspidium* Presl. Epim. Bot. 63. Mett. *Aspidium* 283.

Leyte, *Cuming* 310.

A very peculiar plant known only from *Cuming's* collection. Mettenius gives an excellent figure (l. c. Plate XVII). Some of the indusia are peltate, others not so.

12. **Tectaria devexa** (Kze.) Copel. nom. nov.

Aspidium Kze. Bot. Zeit. (1848); 259. *A. membranaceum* Hook. Sp. Fil. **5**; 105.

Cuming s. n.; Mindanao, Davao, *Copeland* 898.

This is certainly *A. membranaceum* Hook.; I have followed Christensen in taking Kunze's name.

13. Tectaria melanocaulon (Bl.) Copel. nom. nov.

Aspidium Blume, Enumeratio 161, 1828.

(*Cuming* 57): Luzon, Cavite, *Copeland* 1788, *Mangubat* B. S. 1295; Los Baños, *Matthew*; Mindanao, Camp Keithley, *Clemens* 381; Mt. Apo, *Copeland* 1466A. A most clearly marked species here, but connected with the *Cicutariæ* by *T. latifolia*.

14. Tectaria malayensis (Christ) Copel. nom. nov.

Aspidium Christ, Phil. Journ. Sc. **2** C. (1907); 187; *A. cicutarium* of Christ and myself, not of Swartz.

Luzon (Laguna, *Loher*); Rizal, *Ramos*, B. S. 1033; Mt. Mariveles, *Copeland* 217, 1396; *Topping*, 533, *Borden*, F. B. 1959; Mindanao, Surigao, *Bolster* 386; Mt. Apo, *Copeland* 1468.

15. Tectaria Christii Copel. nom. nov.

Aspidium coadunatum Wall. non Kaulfuss. For a diagnosis of this plant see Christ. Phil. Journ. Sc. **2** C. (1907) 187.

Luzon, Bontoc, *Copeland* 1899; Benguet, *Topping* 229; Mindoro, Merrill 5869.

16. Tectaria irregularis (Presl) Copel. nom. nov.

Polypodium Presl, Rel. Haenk. 25. 1825, Plate IV Fig. 3: *P. pteroides* Presl. l. c. fig. 4: *Aspidium difforme* Bl., Enumeratio 160, 1828; *Polypodium Brogniartii* Bory. Dup. Voy. Bot. **I** (1828) 263; *Phegopteris macrodonta* (Reinw.) Mett. Pheg. & Aspid. 1858. No. 68: *Polypodium Cumingianum* (Presl.) Hooker Sp. Fil. **5** (1864) 103; *Aspidium Whitfordi* Copel. Perkins' Fragmenta 176, 1905.

Easy as it is to distinguish the extremes of the forms combined here, and to find diagnostic characters which look valid when only the extremes are considered, the differences are found inconstant and the gaps are bridged over with the accumulation of many specimens. In our herbaria this accumulation has reached the point where I can no longer see a good specific distinction between any one of them and some other. In place of the hitherto recognized species, we will better recognize 3 or 4 inconstant varieties.

a. Var. euirregularis.

Frond simply pinnate: sori small and very numerous, sometimes much more so even than Presl figured, and irregularly scattered. Presl, surely in error, credits the type to Mexico.

Culion, *Merrill* 665; Mindanao, Surigao, *Bolster* 220; Camp Keithley, *Clemens* s. n.; San Ramon, *Copeland* 1569.

Simply pinnate fronds with the sori more like those of the next variety are: *Foxworthy* B. S. 1962, Pampanga; *Topping* 424, *Elmer* 6684, Mt. Mariveles; *Copeland* s. n. Mt. Maquiling. *Elmer* 6230, is simple, but has the slender upper pinnæ of var. *Brogniartii*.

b. Var. **macrodon.**

Frond bipinnate: sori less numerous and forming irregular (usually very irregular) rows parallel to the costæ and main veins.

Luzon, *Cuming* 9; Mt. Maquiling, *Matthew* s. n., *Topping* 657, 697: Mt. Mariveles, *Topping* 447; Tayabas, *Whitford* 668, *Gregory* 144; Mindoro, *Merrill* 5873; Samar, *Merrill* 5198.

Intermediate forms between this and the next variety are: *Mearns*, B. S. 2986, Casiguran, Luzon; *Topping* 335, Benguet; *Merrill* 1806 Mindoro.

c. Var. **Brogniartii.**

Bipinnate, the sori restricted to a single, more or less regular, closely marginal row.

Luzon, *Cuming* 171: Negros, *Whitford* 1665, *Copeland* 2076, typical; *Copeland* s. n. Negros, has the sori of this var., but the form of var. *euirregularis.*

Aspidium Whitfordi is probably a form, rather than a good variety; characterized by having the veins conspicuous on the upper surface; in other respects the plants are mostly referable to var. *macrodon.* Luzon, Mt. Mariveles, *Whitford* 201; Rizal, *Foxworthy* B. S. 76, *Ramos* B. S. 957; Cavite, *Copeland* s. n., *Mangubat*, B. S. 1285.

This species as a whole is a very natural one, distinguished from the preceding by the color, texture, margins, naked and often irregular sori, absence of free included veinlets, and coarse, chocolate-colored scales crowning the rhizome and burying the bases of the stipes. It is the type of *Arcypteris,* but is not isolated enough to be given generic rank by itself, and is no near relative of our other exindusiate species, *T. Bryanti* and *T. ambigua.* On the other hand is nearly related to *Pleocnemia,* as is shown by the venation, the shape of the sori, the texture and the pubescence. But for these many points of resemblance I should not include *Pleocnemia* in *Tectaria.*

17. **Tectaria leuzeana** (Gaud.) Copel. nom. nov.

Polypodium Gaud. Freyc. Voy. Bot. 361, 1827.

As distinct as *Aspidium angilogense* Christ, Bull. Herb. Boiss, II **6** (1906) 1003, appears in its description, and as different as its huge typical form is from the more common small forms, I am unable to find any constant character distinguishing them in the field. Even the few specimens collected by *Cuming* seen to me to have been more safely treated as a single species. At any rate, our specimens of No. 33 (*P. Cumingiana*) and No. 289 (*P. leuceana*), of Epimeliae Bot. p. 50, are alike. However, I am not ready to pass positive judgment on the question. *T. leuzeana* remains a very variable species even if the arborescent form is removed from it. We have *T. leuzeana, s. s.,* as follows:

Luzon, *Cuming* 34; Rizal, *Ramos,* B. S. 1046, 1091; Cavite, *Copeland*

1795; Mindanao, *Cuming* 289, *Copeland* 1696; Balabac, *Merrill* 5378. Specimens referable to *Pleocnemia cumingiana* Presl, but acaulescent, are: *Merrill* 5876, Mindoro, and *Elmer* 7062, Leyte.

Aspidium angilogense Christ is represented by: (?) Cuming 33, 107, *Luzon; Matthew s. n.*, Mt. Maquiling; *Copeland* 1698, San Ramon, Mindanao; (?) *Clemens s. n.* Camp Keithley.

EXCLUDED SPECIES.

Aspidium membranifolium (Pr.) Kze. This is *Dryopteris dissecta* (Forst.) O. K.

Aspidium heterophyllum Hooker, collected in Samar by *Cuming,* No. 322. The nearest affinity of this peculiar plant is to *Dryopteris canescens.*

Aspidium latifolium J. Sm. (*Aspidium microsorum Pr.*) ? *T. melanocaulon,* vel potius *T. siifolia.*

Aspidium giganteum Bl. This Javan plant is unknown to me in the Philippines, and is probably credited to these Islands on the strength of wrong determinations.

TWO NEW PHILIPPINE GRASSES.

By E. HACKEL.
Graz, Austria.

ARUNDINELLA Raddi.

Arundinella pubescens Merr. & Hack. n. sp.

Annua. Culmi erecti, 20 ad 30 cm. alti, graciles, teretes, in parte superiore nuda laxe hirsuti, 3 ad 5-nodes, nodis omnibus in culmi parte $\frac{1}{4}$ vel $\frac{1}{3}$ sitis reverso-barbatis, simplices vel ramo uno alterove aucti. Folia undique hirsuta, pilis patulis basi plerumque tuberculatis vestiti. Vaginae internodia superantes, teretes, laxiusculae, ligula series pilorum brevium. Laminae e basi rotundata lineari-lanceolatae, acutae, 6 ad 10 cm. longae, 4 ad 7 mm. latae, erectae, rigidulae, nervis crassiusculis parum prominulis percursae. Panicula elongato-lineari-oblonga, 15 ad 30 cm. longa, basi subinterrupta et laxa, superne laxiuscula vel apice densiuscula, contracta, angusta (vix 2 cm. lata), rhachi scabra et laxe pilosa, ad ramorum ortum parce barbata, internodiis inferioribus usque ad 9 cm., superioribus vix 2 cm. longis, ramis inferioribus solitariis vel binis, superioribus binis usque ad quaternis, omnibus tenuiter filiformibus scaberrimis suberectis, primariis in $\frac{1}{4}$ inferiore nudis, secundariis plerisque basilaribus brevissimis a basi spiculiferis, spiculis secus ramos geminis, paribus in inferiore parte rami valde distantibus, in superiore subcontiguis, pedicellis inaequalibus, altero spicula longiore, altero ca. 2-3-plo breviore. Spiculae lanceolatae, 4 mm. longae, pallide viridulae; gluma I, 2.5 mm. longa, ovato-lanceolata, acuta, mucronata, 5-nervis, carina scabra; II, spiculam aequans, ovato-lanceolata, in rostrum subulatum apice anguste obtusum acuminata, 5-nervis, laevis; III, 3 mm. longa, ovato-lanceolata, acuta, mucronulata, chartacea, tenuiter 3-nervis, glaberrima (etiam callo), paleam paullo breviorum floremque ♂ fovens; IV 2 mm. longa, lanceolato-oblonga, ex apice obtusiusculo aristam emittens 5 mm. longam geniculatam, tenuiter papillosa, subenervis, callo minute barbulato. Palea glumam aequans, binervis.

PALAWAN; prope Iwahig (B. of S. 856 *Foxworthy*) Majo, 1906.

Affinis *A. hispidae* O. Kuntze (*A. nepalensi* Trin.) quae a nostra differt radice perenni, culmo robusto multinodo glabro, nodis glabris, laminis elongatis, panicula oblonga dense haud interrupta, gluma I. III^{am} subaequante IV callo pilis $\frac{1}{3}$ glumae aequantibus barbata.

SCHIZOSTACHYUM Nees.

Schizostachyum hirtiflorum Hack, n. sp.

Ramuli foliiferi fasciculati, folia 5–6 gerentes. Vaginae pilis rigidulis basi tuberculatis hispidae, ore utrinque auriculatae, auriculis subfalcatis amplexicaulibus, longe fimbriatis. Ligula interiore margo membranaceus angustissimus, externa nulla. Laminae brevissime pedunculatae, penduculo 1 ad 2 mm. longo fere latiore quam longiore, e basi cordata lineari-lanceolatae, longe tenuiterque acuminatae, apice setaceo-involutae, 12 ad 20 cm. longae, 1.6 ad 2.2 cm. latae, tenues, utrinque virides, supra glabrae, subtus brevissime parceque puberulae, margine scaberrimae, tenuiter nervosae, venulis transversis nullis. Panicula (quae incompleta adest) magna vel aphylla vel ramus unus alterve apice breviter foliatus, rhachi glaberrima, ramis dense fasciculatis 20 ad 60 cm. longis glaberrimis ramulosis, ramulis solitariis, inferioribus 6 ad 10 cm. longis, superioribus brevibus, omnibus bractea subcymbiformi 1 ad 2 cm. longa lamina brevissima vel mucroniformi terminata fultis, spiculas 4 ad 6 saepius remotiusculas, rarius basi plus minus fasciculatis sed nunquam glomeratis gerentibus, in parte superiore rami vel ramuli spiculae solitariae, singulis bracteis fultae. Spiculae lineari-fusiformes acutae, ca. 12 mm. longae, hirsutae; glumae steriles 3 ad 4, inferiores breves, ovales, obtusae, mucronatae, 7 ad 9–nerves, infra apicem tantum hirsutae, singulae in axilla gemmulam foliaceam parvam foventes, I, circiter 2.5 mm.; II, 4 mm.; III, 5 mm. longa: IV, ovali-oblonga, 7 mm. longa, obtusiuscula, mucronata, involuta, 13–nervis, in ½ superiore hirsuta, vacua: gluma fertilis (V) spiculam aequans, lanceolata, acuta, involuta, apice subuliformis, 9–nervis, in ¼ superiore hirsuta. Palea nulla. Stamina 6, antheris 5 mm. longis apice obtusis. Ovarium (immaturum) parvum, glabrum, stylis in rostrum circiter 10 mm. longum connatis, stigmatibus 3 brevibus, breviter plumosis.

LUZON; prope Sablan, Prov. Benguet (6173 *A. D. E. Elmer*).

Affinis *S. acutifloro* Munro, quod differt a nostro laminibus basi rotundatis nec cordatis, spiculis fertilibus (quibus steriles crebrae immixtae sunt) 6 ad 8 mm. longis glaberrimis.

SOME GENERA AND SPECIES NEW TO THE PHILIPPINE FLORA.

By Elmer D. Merrill.

(From the botanical section, Biological Laboratory, Bureau of Science, Manila, P. I.)

In recently published papers on Philippine botany many genera and species have been recorded from the Philippines that were previously not known to occur in the Archipelago, one of the striking proofs of the present comparatively limited knowledge that we have of the flora of the Philippines. As collections of botanical material are made in various islands of the group, genera and species previously known from surrounding regions are constantly being found, and a few of these, mostly observed in recently collected material, are recorded in the following paper.

ALISMACEÆ.

SAGITTARIA Linn.

Sagittaria sagittifolia Linn. Sp. Pl. (1753) 993; Buchenau in Engler's Pflanzenreich **16** (1903) 46. *S. sagittæfolia* var. *diversifolia* Micheli in DC. Monog. Phan. **3** (1881) 66.

MINDANAO, Lake Lanao, Camp Keithley (888 Mrs. *Clemens*) November, 1906.

Europe and Asia, extending to Hainan, Formosa, Japan and Java.

This widely distributed species has previously been known as a Philippine plant only by the doubtful record given by *Naves*,[1] who states that he saw living specimens in the Island of Panay, and dried specimens in *Vidal's* herbarium, although *Vidal* does not record the species in any of his published works on the Philippine flora.

GRAMINEÆ.

Phalaris minor Retz. Obs. **3**: 8; Hook. f. Fl. Brit. Ind. **7** (1897) 221.

LUZON, Province of Benguet, near Baguio (Major *E. A. Mearns*) April, 1907, probably introduced.

Southern Europe to British India, South Africa and Australia.

[1] (Blanco's Flora de Filipinas) Nov. App. (1883), 298.

CYPERACEÆ.

MAPANIA Aubl.

Mapania macrocephala (Gaudich.) K. Sch. ex Warb. in Bot. Jahrb. **13** (1891) 265. *Hypolytrum macrocephalum* Gaudich. In Freyc. Voy. Bot. (1826) 414. *Lepironia macrocephala* Miq. Ill. (1871) 64. *pl.* 27.

BALUT ISLAND (5409 *Merrill*) October 8, 1906. In wet forests at 700 m. alt.

Moluccas and the Bismarck Archipelago.

Mapania kurzii C. B. Clarke in Hook. f. Fl. Brit. Ind. **6** (1904) 681.

LUZON, Province of Tayabas, Atimonan (4001 *Merrill*) March, 1905. In forested ravines.

Malayan Peninsula.

This species was so identified by the late *C. B. Clarke*, but omitted from his list of Philippine Cyperaceae.[2] Mr. *Clarke* observes that the specimen might be the closely allied *M. multispicata*, but it agrees well with No. 11476 *Ridley* from Singapore, determined by *Clarke* as *M. kurzii*, and also with the description of that species.

SCIRPODENDRON Kurz.

Scirpodendron costatum Kurz in Journ. As. Soc. Beng. **38**[2] (1869) 85; Clarke in Hook. f. Fl. Brit. Ind. **6** (1894) 684.

PALAWAN, San Antonio Bay (5257 *Merrill*) October 17, 1906. Forming dense thickets along a small stream at about 10 m. above sea level. The genus new to the Philippines.

Ceylon, Malayan Peninsula, Java, Australia and Samoa.

ARACEÆ.

CYRTOSPERMA Griff.

Cyrtosperma griffithii (Hassk.) Schott. in Oest. Bot. Wochenbl. (1857) 61; Engler in DC. Monog. Phan. **2** (1879) 271. *Lasia merkusii* Hassk. Cat. Bog. (1844) 59; Pl. Jav. Rar. (1848) 161; Miq. Fl. Ind. Bat. **3** (1855) 177.

MINDORO, Bulalacao (B. S. 1515 *Bermejos*) September, 1906. SAMAR, Borongan (5218 *Merrill*) October, 1906.

Java, Borneo and the Fiji Islands.

This species is rather abundant in the Visayan Islands and was observed by the author at several localities on the east coasts of Samar and Mindanao in October, 1906. It is extensively cultivated in some places, notably Borongan, and probably does not occur strictly wild in the Philippines. At Borongan and other places where it was observed it was grown in ravines in coconut groves, the petioles often being 8 feet in length and 3 to 4 inches in diameter and the leaf-blades 5 feet in length. The genus is new to the Philippines.

MAGNOLIACEÆ.

KADSURA Kaempfer.

Kadsura scandens Blume Fl. Jav. Schizandreae (1830) 9 *t. 1;* Miq. Fl. Ind. Bat. **1**[2] (1859) 19; King in Journ. As. Soc. Beng. **58**[2] (1889) 375; Ann. Bot. Gard. Calcutta **3** (1891) 221. *pl. 71. Sarcocarpon scandens* Blume Bijdr. (1825) 21.

[2] *This Journal, Bot. Sec. C.* (1907), **2**, 109.

MINDANAO, Lake Lanao (683 Mrs. *Clemens*) September–October, 1906. Malayan Peninsula to Java and Sumatra, and probably other islands in the Malayan Archipelago.

The specimens from Mindanao are apparently typical, and with the exclusion of *Kadsura blancoi* = *Phytocrene! (Icacinaceæ)* from the *Magnoliaceæ*, the above species is the first one of the genus to be recorded from the Philippines.

ICACINACEÆ.

CARDIOPTERYX Wall. (*Cardiopteris*).

Cardiopteryx moluccana Blume in Rumphia 3 (1837) 277. *t.* *177.* *f.* 2. *C. lobata* R. Br. var. *moluccana* Mast. in Hook. f. Fl. Brit. Ind. 1 (1875) 597; F.-Vill. Nov. App. (1883) 46. *C. rumphii* Baill., var. *integrifolia* Baill. in DC. Prodr. 17 (1873) 26.

MINDANAO, Lake Lanao, Camp Keithley (137 Mrs. *Clemens*) 1906, in fruit in February, in flower in July and September.

Baillon gives the distribution of the variety *integrifolia*=*C. moluccana*, as from British India through Malaya to New Guinea, but *Engler* in Nat. Pflanzenfamilien gives its distribution as from the Moluccas, Ceram and New Guinea. *F.-Villar* reported it from Luzon and Panay, but his record of the species as a Philippine plant has previously never been verified.

VITACEÆ.

PTERISANTHES Blume.

Pterisanthes sinuosa Merrill n. sp.

Glabra; foliis ovatis vel oblongo-ovatis, membranaceis, 11 ad 20 cm. longis, 6 ad 11 cm. latis, apice acuminatis, basi leviter cordatis, margine grosse distanter sinuato-dentatis; receptaculo oblongo-lanceolato, longe pedunculato, 11 ad 20 cm. longo, 1 ad 1.8 cm. lato, floribus marginalibus pedicellatis, floribus sessilibus immersis, 4-meris.

Nearly glabrous throughout when mature, the younger parts slightly ferruginous-pilose. Branches slender. Leaves ovate to oblong-ovate, simple, 11 to 20 cm. long, 6 to 11 cm. wide, membranaceous, the apex sharply acuminate, the base somewhat cordate, broad, the margins distantly and coarsely sinuate-dentate, dull or shining; nerves prominent, curved-ascending, about 5 on each side of the midrib; petioles 3 to 5 cm. long. Tendrils bifid. Receptacles red (?) apparently somewhat fleshy when fresh, oblong-lanceolate, 11 to 20 cm. long, 1 to 1.8 cm. wide, long pedunculate, the marginal flowers rather numerous, pedicellate, the pedicels about 1.5 cm. long, those on the surface of the lamina sessile, immersed, very numerous, 4-merous, the petals triangular-ovate, acute, 1 mm. long, the calyx disciform, truncate, about 1.7 mm. in diameter.

MINDANAO, Lake Lanao, Camp Keithley (647 Mrs. *Clemens*), July–October, 1906, four collections.

The first species of this very characteristic Malayan genus to be found in the Philippines and apparently closely related to *Pterisanthes polita* Miq., of the Malayan Peninsula, Sumatra and Borneo, differing from that species in its coarsely sinuate-toothed larger leaves and other characters.

STERCULIACEÆ.

TARRIETIA Blume.

Tarrietia riedeliana Oliver in Journ. Linn. Soc. Bot. **15** (1887) 98.

MINDANAO, Lake Lanao, near Camp Keithley (Mrs. *Clemens*) June, 1906.

The second species of the genus to be found in the Philippines, and an addition to the Celebes element in the Philippine flora now known to be very prominent.

Celebes.

LEGUMINOSÆ.

STRONGYLODON Vog.

Strongylodon lucidus (Forst.) Seem. Fl. Vit. (1865–68) 61. *Glycine lucida* Forst. f. Prodr. (1786) 51. *Rhynchosia lucida* DC. Prodr. **2** (1825) 387. *Strongylodon ruber* Vogel in Linnaea **10** (1836) 585; A. Gray Bot. Wilke's U. S. Explor. Exped. (1854) 446. *t. 48;* Baker in Hook. f. Fl. Brit. Ind. **2** (1876) 191; Prain ex King in Journ. As. Soc. Beng. **66**² (1897) 69.

BALUT (south coast of Mindanao) (5411 *Merrill*) October 8, 1906. In thickets along streams at 150 m. Not previously reported from the Philippines.

Ceylon, Andaman Islands, New Guinea to the Fiji Islands and Hawaii.

RUTACEÆ.

FAGARA L.

Fagara torva (F. Muell.) Engl. in Engl. und Prantl. Nat. Pflanzenfam. **3**⁴ (1895) 119. *Xanthoxylum torvum* F. Muell. Fragm. **7** (1871) 140; Hochr. Pl. Bog. Exsicc. (1904) 18, No. 28. *Zanthoxylum glandulosum* T. et B. Cat. Hort. Bog. (1866) 234, nomen.

MINDANAO, Lake Lanao, Camp Keithley (667 Mrs. *Clemens*) September, 1906.

The identification of the above plant is based largely on a specimen of No. 28 of *Hochreutiner's* Plantae Bogoriensis Exsiccatae in our herbarium. The Mindanao plant seems to be quite the same. *Hochreutiner* is authority for the reduction of the Javan *Zanthoxylum glandulosum* T. et B., to the Australian *Xanthoxylum torvum* F. Muell. = *Fagara torva* (F. Muell.) Engl. An interesting addition to our knowledge of the Philippine flora, the species previously being known from Java and Australia.

MALASTOMATACEÆ.

OCHTHOCHARIS Blume.

Ochthocharis javanica Blume in Flora **2** (1831) 525 et Mus. Bot. **1** (1849) 40; Naud. in Ann. Sc. Nat. III. **15:** 307; Triana Melast. 74. *t. 6. f. 67;* C. B. Clarke in Hook. f. Fl. Brit. Ind. **2** (1879) 528; King in Journ. As. Soc. Beng. **69**² (1900) 14; Miq. Fl. Ind. Bat. 1¹, (1855) 556; Cogn. in DC. Monog. Phan. **7** (1891) 480.

MINDORO, Baco River (F. B. 5518 *Merritt*) November, 1906, in mangrove swamps.

Var. **longipetiolata** Merrill n. var.

Petiolo 1.5 ad 4.5 cm. longo, ceteroquin ut *O. javanica* Bl.

MINDORO, Subaan (6226 *Merrill*) December, 1906; Baco River (F. B. 5488 *Merritt*) November, 1906, in mangrove swamps.

The genus is new to the Philippines, the five known species being Malayan, the genus therefore a characteristic Malayan one. *O. javanica* is found along the seashore from Tenasserim through the Malayan Peninsula to Borneo, Java, Banka and Billeton.

VERBENACEÆ.

PETRAEOVITEX Oliver.

Petraeovitex trifoliata Merrill n. sp.

Frutex scandens; foliis oppositis, trifoliatis, glabris vel parce puberulis; petiolo 3 ad 4 cm. longo; foliolis ovatis vel oblongo-ovatis, acuminatis, usque ad 7 cm. longis; paniculis terminalibus, 20 ad 40 cm. longis, puberulis; flores 8 mm. longi; calycis fructiferis lobis 1.5 ad 2 cm. longis.

A scandent shrub. Branches brown or gray, quadrangular, puberulent, becoming glabrous, 1.5 to 2 mm. in diameter. Leaves opposite, trifoliate, the petiole 3 to 4 cm. long, puberulent, the petiolules 4 to 10 mm. long; leaflets ovate to oblong-ovate, 7 cm. long, 2.5 to 5 cm. wide, subcoriaceous, glabrous or nearly so, somewhat shining, paler beneath, the apex acuminate, the base rounded, often somewhat inequilateral; nerves rather prominent beneath, about 4 on each side of the midrib, the reticulations lax. Panicles terminal, the lower branches subtended by leaves, 20 to 40 cm. long, puberulent, the axis and branches quadrangular. Flowers purple, about 8 mm. long. Calyx puberulent, the tube 4 mm. long, the lobes in anthesis 4 mm. long, oblong-lanceolate. Corolla equalling the calyx, the tube 5 mm. long, narrowly funnel-shaped, puberulent outside, slightly so at the throat inside, the lobes 3 mm. long, oblong-ovate, rounded. Stamens 4; filaments about 4 mm. long, slightly puberulent; anthers 1 mm. long. Calyx in fruit accrescent, the lobes 1.5 to 2 cm. long, oblanceolate, spatulate, obtuse.

PALAWAN, Victoria Peak (B. S. 708 *Foxworthy*) March, 1906, on open steep slopes at 900 m. alt. Nearly or quite the same species is represented by a specimen collected by *Hallier* in February, 1904, at San Ramon, near Zamboanga, MINDANAO.

The genus new to the Philippines, but two species previously known, one from Buru Island and one from New Guinea. The present species is apparently most closely related to the Buru species, *Petraeovitex riedelii* Oliv. in Hook. Icon. Pl. III. 5 (1883) 15. *Pl. 1420*, the type of the genus, distinguished at once from that species by its trifoliate leaves.

LABIATÆ.

POGOSTEMON Desf.

Pogostemon heyneanus Benth. in Wall. Pl. As. Rar. **1** (1830) 31; Lab. (1832–36) 154; Benth. in DC. Prodr. **12** (1848) 153; Wight Icon. *t. 1440.* F.-Vill. Nov. App. (1883) 164; Miq. Fl. Ind. Bat. **2** (1856) 961. *P. Patchouly* Pellet. in Mem. Soc. Sc. Orleans. **5** (1845) 277. *t. 7;* Benth. l. c.; Miq. l. c.; F.-Vill. l. c., Hook f. Fl. Brit. Ind. **4** (1885) 633.

LUZON, Province of Pampanga, Mount Arayat (5025 *Merrill*) February, 1906, det. *Rolfe* as *P. patchouli* Pellet.: Province of Rizal, Montalban (2442 *Ahern's collector*) January, 1905. In forests and thickets, perfectly wild, not cultivated.

The only record of this plant as a Philippine species that I have seen is *F.-Villar's,* who states that he saw living specimens in Luzon, this record, like so many of *F.-Villar's* and *Naves',* being subject to doubt, and accordingly it has been thought best again to record the species as a Philippine one, with citation of specimens. I have followed *Hooker f.,* in considering *Pogostemon heynianus* Benth., identical with *P. patchouli* Pellet., but the former name being the earlier is retained. In consideration of the fact that *Hooker f.* states "perhaps only a var. of *P. parviflorus*", it seems probable that the plant recorded from Luzon by *F.-Villar,* l. c., as *Pogostemon parviflorus* Benth., was only a form of *P. heynianus.* I have not seen the species cultivated in the Philippines and the specimens collected on Mount Arayat were growing on steep forested slopes at an altitude of about 400 m., remote from any dwelling or settled region, while *Ahern's collector* informs me that the Montalban specimens were from open forests.

British India to the Malayan Peninsula, Sumatra and Borneo.

ACANTHACEÆ.

THUNBERGIA Linn. f.

Thunbergia alata Boj. in Hook. Exotic Fl. (1823–27) *t. 177;* Nees in DC. Prodr. **11** (1857) 58; Clarke in Hook. f. Fl. Brit. Ind. **4** (1884) 391.

LUZON, Manila (14 *Merrill*) April, 1902, in waste places; Province of Bataan, Lamao River (B. S. 1612 *Foxworthy*) October, 1906.

A native of tropical Africa, now widely distributed in the tropics of both hemispheres.

The species has apparently been distributed by cultivation as an ornamental plant, and undoubtedly was so introduced into the Philippines, although I have not seen specimens in cultivation in the Archipelago, where it is perfectly spontaneous, although not common. It has not previously been reported from the Philippines.

RUBIACEÆ.

PETUNGA DC.

Petunga racemosa (Roxb.) K. Sch. in Engl. und Prantl. Nat. Pflanzenfam. 4[4] (1891) 80. *Randia racemosa* Roxb. Hort. Beng. (1814) 15: Fl. Ind. **1**, (1820) 144. *Petunga roxburghii* DC. Prodr. **4** (1830) 399; Hook. f. Fl. Brit. Ind. **3** (1880) 120; King & Gamble in Journ. As. Soc. Beng. **72**[2] (1903) 223.

BALABAC (B. S. 447 *Mangubat*) March, 1906, a shrub in forests, no representative of the genus having previously been reported from the Philippines. An Indo-Malayan type.

Northern India to Burmah, Malayan Peninsula, Java, Sumatra and Borneo.

RANDIA Linn.

Randia auriculata (Wall.) K. Sch. in Engl. und Prantl. Nat. Pflanzenfam. **4**[4] (1891) 75; King & Gamble in Journ. As. Soc. Beng. **72**[2] (1903) 207. *Webera auriculata* Wall. in Roxb. Fl. Ind. ed Carey & Wall. **2**: 537. *Stylocoryna auriculata* Wall. Cat. (1828) No. 8402. *Cupia auriculata* DC. Prodr. **4** (1830) 394. *Pseudixora ? auriculata* Miq. Fl. Ind. Bat. **2** (1856) 210. *Anomanthodia auriculata* Hook. f. in Benth. et Hook. f. Gen. Pl. **2**: 87; Fl. Brit. Ind. **3** (1880) 108. *Randia corymbosa* Boerl. in Koord. & Val. Bijd. Boomsoort. Java **8** (1902) 88, non Wight. & Arn.

NEGROS, Gimagaan River (B. F. 4265 *Everett*); (1624 *Whitford*) May, 1906. A Malayan type, new to the Philippines.

Malayan Peninsula and Archipelago.

BIKKIA Reinw.

Bikkia grandiflora Reinw. in Blume Bijdr. (1826) 1017; Miq. Fl. Ind. Bat. **2** (1856) 156; K. Sch. und Lauterb. Fl. Deutsch. Schutzgeb. Südsee (1901) 549.

SIBUTU (Sulu Archipelago) (5297 *Merrill*) October 13, 1906. In thickets on rocky seashores, the genus new to the Philippines, an eastern Malayan and Polynesian type.

Eastern Malaya to New Guinea and Polynesia.

IXORA Linn.

Ixora congesta Roxb. Fl. Ind. **1** (1820) 76; DC. Prodr. **4** (1830) 486; Hook. f. Fl. Brit. Ind. **3** (1880) 146; King and Gamble in Journ. As. Soc. Beng. **73**[2] (1904) 76. *Pavetta congesta* Miq. Fl. Ind. Bat. **2** (1856) 76.

MINDANAO, Lake Lanao, Camp Keithley (237 Mrs. *Clemens*) February, April and May, 1906; Mount Malindang (B. F. 4757 *Mearns & Hutchison*) May, 1906. An Indo-Malayan type new to the Philippines.

Burma through the Malayan Peninsula to the Malayan Archipelago.

GOODENIACEÆ.

SCAEVOLA Linn.

Scaevola minahassæ Koord. in Meded. 's Lands Plant. **19** (1898) 513, 628.

MINDANAO, Lake Lanao, Camp Keithley (690 Mrs. *Clemens*) September–October, 1906.

North-east Celebes.

I have made the above identification entirely from *Koorders'* rather short description which applies closely to our specimens. According to Mrs. *Clemens's* notes, the plant is a vine 30 to 40 feet in height, while *Koorders* describes the species as a shrub 1.5 to 2 m. high, the inference being, from his statement "frutex 1½–2 m. alta", that the plant is erect, but on page 513 he speaks of it as a scandent shrub.

The first species of the section *Enantiophyllum* to be found in the Philippines, a second apparently undescribed species being found also in Jolo which is described below.

Scaevola dajoensis Merrill n. sp. § *Enantiophyllum*.

Herbacea, scandens, ramulis foliisque oppositis, axillis barbatis; foliis ovato-lanceolatis vel oblongo-lanceolatis, acuminatis, basi acutis, margine irregulariter subrepando glanduloso-dentatis; cymae axillares, pauciflorae; corolla flava, 13 mm. longa.

A scandent herbaceous plant reaching a height of 5 m. Branches terete, glabrous, slender. Leaves opposite, membranous, glabrous or somewhat pubescent along the midrib beneath, ovate-lanceolate to oblong-lanceolate, the base acute, the apex long slender acuminate, the margins irregularly rather coarsely subrepand glandular-dentate, 6 to 9 cm. long, 2 to 3 cm. wide; nerves 6 to 7 on each side of the midrib, not prominent; petioles 2 to 5 mm. long, bearded in the axils. Cymes axillary, few-flowered, the peduncles 1.5 cm. long or less, pubescent. Calyx about 5.5 mm. long, the tube slightly hirsute, narrowly ovoid, 2.5 mm. long, the lobes lanceolate, 3 mm. long, persistent. Corolla yellow, 13 mm. long, slightly hirsute outside, densely so within, the lobes 5 to 6 mm. long, hyaline margined. Stamens glabrous. Style glabrous; stigma ciliate-fringed. Fruit fleshy, ovoid, dark-purple, about 8 mm. long.

JOLO, Mount Dajo (5324 *Merrill*) October 11, 1906, scandent in thickets on exposed ridges at an altitude of 650 m., in the Moro stronghold on Mount Dajo which was reduced by the American troops in March, 1906.

Apparently most closely related to *Scaevola minahassæ* Koord., from Celebes and Mindanao, differing from that species in being more glabrous and with larger leaves which are acute at the base and not pubescent beneath.

COMPOSITÆ.

BLUMEA DC.

Blumea sericans Hook. f. Fl. Brit. Ind. **3** (1881) 262; Forbes & Hemsl. Journ. Linn. Soc. Bot. **23** (1888) 422.

MINDANAO, Lake Lanao, Camp Keithley (894 Mrs. *Clemens*) January, 1907, common in open grass lands; not previously recorded from the Philippines.

Chittagong, Burma and Martaban to southern China and Formosa.

ADDITIONAL IDENTIFICATIONS OF THE SPECIES DESCRIBED IN BLANCO'S "FLORA DE FILIPINAS."

By Elmer D. Merrill.

(*From the botanical section, Biological Laboratory, Bureau of Science, Manila, P. I.*)

In April, 1905, I published a work entitled "A Review of the Identifications of the Species Described in Blanco's Flora de Filipinas"[1] in which an attempt was made to correlate the species considered by *Blanco* in the different editions of his "Flora de Filipinas," and to summarize what was known regarding them, indicating those that were referable to known and previously described species, those that were apparently valid and those that were unknown or doubtful. Since the publication of that paper, one of the objects of which was to serve as a guide in the collection of material and data that might serve to clear up doubtful points in *Blanco's* work, extensive collections have been made and our knowledge of the Philippine flora has been greatly extended, while from time to time certain points regarding *Blanco's* work have been cleared up. Accordingly it is proposed occasionally to publish notes regarding *Blanco's* species, as the data available seems to warrant such action.

In the preparation of the previous work too much dependence was placed on the work of *Fernandez-Villar,* and his identifications of some of *Blanco's* species were then accepted which have since been found to have been erroneous. Doubtless in the future as various groups are carefully monographed, numerous changes will have to be made in his identifications that were previously accepted by me.

Through the kindness of Dr. *C. B. Robinson,* of the New York Botanical Garden, I have received copies of two papers in which references are made to previous attempts at clearing up *Blanco's* species, which seem to have been overlooked by most, if not all, recent investigators of Philippine botany. The first paper is that of *Walpers,* a summary of the first edition of Blanco's "Flora de Filipinas" published in *Linnæa* (1842), vol. **16**, Litteratur-Bericht, pages 1 to 68, in which the first

[1] *Publications of the Bureau of Government Laboratories,* Manila (1905) No. **27.**

447 species described by *Blanco* are enumerated, with Latin translations of the species that *Blanco* described as new, and accepting those that *Blanco* ascribed to other authors without question. The paper is of little value and adds but very little to our actual knowledge of *Blanco's* species. The next paper is by Hasskarl, published in Flora, vol. **47** (1864), pages 17–23, 49–59; this was intended to be a critical review of the first edition of *Blanco's* work, but was apparently discontinued after the first thirty-three species described by *Blanco* were considered. Latin translations of *Blanco's* descriptions are given and some critical notes, while some new names appear, most of which must fall as synonyms. Still another reference supplied me by Dr. *Robinson,* is a review of *Blanco's* "Flora de Filipinas" by *George Tradescant Lay* in the Chinese Repository **7**: 422–437, 1838. Of this I have seen no copy, but Dr. *Robinson* informs me that it is of no scientific importance, data regarding about 15 species only being abstracted, with additions from the author's observations.

In the following paper notes on a number of *Blanco's* species are included, the arrangement following my previous publication,[2] the page references following the family names referring to that paper.

MAGNOLIACEÆ (p. 15).

Kadsura blancoi Azaola is excluded from the *Magnoliaceæ* and referred to *Phytocrene* (p. 423).

ANONACEÆ (p. 16).

Uvaria lanotan Blanco, ed. 1, 464. **Unona latifolia** Blanco, ed. 2, 324= *Mitrephora lanotan* (Blanco) Merr. in Govt. Lab. Publ. **35** (1905) 71, with description, synonomy and citation of specimens.

NYMPHÆACEÆ (p. 17).

Nymphæa lotus Blanco, ed. 1, 456; ed. 2, 317; ed. 3, **2** (1878) 222; F.-Vill. Nov. App. (1880) 9, non Linn.

Following Conard[3] true *Nymphæa lotus* is found in Africa and Madagascar only, while the Asiatic-Malayan-Australian form treated by various authors as *N. lotus* is *N. pubescens* Willd., which name should be accepted for the Philippine plant.

PITTOSPORACEÆ (p. 18).

Bursaria inermis Blanco, ed. 2, 124; ed. 3, **1**: 122, previously considered, after *F.-Villar*, to be probably identical with *Pittosporum ferrugineum* Ait., is more probably identical with *Pittosporum pentandrum* (Blanco) Merr. The species was really described by *Azaola* and not by *Blanco*, according to the latter's statement. See *Merrill* in Govt. Lab. Publ. **35** (1905) 18.

[2] *Ibid.*

[3] *Carnegie Inst. Pub.* (1905) No. 4, 198.

ELATINACEÆ (p. 19).

Bergia serrata Blanco Fl. Filip., ed. 1 (1837) 387. *Spergula serrata* Blanco l. c., ed. 2 (1845) 271; ed. 3, 2: 140. *Bergia glandulosa* Turcz. in Bull. Soc. Nat. Mosc. 27 [2] (1854) 371: Rolfe in Journ. Bot. 23 (1885) 210; Vid. Phan. Cuming. Philip. (1885) 95: Rev. Pl. Vasc. Filip. (1886) 51. Mats. and Hayata. Enum. Pl. Formosa. (1906) 40. *Bergia verticillata* F.-Vill. Nov. App. (1880) 15, non Willd.

LUZON, without locality (1058 *Cuming*), duplicate type of *Bergia glandulosa* Turcz: (138 *R. Marare*) 1894–95. Manila (Normal School Students) 1904; Province of Ilocos Norte (B. S. 2304 *Mearns*) January, 1907; Province of Zambales, Subic (*Hallier*) December, 1903; Province of Rizal, Bosoboso (B. S. 2058 *Ramos*) February, 1907.

In my treatment of *Blanco's* species I followed *F.-Villar* in considering this species the same as *Bergia verticillata* Willd. From the description, however, it can not be *Willdenow's* species, but I can not distinguish it from *Bergia glandulosa* Turcz., and *Blanco's* name being the earlier is here retained. The species, so far as is known, is confined to Luzon and Formosa.

GERANIACEÆ (p. 26).

Oxalis acetosella Blanco ed. 1, 388; ed. 2, 272, non Linn.

Following *B. L. Robinson* [1] *Oxalis corniculata* Linn., to which *Blanco's* species has been reduced, is confined to Europe and the eastern United States, while the widely distributed form found in southern Europe, the southern United States and in subtropical and tropical regions of both hemispheres is a distinct species, *Oxalis repens* Thunb. Oxal. (1781) 16. In case Robinson's distinctions hold good, this name should be applied to the common Philippine form of Oxalis.

RUTACEÆ (p. 27).

Fagara octandra Blanco ed. 1, 67; ed. 2, 48, non Linn. = *Melicope luzonensis* Engl. in Perk. Frag. Fl. Philip. (1905) 161. See *Merrill* in Govt. Lab. Publ. **35** (1905) 24.

Limonia linearis Blanco, ed. 1, 357; **Limonia monophylla** Blanco, ed. 2, 252, non Linn.=*Atalantia linearis* (Blanco) Merr. in Philip. Journ. Sci. **1** (1906) Suppl. 200 ! A characteristic endemic species.

Cookia anisum-olens Blanco, ed. 1, 359; **Cookia anisodora** Blanco ed. 2, 253. This is a *Clausena* as indicated by the author in Govt. Lab. Publ. **17** (1904) 21, and later described by *Perkins* as *Clausena warburgii*, Frag. Fl. Philip. (1905) 162. There is no doubt whatever as to the identity of the material cited with *Blanco's* species. In case of objection to *Blanco's* poorly constructed specific name, his second specific name, *anisodora*, still has priority over *Perkin's* name.

SIMARUBACEÆ (p. 29).

Ailanthus pongelion Blanco ed. 1, 380; ed. 2, 268, non Gmel., is not *A. malabarica*, DC., as determined by *F.-Villar*, but is a distinct species, *A. philippinensis* Merr. in Govt. Lab. Publ. **35** (1905) 25, with synonomy, description and citation of specimens.

[1] *Journ. Bot.* **44**, (1906) 391.

BURSERACEÆ (p. 30).

Guiacum abilo Blanco ed. 1, 30; *Icica abilo* Blanco ed. 2, 256=*Garuga abilo* (Blanco) Merr., in Govt. Lab. Publ. **35** (1905) 73 ! *G. mollis* Turcz., is a synonym, and *G. floribunda* Decne., ex descr., a quite different species.

CHALLETIACEÆ (p. 32).

Riana tricapsularis Blanco ed. 1, 850; ed. 2, 126=*Dichapetalum tricapsularc* (Blanco) Merr. in Govt. Lab. Publ. **35** (1895) 35. Apparently a very distinct endemic species.

OLACINEÆ (*Icacinaceæ*).

PHYTOCRENE Wall.

Phytocrene blancoi (Azaola) *Kadsura blancoi* Azaola in Blanco Fl. Filip. ed. 2 (1845) 594; ed. 3, **3** (1879) 118; Merr. in Govt. Lab. Publ. **27** (1905) 15. *Schizandra elongata* F.-Vill. Nov. App. (1880) 4, non Hook. f. et Th. *Phytocrene luzoniensis* H. Baill, in Adansonia **10** (1872) 28, et in DC. Prodr. **17** (1873) 10. *Gynocephalum luzoniense* Llanos ex Baill. ll. cc. as syn.

LUZON, Province of Rizal (1661 *Merrill*) March, 1903; 2439 *Ahern's collector*) January, 1905: Province of Laguna (*Alberto*) May, 1905.

MINDANAO, Lake Lanao, Camp, Keithley (447 Mrs. *Clemens*) March, 1906.

In my treatment of Blanco's species[5] I considered *Kadsura blancoi* as a doubtful species, following *F.-Villar* in treating it as a *Magnoliaceous* plant. However, a careful examination of *Blanco's* description shows that the plant can not be a *Kadsura* or a *Schizandra*. The description is very imperfect, but from the gross characters and the fruit description the species can belong to no other genus than *Phytocrene* "fruto en una cabezuela ó capítulo que contiene más de setenta frutos, de tres lados, á manera de los del plátano, musa, apiñados ó reunidos sobre un receptáculo que pesaba 25 á 30 libras." In the one specimen that I have seen in fruit, the heads weighed about 15 pounds. The locality from which the material came, on which the description of *Kadsura blancoi* was based, is not given, but the specimens undoubtedly came from one of the provinces near Manila, Rizal, Laguna or Bulacan. After a careful consideration of the matter I do not hesitate to refer the species to *Phytocrene*, adopting *Azaola's* name as the earliest one for it and reducing to it *Phytocrenc luzoniensis* H. Baill.

A second species of the genus, perhaps *Phytocrene macrophylla* Blume, is represented by material collected near Davao, Mindanao (Nos. 2765, 2995 *Williams*).

AMPELIDACEÆ (p. 33).

Cissus pedata Blanco, ed. 1, 71; ed. 2, 52, non Lam = *Tetrastigma lanceolarium* (Roxb.) Planch. ! *Blanco's* description is entirely too short and imperfect to warrant the above identification from the description alone, but the Tagalog name *Ayo*, cited by him, is almost universally and quite consistently applied to *Planchon's* species which is common in the Philippines.

[5] *Publications of the Bureau of Government Laboratories*, Manila, (1905), No. **27**, 15.

LEGUMINOSÆ (p. 37).

Cylista piscatoria Blanco, ed. 1, 589; **Galactia ? terminiflora** Blanco ed. 2, 411, previously considered by the author to be a distinct species of *Millettia*. *M. piscatoria* (Blanco) Merr. is certainly identical with *Derris elliptica* (Wall.) Benth., a species widely distributed from Martaban to the Malayan Peninsula and Archipelago. (See *Merrill* in Philip. Journ. Sci. **1** (1906) Suppl. 66.)

Cytisus quinquepetalus Blanco ed. 1, 598; ed. 2, 581. Following *F.-Villar* this was considered to be the same as *Desmodium cephalotes* Wall., but on securing specimens it was found to be quite distinct from *Wallich's* species and to represent a distinct species of *Desmodium*, *D. quinquepetalum* (Blanco) Merr. in Govt. Lab. Publ. **35** (1905) 20.

Negretia mitis Blanco ed. 1, 588; ed. 2, 410, non Beauv., has been considered by the author as *Mucuna lyonii* Merr. in Philip. Journ. Sci. **1** (1906) Suppl. 197. However the validity of the latter species seems doubtful, and it may prove only a form of *Mucuna nivea* DC., to which *Blanco's* species was reduced by *F.-Villar*.

Mimosa membranulacea Blanco ed. 1, 739; **Reichardia pentapetala** Blanco, ed. 2, 233. This was considered to be the same as *Pterolobium indicum* A. Rich., after *F.-Villar*, but was later considered by the author to represent a distinct species. *Pterolobium membranulaceum* (Blanco) Merr., in Govt. Lab. Publ. **35** (1905) 22, where the species is redescribed.

Bauhinia grandiflora Blanco, ed. 1, 332; ed. 2, 231, non Juss. This is apparently identical with *Bauhinia acuminata* Linn., and not at all *B. variegata* Linn., to which *F.-Villar* reduced it. *B. acuminata* Linn., seems to be widely distributed in Luzon but is nowhere abundant.

MYRTACEÆ (p. 45).

Metrosideros pictapetala Blanco, ed. 2, 295. This species was described by Blanco in the first edition under the name *Legnotis lanceolata*, p. 445, as pointed out to me by Dr. *C. B. Robinson* in lit. *F.-Villar* failed to connect *Legnotis* of the first edition with *Metrosideros* of the second edition, and made no attempt to reduce the former. I have as yet not succeeded in connecting *Blanco's* species with any known one.

Eugenia lobas Blanco ed. 1, 857; **Eugenia cauliflora** Blanco ed. 2, 291. Considered by the author to represent a distinct species and redescribed under the former name in Govt. Lab. Publ. **35** (1905) 48.

ONAGRACEÆ (p. 48).

Balingayum is excluded from this family and referred to *Calogyne* (*Goodeniaceæ*) which see (p. 434).

ARALIACEÆ (p. 51).

Nauclea digitata Blanco, ed. 2, 102=*Schefflera blancoi* Merr. in Philip. Journ. Sci. **1** (1906) Suppl. 109.

RUBIACEÆ (p. 52).

Remijia odorata Blanco ed. 2, 115. I am now of the opinion that *F.-Villar* was correct in referring this species to the widely distributed *Randia densiflora* (Wall.) Benth.

Serissa pinnata Blanco, ed. 1, 163; **Remijia oscura** Blanco, ed. 2, 116 Previously referred by me to *Gardenia* as a distinct species, *G. pinnata* (Blanco) Merr. The species is, however, identical with *Hypobatherum glomeratum* (Bartl.) K. Schum. as pointed out by *K. Schumann* in Engl. und Prantl Nat. Pflanzenfam. IV. 4 (1891) 156, the synonomy being as follows: **Hypobatherum glomeratum** (Bartl.) K. Sch. in Engl. und Prantl Nat. Pflanzenfam. IV. **4** (1891) 156; Elmer Leaflets Philip. Bot. **1** (1906) 8. *Platymerium glomeratum* Bartl. in DC. Prodr. **4** (1830) 619; Miq. Fl. Ind. Bat. **2** (1856) 200; F.-Vill. Nov. App. (1883) 113. *Serissa pinnata* Blanco Fl. Filip., ed. 1 (1837) 163. *Remijia oscura* Blanco l. c. ed. 2 (1845) 116; ed. 3, 1 (1877) 207. *Randia obscura* F.-Vill. Nov. App. (1883) 108: Vid. Sinopsis Atlas (1883) 29. *t. 57 f. B. Gardenia pinnata* Merr. in Govt. Lab. Publ. **27** (1905) 53. *Gardenia obscura* Vid. Phan. Cuming. Philip. (1885) 119; Rev. Pl. Vasc. Filip. (1886) 153; Ceron. Cat. Pl. Herb. (1892) 95; Merr. in Forestry Bureau Bull. **1** (1903) 54.

Pavetta membrenacea Blanco ed. 1, 59; **Pavetta sambucina** Blanco, ed. 2, 41, non DC. Erroneously reduced by *F.-Villar* to *Pavetta angustifolia* R. et S. A distinct species represented by No. 1584 *Merrill;* Nos. B. S. 996, 1834 *Ramos;* Nos. 1862, 3309 *Ahern's collector. Pavetta manillensis* Walp. (1843) is a synonym. *Blanco's* name *P. membrenacea* (1837) being the first published for the species.

Coffea volubilis Blanco, ed. 1, 157; ed. 2, 111 = *Morinda volubilis* (Blanco) Merr. Philip. Journ. Sci. **1** (1906) Suppl. 137, with description.

GOODENIACEÆ (p. 56).

Balingayum decumbens Blanco ed. 1, 187; ed. 2, 132 = *Calogyne pilosa* R. Br., or a closely related species, see *Merrill* Govt. Lab. Publ. **35** (1906) pp. 66–68, for a discussion of this previously doubtful genus, which was considered by *Bentham* and *Hooker* as belonging in the *Olacaceæ* and by *F.-Villar* as belonging in the *Onagraceæ*.

SAPOTACEÆ (p. 57).

Sideroxylon duclitan Blanco ed. 1, 129; ed. 2, 92. *Sideroxylon ramiflorum* Merr. in Govt. Lab. Publ. **17** (1904) 43, should be reduced to *Blanco's* species.

OLEACEÆ (p. 57).

Mogorium aculeatum Blanco ed. 1, 9; ed. 2, 7 = **Jasminum aculeatum** (Blanco) Walp. in Linnaea **16** Litt.-Bericht 3, 12, Hassk. in Flora **47** (1864) 50; Merr. in Govt. Lab. Publ. **35** (1905) 76. The transfer of the specific name to *Jasminum* was first made by *Walpers* fide *Hasskarl*. The combination is not given in Index Kewensis. A full description of the species with synonomy is given by *Merrill* l. c.

APOCYNACEÆ (p. 58).

Echites repens Blanco ed. 1, 109 non Jacq.; **Echites procumbens** Blanco ed. 2, 78 = *Aganosma marginata* G. Don. (*Holarrhena procumbens* (Blanco) Merr., *H. macrocarpa* F.-Vill.)

Echites spiralis Blanco ed. 1, 110; ed. 2, 79, non Wall. = *Parsonsia confusa* Merr. in Philip. Journ. Sci. 1 (1906) Suppl. 118. *P. rheedii* F.-Vill, non *Heligme rheedii* Wight.

CONVOLVULACEÆ (p. 62).

Convolvulus dentatus Blanco ed. 1, 89; ed. 2, 66, non Vahl. = *Ipomoea triloba* Linn. (*I. blancoi* Choisy). Dr. *C. B. Robinson*, in lit., states that he is of the opinion that *Blanco's* species is identical with *Ipomoea triloba* Linn., and I can not but agree with him after a careful examination of the various descriptions and the tropical American material of the Linnean species in our herbarium.

ACANTHACEÆ (p. 66).

Antirrhinum molle Blanco ed. 1, 503; ed. 2, 353 non Linn. = *Hygrophila phlomoides* var. *roxburgii* Hook. f.!

VERBENACEÆ (p. 67).

Premna serratifolia Blanco, ed. 2, 342, non Linn. = *Premna odorata* Blanco, see Merrill in Philip. Journ. Sci. 1 (1906) Suppl. 232.

Premna cordata Blanco ed. 1, 489; non R. Br. *Premna tomentosa* Blanco ed. 2, 342, non Wall. = *Premna cumingiana* Schauer, see Merrill l. c. 230.

EUPHORBIACEÆ (p. 75).

Phyllanthus niruri Blanco ed. 1, 690, non Linn.; **Phyllanthus tetrander** Blanco ed. 2, 480, non Roxb. = *Phyllanthus blancoanus* Muell. Arg., *Mueller's* species being based entirely on *Blanco's description*. To this species I have referred various specimens in Philip. Journ. Sci. 1 (1906) Suppl. 74.

MORACEÆ (p. 78).

Ficus payapa Blanco ed. 1, 683; ed. 2, 475. *Blanco's* description is quite too short and indefinite from which to determine this species. Material received under the Tagalog name *Payapa* seems to be referable to *Ficus forstenii* Miq. See Merrill in Philip. Journ. Sci. 1 (1906) Suppl. 47.

Ficus laccifera Blanco, ed. 1, 673; ed. 2, 468, non Roxb. Material received from the Visayan Islands under the Visayan name *lagnob* cited by *Blanco* for this species is identical with *Ficus hauili* Blanco, known to the Tagalogs as *hauili*, which in turn is perhaps not distinct from *F. leucantotoma* Poir.

SCITIMINEÆ (p. 83).

Costus nigricans Blanco ed. 1, 3; ed. 2, 3, = *Curcuma zeodaria* (Berg.) Rosc.! (See Merril in Philip. Journ. Sci. 1 (1906) Suppl. 36.

DIOSCOREACEÆ (p. 86).

Dioscorea divaricata Blanco ed. 1, 797; ed. 2, 550. Apparently a distinct species; see Merrill in Philip. Journ. Sci. **1** (1906) Suppl. 35.

LILIACEÆ (p. 87).

Smilax latifolia Blanco ed. 2, 548, non R. Br. = *Smilax vicaria* Kth.! Apparently a distinct species; see Merrill in Philip. Journ. Sci. **1** (1906) Suppl. 35.

PANDANACEÆ (p. 89).

Pandanus gracilis Blanco ed. 1, 778; ed. 2, 536. The species recently described by the author, *Pandanus whitfordii* Merr. in Govt. Lab. Publ. **17** (1904) 8, may prove to be identical with *Blanco's* species although it does not agree in habit with the form described by *Blanco*.

Pandanus exaltatus Blanco ed. 1, 778; ed. 2, 536. Manifestly two species are included in the description. The mountain form is doubtless the one I have described as *Pandanus arayatensis*, and I have identified as *Blanco's* species a coast form from Semerara Island.

Pandanus radicans Blanco ed. 1, 780; ed. 2, 537. This is not the same as *P. dubius* Spreng., but apparently a valid species as redescribed by Elmer, Leaflets Philip. Bot. (1906) 74.

GRAMINEÆ (p. 91).

Paspalum villosum Blanco ed. 1, 40; ed. 2, 28, non Thunb. I am now of the opinion that this is a form of *Paspalum scrobiculatum* Linn.

Andropogon ramosus Blanco, ed. 1, 37; ed. 2, 25, non Forsk. = *Ischaemum rugosum* var. *distachyum* (Cav.) Merr. in Philip. Journ. Sci. **1** (1906) Suppl. 330.

Cenchrus hexaflorus Blanco ed. 1, 36; ed. 2, 24. This was previously reduced by me to *Pennisetum macrostachyum* Brongn., but I am now of the opinion that it is the same as *Pennisetum compressum* R. Br.

Andropogon schoenanthus Linn., Blanco ed. 1, 39; ed 2, 27. I have considered (Philip. Journ. Sci. **1** (1906) Suppl. 339) that *Blanco* described the *Linnean* species, but it is impossible to be quite sure of this until flowering specimens are received.

INDEX TO PHILIPPINE BOTANICAL LITERATURE, III.

By Elmer D. Merrill.

(*From the botanical section, Biological Laboratory, Bureau of Science, Manila, P. I.*)

Berkeley, M. J. Contributions to the Botany of H. M. S. Challenger, XXXVIII, Enumeration of the Fungi collected during the Expedition of H. M. S. Challenger, 1874–75. (*Journ Linn. Soc. Bot.* **16** (1878) pp. 38–54.)

On pages 45 to 48, thirty-five species and varieties of Philippine fungi are enumerated from "Camiguin, Malanipa and Malamon (Philippines)." Several species of fungi are described as new.

Britton, N. L. Botanical Exploration of the Philippines. (*Journ. N. Y. Bot. Gard.* **5** (1904) pp. 40–41.)

A short account of R. S. Williams' collecting trip in the Philippines.

Conard, Henry S. The Waterlilies, a Monograph of the Genus Nymphaea. (*Carnegie Inst. Publ.* **4** (1905) pp. 1–279.)

In the genus *Nymphaea* 34 species are recognized, two of which are credited to the Philippines, *Nymphaea pubescens* Willd. (*N. lotus* Linn., in part and Philippine authors), British India to the Philippines, Java and Australia, and *N. stellata* Willd., south and southeast Asia, the Philippines, Java and Borneo.

Engler, A. Beiträge zur Kenntniss der Araceae, X. (*Engl. Bot. Jahrb.* **3** (1906) pp. 110–143.)

The following Philippine species are described: *Rhaphidophora perkinsiae, R. copelandii, R. merrillii, R. warburgii; Aglaonema densinervium, A. latifolium*, and *Alocasia culionensis*, while *Epipremnum mirable* Schott., is credited to the Archipelago.

Haviland, G. D. A Revision of the Tribe Naucleeae (Nat. Ord. Rubiaceae). (*Journ. Linn. Soc. Bot.* **33** (1897) pp. 1–94, plates 4.)

Seven genera are recognized of which five are represented in the Philippines by the following species: *Sarcocephalus cordatus* Miq., India to Australia, *S. glaberrimus* Miq., Celebes and Philippines, *S. junghuhnii* Miq., Malaya; *Adina multifolia* n. sp., endemic; *Nauclea gracilis* Vid., endemic, *N. philippinensis* (Vid.) Hav., endemic, *N. strigosa* Korth., Borneo, Philippines, *N. nitida* n. sp., endemic, *N. media* n. sp., endemic, *N. forsterii* Seem., Philippines to Samoa, *N. purpurascens* Korth., Java, Borneo, Celebes, *N. bartlingii* DC., endemic, *N. reticulata* n. sp., endemic; *Mitragyna speciosa* Korth., Borneo and New Guinea, *M. diversifolia* (Wall.) Hav., India to Malaya; *Uncaria pedicellata* Roxb., India to New Guinea, *U. insignis* DC., Borneo, *U. velutina* Hav., endemic, *U. setiloba* Benth., *Amboina* and *U. hookeri* Vid., Borneo.

Hose, Bishop. A Catalogue of the Ferns of Borneo and some of the adjacent islands which have been recorded up to the present time. (*Journ. Straits Branch R. A. Soc.* **32** (1899) pp. 31–84.)

In this paper 430 species are enumerated, many of which extend to the Philippines.

Loher, A. Lophopetalum toxicum Loher. (*Icon. Bogor.* 1 (1897) pp. 56–67, plate 16.)

Lophopetalum toxicum, from Luzon, figured and described, with a note regarding the use of its bark by the Negritos as a source of arrow poison.

Massee, George. Fungi Exotici, II, Philippine Islands. (*Kew Bull.* (1899) p. 176.)

Nine species of fungi are recorded from Loher's Philippine collections, of which one, *Favolus purpureus*, is described as new.

Masters, Maxwell T. A General View of the Genus Pinus. (*Journ. Linn. Soc. Bot.* **35** (1904) pp. 560–659, plates 4.)

In this paper 73 species are considered, two of which are Philippine, *Pinus insularis* Endl., endemic, and *P. merkusii* Jungh. et DeVr., Luzon, Sumatra, Borneo and (?) the Shan States.

Pearson, H. H. W. On some Species of Dischidia with Double Pitchers. (*Journ. Linn. Soc. Bot.* **35** (1902) pp. 375–390 with one plate.)

On page 377 *Dischidia pectenoides* Pearson, is described from Luzon.

Rehder, Alfred. Synopsis of the Genus Lonicera. (*Rept. Mo. Bot. Gard.* **14** (1903) pp. 27–232, plates 20.)

Of this genus, 150 species, with many varieties and forms, are recognized, none of which, however, occur in the Philippines. Since the publication of the work, however, one or two species of *Lonicera* have been found in northern Luzon.

Ridley, H. N. New or little known Malayan Plants. (*Journ. Straits Branch R. A. Soc.* **44** (1905) pp. 189–211.)

Many species are described from different parts of the Malayan Peninsula, Borneo, etc., including one from the Philippines, *Calamus lindeni* Hort., page 200, based on a specimen from the Philippines cultivated in the Botanic Gardens, Singapore. On page 199 *Joinvillea malayana* is also described from material collected in Perak, Selangor and Sarawak, also being found in Palawan. (See Merrill in *Philip. Journ. Sci.* **1** (1906) Suppl. 181.) In the same work two other papers by the same author are published, both bearing more or less on Philippine botany, "The Gesneraceæ of the Malay Peninsula" **43** (1905) pp. 1–92, and "The Aroids of Borneo" **44** (1905) pp. 169–188.

Ridley, H. N. The Flora of Singapore. (*Journ. Straits Branch R. A. Soc.* **33** (1900) pp. 27–196.)

An enumeration of all the flowering plants and vascular cryptogams known to occur on the Island of Singapore, over 1,900 species being recorded from an area of a little over 200 square miles. Many of the species enumerated extend to the Philippines.

Ridley, H. N. Grasses and Sedges of Borneo. (*Journ. Straits Branch R. A. Soc.* **46** (1906) pp. 215–228.)

An enumeration of the *Cyperaceæ* and *Gramineæ* in recent Bornean collections, 87 species of grasses and 99 species of sedges being enumerated, both numbers much smaller than in the corresponding groups in the Philippines. Many of the species enumerated extend to the Philippines.

Ridley, H. N. Scitamineæ of Borneo. (*Journ. Straits Branch R. A. Soc.* **46** (1906) pp. 229–246.)

Including *Zingiberaceæ, Marantaceæ, Musaceæ* and *Lowiaceæ* 86 species are enumerated, many described as new and a few extending to the Philippines.

Ridley, H. N. The Scitamineæ of the Malay Peninsula. (*Journ. Straits Branch R. A. Soc.* **32** (1899) pp. 85–184.)

A paper of the same scope as the same author's Scitamineæ of Borneo, above, about 140 species being enumerated, including many new ones and a few that extend to the Philippines.

Skan, S. A. Skimmia japonica Thunb. (*Curtess' Bot. Mag.* IV. **1** (1905) Tab. 8038.)

This Japanese species figured and described; recently found also in northern Luzon.

Sydow, H. et P. Novae Fungorum species III. (*Ann. Myc.* **4** (1906) pp. 343–345.).

Among species described from various parts of the world is one from the Philippines, *Auerswaldia copelandi*, on leaves of *Caryota*.

Sydow, H. et P. Neue und kritische Uredineen IV. (*Ann. Myc.* **4** (1906) pp. 28–32.)

The following new species are described from Philippine material: *Uromyces hewittæ, Uredo davaoensis, U. hygrophilæ, U. philippinensis*, and *U. wedeliæ-bifloræ*.

Tavera, T. H. Pardo de. The Medicinal Plants of the Philippines (1901) pp. 1–269+XVI. English translation by Jerome B. Thomas.

A compilation of notes regarding the medicinal uses of various native plants, arranged according to Bentham and Hooker's Genera Plantarum, with descriptions of the species considered. The original work "Plantas medicinales de Filipinas" was published in Manila in 1892.

Underwood, L. M. The genus Alcicornium of Gaudichaud. (*Bull. Torr. Bot. Club.* **32** (1905) pp. 567–596.)

The generic name *Alcicornium* Gaudich., is accepted for *Platycerium*, and 13 species are recognized, of which two are found in the Philippines, *Alcicornium coronarium* (Müller) Underw., and *A. grande* (J. Sm.) Underw.

Wright, C. H. Pinanga maculata Porte. (*Curtiss' Bot. Mag.* IV. **1** (1905) Tab. 8011.)

This previously imperfectly known Philippine species figured and described.

INDEX.[1]

[Synonyms in *italics*.]

[1]For Index to No. 5, see pp. 397–404.

D

M

U

V

Lightning Source UK Ltd.
Milton Keynes UK
UKHW020648060119
334855UK00009B/1499/P

9 780265 453414